Perspectives in Physiology

Published on behalf of The American Physiological Society
by Springer

This fascinating series seeks to place biomedical science inside a greater historical framework, describing the main pathways of development and highlighting the contributions of prominent investigators. This book series is published on behalf of the American Physiological Society by Springer. Access to APS books published with Springer is free to APS members.

Guido Ferretti
Editor

Exercise, Respiratory and Environmental Physiology

A Tribute from the School of Milano

Springer

american physiological society

Editor
Guido Ferretti
Department of Molecular and
Translational Medicine
University of Brescia
Brescia, Italy

ISSN 2625-2813 ISSN 2625-2821 (electronic)
Perspectives in Physiology
ISBN 978-3-031-19196-1 ISBN 978-3-031-19197-8 (eBook)
https://doi.org/10.1007/978-3-031-19197-8

Jointly published with The American Physiological Society

This Springer imprint is published by the registered company Springer Nature Switzerland AG
The registered company address is: Gewerbestrasse 11, 6330 Cham, Switzerland

Foreword

Dear Reader.

The book you are holding is truly a goldmine of exercise, respiratory and environmental physiology, and if you had not realized the greatness of the Milano School before, you will certainly do so after having read it. The term "School" can only be used when a continuum of academic achievements spans over several generations and has maintained its scientific relevance also in a historical perspective. This is certainly true for the scientific work described here and starting in Italy more than a century ago. That such important work was done outside Scandinavia was not evident to me at the time I started my doctoral studies at the Department of Physiology at Karolinska Institute in Stockholm, Sweden in the early 1970s. The perspective then was Scandinavian, and the great names in exercise physiology were E. Asmussen and M. Nielsen in Denmark and E. Hohwu-Christensen and P. O. Åstrand in Sweden. My supervisors C.M. Hesser and H. Bjurstedt had developed advanced techniques for continuous blood gas and ventilatory measurements during non-steady state situations such as apnoea and hypergravity stress and were among the first to install computers (at the time analogue) in their laboratory. They encouraged me to apply such techniques to non-steady state exercise in humans and analyse the data in terms of control engineering. As my work progressed, my limited perspective changed dramatically as shown by the following quote from my thesis:

Recalling the study of Margaria et al. (1933), their approach was in essence that of a systems analysis, although the terminology of control engineering was not used. Thus, a step function was used as the input function and the relationship between the time courses of the output and input functions was studied. The dynamic properties of the unknown system were described in terms of a mathematical model, which was found to be consistent with the hypothesis that there are two compartments of the O_2 debt.

Another key paper that inspired me greatly was by P.E. di Prampero et al. (1970) describing the kinetics of oxygen uptake during the onset of submaximal exercise, and I was fascinated by the fact that R. Margaria as a coauthor was still active 40 years after his seminal work in the 1930s.

A typical feature in papers from members of the Milano School is the clarity, even beauty, of the diagrams, for example that reproduced as Fig. A1 on page 85 in this book and originating from the classical paper by Margaria et al. (1933). Typically, the diagrams report original data without becoming too crowded, and they are usually accompanied by the underlying equations in the text, with explanations, so that the reader can understand how they were derived. Many further examples of such diagrams and informative equations are found in Chaps. 5 and 6 dealing with the energetics of various forms of locomotion. Present technologies with watches and cellphones that estimate caloric expenditure would not have been possible without the data summarized in these two chapters.

The book is by definition a subjective account on the development in central research areas in exercise and environmental physiology, as seen with the eyes of a group of senior members of the Milano school and does not pretend to be neutral. Therefore, the expert reader has no reason to be disappointed if his/her own work is not mentioned or not recognized as expected. Hopefully, other schools will write their own stories in their way, and I will be the first to read and enjoy their work.

In a similar way, this foreword is not objective, but instead influenced by my personal knowledge of the authors. During my postdoctoral period at the Department of Physiology, State University of New York at Buffalo, I was happy to meet members of the Milano School such as P. Cerretelli and P. E di Prampero, who were frequent visitors to the laboratory. Later I collaborated with several members of the Milano School during studies of exercise in weightlessness on the US Space Shuttle and on the Russian space station Mir, where members of the Milano School such as G. Ferretti, C. Capelli, G. Antonutto, P. Zamparo, D. Negrini and G. Miserocchi played important roles. As before, I was impressed by the ingenuity of their experimental designs and the quality of their work. I am certain that the reader of this book will agree.

Department of Physiology, Karolinska Institute, Stockholm, Sweden

Dag Linnarsson, MD PhD

References

Margaria R, Edwards HT, Dill DB (1933). The possible mechanism of contracting and paying the oxygen debt and the role of lactic acid in muscular contraction. Am J Physiol 106: 689–714.

di Prampero PE, Davies CT, Cerretelli P, Margaria R (1970). An analysis of O_2 debt contracted in submaximal exercise. J Appl Physiol 29: 547–551. doi: 10.1152/jappl.1970.29.5.547. PMID: 5474842.

Acknowledgements

We are grateful to Guglielmo Antonutto (Udine), Egidio Beretta (Milano), Carlo Capelli (Verona), Paolo Cerretelli (Genève), Fiorenzo Conti (Ancona), Mario Costa (Siracusa), Antonio Dal Monte (Roma), Fabio Esposito (Milano), Giorgio Fanò (Chieti), Raffaello Grandi (Milano), Bruno Grassi (Udine), Dag Linnarsson (Stockholm), Alberto Minetti (Milano), Igor Mekjavic (Ljubljana), Jacopo Mortola (Montreal), Daniela Negrini (Varese), Dave Pendergast (Buffalo), Corrado Poggesi (Firenze), Carlo Reggiani (Padova), Giulio Sergio Roi (Bologna), Roberta Samaja (Milano), Paola Zamparo (Verona) for fruitful discussions, especially on historical matters, and/or for having provided some iconographic material.

Contents

Abbreviations

3D	Three-dimensional
A	Ratio between the metabolic power at altitude and at sea level
a	Proportionality constant relating phosphate concentration to muscle oxygen consumption
a	asymptotic (suffix, in Chapter 7)
A_1	Phase I amplitude of a double or triple exponential equation
A_2	Phase II amplitude of a double or triple exponential equation
A_3	Phase III amplitude of a triple exponential equation
A_c	Centrifugal acceleration
ADP	Adenosine-di-phosphate
A_f	Area projected on the frontal plane
a_f	Forward acceleration
alt	Altitude (suffix, used for cycling at altitude)
AMREE	American Medical Research Expedition to Everest
AT	Anaerobic threshold
A_{tot}	Total body surface
ATP	Adenosine-tri-phosphate
$\dot{Al}$	Anaerobic alactic power
$\dot{Al}^{max}$	Maximal anaerobic alactic power
$A\dot{T}P$	Rate of ATP resynthesis
$\overrightarrow{A\dot{T}P}$	Rate of ATP hydrolysis
$\overleftarrow{A\dot{T}P}$	Rate of ATP resynthesis
B	ratio between constant k_e at altitude and at sea level
b	proportionality constant, corresponding to the moles of ATP resynthesized per mole of lactate accumulated
C	Energy cost (generic, concerning locomotion)
C	Compliance (distensibility, generic, concerning lung mechanics)

c	proportionality constant, corresponding to the moles of ATP resynthesized per mole of oxygen consumed
C_a	Aerodynamic energy cost
C_{acc}	Energy cost of accelerated running
C_aO_2	Oxygen concentration in arterial blood
C_c	Energy cost of cycling
C_{cw}	Compliance of the chest wall
C_h	Hydrodynamic energy cost
C_L	Compliance of the lungs
C_{n-a}	Non-aerodynamic energy cost
CO_2	Carbon dioxide
COPD	Chronic obstructive pulmonary disease
Cr	Creatine
C_r	Energy cost of running
CRESE	Center for Research in Special Environments, Buffalo NY, USA
C_{rs}	Compliance of the overall respiratory system (chest plus lung)
C_{sk}	Energy cost of skiing
C_{sw}	Energy cost of swimming
CT	Computerized tomography
C_{tot}	Total creatine concentration
$C_{\bar{v}}O_2$	Oxygen concentration in mixed venous blood
C_w	Energy cost of walking
CWT	Critical water temperature
C_x	Aerodynamic coefficient
D	Drag
d	Time delay
D_A	Added drag
d_A	Thickness of the air–blood barrier
dCO_2	Infinitesimal increase in oxygen concentration at distance x in a lung capillary
$DefO_2$	Overall oxygen deficit
D_L	Lung diffusing capacity, generic
D_LCO	Lung diffusing capacity for carbon monoxide
D_LNO	Lung diffusing capacity for nitrogen monoxide
D_LO_2	Lung diffusing capacity for oxygen
D_m	Diffusing capacity of the alveolar-capillary membrane
$d\dot{M}O_2$	Infinitesimal blood oxygen uptake from alveoli
DO_{2M}	Muscle oxygen deficit
D_s	Time delay of the slow component (phase III) of the $\dot{V}O_2$ kinetics
D_t	Peripheral tissue diffusing capacity, generic
D_tO_2	Peripheral tissue oxygen diffusing capacity
E	Metabolic energy
E_BO_2	Aerobic energy stores in the blood
ECG	Electrocardiogram

E_k	Kinetic energy
E_{La}^{max}	Maximal anaerobic lactic capacity
E_LO_2	Aerobic energy stores in the lungs
EM	Equivalent body mass in accelerated running
EO_2	Energy from aerobic sources stored in the body
E_p	Potential energy
E_{PC}^{max}	Maximal anaerobic alactic capacity
EPOC	Excess post-exercise oxygen consumption
ES	Equivalent slope during accelerated running
E.S.	Extracellular space
ESA	European Space Agency
ES_a	Aerodynamic component of the equivalent slope
ES_{n-a}	Non-aerodynamic component of the equivalent slope
$E_{st,rs}$	Standard elastance of the respiratory system (chest plus lung)
E_tO_2	Aerobic energy stores in tissues
$\dot{E}$	Metabolic power
$\dot{E}_0$	Intercept of the linear relationship between $\dot{E}$ and $\dot{w}$ on a cycle ergometer
$\dot{E}_a$	Metabolic power spent to overcome aerodynamic forces
$\dot{E}_c$	Metabolic power in cycling
$\dot{E}_i$	Internal metabolic power
$\dot{E}_{max}$	Maximal metabolic power
$\dot{E}_{n-a}$	Metabolic power spent to overcome non-aerodynamic forces
$\dot{E}_r$	Metabolic power at rest
F	Force
f	Frequency, generic
F_a	Aerodynamic component of the force opposing motion
F_B	Body weight
F_c	Overall force opposing motion in cycling
FEV1	Forced expiratory volume after 1 s
FFA	Free fatty acids
f_H	Heart frequency, or heart rate
f_H^{max}	Maximal heart rate
F_i	Fractional limitation to $\dot{V}O_2^{max}$ imposed by the ith resistance to oxygen flow, where i indicates the position of the resistance at stake in a series of resistances
F_L	Fractional limitation to $\dot{V}O_2^{max}$ imposed by the lung resistance to oxygen flow
F_n	Fractional limitation to $\dot{V}O_2^{max}$ imposed by the nth resistance to oxygen flow (last downstream in a series of resistances).
F_{n-a}	Non-aerodynamic component of the force opposing motion
F_{n-a0}	Non-aerodynamic component of the force opposing motion on flat terrain

F_p	Fractional limitation to $\dot{V}O_2^{max}$ imposed by the lumped peripheral resistance to oxygen flow
f_p	Pedalling frequency
F_Q	Fractional limitation to $\dot{V}O_2^{max}$ imposed by the cardiovascular resistance to oxygen flow
Fr	Froude number
FRC	Functional residual capacity
F_{tot}	Total force acting on a body in water in the vertical direction
F_V	Fractional limitation to $\dot{V}O_2^{max}$ imposed by the ventilatory resistance to oxygen flow
F_v	Vertical force opposing motion, or work per unit distance on the vertical axis
F_w	Buoyant force
G	Free energy
g	Gravity acceleration
g'	Vectorial sum of forward (or centrifugal) acceleration plus gravity acceleration
G_L	Conductance to oxygen flow related to lung diffusion/perfusion
G_m	Mitochondrial conductance to oxygen flow
G_p	Lumped peripheral conductance to oxygen flow
G_Q	Cardiovascular conductance to oxygen flow
$\dot{G}l_{ox}$	Rate of glycogen oxidation
G_T	Overall conductance to oxygen flow
G_t	Peripheral tissue conductance to oxygen flow
G_V	Ventilatory conductance to oxygen flow
h	Heat (chapter 2)
h	Height (whenever expressing potential energy as Mgh, as in chapters 5, 6 and 11)
H^+	Hydrogen ions
H_2O	Water
HAPE	High altitude pulmonary oedema
Hb	Haemoglobin
[Hb]	Blood haemoglobin concentration
Hct	Haematocrit
h_{irr}	Irreversible heat
HIF-1	Hypoxia inducible factor 1
h_{rev}	Reversible heat
h_{tot}	Total heat ($= h_{rev} + h_{irr}$)
i	Incline of the terrain
IA	Iodo-acetamide
IL	Index Locomotorius (locomotor index)
iPWel	Iso-inspiratory power
I.S	Intracellular space
J_l	Lymph flow

J_v	Trans-epithelial water flow
k	Proportionality constant, generic
k_1	Phase I rate constant of a double or triple exponential equation
k_2	Phase II rate constant of a double or triple exponential equation
k_3	Phase III rate constant of a triple exponential equation
k_e	Proportionality constant relating the energy cost C to the square of speed
$k_{e\,alt}$	Proportionality constant relating the energy cost C to the square of speed at altitude in cycling
K_f	Filtration coefficient
k_i	Proportionality constant of the relationship between the internal metabolic power in cycling and the pedalling frequency
K_l	Lymphatic conductance
K_p	Dimensionless constant relating $P_{\bar{v}}O_2$ to $P_{\bar{c}}O_2$.
K_{Wa}	Wagner's constant, equal to the product between D_tO_2 and K_p
k_w	Proportionality constant relating the force opposing motion to the square of speed
L	Length, distance
LA	Los Angeles
La_b	Blood lactate concentration
$\hat{La}_b$	Peak or maximal blood lactate concentration attained in recovery after exercise
La_{b0}	Blood lactate concentration at recovery time 0 (end of exercise)
LBNP	Lower body negative pressure
L_{eq}	Lung equilibration coefficient
L_{ho}	Distance (length) covered on the horizontal plane
L_p	Water specific conductance
L_t	Distance (length) covered along the treadmill's axis
$\dot{La}$	Rate of lactate accumulation
$\dot{La}_b$	Rate of blood lactate accumulation
$\dot{La}_p$	Rate of lactate production
$\dot{La}_r$	Rate of lactate removal
M	Mass
M_b	Body mass
MEP	Maximal explosive power
MFT	Mean (subcutaneous) fat thickness
MLSS	Maximal lactate steady state
n	Number (e.g. of moles)
N_{alv}	Number of alveoli
NASA	National Aeronautics and Space Agency, USA.
NIRS	Near-infrared spectroscopy
NMR	Nuclear magnetic resonance
O_2	Oxygen
O_t	Oarlock reaction force

P	Phosphate
P	Pressure
P	Reaction force of water against the blade of the oar (chapter 6, rowing)
$\overline{P}$	Mean arterial pressure
$P_{0.1}$	Pressure measured at 0.1 s
p_a	Component of the reaction force of water against the blade of the oar, acting along the same direction of the shell of the boat
P_ACO_2	Carbon dioxide partial pressure in alveolar air
P_aCO_2	Carbon dioxide partial pressure in arterial blood
P_{alv}	Alveolar pressure
P_AO_2	Oxygen partial pressure in alveolar air
P_aO_2	Oxygen partial pressure in arterial blood
PB	Barometric pressure
PB_0	Barometric pressure at sea level
PCO_2	Carbon dioxide partial pressure, generic
$P_{\bar{c}}O_2$	Mean capillary oxygen partial pressure
PCr	Phosphocreatine
PCr_0	Phosphocreatine concentration at exercise start (exercise time 0)
PCr_{ss}	Phosphocreatine concentration at exercise steady state
P_{cw}	Elastic recoil pressure of the chest wall
P_d	Distending pressure
PEEP	Pulmonary end-expiratory pressure
P_{el}	Elastic recoil pressure, generic
PFK	Phospho-fructo-kinase
pH_m	Muscle pH
Pi	Inorganic phosphate
P_IO_2	Oxygen partial pressure in inspired air
P_{ip}	Pulmonary interstitial pressure
P_L	Elastic recoil of the lung
P_{liq}	Pleural liquid pressure
P_mO_2	Oxygen partial pressure in mitochondria
PO_2	Oxygen partial pressure, generic
$\tilde{}PO$	Overall high energy phosphate concentration
P_{pl}	Intrapleural pressure
P_{rs}	Overall elastic recoil pressure of the overall respiratory system (chest plus lung)
P_t	Component of the reaction force of water against the blade of the oar, acting along a perpendicular direction to that of the shell of the boat
P_tO_2	Oxygen partial pressure in peripheral tissues
$P_{\bar{v}}O_2$	Oxygen partial pressure in mixed venous blood
$P\gamma$	Pressure reflecting surface tension
P/O_2 ratio	Amount (moles) of high-energy phosphates resynthesized by the consumption of one mole of oxygen

$\overrightarrow{\dot{P}Cr}$	Rate of phosphocreatine splitting
$\dot{Q}$	Total blood flow (cardiac output)
$\dot{Q}_aO_2$	Oxygen flow in arterial blood (systemic oxygen delivery)
$\dot{Q}_aO_2{}^{max}$	Maximal oxygen flow in arterial blood
$\dot{Q}_{max}$	Maximal cardiac output
Q_{st}	Stroke volume of the heart
R	Resistance
r	Rest (suffix)
RBC	Red blood cell
R_g	Radius of gyration
R_i	i^{th} resistance to oxygen flow along the respiratory system
R_i	i^{th} resistance to oxygen flow, where i indicates the position of the resistance at stake in a series of resistances.
R_L	Lung resistance to oxygen flow
R_m	Mitochondrial resistance to oxygen flow
$R_{min,rs}$	Minimal resistance of the respiratory system (chest plus lung)
R_n	n^{th} resistance to oxygen flow (last resistance downstream in a series of resistances).
R_p	Lumped peripheral resistance to oxygen flow
RPE	Rate of perceived exertion
RQ	Respiratory quotient, generic
R_Q	Cardiovascular resistance to oxygen flow
R_{rs}	Resistance of the respiratory system (chest plus lung)
R_T	Total resistance to the oxygen flow along the respiratory system
R_t	Tissue resistance to oxygen flow
R_{tr}	Overall traction resistance
RV	Residual volume
R_V	Ventilatory resistance to oxygen flow
S_A	Overall surface of the air–blood barrier
S_aO_2	Arterial oxygen saturation
sl	Sea level (suffix)
ss	Steady state (suffix)
STPD	Standard temperature and pressure, dry
T	Temperature
t	Time
T_c	Temperature of the heat sink
T_h	Temperature of the heat source
TLC	Total lung capacity
TPR	Total peripheral resistance
T_r	Traction force exerted by the rower on the oar
t_t	Transit time
TTI	Tension time index
T_u	Underwater torque

UCI	Union Cycliste Internationale, the governing body of international cycling competitions
UCP-1	Uncoupling protein 1
V	Volume
v	Velocity or speed
V_A	Alveolar volume
V_c	Blood volume in the alveolar capillary network
V_LO_2	Volume of oxygen taken up through the lungs
V_LO_{2ss}	Volume of oxygen taken up through the lungs during steady state exercise of any duration
V_m	Mitochondrial volume
v_m	Mean velocity of locomotion
v_{max}	Maximal speed
$v_{max,sl}$	Maximal speed at sea level (in cycling)
v_{min}	Minimal speed
VO_2	Oxygen volume (oxygen consumption, generic)
${VO_2}^{eLa}$	Oxygen deficit due to early lactate accumulation, in oxygen equivalents
${VO_2}^{PCr}$	Oxygen deficit due to PCr hydrolysis, in oxygen equivalent (alactic oxygen deficit)
v_{opt}	Optimal walking velocity
v_s	Shortening velocity
VT	Tidal volume
v_v	Vertical velocity
V_{vb}	Venous blood volume
$\dot{V}$	Gas flow
$\dot{V}_A$	Alveolar ventilation
$\dot{V}_A/\dot{Q}$	Ventilation/perfusion ratio
$\dot{V}CO_2$	Carbon dioxide output, (carbon dioxide flow, generic)
$\dot{V}_E$	Expired ventilation (total expiratory gas flow)
${\dot{V}_E}^{max}$	Maximal expired ventilation
$\dot{V}_LO_2$	Lung oxygen flow, or oxygen uptake
$\dot{V}_LO_{2r}$	Lung oxygen flow at rest
$\dot{V}_LO_{2ss}$	Lung oxygen flow at the exercise steady state
$\dot{V}_LO_{2t}$	Lung oxygen flow at time t
$\dot{V}_mO_2$	Muscle oxygen flow, or rate of muscle oxygen consumption
$\dot{V}_mO_{2ss}$	Rate of muscle oxygen consumption at steady state
$\dot{V}_mO_{2t}$	Muscle oxygen flow at time t
$\dot{V}O_2$	Rate of oxygen consumption or oxygen flow (generic)
${\dot{V}O_2}^{max}$	Maximal oxygen consumption (maximal oxygen flow). Suffixes *0, a* and *t* indicate that ${\dot{V}O_2}^{max}$ is measured at time 0, at the asymptote and at time t of an exponential kinetics of a ${\dot{V}O_2}^{max}$ change with time.
$\dot{V}O_{2_{ss}}$	Oxygen flow at steady state (generic)
v_{opt}	Optimal walking speed

v_t	Tangential velocity
w	Mechanical work
w_D	Work done for overcoming drag
w_e	External work
w_i	Internal work
w_m	Energy generated by muscle contraction during walking, in order to keep speed constant
w_r	Work of rowing
w_{re}	Work of breathing, or respiratory work
w_{tot}	Total mechanical work
w^*_{ATP}	Work done per mole of ATP split
W/D	Wet weight to dry weight ratio
$\dot{w}$	Mechanical power
$\widehat{\dot{w}}$	Peak power (maximal absolute mechanical power during standing high jump off both feet)
$\dot{w}^{max}$	Maximal mechanical power
$\dot{w}_D$	Mechanical power necessary to overcome drag
$\dot{w}_e$	External mechanical power
$\dot{w}_{el}$	Elastic power of the lungs
$\dot{w}_i$	Internal mechanical power
$\dot{w}_k$	External power transmitting kinetic energy to water, but useless to overcome drag
$\dot{w}_{n-D}$	Mechanical power against non-aerodynamic or hydraulic forces (non-drag forces)
$\dot{w}_r$	Mechanical power of rowing
$\dot{w}_{tot}$	Total mechanical power
x	Distance along a lung capillary
X_{rs}	Reactance of the respiratory system (chest plus lung)
Z_{rs}	Impedance of the respiratory system (chest plus lung)
α	Angle between the mean body sagittal axis and the terrain
β	Angle between the terrain and the forward direction of the treadmill's movement
β_b	Oxygen transfer coefficient in blood
β_g	Oxygen transfer coefficient in a gas phase
γ	Angle between the terrain and the horizontal plane
Δ	Difference (delta) between two values, generic
ΔE_k	Kinetic energy change
ΔE_p	Potential energy change
ΔG	Free energy changes
ΔG^*_{ATP}	Molar free energy change of ATP hydrolysis
ΔG^*_{gl}	Free energy liberated by the oxidation of a glycosidic unit
ΔH	Enthalpy change

ΔP_i	Pressure gradient across the ith resistance to oxygen flow, where i indicates the position of the resistance at stake in a series of resistances.
ΔP	Pressure difference
ΔPCr_{ss}	Amount of PCr split in the transition from rest to exercise steady state
ΔP_n	Pressure difference across the nth (last downstream) resistance to oxygen flow in a series of resistances.
ΔP_T	Total partial pressure difference along the respiratory system
ΔS	Entropy change
ΔU	Overall energy change (in thermodynamics)
ΔV	Volume difference
Δv	Velocity change during a stride (or step)
ΔVO_2^{st}	Oxygen deficit due to oxygen store changes
ε	Thermodynamic efficiency
ε_R	Thermodynamic efficiency of oxidative ATP resynthesis
ε_S	Thermodynamic efficiency of ATP splitting
ε_{Tot}	Overall thermodynamic efficiency of the ATP cycle ($=\varepsilon_R\ \varepsilon_S$)
η	Mechanical efficiency
η_D	Drag efficiency
η_H	Hydraulic efficiency
η_O	Overall efficiency of locomotion
η_P	Propelling efficiency
η_v	Mechanical efficiency of vertical work
θ	Chemical reaction rate of carbon monoxide with haemoglobin
μ	Ratio between the non-aerodynamic component of the force opposing motion ($F_{n\text{-}a}$) and the overall efficiency of locomotion (η_O)
μ	Coefficient of kinetic friction
μ'	Dimensionless friction coefficient of skiing, depending on snow temperature
ν	Ratio between constant k_w (see above) and the overall efficiency of locomotion (η_O)
π	Oncotic pressure
ρ	Density of a fluid
ρ_0	Gas density at $P_B = 760$ mmHg and $T = 273$ °K
ρ_B	Body density
ρ_b	Blood density
ρ_w	Water density
σ	Protein reflection coefficient
τ	Time constant
τ_1	Phase I time constant of a double exponential equation
τ_2	Phase II time constant of a double exponential equation
τ^{al}	Time constant of the exponential decrease of PCr concentration
τ_L	Time constant of the mono-exponential lung oxygen flow increase
Φ	Ratio between body density and water density

φ	Fraction of $\dot{V}O_2^{max}$ that can be sustained to cover a given distance at the maximal possible speed
φ_1	First phase of the cardiovascular response to apnoea
φ_2	Second phase of the cardiovascular response to apnoea
φ_3	Third phase of the cardiovascular response to apnoea

Chapter 1
Before Margaria: Mosso and Herlitzka

Guido Ferretti

Abstract In this chapter, after a short synopsis of the evolution of the concept of natural philosophy and of its transformation into modern physiology at the beginning of the nineteenth century, an instant picture of the state of physiology when Angelo Mosso started his scientific career is portrayed. Then the personality and scientific contributions of Mosso are analyzed. Further, the role of Mosso's pupils and cultural heirs is summarized, with special attention paid to Amedeo Herlitzka, who was the master of Rodolfo Margaria. The cultural and political differences between the times of Mosso and those of Margaria are highlighted. The peculiar fate of the Jewish Herlitzka, a supporter of the fascist regime compelled to exile after the promulgation of the racial laws, is evidenced.

G. Ferretti (✉)
Department of Molecular and Translational Medicine, University of Brescia, Brescia, Italy
e-mail: Guido.Ferretti@unibs.it

G. Ferretti (ed.), *Exercise, Respiratory and Environmental Physiology*, Perspectives in Physiology, https://doi.org/10.1007/978-3-031-19197-8_1

Mosso's ergograph. With the kind permission of Accademia di Medicina, Torino

1.1 Introduction

This book overviews exercise, respiratory and environmental physiology from the viewpoint of a School, the School of Milano, established by Rodolfo Margaria during his entire, long career. The seed launched by Margaria in 1933 and sprayed by his mastery during some 40 years has generated a powerful and solid tree, which, in the third generation after him, still represents one of the pillars sustaining the evolution of factual and theoretical knowledge in integrative physiology.

The aim of this chapter is to introduce the men who forged Margaria as a scientist, set the Institute where he trained in Torino, and in a way can be considered as his predecessors: Angelo Mosso and Amedeo Herlitzka. There would not be a productive exercise physiology in Italy, there would not have been Margaria's epopee in the form which we know nowadays, without Mosso and Herlitzka. In turn, these two men are typical cultural fruits of an era, which we need to outline at least in its basic characteristics, if we wish to understand their parabola.

Thus, in this chapter, we firstly present a synopsis of the cultural climate in which the ancient Academies were transformed into modern Universities, within which the noble science of Physiology underwent a dramatic development, as did other fundamental branches of basic medical sciences. These transformations gave origin to the extraordinarily live and stimulating world of basic medical sciences in the second half of the nineteenth century, which affected the Italian academic environment deeply after the national unification and the proclamation of the Kingdom of Italy in 1861. The new unified State organized a modern national University system, following the example of the Middle Europe countries.

That was the time when Angelo Mosso started his academic and scientific life and became Professor of Physiology at the University of Torino. Amedeo Herlitzka, one of his successors in Torino, was a pupil of him, and was the mentor of Margaria. Thus, Herlitzka represents the chain connecting Mosso to Margaria. The personality and the scientific work of Mosso and Herlitzka are analyzed briefly in the remainder of this chapter.

1.2 Some Considerations on Natural Philosophy

In ancient times, the study of physiology was included in the more general category of natural philosophy. Natural philosophy is the branch of philosophy overlapping with natural science, inasmuch as it concerns the philosophical understanding of nature, whether quantitative and experimental, as in Alexandrian science in Hellenistic times, or after the scientific revolution of the seventeenth century, or qualitative and metaphysical, as in the Presocratics or in Middle Age philosophers. It included the study of inanimate objects, whence modern physics stemmed, as well as of living systems, whence modern biology and physiology were born.

The study of nature as a philosophical inquiry started in Greece, in the Ionian School, around the sixth century BC. Contrary to the Middle East theological empires, in which power relied on a coherent, comprehensive, and pervasive theology and was guaranteed by a powerful class of clergymen, Greek theology consisted of a system of humanlike infantile Gods watching on our destiny, but lacked a coherent arrangement explaining the relationship between Humans and the World. The door to the free inquiry of nature was opened, the curiosity of free Greek spirits trespassed that door, and the philosophical study of nature, as opposed to the theological explanation of nature, started.

When Thales of Miletus (≈625–≈545 BC) identified water as the basic principle of nature, through its changes of state, he formulated a conjecture open to criticism, and, perhaps inadvertently, set the conditions for the deep metaphysical discussions, which further development of Greek philosophy consists of: natural philosophy was born. Thales' pupil Anaximander (610–547 BC) introduced refutation: when he argued that a thing cannot be the contrary of itself, and thus that assuming water, which is essentially wet, as the principle of nature would preclude the existence of dry objects, he implied that Thales' conjecture was false. Criticism of philosophical conjectures played a fundamental role in the development of natural philosophy ever since. After a couple of centuries of even stirred debates, the philosophical theory of the four elements was set and generally accepted: earth (cold and dry), fire (hot and dry), water (cold and wet), and air (hot and wet). All things in nature would consist of different combinations of these four elements. This is definitely a theory, but not a scientific theory, as long as it does not foresee the interaction between theory and experiment: neither Thales nor Anaximander and their successors performed experiments. It is the outcome of metaphysical speculation about nature, a kind of speculation that characterized most of the theoretical settings of natural philosophy ever since.

Another highly influential philosophical evolution, which had a huge impact on subsequent natural philosophy, was the creation of the concept of teleology by Aristotle (383–322 BC). He inherited from Plato the idea that the world tends to perfection, and further developed that vision, in so far as perfection requires a kind of divine, immanent will, driving the world toward it. So, in his natural philosophy, concerning inanimate things, Aristotle conceived the world (κόσμος) as consisting of round bodies (stars, planets, and whatever, σφαῖραι) rotating along a perfect trajectory, and thus describing the circumference of a circle, the most perfect of all geometrical forms (Ross 1936). Concerning living systems, his main question was not: how does it work. It rather was: what is its purpose in the strive toward perfection. This being the context, experimental science was useless to his mind, because experiment alters nature and prevents observation of its perfection (Lanza and Vegetti 1971).

These views had an incredible impact on the developments of Natural philosophy in the subsequent centuries. Indeed, when, on the ashes of the Roman Empire, new theocratic reigns based on monotheistic religions took form, all found in Aristotle's teleological vision of natural philosophy a very useful instrument, allowing integration of natural philosophy in the dominating theological construction. Aristotle's natural philosophy became a dogma, with several consequences, among which the survival of almost all his writings to present times, the hindering of other approaches to the study of nature, the banning of any other discording theory, and the criminalization of dissection as a scientific practice.

Another major effect was the disappearance of most of the science that flourished in early Alexandria in the third century BC. The extraordinary fertility of freethinking in the newly constructed town, unbound from religious and psychological prejudices, generated a short yet golden era of scientific and experimental approach to the study of natural philosophy. In the fields of interest to us, men like Erasistratus and Herophilus, who first practiced dissection as a normal tool for the study of living bodies, humans included, have an extraordinary value. Unfortunately, their works did not survive and only fragments, reported in later writings, are available, largely for the aforementioned reasons. What appears from these fragments is that they attained an impressively detailed gross anatomical knowledge, which they tried to translate into functional knowledge. So, Herophilus (≈325–≈255 BC) distinguished sensory and motor nerves, identified skeletal muscles as the site of voluntary movement, and their contraction, stimulated by motor nerves, as the cause of it, developed a theory of vision as a brain activity, created a mechanical vision of the arterial pulse, which he used as a diagnostic tool (Von Staden 1989). The best-known *corpus* subsuming the anatomical and functional knowledge of antiquity is Galen's, which partly acknowledged the work carried out in Alexandria 400 years earlier.

A scientific approach to the study of nature developed, or was resumed, at the time of the scientific revolution of the sixteenth to seventeenth centuries AD, giving raise to the birth of modern science. In parallel, the natural philosophers, prone to the new scientific approach, felt the need of increasing the dissemination of the new knowledge that they were generating. The first modern philosophical, and thus scientific, Societies were founded in that period to this aim.[1]

Those events have been so widely and deeply analyzed that it is useless to summarize them in the few lines that we could reserve to them in this context. Suffice it to say that the scientific revolution overthrew Aristotle's dogmas and encompassed all fields of natural philosophy. Suffice it to say that Isaac Newton entitled his masterpiece *Philosophiae Naturalis Principia Mathematica (Mathematical Principles of Natural Philosophy*), to understand the philosophical context in which he operated.

The natural philosophy of living systems was fully part of the scientific revolution of those times. That was the golden age of classical gross anatomy, when Andreas Vesalius (1514–1564) criticized Galen's anatomy and, with a typical scientific approach, corrected it under several respects (Vesalius 1543). Shortly afterward, William Harvey (1578–1657), based on theoretical reasoning after gross experimental evidence, proposed that blood must circulate in the body, although no anatomical evidence of a communication between arteries and veins was available yet (Harvey 1628). The subsequent invention of the microscope led Marcello Malpighi (1628–1694) to the discovery of capillaries, thus providing the first experimental evidence of a structural basis for blood circulation (Malpighi 1661). In the field of interest of this book, we remind also the work of Santorio Santorio (1561–1636), who first tried to measure heat production by muscles, initiating in a way the scientific study of metabolism (Santorio 1612), and that of Giovanni Alfonso Borelli (1608–1679), who first interpreted the movements of animals on mechanical laws and laid the foundations of human flight (Borelli 1681). Moreover, Nils Stensen (1638–1686) formulated a theory of muscle contraction based on microscopic observations (Stensen 1667). On Santorio and Borelli, see respectively Reggiani (2012), and Maquet (1989) or Derenzini (1958); on Stensen, we remind the nice analysis of his theory of contraction by Kardel (1990).

An important aspect of the work of these scientists was the attempt at applying the new physical laws to the study of living systems, along the principles of what is now called the iatromechanical school. In fact, they were natural philosophers who combined the nature of inanimate objects (physics) to that of living systems (biology) as the starting axiom of their work. These attempts represent a major step on the pathway that led to the establishment of physiology as an independent branch of

[1]To the best of our knowledge, the first modern philosophical Society was the *Accademia dei Lincei*. It was established in 1603 by Federico Cesi (1586–1630) in Rome, with the aim of collecting, exchanging, and disseminating knowledge in all fields of natural philosophy, beyond any bond of tradition and authority. Its most illustrious affiliate in those times was Galielo Galilei, who was admitted in 1611. *L'Académie Française* and the Royal Society of London followed, respectively, in 1635 and in 1660.

science. Another important step on this pathway, about one century later, consisted of the coupling of chemistry to biology, as happened after the identification of oxygen and carbon dioxide, which led to the demonstration that oxygen is consumed and carbon dioxide is produced in animal metabolism (see Chap. 2).

1.3 The Transformation of Ancient Academies in Modern Universities

At the beginning of the eighteenth century, the technical developments, made possible after the birth of modern physics, progressively generated an industrial revolution in the United Kingdom, in the German states and in France. A British philosopher, John Locke (1632–1703), designed new forms of state organization, which we call liberal. The liberal state started to appear more apt to govern societies, in which the industrial revolution was moving economic power from the rural aristocracy to the industrial bourgeoisie, which however felt deeply underrepresented in the absolute monarchic states of the *ancien régime*. As a consequence, in the second half of the eighteenth century, gorgeous social, cultural, and political changes shattered Europe and revolutionized it. In France, these changes led the Monarchy to crumble in the French Revolution and exited in the deep political reforms introduced by Napoleon Bonaparte. Meanwhile, the liberal regime consolidated in the United Kingdom and in the Netherlands, whereas the revolt of the British colonies on the East Coast of North America and their declaration of independence led to the creation of the United States of America, ruled by a Constitution based on John Locke's liberal principles.

At the same time, new cultural energies were liberated. Along the same critical line of Locke, Voltaire (1694–1778) in France created the Enlightening movement, which put Reason at the center of philosophical thinking, promoted freedom as a necessary state of mind for cultural development, and supported a liberal reformation of the State. In Germany, a less radical movement than the Enlightenment, called *Aufklärung* (Elucidation or Clarification), took over some aspects of the Enlightenment to promote moderate reforms in support of the common people across German courts. It took not the strong political orientation of the Enlightenment and of Locke's political philosophy; it maintained the central role of absolute monarchies in the political systems; nevertheless, it promoted some reforms by the princes and the kings in support of the cultural growth of their subjects. Amongst others, the *Aufklärung* gave education a central role in its political reform program, promoted "philanthropism" and sustained the idea of a popular school, supported by the Courts, aimed at improving the cultural status of men through the use of "reason," pursuing happiness and popular solidarity. The pedagogic ideas of the *Aufklärung* played an important role in the transformations of ancient Academies in modern State Universities at the beginning of the nineteenth century.

Moreover, the recent archaeological discoveries shed new light on the greatness of past civilization. New interests in the study of history arose and generated a novel cultural environment tending to the valorization of classical harmony of ancient artworks in opposition to the baroque style: Neoclassicism, especially promoted in Germany by Johann Joachim Winckelmann (1717–1768) and taken over in France by the Empire style at Bonaparte's times. Neoclassicism gave an intrinsic cultural value to art pieces as such, the fruition of which was to be opened to as many people as possible in public establishments (Museums), along the educational principles of the *Aufklärung*. These principles were not foreign to the creation of State Universities as sites of fruition of philosophy, history, and science, by the largest possible audience.

Neoclassicism contained also the seeds of another, deeper, and more pervasive cultural movement, Romanticism, which contained the germs of nationalism in it. Romantic ideas were not foreign to the birth of independent Greece, to the *Risorgimento* (Resurgence) in Italy, whence the political unification of Italy stemmed, to the naissance of the unified Kingdom of Germany under the Prussian Court. Moreover, Romanticism pervaded also the ideological construction of the new French state after the revolution, and a new competition arose among Nations in Europe, which consisted of not only political confrontation and wars, but also of economic competition and, most important to us, cultural competition, through the richness of their Museums and the cultural power of their Universities.

Before the French Revolution, the educational and cultural power was tightly controlled by religious orders in most European countries, either directly or, as in many reformed cities in Germany and Switzerland, indirectly, through the control exerted by theocratic states. Exceptions to these, since the sixteenth century, were the free Colleges in England, Scotland, and Holland, and the relative tolerance allowed by the Republic of Venice to its academic institutions, primarily the University of Padova. The fall of the cultural monopole of the Church of Rome in catholic countries at the time of the French Revolution and of Bonaparte's empire, [2] and the pervading influence of the *Aufklärung*, not only in Germany, but also in the Austrian Empire, created the new political conditions wherein the State, either revolutionary, as in France, or absolutist, as in the Austrian Empire and in the German kingdoms, focused on Academia and transformed it into modern State Universities, creating new Faculties. In this context, science started to be supported by the States as an instrument of glory and power to them.

---

[2]Napoleon Bonaparte himself promoted the creation of a new Academy in Milano, then capital town of the France satellite state known as *Repubblica Cisalpina*. That Academy, established in 1797, survived the remarkable and turbulent political events of the following two centuries and is still lively active in Milano as *Istituto Lombardo – Accademia di Scienze e Lettere*. Several physiologists issued from the School of Milano were or are members of it, namely Rodolfo Margaria, Emilio Agostoni, Giovanni Cavagna, Paolo Cerretelli, Edgardo D'Angelo and, as corresponding members, Pietro Enrico di Prampero and Guido Ferretti, who, incidentally, are two of the authors of this book.

1.4 From Natural Philosophy to Physiology

Some of the natural philosophers of the sixteenth and seventeenth centuries, actually scientists in the modern sense, operated, at some times, in Academic establishments (Vesalius, Santorio and Harvey in Padova, Borelli in Messina and Pisa, Malpighi in Bologna and Messina), but not necessarily. Science was not a recognized profession, was a hobby, an intellectual pleasure. Many scientists of those days were wealthy gentlemen, some earned their wages from other related jobs. The concept of science as a profession was foreign to the minds of scientists in those times.

Nevertheless, the process leading to the creation of modern State Universities, which we outlined in Sect. 1.2, was close to start. By 1815, after Napoleon had been definitely defeated in Waterloo, most of the ancient academic institutions were already within the new educational system. New Universities were also created, and science became a profession, a rewarding profession. In medical schools, a newly structured program of medical training was progressively consolidating. This prompted a huge development of knowledge in many fields, implying the definitions of new boundaries among different branches of science, i.e., of natural philosophy. The branch of natural philosophy concerning the functional studies of living systems along the principles of physics and chemistry was named **Physiology**.

The first scientific journal formally dedicated to physiology, the *Journal de Physiologie Expérimentale*, started to be published in 1821, under the auspices of François Magendie (1783–1855), who was Professor of Medicine at the *Collège de France* in Paris. Jan Evangelista Purkyně from Bohemia (1787–1869), who first described the so-called Purkyně fibers of the heart, was the first natural philosopher to be formally appointed professor of physiology by a university establishment: it happened at Breslau, Prussia (now Wrocław, Poland) in 1823. The first textbook containing the word physiology in its title appeared in 1837: it is Johannes Müller's *Handbuch der Physiologie des Menschen für Vorlesungen*. Müller (1801–1858), who studied natural philosophy at the medical school in Bonn, and then trained in anatomy at the University of Berlin under Karl Rudolph (1771–1832), obtained a lectureship in physiology at Bonn in 1824, where he became the first full professor of physiology in 1830. In 1850, most medical schools had at least one Professor of Anatomy (structure), Physiology (function), Pathology (alteration of structure and function), Medicine, and Surgery. This structure of medical schools still represents the core of medical training nowadays, although the subdivision of branches never stopped, prompted by the continuously growing body of medical and biological knowledge. The era of natural philosophy was over; the era of physiology had risen and was shining splendidly.

1.5 An Instant Picture of Physiology in the Second Half of the Nineteenth Century

The birth of modern universities, the creation within them of chairs of physiology, and the increasing availability of public resources for free experimentation in university laboratories set the conditions for an impressive development of physiological knowledge. Many of the basic principles on which contemporary physiology still relies nowadays were set in those times.

Just to stay on concepts pertinent to the future development of the School of Milano, when young Angelo Mosso joined the laboratory of physiology in Florence to train with Moritz Schiff (1823–1896)—Italian universities in those times were much more open to hiring foreign scientists than they are nowadays—an instant picture of contemporary physiology shows a bright panorama. Claude Bernard (1813–1878) had created the concept of *milieu intérieur* (Bernard 1859, 1865). Carl Ludwig (1816–1895) had redefined the principles of the old iatrochemical school in a more solid and comprehensive theory of physiology as the science of physics applied to living systems (Ludwig 1858), providing the basis for a new interpretation of respiratory gas exchange and blood circulation in terms of pressure–flow relationships. Along this line, he obtained the first measurements of pressure in the pleural space. He also demonstrated the effect of excitatory (sympathetic) and inhibitory (parasympathetic) nerves on the sinus node of the heart. Moreover, Adolf Fick (1829–1901) had formulated the laws of diffusion (Fick 1855) and the law describing the relationship between oxygen flow and total blood flow (cardiac output) (Fick 1870). Felix Hoppe-Seyler (1825–1895) had demonstrated the nature of the pigment of the blood, which we now call hemoglobin (Hoppe-Seyler 1862). Paul Bert (1833–1886) had already published his classical textbook on the comparative physiology of respiration (Bert 1870). And Hermann von Helmholtz (1821–1894) had demonstrated that the law of conservation of energy applies also to living systems (Helmholtz 1847).

It was in that context that Angelo Mosso entered the game of physiology. He threw a fertile seed in a fertile land. The ground was rich enough to sustain the roots that developed from that seed.

1.6 Angelo Mosso and his Pupils

Angelo Mosso (1846–1910) (Fig. 1.1), from Torino, graduated in medicine in the same town in 1870. After serving the army as a medical officer, he joined Moritz Schiff in Florence, then, in 1875, he moved to Leipzig, in Carl Ludwig's laboratory, where he spent a couple of years. There, he started working on respiration and he was among the first to use the kymograph for recording respiratory movements. In those years, he was immersed in the core of contemporary physiology and he got acquainted with several of the most prominent physiologists of that time.

Fig. 1.1 Portrait of Angelo Mosso. From Sandrone et al. (2012)

Shortly after returning to Torino, when Jakob Moleschott (1822–1893) left Torino for Rome in 1879, Mosso replaced him as Professor of Physiology in his hometown, where he remained until his death. Under his leadership, the Institute of Physiology of the University of Torino took an international allure. It became a pole of attraction for physiologists from all over Europe, especially after he promoted the establishment of a high-altitude laboratory at the Capanna Margherita (m. 4560 above sea level, inaugurated by Queen Margherita di Savoia in 1893, see Fig. 1.2), and a permanent altitude laboratory at Col d'Olen (m. 2900 above sea level, established in 1907), which after his death was entitled to him. Both are located on the Mount Rose massif, in the Pennine Alps (Cogo et al. 2000).

He left numerous pupils, who continued his teaching and his mission. Let us cite, amongst others, Vittorio Aducco (1860–1937), Professor of Physiology at Siena from 1891 to 1894, then at Pisa from 1894 to 1937, who experimented on muscle

Fig. 1.2 Picture taken on the inauguration day at Capanna Margherita, Mount Rose, August 18, 1893. The two red arrows indicate, from left to right, Angelo Mosso and Margherita of Savoy, Queen of Italy. With the kind permission of the University of Torino

physiology along Mosso's line of thinking and studied neural control of breathing, identifying coordinated inspiratory and expiratory centers in the bulb; and Mariano Luigi Patrizi (1866–1935), Professor of Physiology at Modena from 1898 to 1924—with the exception of 3 years, from 1909 to 1912, which he spent in Torino to teach criminal anthropology as the first successor of Cesare Lombroso—then in Bologna until 1935, who was among the first to study psychophysiology in Italy.

As his successors in Torino, Mosso left, on one side, Piero Giacosa (1853–1928), who became professor of *Materia Medica*—we would call it pharmacology nowadays—and who studied the physical chemistry of blood, eyes, and eggs and the therapeutic properties of several alkaloids extracted from various herbs. On the other side, at the Institute of Physiology, Mosso left first of all Amedeo Herlitzka (1872–1949), who succeeded him directly as Professor of Physiology. Nevertheless, we owe a special mention also to Alberto Aggazzotti (1877–1963) (see Fig. 1.3), who chaired the Angelo Mosso laboratory at Col d'Olen (1915–1923), then became Professor of Physiology, firstly in Catania, then in Modena, from 1924 to 1952. In the latter town, he was Dean of the Medical School from 1935 to 1944. Aggazzotti continued the work of Mosso on respiration and was one of the first to investigate acute mountain sickness. His seminal and precursor works on gas exchange of chicken eggs (Aggazzotti 1913, 1914) were highly praised by Hermann Rahn in

Fig. 1.3 Portrait of Alberto Aggazzotti. From Rahn (1988). Picture forwarded by Oreste Pinotti, Torino, to Hermann Rahn and Paolo Cerretelli, reproduced with the kind permission of Paolo Cerretelli

Buffalo (Rahn 1988), and this established a further, indirect and unexpected ideal link between the Schools of Milano and Buffalo.

We finally remind Carlo Foà (1880–1971), perhaps the youngest pupil of Angelo Mosso, whose reputation as a physiologist highly relies mostly on his work on the sexual glands. Foà became Professor of Physiology in 1914 at Messina, whence he moved to Parma in 1918, then to Padova in 1922. In 1924, he was called to the newly established University of Milano, where he created the first laboratory of physiology. When, being Jewish, he was forced to quit in 1938, because of the racial laws of the fascist regime, despite his support to the regime (Troiani and Manni 2007), his position was taken by Margaria. Foà took refuge in Saint Paul, Brazil, where he remained until the end of World War II, when he returned and was reintegrated in the professorial roles of the University of Milano.

Many articles have been devoted to analyze Mosso's vast scientific activity (amongst others, see Cogo et al. 2000; Di Giulio et al. 2006; Di Giulio and West 2013; Gori 2002; Herlitzka 1947; Foà 1957; Lankford 2015; Losano 1996). His pioneering work on altitude physiology was mostly analyzed. In 2016, pH, the

Journal of the Italian Physiological Society, dedicated a special issue to Mosso (Ferretti and Conti 2016).

The importance and popularity of Angelo Mosso went well beyond his academic role as a teacher and a scientist, for his societal engagement. He promoted physical education and sport practice as a tool to improve soldiers' and workers' health, in opposition to the static German gymnastics, which predominated in Italy in those times. We owe much to him, if physical education became a compulsory subject in the educational programs of public Italian school system. We owe much to him if physical training was introduced in the Italian army. He summarized his vision of exercise and health in an important book on physical education (Mosso 1893). In the last years of his life, when he was already affected by the disease that would have brought him to death, and he was counseled to escape altitude for the seaside for health reasons, he devoted himself to the practice of archeology, in which, as a neophyte, he nevertheless obtained remarkable successes by applying his cute scientific attitude. He visited Greece several times, he spent long months in Crete to study Minoan excavations and artifacts, he reported his results in remarkable publications (Mosso 1907, 1910). Mosso's popularity was so high that, when he died, *La Stampa*, the main newspaper of his hometown, Torino, dedicated an entire page to his obituary.

The international reputation of Angelo Mosso was also very extensive and widespread. He was a member of numerous international academies and societies, including l'*Accademia dei Lincei* in Italy, the *Institut de France*, and the Swedish Royal Academy of Science. In 1904, King Vittorio Emanuele III appointed him Senator of the Kingdom of Italy. He died on November 24, 1910. On December 10, 1910, he was solemnly commemorated at the Senate by Luigi Luciani (1840–1919) (Fig. 1.4), colleague Senator and Professor of Physiology himself at the University of Rome.

1.7 An Overview of Mosso's Scientific Activity

This overview makes use of Luciani's words at the Senate. It is not a random choice. The wish is to give a lively idea of how the value of Mosso's scientific activity was perceived in his time. In his intervention, Luciani cited two of the most prominent works of Mosso (Mosso 1880, 1897) and a subsequent German textbook on altitude physiology (Zuntz et al. 1906). The text is the English translation made by Guido Ferretti for pH (Ferretti and Conti 2016). It is reproduced with the kind permission of the Italian Physiological Society.

> *...Mosso combined the figures of experimental physiologist, scientific divulgator, promoter of physical education among the youngsters and, in the last period of his life, of archaeologist explorer. As a physiologist, he devoted his experimental studies to a variety of subjects, and he left traces of his work in various chapters of physiology. The most fortunate of his studies were aimed at obtaining the graphical expression and the mechanical explanation of several physiological movements. This is not the site for mentioning them all and analyzing their intrinsic value and the originality of the methods that he devised. I would mention only*

Fig. 1.4 A portrait of Luigi Luciani in his office at the University of Rome in 1913. Property of the National Library of Medicine, Bethesda MD, USA, in public domain

that, with his plethysmograph, he obtained the curves of volumetric oscillations either of organs, which, after they had been recently detached from a living body, were exposed to artificial circulation, or of the limbs of a normal man, due to changes of vascular tone and to arterial pulsation. His numerous studies "On the blood circulation in the human brain" led him to be awarded the Royal prize of the Lincei Academy in 1879, because the ensemble of the phenomena that he described raised the hope that he had opened a new way to elucidate the physiological changes (at least the most striking) associated with psychical activity.

Among the most ingenious works by Mosso, I count those wherein he analysed the various forms of pulsatile oscillations, either positive or negative, that depend on heart activity and are transmitted to the arteries, the veins, the alveolar air, the thoracic and abdominal walls. These are long-lasting contributions, because they are based on studies carried out with perfect technique and they are driven by brilliant critical discernment. In the study of muscular contractions, he introduced another special instrument that he invented, the ergograph, with which he could obtain the graphical representation of the time course of fatigue of the flexor muscles of the hand, and study how fatigue was affected by a variety of intrinsic and extrinsic conditions. Following the path opened in France by Marey, he contributed, with his sphygmomanometer, to the perfection of the method for obtaining an indirect yet operator-independent measure of the mean pressure of the aorta, thus making the study of blood pressure changes in a variety of diseases a real clinical possibility. These works, together with other, yet less important works on physiological mechanics, certainly constitute the main core of the scientific opus of Mosso, the reason why his name will remain honorably in the hall of fame of physiology.

Less fortunate were his researches on the genesis and metamorphosis of blood corpuscles, in which he defended hazardous ideas, which were not supported by further results. Yet also these attempts proved the originality of his genius, as long as he could stimulate a process of revision of the ongoing doctrines on hematopoiesis, which was useful to the progress of science, the promotion of which, as is well known, may profit more of a fruitful error than of a sterile truth.

His works on the mechanism and the chemistry of respiration and on respiratory innervation were most numerous and altogether most important. His doctrines on periodic breathing, on superfluous breathing, on apnea, on the so-called acapnia, proposed to be one of the main causes of mountain sickness, were and still are a matter of discussion and generated a wide literature, both in Italy and abroad. One may rightly observe that his studies heralded rapidly a very vast field of scientific exploration, without exhausting the single subjects. He certainly threw a lot of seeds along his pathway, some of which produced remarkable fruits. In 1897 he published a richly illustrated book entitled "The physiology of man on the Alps", in which he collected in elegant form, largely accessible to cultivated persons, foreigners to physiological studies, the sum of the researches by his collaborators and himself on the physiological effects of climbing onto the Mount Rose.

If one considers that some years later, in 1906, four German physiologists, Zuntz, Loewy, Fr. Mueller, W. Gaspari, published another bigger volume "On the ascensions and the climate in the mountains", in which all subjects already treated by Mosso underwent rigorous experimental control and were more deeply investigated and discussed, we can well conclude that the genial initiatives of our mourned colleague were very remunerative for science.

We should not forget that the Istituto Scientifico Internazionale, solidly built at Col d'Olen, at the foot of the Mount Rose, at an altitude of about 3000 meters, was honorably given his name, because he promoted its construction, with the help of the Patrons of science, led by our beloved King and by the Mother Queen. Many scientists in the fields of geophysics, meteorology, zoology, physiology and pathology every year convene there for carrying out their studies…

Beside his scientific achievements, Mosso deserves consideration as a constructor of new tools and instruments for physiological measurements on humans. Apart from the kymograph, which he brought back from Leipzig, he constructed, under the impulse of Carl Ludwig, the first plethysmograph, allowing measurements of volumes and volume changes of complex geometrical parts of bodies. A kind of balancing bed allowed him to analyze the distribution of blood in the body (Mosso 1884). Great importance has the construction of the ergograph, an instrument for measuring muscle force, which he used not only to study fatigue (Mosso 1889, 1891) but also to investigate the effects of altitude exposure on muscle contraction (Mosso 1897). He finally realized a pneumatic chamber, similar to that developed in Paris by Paul Bert, in which he could study the effects of changes in barometric pressure on respiration. A nice report of Mosso's instrumentation was provided by Giovanni Losano and Marco Galloni for the special issue of pH, the journal of the Italian Physiological Society, on Mosso (Ferretti and Conti 2016).

1.8 Angelo Mosso on Altitude

At the Capanna Margherita, Angelo Mosso investigated periodic breathing and the ventilatory responses to hypoxia. Using the ergograph, he showed that local muscle fatigue followed the same patterns at sea level (Torino) as at altitude (Mount Rose). Moreover, he demonstrated the existence in nature of anaerobic energy sources, well before the meaning of anaerobic glycolysis, lactate accumulation was understood, and phosphagen discovered (Mosso 1897).

Concerning breathing, Mosso was involved in a controversy, as long as he was convinced that acapnia, or better hypocapnia, rather than hypoxia, was the main cause of respiratory alterations at altitude. Hugo Kronecker (1839–1914) shared the same view, after he reported on the beneficial effects of carbon dioxide breathing on a monkey exposed to extreme hypoxia (equivalent altitude 8800 m) in a pneumatic chamber (Kronecker 1894). The acapnia hypothesis was challenged, mostly by Adolf Löwy (1862–1937) and Nathan Zuntz (1847–1920) (Löwy et al. 1897; Zuntz et al. 1906), who were in favor of the hypoxic hypothesis of Paul Bert (Bert 1878), and particularly by Joseph Barcroft (1872–1947), who reported values of arterial oxygen partial pressure (P_aO_2) in hypoxia (Barcroft 1914). In subsequent years, the experimental evidence led to a clear refutation of the acapnia hypothesis, in particular after the demonstration that the carotid body was an oxygen sensor (Heymans et al. 1934). In fact, the alveolar gas equation provided an explanation of Mosso's result: hypoxia is the main stimulus of hyperventilation at altitude, but hypercapnia reinforces hyperventilation, thereby further elevating P_aO_2. The symbiotic interaction between central and peripheral chemoreceptors may explain this issue (Löschke et al. 1958; Nielsen and Smith 1952). Kellogg (1978) attributed Mosso's error to possible carbon dioxide accumulation in the pneumatic chamber.

Angelo Mosso was interested also in understanding muscle performance during altitude exposure, stimulated by the long history of altitude explorers, and more recently climbers, reporting an impairment of performance in long-duration exercise with increasing altitude. In 1880, Edward Whymper (1840–1911), the first climber of the Matterhorn, when he was in Quito (3100 m), Ecuador, preparing the first ascent of Chimborazo (6420 m), measured the time needed to run a mile at the maximum possible travel speed and compared the results with previous determinations at sea level. He reported a performance decrease of about 8% at altitude.

In 1894, at Capanna Margherita, as compared to Torino, Mosso performed force and work measurements on small muscle groups. He summarized his results as follows (Mosso 1897) [3]: *I made the experiments with the ergograph on August 15 and 16, several days after my brother had come to the Capanna Regina Margherita, and when he was already acclimatized, because the early days he did not feel completely well. The amount of mechanical work done in the path B of Fig. 3* (current Fig. 1.5) *is a little less. The curve (except toward the end, where there is a slight rise) reproduces the pattern of the previous figure, that is A, made in lowland. I repeated similar experiments on myself, Beno Bizzozzero, and all the soldiers of the expedition; I found none on which the barometric depression at 4560 meters had produced a very considerable decrease in strength. Of course, I am talking of the paths taken after the fatigue produced by the ascent had disappeared. So it can be*

[3] In the paragraphs on Angelo Mosso's scientific activity, the parts in italic are English translations of the original reports by Mosso himself, made by Guido Ferretti for the special issue of pH on him. The Figures illustrating these paragraphs are from the same sources.

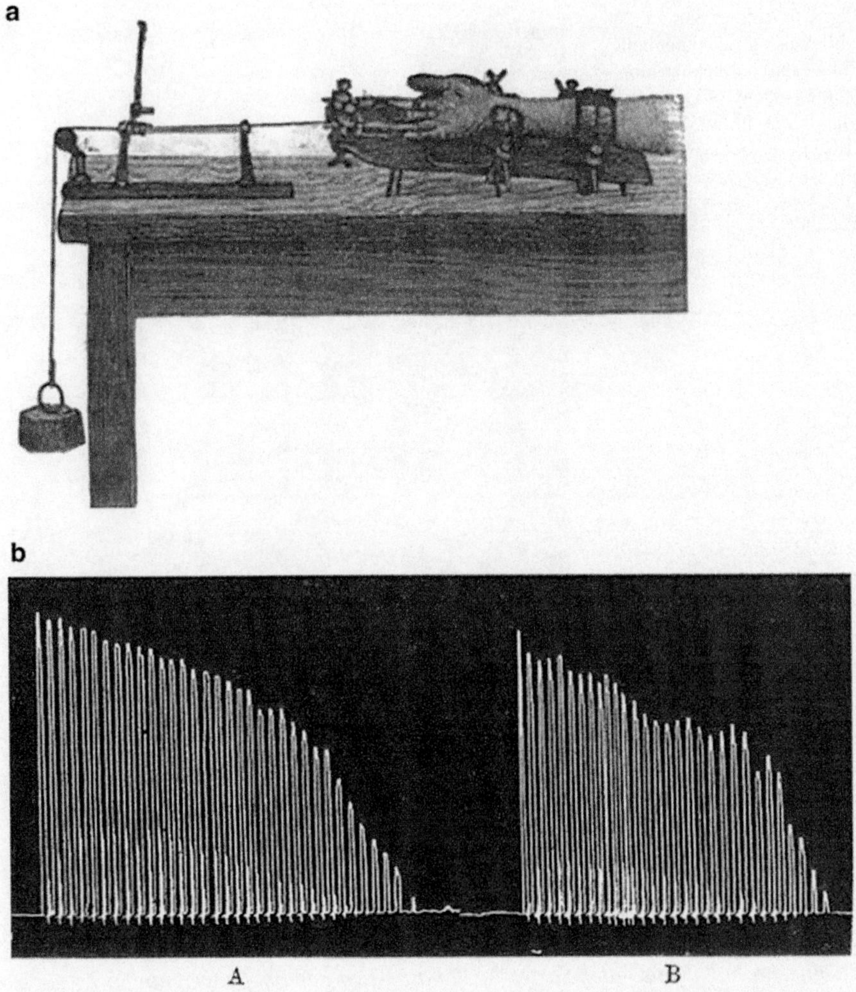

Fig. 1.5 Angelo Mosso's digital ergograph (top panel) and tracings recorded on his brother Ugolino (lower panel) during the exhaustive exercise of lifting 4 kg with the middle finger every two seconds, in Turin (**a**) and at Capanna Margherita (4560 m) (**b**). From Mosso (1897)

said that on Mount Rose at 4560 meters, muscle strength has slightly decreased, and the type and characteristics of the individual effort somewhat changed, as you can see with the ergograph. In some people, the differences between these recordings, concerning the work done, were so small that these variations can be included in the diurnal oscillations.

Mosso's ergographic data referred to very light loads. He was surprised by these results, and he challenged them by performing tests during arm exercises with bigger muscle masses (Fig. 1.6).

Fig. 1.6 Soldier Sarteur while raising two dumbbells (5 kg each) until exhaustion, on the balcony of Capanna Margherita. From Mosso (1897)

He described the results as follows: *In Fig. 4* (current Fig. 1.6) *we see one of these experiments made on a day of good weather on the balcony of the Regina Margherita hut. The soldier Sarteur is holding two dumbbells weighing 5 kg each. A metronome beats the seconds, and he raises the dumbbells over the head every 4 seconds. Assuming that this is the starting position, after one second, by flexing his arms, the dumbbells will be at chest height. In the next second, after stretching his arms, the dumbbells will drop close to the knees. From this position after one more second, he will return to flex his arms and bring the dumbbells at chest, and then the last second, by stretching his arms, he will lift the dumbbells over the head, returning to the position indicated by the figure. In that position, he waits one second and then will put the dumbbells close to his chest, and so continuing until he can endure. By counting the number of lifts before exhaustion (determined by his inability to sustain the initial load), it is possible to determine the overall work and the power developed in various exercises. We repeated these tests in the camps as we were going onto the summit of Mount Rose, and to my surprise, I found that at 4560 meters far more work had been done than in Turin.*

Apart from the impressive clarity of the exposition, demonstrating the vivid and plain description of scientific data in those times as compared to present days, these results were even more surprising than the previous ones to Mosso's mind. He hypothesized, quite correctly, that the work increase at altitude was at least partly due to a training effect, inasmuch as the soldiers competed among them in lifting weights for fun during the climbing pauses. Mosso's enigma can be formulated this way: why running a mile at Quito implies a fall of performance, whereas lifting weights at Capanna Margherita does not?

Several considerations could be done to this purpose. Firstly, the exercise intensity of Mosso's weightlifting experiments was probably submaximal rather than maximal: the soldiers did not lift the dumbbells as fast as possible but were asked to follow a predetermined relatively slow rhythm dictated by a metronome. Secondly, current models of maximal oxygen consumption limitation establish a clear difference between whole-body exercise, like running, and exercise with small muscle masses, like the one performed by Mosso's soldiers (see Chap. 7). In the former case, limitation is central and is largely imposed by oxygen delivery, which is impaired at altitude; in the latter case, limitation is largely peripheral, being imposed more by muscle fatigue than by oxygen delivery. Last, but not least, when Mosso performed his experiments, the glycolytic pathway was not completely elucidated, Hill and Meyerhof's theory of the energetics of muscular exercise was far from being formulated, and even farther were Margaria's refutation of it and the discovery of high energy phosphates. Margaria's revolution was yet to come: Mosso was a necessary precursor of Margaria, who operated amidst the knowledge of his times.

Paolo Cerretelli, whose collaborators repeated Mosso's experiment, had a different view of the issue. According to him, Mosso raised the issue of blood flow partition and regulation in muscles during exercise. In fact, Cerretelli argued that the performance of maximal exercise with a small muscle mass implies a higher maximum blood flow per unit of active muscle mass, than that observed in the same muscles during exercise involving large muscle masses. This may be of little importance in normoxia, yet in hypoxia, an extra increase in the absolute blood flow in a given limited muscle mass may improve muscle oxygen availability, thus leading to a greater capacity for work (Kayser et al. 1994). To our opinion, however, Cerretelli underscored the fact that Mosso's soldiers performed submaximal exercise.

A few years later, Zuntz et al. (1906), with Mosso's logistic support, measured oxygen consumption and carbon dioxide production in humans during uphill walking at various altitudes up to Capanna Margherita. These authors filled a gap between Whymper's anecdotal observation at Quito on running the mile and a rigorous experimental demonstration of the deterioration of exercise capacity at altitude, along the theoretical thinking of Bert. Moreover, this data allowed the earliest computation of the energy cost of vertical work at altitude, suggesting that the mechanical efficiency of walking is the same at altitude as at sea level. Of course, Zuntz had no acquaintance with the concept of energy cost of locomotion, yet by that study, he established a new unpredictable bridge between Mosso and Margaria (Margaria 1938).

1.9 Angelo Mosso and the Firefly

In the time of Mosso, it was a common belief that the chemical energy sustaining muscular work was to come from the oxidation of energetic substrates, particularly carbohydrates. So, Mosso could not believe the reports of his old guide and friend Jean-Joseph Maquignaz[4] (1829–1890) from Valtournenche, Aosta Valley, Italy—Valtournenche is located on the Italian side of the Matterhorn –, who claimed that at an altitude above 4000 m it was hard to work, because whenever men tried to break rocks, they had to stop after a few shots for breathlessness. There was no clear connection between ventilation and muscle performance: fatigue could be only of muscular origin. Although it was known that both metabolism and ventilation increase at exercise, the effects of altitude on maximal oxygen consumption were yet to be determined.

However, Mosso performed an observational experiment on fireflies, which suggests that Maquignaz might not be fully wrong, and Mosso acknowledged this, and which he reported as follows (Mosso 1897): *I made tests on fireflies, which, by means of phosphorescence, let us know better about the excitation of the nervous system. The phosphorescence of these insects is a chemical process, such that the light irradiated from phosphorus, although the process was incomparably different in its intimate nature. I was very interested in seeing whether oxygen reduction in rarefied air would modify the vital phenomenon of light. For this reason, I took a few fireflies and put them under a large glass bell, in which I slowly made air rarefied, until the pressure was only 30 centimeters* (of mercury), *as it is on the highest peak of the Himalayas. I was amazed to see that their brightness grew, the more oxygen diminished as a consequence of air rarefaction. We all know that the firefly shines with intermittent periods of light and darkness. As soon as air became rarefied, such periods ceased, and the fireflies kept constantly bright. Even if after half an hour, or an hour, they had been in such a rarefied air that a dog or a man would die, these beetles were shining with a brighter light than they show in our meadows at the time of their loves. The color of the light was less yellow and took a more fiery and red tint on the edge of the segments. During summer nights, a man who picks up a firefly sees that only the two segments in the back of the body are bright. In the rarefied air, the body of the fireflies stretches by about three millimeters in the bright posterior part. For all of them, the period of rest, or of darkness, of light organs becomes shorter, so that the light is continuous, with periods of more intense and fainter glow. We believed from physiologists that the light these animals emit was a life phenomenon closely linked to the oxygen in the air. Now we see that this is not true.*

[4] Jean-Joseph Maquignaz, head of the alpine guides in Valtournenche, was one of the best climbers from the Aosta Valley in those times. In 1865, he was part of the group of Italian climbers, led by Jean-Antoine Carrel, who attained for the first time the summit of Matterhorn from the Italian side, three days after the first ascent by Whymper's group. In 1882, he was the first to attain the summit of the Dente del Gigante, in the Mont Blanc massif, the last inviolate summit higher than 4000 m in the Alps. He disappeared in 1890 during a climb on the Mont Blanc. His body was never found. Most of the greatest English climbers of that era had been his clients.

Therefore, Maquignaz could be right, sure, but there is much more in this: the first demonstration of anaerobic energy production in living systems. Moreover, as Cerretelli pointed out (Cerretelli 2014), this observation opens the way to admitting that even in humans cell metabolism may be less closely linked to oxygen than was believed up to then. The fireflies experiment provided evidence that the cells of light organs must contain some substances allowing them to produce light without the combustion of oxygen.

Only some 30 years later, when Karl Lohmann (1898–1978) discovered "phosphagen" and Einar Lundsgaard (1899–1968) performed his celebrated experiment on the contraction of muscles poisoned with monoiodoacetic acid (see Chap. 3), it was understood that high energy phosphates actually provide the energy for muscle contraction. Mosso was obviously unaware of the existence of anaerobic energy sources. Notwithstanding this, he might have perceived the possible existence of energy primary sources in the cell, presumably also in the muscle fiber, capable to develop high power levels, but for a short time. Mosso showed the muscle ability to develop very high powers, sustainable only for a few seconds, without any increase in oxygen consumption, thanks to the energy generated by the transformation of a specific unknown substance, which is now known under the name of phosphocreatine. It is at the same time tempting and intriguing to conclude that the first hint toward Margaria's revolutionary concept of oxygen debt (Margaria et al. 1933) might have come from a minimalistic experiment of Angelo Mosso on little fireflies.

1.10 Angelo Mosso on Circulation

In the preceding paragraphs, we highlighted in some detail a few aspects of Mosso's scientific activity directly pertaining to the main subject of this book. Yet, as underlined by Luciani in his commemoration at the Italian Senate, the scientific interests of Mosso were widespread. They were also intermingled with philosophical thinking. He was aware of the role that the Enlightenment played in the development of his contemporary cultural milieu, yet he was aware that Reason was not an abstract concept, so that we need solid grounds on which we may exert our drive toward a better knowledge of the world in the light of Reason. Science was the main field on which this could be done, because a scientific approach to knowledge was the necessary path allowing the scrutiny of factual observation to the "infallible" judgment of Reason. This vision of science establishes a direct link between Enlightenment and Positivism and its mechanistic interpretation of Nature. Moreover, it places Mosso in continuity with the iatromechanichal school of the XVII century and explains Mosso's materialistic approach to science and the central role that the development of new measurement tools played in his activity.

We already mentioned the use of the ergograph at Capanna Margherita for the investigation of the effects of altitude on fatigue during exercise with small muscle masses. In fact, he used the ergograph to study fatigue in a variety of conditions, including the effects of aging, and of the type of work. From this, it was an easy step to introduce the concept of training, which was already governing the British approach to exercise and sport, and to investigate the effects of specific training maneuvers on muscle performance. His results were summarized in his famous book on fatigue (Mosso 1891).

Another key instrument in his vision was the plethysmograph, which was not only useful to measure volume changes, but also to infer therefrom blood volume and blood flow changes in various organs. The effects of artery pulsation, respiration, and stress on blood flow could be investigated as well, although the inertia of the recording tools was such as to prevent the follow-up of rapid phenomena. To get rid of this inconvenience, he devised the construction of the hydrosphigmograph, combining a pneumatic and a hydraulic transmission system. Using this instrument, Mosso obtained the first recordings of cerebral blood flow (Mosso 1880). In fact, he was able to profit of an accident, from which two individuals exited with an opening of the skull, making the brain accessible through the skin. Mosso placed a tympanum on the opening, which transmitted brain oscillations to another tympanum through a hydraulic system, and the oscillations of the second tympanum could be recorded by a kymograph. So, the kymograph recorded what the meninx and the skin transmitted, id est the pulsation of cerebral arteries. Since brain blood flow varies locally depending on the degree of activity of the underlying brain parts, he could, under some circumstances, investigate brain activity *sub specie* of brain blood flow changes.

There is something extraordinary in these experiments, as long as these same principles are applied nowadays to study brain activation during mental tasks and to identify which areas of the brain are excited during specific body activities, by means of functional magnetic resonance imaging. These are key issues in modern system neuroscience (Le Bihan 1996; Norris 2006). In the Pubmed database, we reckoned 135,793 publications in the last ten years, after typing “brain functional magnetic resonance imaging” as keyword.

In sum, Mosso can be considered also as one of the initiators of the Italian tradition in neuroscience. A direct link, parallel to the Mosso—Herlitzka—Margaria connection, is clearly identifiable with Pisa, Parma and the development of major Italian neurophysiological Schools, through his pupil Vittorio Aducco and Aducco’s pupil Mario Camis (1878–1946). Incidentally, we note that the early Giuseppe Moruzzi (1910–1986), before his departure to Brussels, trained with Camis at Parma. Moruzzi had, for Italian neurophysiology, a role at least equivalent to that of Margaria for the School of Milano.

1.11 From Mosso to Herlitzka

Mosso was a full professor of physiology and chairman of the Institute of Physiology at the University of Torino until his death, although in the last years, because of his health problems and his new passion for archaeology, he was less and less present.

After his death, Herlitzka, his older pupil in Torino, took over Mosso's role at the Institute of Physiology. Herlitzka, who was perhaps the most direct heir of Mosso's physiological approach and experimental tradition, continued several of the research lines initiated by Mosso, in a different political context (Margaria and Rowinsky 1950). He kept working on altitude physiology, a passion which he transmitted also to Margaria, but, attracted by the technological advancements of his times and the bombastic development of aviation, he interpreted it more in the sense of aviation medicine than of altitude physiology. Nevertheless, he chaired the Angelo Mosso laboratory at Col d'Olen from 1923, after Aggazzotti moved to Catania, to 1938, when he was forced to quit University because of racial laws (Troiani and Manni 2007).

Rodolfo Margaria (1901–1983) was the most eminent of Herlitzka's pupils. We will talk of him extensively in the next chapters, as the actual initiator of the School of Milano. Here we remind only that it was Herlitzka who trained young Margaria in physiology and put him in contact with Otto Meyerhof and with Archibald Vivian Hill. Like every scientist, Margaria had several masters and one academic father: Herlitzka. The next paragraphs are devoted to an outline of Herlitzka's contribution to physiology, covering the gap between Mosso and Margaria.

Gaetano Viale (1889–1934) was another important pupil of Herlitzka, with whom he graduated in 1913 and trained in the laboratory of physiology in Torino. He was still a medical student when he collaborated with Herlitzka on the conservation of living tissues in liquid solution, by studying the action of lipid solvents (Viale 1910). We owe to him also one of the first measurements of energy expenditure during level walking at altitude (Barkan et al. 1914, see Chap. 5), which resulted equal to that obtained at sea level at the same speed. In 1925 he moved to Rosario, Argentina, where he taught physiology at the Medical School for four years. He returned to Italy in 1929, to succeed Valentino Grandis in Genova. His premature death put an end to his career when he was 45.

1.12 Amedeo Herlitzka

Born in Trieste, which was under the Austrian Empire in those times, in 1872 from a wealthy Jewish family of Italian lineage, Amedeo Herlitzka (Fig. 1.7) joined the medical school of Pisa in 1891, whence he moved to Florence in 1894, and finally to Rome, where, under the guidance of Luciani, graduated in Medicine in 1897. In fact, he took a longer time to graduate than usually in those days for a very peculiar reason: his acquaintance with irredentist movements, promoting the association of Trieste to the Kingdom of Italy. The Austrian police arrested him in 1892; on trial, the court found him guilty of high treason and sentenced him to prison, where he spent almost seven months, from September 1892 to April 1893.

When he was in Florence, Herlitzka started his scientific activity as an embryologist at the Institute of Physiology chaired by Giulio Fano (1856–1930), working on the differentiation of newt's eggs (Herlitzka 1897). Then, when he moved to

Fig. 1.7 Amedeo Herlitzka in 1937. Picture taken at Col d'Olen, at the celebration of the 30th anniversary of Angelo Mosso's laboratory. From: Abrahamson (2019)

Luciani's laboratory in Rome, he studied cells' differentiation (Herlitzka 1898), contributing to the definition of the law of the constancy of cell volume in epithelia and analyzing its physiological significance.

After graduation, in 1898, he moved to Torino to join Mosso, where he remained for 40 years. Initially, he led the laboratory of physical chemistry (now biochemistry, then a branch of physiology). As already pointed out, he succeeded Mosso in 1910. During World War One, he was enrolled as a volunteer in the Italian army, serving as medical officer. He happened soon to treat soldiers exposed to toxic gases. This prompted him to investigate their effects on rabbits, in particular with regard to carbon dioxide and bromine. He introduced the use of soda lime to absorb carbon dioxide and tested a specific facemask first on animals and then on humans (Herlitzka 1915), which however the Army never used. In 1917, he was charged to organize a medical center for the clinical and physiological evaluation of aircraft pilots and for the study of the psychophysiology of human flight. This is when he developed his passion for aviation, perceiving the important physiological and medical problems related to it (Galloni and Musso 2017). It is thanks to his efforts that the Institute of Physiology of Torino became a psycho-physiological office of the Royal Air Force.

Herlitzka's connection with the army continued after the war. In 1919, he became chairman of the Army cabinet for war gases. He established a second center for psychological testing of military pilots in Rome, initially chaired by Aggazzotti, but soon taken over by Brother Agostino Gemelli (1878–1959), a Franciscan friar, who graduated in medicine, served the army during the war and founded the Catholic University in Milano in 1921 on behalf of the Pope. Herlitzka's studies on aviation medicine carried out in that period were summarized in a fine book in 1923 (Herlitzka 1923). In the same year, he founded the Centre for Physiological Research on deep-sea divers, which he chaired until its closure in 1933.

Under the fascist regime, which he strongly supported, Herlitzka further contributed to the development of aviation medicine in Italy together with Agostino Gemelli. He became a friend of Italo Balbo (1896–1940), a prominent hierarch of the regime, Vice Minister, then Minister of the Royal Air Force between 1926 and 1934, and an eminent pilot himself. He was consultant of the Ministry of the Navy on physiological problems of deep sea diving. In 1929, he was admitted to the *Accademia Nazionale dei Lincei*. He never stopped in promoting altitude research at the Col d'Olen, as a tool for understanding aviation respiratory problems. Nowadays, we will call him a physiologist of adaptation to special environments.

In 1938, after the promulgation of the racial laws, he was obliged to quit all his positions and roles in the Italian academic and military systems. He escaped in exile. After a short stay in Paris, where he tried to obtain a position in England, but unsuccessfully for his acquaintance to fascism, he took refuge in Argentina, where he had excellent relationships after a travel there in 1927 to support Italian culture on behalf of Balbo, and where his brother, an electric engineer, chief executive officer of an electrical company in Buenos Aires, was living since long. He remained there until 1946, working at the Centre of aviation studies of the Argentinian army (Abrahamson 2019; Troiani and Manni 2007). When he returned to Italy, he reobtained his academic roles at the University of Torino but was purged from the *Accademia dei Lincei* for his strong contiguity with the fascist regime, despite his destitution for racial reasons. He was one of the founders and the first President of the new Italian Physiological Society after the War (Fanò-Illic 2021). He died on July 12, 1949, at age 77.

Herlitzka was a man of his times, which were by far different from the times of Mosso. The international environment in which Mosso operated was restricted after World War One by the advancement of nationalistic regimes. In Italy, the liberal state, which had been created after the unification of the country in 1861, collapsed after the war and the fascist dictatorship took over. Herlitzka, whose nationalism developed in his early age under the Austrian imperial regime in Trieste, adhered to it enthusiastically. The regime was not racist at the beginning, yet it undertook a racist policy after Hitler went to power in Germany and the alliance between Italy and Germany was signed. Herlitzka's acquaintance to the regime revealed useless to him in those circumstances: when the racial laws were introduced, he just had to quit his functions, as every other Jewish professor, as any other Jewish professional. This was one of the most shameful pages in the history of fascism and of modern Italy altogether.

Nevertheless, fascism's fascination for aviation, which the regime pursued as a powerful propaganda tool, partly prompted by its attraction for the futuristic avant-garde, helped Herlitzka develop a passion that stemmed during World War One. His love for aviation medicine and physiology, associated with a deep interest for the solution of physiological problems of practical and social interest, which he inherited from Mosso, characterized his scientific history between the two world wars. Yet, he did it in a different way from Mosso's. Mosso promoted physical education for the improvement of the health of workers. Fascism militarized physical education and placed it in a different, propagandistic context. Times had changed, priorities had changed, and the psychological and political attitude of Herlitzka was different from Mosso's. Herlitzka addressed his efforts to a niche field, highly technological, connected to the army more than to the society, strongly related to one of the main interests of the regime: aviation.

Notwithstanding his political relations, Herlitzka was a bright physiologist, imprinted by Mosso's positivism, with widespread physiological interests, including exercise physiology, as will appear from the next paragraphs.

1.13 Herlitzka's Scientific Activity

Although of less apparent impact and international recognition than Mosso's work, partly because of Italy's closure on itself in the fascist era, as pointed out above, Herlitzka's contribution to the advancement of physiological knowledge was of great importance, especially in the fields of applied physiology and aviation medicine. Herlitzka inherited from Mosso his rigor in conceiving and realizing experiments and his great attention to the technical aspects of physiological research. He inherited also the sense for a practical application of scientific studies, which in his mind were aimed at providing, through a rigorous scientific approach, novel solutions to prominent practical problems of public and political interest.

When he moved to Torino, he initially pursued his former studies on embryo development and investigated the possibility of gonad transplantation. He was among the first to demonstrate tissue degeneration after organ transplantation in animals, and he performed a remarkable series of experiments on the effects of various types of saline solutions on organ and tissue conservation.

However, his scientific activity took soon different directions. Biochemistry was booming. New substances were continuously identified. The various steps of the glycolytic pathway were on the way of being recognized, setting the terrain for Embden and Meyerhof's synthesis, which we call glycolysis. Herlitzka was attracted by that cultural climate and started investigating glycogenolysis. He demonstrated the possibility of glycogenolytic activity by nucleohistones in several tissues, including the liver and the kidney epithelia. Another important contribution of him was the demonstration that pepsin is a protein (Herlitzka 1904). He studied muscle thermodynamics, investigating heat production of isolated rabbit hearts (Herlitzka 1905a, b; Herlitzka 1912).

It is a matter of fact that Herlitzka was a restless man. His curiosity was extremely wide and led him to encompass numerous fields of biological research. He demonstrated the additive effect of refraction indexes of proteins in ionic solutions (Herlitzka 1906). This was the first important step, on the path leading to the realization of methods for the refractometric determination of plasma proteins (Sunderman 1944). However, he also investigated chlorophyll. He studied taste, testing several types of salts. He proposed that the salty taste may be a consequence of anion liberation and of a negative charge on colloids (Herlitzka 1909). He demonstrated the possibility of restoring the heart beat by injecting small quantities of adrenaline in the coronary artery of the dog. He showed that the heart function is optimal within given blood pressure ranges (Herlitzka 1905a). His practical mood led Herlitzka to go on with his investigations on how to conserve living tissues, a field of great importance to those who wish to perform organ transplantation. Concerning neurophysiology, he proposed that a unique center, located in the left-brain hemisphere, to which the right motor areas are completely subjected, controls the genesis of voluntary motor impulses. And this is only a representative yet tiny account of his widespread scientific interests.

The war implied a dramatic shift in Herlitzka's scientific work. What we now call "applied physiology" became predominant in his interests. Muscle physiology took a central place. He investigated the heat production by the heart and during muscular contraction. He was among the first to perform exercise testing on and to carry out a physiological evaluation of athletes. His practical mood drew him to the investigation of the functional characteristics of athletes and sportsmen. His results were summarized in a remarkable booklet on the physiological evaluation of athletes (Herlitzka 1931).

This booklet deserves a closer inspection, because it is indicative of Herlitzka's approach to science. Each chapter is on a specific subject, analyzed along a clear line, which we can describe as follows: (i) set a practical problem to be solved; (ii) define the physiological context of its possible solutions, and thus the necessary experiments; (iii) describe the equipment to be used and the experiments to be done, and why; (iv) describe the data in detail; (v) draw physiological interpretations of the data; and (vi) describe the practical impact and the possible solutions of the practical problem at stake. To our eyes, this appears as a simple, rigorous description of scientific methodology and of its connections to actual life. As an example, fatigue is an important practical problem to athletes and in sport practice; physiology can help understand fatigue, but this is not what counts most to him: what counts is what we could do to prevent and treat fatigue in athletes.

During the war, aviation physiology and medicine became his greatest scientific and practical passion. The fact of collaborating closely with the Air Force and having implemented the national center for the psychophysiological evaluation of pilots placed him in a privileged condition for performing unique studies. In those times, the aircrafts were not pressurized, so hypoxia tolerance was a great problem to deal with. The study of altitude physiology could provide several answers to questions raised by the Air Force Ministry. So, privileged access to the Angelo Mosso laboratory at Col d'Olen, which he chaired, allowed him the performance of pioneering studies on altitude physiology and mountain sickness, namely concerning the chemico-physical alterations of blood (Herlitzka 1926), the onset of loss of consciousness, the hyperventilation at altitude. Theoretical considerations on breathing in the stratosphere and on the acceleration effects on pilots can be found in a book that he published during his exile to Argentina (Herlitzka 1945). Fenn, Otis and Rahn revolutionized our understanding of the physiological responses to hypoxia, during World War Two (Farhi 1990; Ferretti 2018). Yet Herlitzka was among those who set the basis for that and asked questions, which the Rochester triad provided convincing answers to.

Unfortunately, his scientific career was suddenly cut by the shame of racial laws. Nevertheless, from the scientific viewpoint, Herlitzka is and remains the ideal *trait d'union* between Angelo Mosso and Rodolfo Margaria, the necessary link between the pioneering, international, positivistic nineteenth-century scientific mood of Mosso, and the revolutionary vision of Margaria. Margaria's revolution would not have been possible if he remained confined in Italy, with the sole Herlitzka as master; nonetheless, it would not have been possible even without Herlitzka's traineeship as well. An ideal physiological link appears clearly between Mosso, Herlitzka, and Margaria, even though the general cultural and political contexts in which they operated differed remarkably. It is this link that sustains the foundation on which Margaria has built the School of Milano.

1.14 Herlitzka on the Energetics of Muscular Contraction

The next chapter of this book treats of Hill and Meyerhof's theory of muscular contraction and its refutation by Margaria et al. (1933). This is a cornerstone in the history of exercise physiology. Herlitzka published his book on the physiological evaluation of athletes in 1931. A chapter of it is dedicated to the energetics of muscular exercise. Hill and Meyerhof, Nobel Prize winners, were still dominating the scene. Yet phosphagen had already entered the game (Lohmann 1928). Hill would have defended his theory against phosphagen one year later (Hill 1932). Margaria had just arrived in Boston, where he started working on the experiments that would have generated the refutation of Hill and Meyerhof's theory of the energetics of muscle contraction (Margaria et al. 1933). The words that Herlitzka used to describe the energetics of muscular exercise, besides reflecting common ideas of those times, tried to accommodate phosphagen in it. They are impressive: they show how unexpected, fulgurating, and dramatically shattering would have been the article by Margaria, Edwards, and Dill 2 years later. This is the reason why we report a few pages of Herlitzka's text on this subject. The translation from Italian is by Guido Ferretti.

> *Recent researches on the chemistry of muscular contraction have evidenced that this is accompanied by the formation of lactic acid coming from a compound of glucose and phosphoric acid, which* Embden et al. *(*1912*) called lactacidogen and is now identified as an exose-mono-phosphoric acid (Embden and Zimmermann* 1927*). This substance, when split,*

liberates, on one side, lactic acid, on the other side, phosphoric acid. We shall describe later, when we will discuss about recovery, the further fate of lactic acid.

For the moment, we just remind that lactic acid formation occurs at the beginning of contraction and is fully independent of the presence of oxygen. This, in turn, determines disappearance of already formed lactic acid, so that the lack of oxygen favors lactic acid accumulation in muscle. When muscular work is intense, and thus lactic acid production is active, this cannot be quickly destroyed by oxygen, so that it accumulates in muscles, where it causes contractures, and moves into blood, from where it is eliminated through the urine. We should not forget that lactic acid is always formed in muscle from glucose, which originates from muscle glycogen, after glucose has bound phosphoric acid to form lactacidogen. In addition, fats or proteins, when the organism works using them instead of sugars, firstly, by splitting and oxidizing, form sugar (glucose), which is the only substance that a muscle can burn.

However, another reaction can occur in muscles, which may intervene in certain types of contractions. The Eggleton spouses (Eggleton and Eggleton 1927*) have recently found that phosphoric acid in muscle can exist, not only bound to glucose, but also bound to another substance, namely a nitrogen containing substance, creatine, which we know since long to be present in muscles. We also know that a side product of it, creatinine, a normal constituent of urine, can increase in muscle in certain types of muscular activities. The compound formed by the binding of creatine to phosphoric acid, called phosphagen, undergoes very likely a splitting since the beginning of muscle contraction, but the formed phosphoric acid disappears very quickly, even from isolated muscle, so that it must be used in some new combination.*

The most credited doctrine nowadays states that muscle contraction depends on the action of lactic acid on muscle fibrils. The chain of events would be the following: after nerve excitation, the lactacidogen is split, thus liberating lactic acid. Lactic acid causes fibril's shortening. Lactic acid neutralization by alkali, with lactate formation, makes fibrils' relaxation possible. As you can see, oxygen does not participate in this process. Its action becomes evident after the end of muscular work, in a successive step, recovery, which we shall describe thereafter.

Very recently, it was shown that (Lundsgaard 1930*), in muscles poisoned with sodium mono-iodo-acetate, contraction occurs without lactic acid formation, that is without sugar splitting. There is however complete splitting of phosphagen. Lactic acid formation should then represent the necessary condition to restore phosphagen.*

*Whether these two processes, phosphagen splitting and lactic acid formation, integrate or take place separately in different forms of muscle shortening is yet to be elucidated. What could these different forms of shortening be? We need to go back to the old researches by Pekelharing and van Hoogenhuyze (*1910*), who studied substances replacement during muscle contractions that generate work and compared it with what happens during tonic contractions, like those occurring when we maintain a posture for long time. In the latter case, according to the two authors, there would be an abundant elimination of creatinine through the urine, which however they did not observe during the execution of external work, i.e. in isotonic contractions. This suggests that muscular tone may be maintained by phosphagen splitting, whereas rapid isotonic contractions be due to lactacidogen splitting.*

*These differences in the chemical process in the two forms of shortening, if confirmed, would be in perfect agreement with the doctrine by Bottazzi (*1897*), which assigns rapid contractions and muscle tone to two distinct parts of muscle fibers, the former to the myofibril, the latter to the sarcoplasm. We understand that, if shortening occurs by modifying the state of two distinct substances, also the chemical processes that induce these modifications may well differ.*

Anyway, we can follow the chemical changes in athletes, during their sport activity, by examining blood and secretions, in various types of sports, which, as we said several times, depend on different forms of muscle contractions. To this purpose, it is noteworthy that a fraction of the formed lactic acid that is not further destroyed, and thus accumulates in muscles, moves into blood, from where it is eliminated through urine and sweat.

The scientific context in which Herlitzka developed his reasoning is clearly affected by Hill and Meyerhof's theory. The role of lactate was undisputed. Phosphagen was breaking the game. Herlitzka strives at accommodating phosphagen into the dominating theory. His attempt is ingenuous, in a way pathetic. He attaches to old studies postulating distinct types of contractions to conceive different yet intermingled energetic processes, to protect the central role of lactate in the energetics of muscular exercise. It is also noteworthy that there is always a practical view in Herlitzka's reasoning: two mechanisms should explain differences in sport practice by athletes. Take a practical problem (sport performance, in this case, pilots at altitude in another context) and find a solution by appealing to science and theoretical physiological thinking based on experiment. In the following chapter of his book, on recovery, Herlitzka fully adheres to Hill's vision of oxygen consumption as the chemical process ensuring lactate washout during recovery: an oxygen debt is paid, but this is a radically different concept from Margaria's oxygen debt.

A page is going to be turned, a new world will soon appear, and a new star will soon rise: Rodolfo Margaria and his new concept of oxygen debt. Herlitzka shaped him, Hill and Meyerhof refined him without altering his critical insight, Dill provided the environment in which he could give free course to his scientific ability. And the revolution exploded.

References

Abrahamson E (2019) Intellettuali in fuga dall'Italia fascista. Firenze University Press, Firenze

Aggazzotti A (1913) Influenza dell'aria rarefatta sull'ontogenesi. Nota I. La perspirazione delle ova di gallina durante lo sviluppo in alta montagna. Arch Entw Mech 36:639–648

Aggazzotti A (1914) Influenza dell'aria rarefatta sull' ontogenesi. Nota III. Le modificazioni che avvengono nei gas della camera d'aria dell'ovo durante lo sviluppo. Arch Entw Mech 40:65–97

Barcroft J (1914) The respiratory function of the blood. Cambridge University Press, Cambridge UK

Barkan O, Giuliani F, Higgins HL, Signorelli E, Viale G (1914) Gli effetti dell'alcool sulla fatica in montagna. Arch Fisiol 13:277–295

Bernard C (1859) Leçons sur les propriétés physiologiques et les altérations physiologiques des liquides de l'organisme. Baillière, Paris

Bernard C (1865) Introduction à l'étude de la médecine expérimentale. Baillière, Paris

Bert P (1870) Leçons sur la physiologie comparée de la respiration. Masson, Paris

Bert P (1878) La pression barométrique, recherches de physiologie expérimentale. Masson, Paris

Borelli GA (1681) De Motu Animalium. Angelo Bernabò, Roma

Bottazzi F (1897) The oscillations of the auricular tonus in the batrachian heart, with a theory on the function of sarcoplasma in muscular tissues. J Physiol Lond 21:1–21

Cerretelli P (2014) Angelo Mosso e lo studio del muscolo ad alta quota. Rivista di Storia dell'Università di Torino III:61–67

Cogo A, Ponchia A, Pecchio O, Losano G, Cerretelli P (2000) Italian high altitude laboratories: past and present. High Alt Med Biol 1:137–147

Derenzini T (1958) Giovanni Alfonso Borelli fisico. In: Celebrazione della Accademia del Cimento nel tricentenario della fondazione. Accademia del Cimento, Pisa, pp 35–52

Di Giulio C, Daniele F, Tipton CM (2006) Angelo Mosso and muscular fatigue: 116 years after the first congress of physiologists: IUPS commemoration. Adv Physiol Educ 30:51–57

Di Giulio C, West JB (2013) Angelo Mosso's experiments at very low barometric pressures. High Alt Med Biol 14:78–79

Eggleton P, Eggleton GP (1927) The inorganic phosphate and a labile form of inorganic phosphate in the gastrocnemius muscle of the dog. Biochem J 21:190–195

Embden G, Kalberlab F, Engel H (1912) Über Milchsäurebildung in Muskelpreßsaft. Biochem Z 45:45–62

Embden G, Zimmermann M (1927) Über die Chemie des Lactacidogens. 5. Mitteilung. Hoppe Seylers Z Physiol Chem 167:114–136

Fanò-Illic G (2021) Imagine yourself at a dinner party in 1947. pH 2:28–46

Farhi LE (1990) World war II and respiratory physiology: the view from Rochester, New York. J Appl Physiol 69:1565–1570

Ferretti G (2018) The O_2-CO_2 diagram: a revolution in respiration physiology. pH 2: 36–57

Ferretti G, Conti F (Eds.) (2016) Angelo Mosso. pH 1: 2–62

Fick A (1855) Über diffusion. Ann Phys Chem 94:59–86

Fick A (1870) Ueber die Messung des Blutquantums in den Herzventrikeln. Physikalisch Medizin Gesellschaft, Wurzburg

Foà C (1957) Angelo Mosso. Sci Med Ital 5:549–567

Galloni MR, Musso S (eds) (2017) L'università di Torino nella Grande Guerra. Hapas, Torino

Gori G (2002) The apostle of Italian sport: Angelo Mosso and English athleticism in Italy. Eur Sport Hist Rev 4:230–252

Harvey W (1628) Exercitatio Anatomica de Motu Cordis et Sanguinis. Guilielmus Fitzerius, Frankfurt

Helmholtz H (1847) Über die Erhaltung der Kraft. G. Reimer, Berlin

Herlitzka A (1897) Sullo sviluppo di embrioni completi da blastomeri isolati di uova di tritone (*Molge cristata*). Archiv für Entwicklungsmechanik der Organismen 4:624–658

Herlitzka A (1898) Ricerche sulla differenziazione cellulare dello sviluppo embrionale. Archiv für Entwicklungsmechanik der Organismen 8:45–103

Herlitzka A (1904) Sull'autodigestione della pepsina. Atti Reale Accad Lincei 13:51–57

Herlitzka A (1905a) Über den Einfluss des arteriellen Druckes auf die Tätigkeit des isolierten Säugetierherzens. Pflügers Arch 107:557–584

Herlitzka A (1905b) Ricerche sull'azione della temperatura sul cuore isolato di mammifero. Z Allg Phys 5:265

Herlitzka A (1906) Ricerche sull'indice di rifrazione delle soluzioni di proteina in presenza di elettroliti. Biologica 1:157–232

Herlitzka A (1909) Contributo all'analisi fisico-chimica del sapore dei sali. Arch Fisiol 7:557–578

Herlitzka A (1912) Ricerche di termodinamica muscolare. Nota prima. Produzione di calore nel cuore isolato di mammifero. Arch Fisiol 10:501–536

Herlitzka A (1915) Sulla difesa dai gas asfissianti. Atti Reale Accad Med Torino, parte 2^a, pp. 277–304

Herlitzka A (1923) Fisiologia e Aviazione. Zanichelli, Bologna

Herlitzka A (1926) Gli equilibri fisico-chimici del sangue nel mal di montagna. Arch Fisiol 24:676–691

Herlitzka A (1931) Valutazione fisiologica degli atleti. Licinio Cappelli, Bologna

Herlitzka A (1945) El factor humano en los vuelos a gran altura y velocidad. Resen Clin Cient 14: 51–58

Herlitzka A (1947) Angelo Mosso e il centenario della sua nascita. Riv Med Aeronaut 10:97–109

Heymans C, Bouckaert JJ, Samaan A (1934) Influence des variations de la teneur du sang en oxygène et en CO_2 sur l'excitabilité réflexe et directe des éléments centraux et périphériques des nœuds cardio-régulateurs. Arch Int Pharmacodyn Ther 48:457–487

Hill AV (1932) The revolution in muscle physiology. Physiol Rev 12:56–67
Hoppe-Seyler F (1862) Über das Verhalten des Blutfarbstoffes im Spektrum des Sonnenlichtes. Arch Path Anat Physiol 23:446
Kardel T (1990) Niels Stensen geometrical theory of muscle contraction (1667): a reappraisal. J Biomech 23:953–965
Kayser B, Narici M, Binzoni T, Grassi B, Cerretelli P (1994) Fatigue and exhaustion in chronic hypobaric hypoxia: influence of exercising muscle mass. J Appl Physiol 76:634–640
Kellogg RH (1978) "La Pression barométrique": Paul Bert's hypoxia theory and its critiques. Respir Physiol 34:1–28
Kronecker H (1894) Über die Bergkrankheit mit Bezug aud die Jungfraubahn. Gutachten, Bern
Lankford HV (2015) Dull brains, mountaineers, and Mosso: hypoxic words from on high. High Alt Med Biol 16:363–370
Lanza D, Vegetti M (1971) Aristotele: opere biologiche. UTET, Torino
Le Bihan D (1996) Functional MRI of the brain: principles, applications and limitations. J Neuroradiol 23:1–5
Lohmann K (1928) Über die isolierung verschiedener natürlicher Phosphorsäure-verbindungen und die Frage ihrer Einheitlichkeit. Biochem Z 194:306–327
Losano G (1996) Angelo Mosso: uno scienziato a cavallo di due secoli. Boll Soc Ital Med Sper 71 (suppl 11–12):10–22
Löschke HH, Köpchen HP, Gertz KH (1958) Über den Einfluss von wasserstoffionenkonzentration und CO_2 druck im Liquor cerebrospinalis auf die Atmung. Pflügers Arch 266:569–585
Löwy A, Löwy J, Zuntz L (1897) Über den Einfluss der verdünnten Luft und des Höhenklimas auf den Menschen. Pflügers Arch 66:477–538
Ludwig C (1858) Lehrbuch der Physiologie des Menschen. Winterschee Verlagshandlung, Leipzig
Lundsgaard E (1930) Untersuchungen über Muskelkontraktionen ohne Milchsäurebildung. Biochem Z 217:162–177
Malpighi M (1661) De Pulmonibus Observationes Anatomicae. Giovanni Battista Ferroni, Bologna
Maquet P (1989) Borelli: De Motu Animalium. Un premier traité de biomécanique. Acta Orthop Belg 55:541–546
Margaria R (1938) Sulla fisiologia e specialmente sul consumo energetico della marcia e della corsa a varia velocità ed inclinazione del terreno. Atti Reale Accad Lincei 7:299–368
Margaria R, Edwards HT, Dill DB (1933) The possible mechanism of contracting and paying the oxygen debt and the role of lactic acid in muscular contraction. Am J Phys 106:689–714
Margaria R, Rowinsky P (1950) Necrologie: Amedeo Herlitzka (1872-1949). Arch Sci Biol 34: 564–576
Mosso A (1880) Sulla circolazione del sangue nel cervello dell'uomo. Atti Reale Accad Lincei 5: 237–358
Mosso A (1884) Applicazione della bilancia allo studio della circolazione sanguigna dell'uomo. Atti Reale Accad Lincei 19:531–543
Mosso A (1889) Le leggi della fatica studiate nei muscoli dell'uomo. Atti Reale Accad Lincei 5: 409–426
Mosso A (1891) La fatica. Treves, Milano
Mosso A (1893) L'educazione fisica della gioventù. Treves, Milano
Mosso A (1897) La fisiologia dell'uomo sulle Alpi. Treves, Milano
Mosso A (1907) Escursioni nel Mediterraneo e gli scavi di Creta. Treves, Milano
Mosso A (1910) Le origini della civiltà mediterranea. Treves, Milano
Nielsen M, Smith H (1952) Studies on the regulation of respiration in acute hypoxia; with an appendix on respiratory control during prolonged hypoxia. Acta Physiol Scand 24:293–313
Norris DG (2006) Principles of magnetic resonance assessment of brain function. J Magn Reson Imaging 23:794–807
Pekelharing CA, van Hoogenhuyze CJC (1910) Die Bildung des Kreatins im Muskel beim Tonus und bei der Starre. Hoppe Seylers Z Physiol Chem 64:262–293

Rahn H (1988) Air cell gas tensions of the chick embryo at sea level and altitude: the contribution of Aggazzotti, 1914. Respir Physiol 72:343–346
Reggiani C (2012) Santorius and the foundation of experimental physiology. pH 1:8–18
Ross WD (ed) (1936) Aristotle's physics: revised Greek text. Clarendon Press, Oxford
Sandrone S, Bacigaluppi M, Galloni MR, Martino G (2012) Angelo Mosso, 1846-1910. J Neurol 259:2513–2514
Santorio S (1612) De Statica Medicina. Polo, Venezia
Stensen N (1667) Elementorum Myologiae Specimen, seu Musculi Descriptio Geometrica. Stellae, Firenze
Sunderman FW (1944) A rapid method for estimating serum proteins. J Biol Chem 153:139–145
Troiani D, Manni E (2007) A tribute to Italian physiologists of Jewish descent evicted during the persecution ordered by the Fascist Regime in 1938. Adv Physiol Educ 31:123–128
Vesalius A (1543) De Humani corporis Fabrica. Ioannes Oporinus, Basel
Viale G (1910) Azione di alcuni solventi dei lipoidi su la sopravvivenza del sistema nervosa nelle rane. Arch Fisiol 8:537–570
Von Staden H (1989) Herophilus. The art of medicine in early Alexandria. Cambridge University Press, Cambridge
Zuntz N, Löwy A, Müller F, Caspari W (1906) Höhenklima und Bergwanderungen in ihrer Wirkung auf den Menschen. Bong, Berlin

Chapter 2
Margaria's Revolution: A Novel Energetic View of Muscular Contraction

Pietro Enrico di Prampero and Guido Ferretti

Abstract This chapter is divided into two parts. The first has a historical approach and reports briefly on the evolution of knowledge on muscle, on oxygen as a gas-sustaining life, on energy transformations, and on lactate as an end-product of metabolism during exercise. The ensemble of this knowledge is the basis on which Archibald Vivian Hill and Otto Meyerhof created the first comprehensive theory of the energetics of muscular exercise, centered around lactate as a main energy source for muscle contraction, for which they were awarded the Nobel Prize in Physiology or Medicine in 1922. The subsequent events, most importantly the discovery of high-energy phosphates, which led Rodolfo Margaria to confute Hill and Meyerhof's theory in 1933, are further summarized. A short biography of Margaria is also provided, with a detailed analysis of the keystone paper by Margaria, Edwards, and Dill, published in 1933, which set the foundation of the School of Milano. In the second part, we provide an analysis of Margaria's approach to the energetics of muscular exercise, the driving principle of which is compatibility of energetic data with the laws of thermodynamics, which are briefly summarized.

P. E. di Prampero
Emeritus Professor of Physiology, University of Udine, Udine, Italy
e-mail: Pietro.Prampero@uniud.it

G. Ferretti (✉)
Department of Molecular and Translational Medicine, University of Brescia, Brescia, Italy
e-mail: Guido.Ferretti@unibs.it

G. Ferretti (ed.), *Exercise, Respiratory and Environmental Physiology*, Perspectives in Physiology, https://doi.org/10.1007/978-3-031-19197-8_2

Archibald Vivian Hill (standing, wearing short trousers) while measuring acceleration of a sprint runner at start. Ithaca, NY, USA, 1927

2.1 Introduction

After some introductory historical notes on the development of knowledge in muscle physiology and the energetics of muscular exercise, the present chapter is devoted to a description of the theoretical synthesis by Hill and Meyerhof that led to the first comprehensive theory of the energetics of muscular contraction. The cultural climate within which these events took place has been briefly outlined in Chap. 1, discussing Herlitzka's views on lactate as the energy source for muscle contraction. Hill and Meyerhof's theory was falsified by the memorable study by Margaria et al. (1933), which we subsequently analyze in detail. This study modified the current view of the energetics of muscular exercise radically, opening the way to the present view of muscle energetics, which is discussed in Chap. 3. A peculiar aspect of Margaria's approach to the energetic processes underlying muscular contraction is their full compatibility with the principles of thermodynamics, as summarized in Sect. 2.3. A brief section on maximal power in humans concludes this chapter.

2.2 Historical Notes

2.2.1 Antiquity to the Nineteenth Century

Herophilus (Alexandrian School) and his younger contemporary Erasistratus, in the early third century B.C., were the first, at least as far as the Western World is concerned, to recognize the importance of muscles for movement and to propose what is very likely the first theory of contraction. Galen (129–201 A.D.), who summarized the theory as follows: "*during contraction the muscles are filled with pneuma,*[1] *increase in breadth, but diminish in length*", clearly defined the muscles as the organs of voluntary movement; they have only two possible states: contraction and relaxation, the latter being regarded as purely passive.

These ideas held, substantially unmodified, until the Renaissance: the great natural philosophers of that era (sixteenth to seventeenth century) viewed muscles as the organs of voluntary movement, like the Alexandrian School, placing however emphasis to varying degrees on "pneuma" as the origin of contraction. René Descartes (1596–1650) replaced the concept of "pneuma" as *primum movens* of the voluntary contraction with the "vital spirits," the behavior of which, at variance with the previous metaphysical concept of "pneuma," was amenable to physical laws.

By the second half of the seventeenth century, the discovery of the microscope, together with the "revolutionary" Galilean climate, allowed substantial advances to be made also in the field of muscle physiology: William Croone (1633–1684), Giovanni Alfonso Borelli (1608–1679), Anton van Leeuwenhoek (1632–1723), Nils Stensen (1638–1686), and Jan Swammerdam (1637–1680) established by observation and experiments the fibrous structure of muscle, its cross-striation, and its contraction with no changes of volume. The cultural interrelations among these scientists, who influenced each other deeply, were nicely discussed by Wilson (1961). Stensen also proposed a geometric theory of muscle contraction based on microscopic observations of its fibrous structure (Kardel 1990).

John Mayow (1643–1679) made a further step forward by proposing, for the first time in history, the idea of the utilization, during contraction, of "nitro-aerial" particles of air with the simultaneous disappearance of a body constituent (the analogy with the oxygen utilization for burning carbohydrates and fats is self-evident). Indeed, in those days, it was already clear that air does not consist of a single pure gas, but is a mixture of different gases. Michal Sedziwój (1566–1636) had already proposed that one component of air may be a "life-giving" substance. In fact, when kalium nitrate is heated, a gas is liberated that can be collected in flasks and used. According to the Dutch engineer Cornelis Drebbel (1566–1625), that gas could sustain up to 12 men in a submarine rowing longer than one hour from

[1]* The Greek word "pneuma" (πνεῦμα), literally "breath of life," indicates here a vital principle midway between soul and spirit on the one side and more concrete "somatic" functions, on the other.

Westminster to Greenwich down the river Thames (Poole et al. 2015). That substance may well correspond to Mayow's "nitro-aerial" particles of air.

However, those pioneering ideas were ignored for more than 100 years. Meanwhile, gas chemistry slowly evolved. Already in 1640, Jan Baptiste van Helmont (1580–1644) had demonstrated that the mass of the remaining ashes after charcoal burning in a sealed vessel was less than that of the charcoal initially put into the vessel. His interpretation was that the burning process generates an invisible substance, which is liberated into air and which he called gas or wild spirit. The concept of a pure gas different from air had been created. Joseph Black (1728–1799) further studied the nature of this gas. In 1764, he demonstrated that limestone, if heated or combined with some acids, liberates a gas, which he called *fixed air*. Then he showed that fixed air was heavier than air, did not sustain burning, and killed animals that were exposed to it, even though he showed also that animals produced it during breathing. In fact, Black identified carbon dioxide.

In the 1770s, Joseph Priestley (1733–1804) and Carl Wilhelm Scheele (1742–1786) discovered oxygen independently. Priestley, who first published his experiment in 1774, called it "dephlogisticated air." This means air without phlogiston. Phlogiston was a kind of metaphysical fire-like element, which was thought to be liberated during combustion. The theory of phlogiston was deeply rooted in eighteenth-century chemistry. Georg Ernst Stahl (1660–1734) was its strongest advocate. According to him, inflammation (combustion) liberates phlogiston, thus transforming the burned substance into ashes. Clearly, Stahl did not follow, or was not aware, of Helmont's interpretation of burning. The identification of oxygen led Antoine-Laurent Lavoisier (1743–1794) to refute the phlogiston theory, demonstrating that combustion implies the combination of a fuel with oxygen (theory of oxidation).

By the end of the eighteenth century, Lavoisier, together with Pierre-Simon de Laplace (1749–1827), had shown the simultaneous occurrence of carbon dioxide production (by an increase in weight of alkali), oxygen consumption (by a decrease in volume) and heat production (by the melting of ice) in guinea-pigs at rest. Moreover, Armand Séguin (1767–1835) and Lavoisier showed, by experiments on men, that oxygen consumption is increased by muscular work (Séguin and Lavoisier 1789). Lavoisier and Séguin can therefore be considered as the first exercise physiologists. The ensemble of these observations generated the theory that the production of mechanical work by muscle contraction is a chemical process of combustion, in which the reaction of oxygen with a yet undefined fuel produces carbon dioxide, in a process implying heat formation. This means that animals (humans) exercise and move thanks to chemical energy transformations into mechanical work and heat, although heat was not yet considered a form of energy. This is a revolutionary concept indeed, so fraught with consequences for future physiological developments, that we can easily recognize it as the seminal starting point of exercise physiology. Without Lavoisier's revolutionary vision, neither Hill and Meyerhof nor Margaria would have been possible. In 1794 the guillotine put an end, together with his life, to the pioneering work of Lavoisier.

As pointed out in Chap. 1, by the second half of the nineteenth century, the belief that the behavior of all living things, muscles included, is amenable to physical laws pervaded the scientific community. The idea was not new, think for instance of the iatromechanical vision of natural philosophy in the seventeenth century, but became hegemonic in the positivistic cultural climate of those times. In this context, Hermann Ludwig von Helmoltz (1821–1894) did show that the law of the conservation of energy applies to living organisms as well (Helmholtz 1847), a finding confirmed later in more detail by Danilewski (1880) and by Rubner (1894) in animals and by Atwater (1904) in man. The question of whether the muscle could be considered a heat engine was also posed and answered positively by Mayer (1845), and later by Engelmann (1895). In contrast, Fick (1893) rejected it, based on the observed efficiency of human and horse muscles (0.20–0.25) that, as he rightly pointed out, would require, on a heat engine hypothesis, temperature gradients physiologically incompatible with living matters. A detailed review of the early studies on muscle thermodynamics was published in 1904 (Frank 1904).

Louis-Antoine Ranvier (1835–1922) recognized the existence of red and white muscles with different morphological and physiological characteristics (Ranvier 1873). By the end of the century, numerous experimental investigations on the energetics of muscular exercise in man had been carried out both in France and Germany, in the groove opened by the pioneering studies of Lavoisier. The conclusions arrived at by several authors (Chauveau and Kaufmann 1887; Chauveau and Tissot 1896; Heidenhain 1864; Heinemann 1901; Pettenkofer and Voigt 1866; Zuntz 1901) are remarkably close to presently accepted knowledge, in terms of fuel of choice for muscular exercise, and of energy expenditure of static contractions and of locomotion in man.

2.2.2 *The Twentieth Century*

At the beginning of the twentieth century, Fletcher and Hopkins (1907, 1917) did show the appearance of lactic acid in muscles during contraction. Along this line, Archibald Vivian Hill (1886–1977) and Otto Fritz Meyerhof (1884–1951) (Fig. 2.1) obtained a substantial body of evidence, collected by the two of them separately (Hill 1913, 1916, 1922; Meyerhof 1920, 1921, 1922), on the appearance of lactic acid, on heat production, on mechanical work and on the efficiency of contraction. On these bases, they proposed the first modern theory of the energetics of muscular contraction, currently subsumed under the term "The Hill and Meyerhof's theory." For this, the two were awarded the 1922 Nobel Prize for Physiology or Medicine.

In Hill and Meyerhof's theory, the breakdown of glycogen to lactic acid is the essential energy-yielding reaction, regardless of the availability of oxygen. In aerobic conditions, it is assumed that about one-fourth of the lactate produced is oxidized, thus yielding the energy for the resynthesis of the remaining three-fourths to glycogen (Meyerhof 1922, 1924). In 1924, an analysis of the time course of oxygen consumption at the onset of, during, and after moderate exercise, prompted Hill et al. (1924) to introduce the term "oxygen debt." By this, they indicated the amount of oxygen utilized in the recovery after exercise for the resynthesis to glycogen of part of the lactic acid formed during the preceding exercise period.

Fig. 2.1 Otto Fritz Meyerhof (left) and Archibald Vivian Hill (right) sitting ahead of the members of the Kaiser Wilhelm Institut für Experimentelle Medizin, in Heidelberg. Picture taken on the occasion of Hill's visit there in 1930. The standing man on the extreme left is Karl Lohmann, who discovered ATP. With the kind permission of the Archives of the Max Planck Society, Berlin

Thus, in this view, the oxygen debt was necessarily associated with the removal of about one-fourth (see above) of the lactate produced in the initial exercise period, wherein the oxygen consumption has not yet reached a constant "steady state" value.

It is noteworthy that, when Hill and Meyerhof defined their theory, the entire series of chemical reactions characterizing the glycolytic pathway was not fully elucidated yet. The first representation of a biochemical process leading to lactate formation from glucose degradation through the reduction of pyruvate was published afterward, although the full glycolytic chain was yet to be identified (Meyerhof and Lohmann 1927). The first systematic description of a biochemical pathway corresponding to most of what we now call glycolysis was proposed only in 1933 by Gustav Georg Embden (1874–1933) (Embden et al. 1933), the same year in which the enzyme catalyzing the reduction of pyruvate into lactate (lactate dehydrogenase) was identified (Andersson 1933). Therefore, Hill and Meyerhof outlined their theory without having a clear idea of the underlying biochemical pathways, although Meyerhof perceived the existence of a pathway degrading glucose to lactate. In fact, the identification of glycolysis was a long and complex process, lasting decades, involving several scientists, requiring the identification of several intermediate compounds and of the enzymes catalyzing the chemical reactions involved, and measurement of the heat produced in each transformation. The complete glycolytic chain as we know it nowadays was published only after World War Two (Meyerhof and Geliazkowa 1947).

However, as time went by, a large number of experimental data emerged that did not necessarily fit into Hill and Meyerhof's theory, which was therefore put under closer scrutiny:

(a) Embden and Lawaczeck (1922) and Stella (1928) showed that the concentration of inorganic phosphate (*Pi*) in muscles increased during a series of contractions.
(b) Eggleton and Eggleton (1927a, b), Fiske and Subbarow (1927, 1928) and Nachmanson (1928) identified Phosphocreatine (PCr) and showed that its concentration in muscle decreases during contraction.
(c) Lohmann (1928)[2] isolated a new type of organic phosphate, Adenosine-Tri-Phosphate (ATP), also present in muscle.
(d) Lundsgaard (1930a, b) showed that muscles poisoned with iodoacetic acid contract repeatedly, albeit for a limited period, without lactic acid accumulation.

Hill (as Herlitzka, see Chap. 1) realized that the appearance of phosphates could undermine his theory. Since he was a real scientist, he accepted the risk of refutation. In 1932, he wrote a remarkable review (Hill 1932), wherein, from a "loser" perspective, he predicted a "revolution" in muscle physiology and tried to defend his theory by integrating the phosphates in it—pretty much as Necker, prime minister of King Louis XVI, tried to do in order to save the *ancien régime* at the onset of the French revolution in 1789. Every physiologist should read Hill's review to take a lesson of humility and intellectual honesty. However, contrary to Hill's prediction, phosphates did not kill Hill and Meyerhof's theory. They could in principle be accommodated by considering them as the link between the biochemical pathways and the contractile unit in muscle fibers. In fact, Margaria actioned the guillotine and his concept of the alactic oxygen debt hammered the nail in the coffin of Hill's theory. The way Margaria himself told this story is reproduced in the **Appendix** to this chapter (Margaria 1975, historical notes, pp. 10–16).

[2] Karl Lohmann (1898–1978) had a very peculiar life. Born in Bielefeld, he studied chemistry at Münster, and, in 1924, he obtained a PhD in chemistry at Göttingen. Then, he joined Meyerhof in Berlin and followed him to Heidelberg, as Meyerhof became chairman of the *Kaiser Wilhelm Institut für Experimentelle Medizin* (see Fig. 2.1). Those were the days of his discovery of ATP and description of the Lohmann reaction. In Heidelberg he also obtained his MD degree. He accepted the Nazi regime and, in 1937, became Professor of Physiological Chemistry at Friedrich-Wilhelms University in Berlin, and in 1944 he became a consultant at the Ministry of Health. Yet there was no proof of his adhesion either to the Party or to the SS. Nevertheless, he was charged for having collaborated with a children neurological clinic during the war. His ambiguous relations to the Nazi regime might explain why he was never awarded the Nobel Prize in Physiology or Medicine. In contrast, after the Soviet occupation of Berlin, he adhered to the communist regime. So, in 1948, he was named Dean of the Faculty in his University, whose name had meanwhile changed into Humboldt University, and in 1949 he became a member of the Berlin Academy of Science (the Acadamy of Science of the DDR, ADW). In 1950, he became vice director of the ADW Institute for Medicine in Berlin-Buch, and in 1951 was named first director of the newly founded Institute for Medicine and Biology, also in Berlin-Buch. In 1957 he became President of the Institute for Nutrition in Potsdam. Although he never became a member of the Communist Party, he was awarded many honors by the communist regime. His peculiarity (mean or great, it depends on the viewpoint) is that, somehow coherently, he always accepted to collaborate with any political regime, independent of its characteristics, without compromising by a formal adhesion to any political party on power.

2.2.3 Rodolfo Margaria

Rodolfo Margaria (Fig. 2.2), born in Chatillon, Aosta Valley, on November 15, 1901, obtained his degree in medicine in 1924 at the University of Torino, where he started working at the Institute of Physiology with Herlitzka, on altitude physiology (Margaria 1928, 1929). Despite the nationalistic closure of Italy in the fascist era, Herlitzka was convinced of the value of international collaborations.

So, he managed in such a way as to send young Margaria to train abroad, initially at University College London (1930–31), where he worked with Joseph Barcroft

Fig. 2.2 Portrait of Rodolfo Margaria (left) with David Bruce Dill. They were co-authors of the fundamental 1933 paper, which provided refutation of Hill and Meyerhof's theory of the energetics of muscular contraction. Dill himself attributed the paternity of the idea to Margaria. With the kind permission of Paolo Cerretelli

(1872–1947) on the effects of carbon dioxide on respiration and on carbon dioxide binding to hemoglobin (Barcroft and Margaria 1931, 1932; Margaria 1931), and where he got acquainted with Hill. Then Margaria attended the Harvard Fatigue Laboratory in Boston (1931–32), chaired by Lawrence Henderson (1878–1942), who first analyzed the buffer power of bicarbonate in biological fluids (Henderson 1908), whence Karl Albert Hasselbalch (1874–1962) derived the logarithmic equation currently known as Henderson-Hasselbalch equation (Hasselbalch 1916). There, he worked with David Bruce Dill (1891–1986) and falsified, as outlined in detail in the next section, Hill and Meyerhof's theory of the energetics of muscular contraction. In 1934, he eventually completed his training at the Kaiser Wilhelm Institut für Experimentelle Medizin in Heidelberg, chaired by Meyerhof, working on muscle pH changes during contraction with Alexander von Muralt (1903–1990) (Margaria and von Muralt 1934).

He became Professor of Physiology in 1934, initially in Ferrara, then in Pavia, where he performed his classical study on the energy cost of walking and running (Margaria 1938). In 1938, he was appointed Professor of Physiology at the University of Milano, where he remained until the end of his career in 1972, after the eviction of Carlo Foà (1880–1971) following the promulgation of the racial laws by the fascist regime. In the same year, he took the chairmanship of the newly created Centro di Studi e Richerche di Medicina Aeronautica at Guidonia, Rome, which he maintained until 1943.

After the war, he was Visiting Professor at Yale University, New Haven for a couple of years (1948–50). His long-lasting interest in sport's physiology led him to create and chair, since 1957, the first School of Specialization in Sport Medicine, in Milano. The participants in the Italian mountaineering expedition to K2, in 1954, underwent exercise testing in the Institute of Physiology of Milano. Several athletes participating in the Olympic Games in Rome in 1960 and in Mexico in 1968 were evaluated by his collaborators on the spot (Cerretelli and Radovani 1960; Cerretelli et al. 1960; di Prampero et al. 1970). In addition, Eddy Merckx was tested in Milano before establishing his celebrated one-hour world record in unaccompanied cycling in 1972. Margaria was a member of the *Accademia Nazionale dei Lincei* and of the board of the International Union of Physiological Sciences.

His scientific activity was at the core of the development of the School of Milano. His papers are widely discussed at various stages in this book. He summarized his contribution to our knowledge of exercise physiology in his small yet rich book entitled *Fisiologia Muscolare e Meccanica del Movimento* (Margaria 1975). Along the line traced by his predecessors and masters, Mosso and Herlitzka, he pleaded for the application of physiological knowledge. According to him, everyone ought to know his maximum muscular power, aerobic and anaerobic. This is necessary for athletes wishing to attain their maximum efficiency and useful to the Nation's medical system as a major tool for preventive medicine. This vision is the background of the law on sports medicine, promulgated in 1982, shortly before his death, and of the ensuing sports medicine certificate of habilitation to sports practice, then made mandatory for everybody who wish to compete in any sport in Italy: a law, which should be taken as a virtuous model all over the world. Margaria died in Milano on January 29th, 1983.

Margaria's legacy goes well beyond the impact of his scientific work and of his publications. He was the founder, on the groove traced by Mosso and Herlitzka, of the physiological School where we come from. He trained people like Emilio Agostoni (1929–2021), Joseph Milic Emili (1931–2022), Paolo Cerretelli, Giovanni Cavagna, Piero Mognoni (1936–2008), Gabriele Cortili (1939–2009), Giuseppe Sant'Ambrogio (1933–2002), Franco Saibene (1932–2008), Pietro Enrico di Prampero, and, in his last academic years, he introduced to physiology Edgardo D'Angelo, Giuseppe Miserocchi and Arsenio Veicsteinas (1944–2017), just to mention those who remained active in physiology all life long. He transmitted to all of them an intellectual mood, a quantitative vision of science and the ambition of pursuing the truth, as far as possible in science. He also transmitted qualities that he inherited from Herlitzka, in particular the importance of including in their scientific activities practical and societal problems. Sports medicine and environmental physiology as related to altitude and microgravity were not, to his mind, only instruments for testing new scientific hypotheses, but also fields in which the physiological knowledge could help solve practical problems. Margaria's legacy is still well present in all the new generation issued and trained by his pupils, 40 years after his death.

2.2.4 *Margaria's "Revolution": The Alactic Oxygen Debt*

Thanks to the detailed series of experiments performed at the Harvard Fatigue Laboratory, in Boston, Margaria, with Edwards and Dill (1933), did show that, over a large range of running speeds in humans, the blood lactate concentration did not increase appreciably above resting, thus indicating that no lactate production had occurred. Hence, under these conditions, the oxygen debt paid in the recovery period could not be attributed to lactate removal. Moreover, lactate could not be the main energy source for muscular contraction in light exercise, as long as it did not increase in proportion with the overall metabolic rate. Therefore, the contraction of the oxygen debt (oxygen deficit, we would call it nowadays) at the beginning of light exercise could not be due to lactate production and the subsequent increase in oxygen consumption could not be due to the aerobic removal of produced lactate. Therefore, Margaria et al. (1933) distinguished two causal mechanisms of the oxygen debt: the one due to lactic acid, which they called the lactacid mechanism (nowadays lactic), the other causing the contraction of an oxygen debt in absence of any formation of lactic acid, which they called the alactacid mechanism (nowadays alactic). The former was shown to intervene only at workloads higher than the 60% of the maximal oxygen consumption of the subject. This implied that the alactacid mechanism occurs by far more frequently in ordinary life than the lactacid mechanism. These facts, as they wrote, *raised the importance of the alactacid mechanism of contracting the oxygen debt to the primary degree*, whereas the lactacid mechanism was downgraded to the level of a mechanism of emergency.

In addition, the amount of the alactacid oxygen debt increased linearly with the increase of the steady-state metabolic rate above the resting level during exercise, whereas the lactacid oxygen debt did not. Thus, they commented as follows: *the alactacid oxygen debt is very probably the expression of an oxidative process, in which the energy liberated* [after the end of exercise] *is spent in repaying the amount of energy set free anaerobically during the muscular contraction* [at the beginning of exercise], i.e., *it is a true oxygen debt in the meaning given by A. V. Hill to this term.* Then, after recognizing that the most important anaerobic process, beyond lactate formation, was phosphagen splitting to form PCr and phosphate, they put forward the hypothesis *that the resynthesis of the phosphagen is the process, which absorbs the energy developed in the oxidative processes attributable to the alactacid oxygen debt payment.*

To sum up, the beautiful theory of the energetics of muscular contraction by Hill and Meyerhof, centered around the primary role of lactic acid as energy source, was slain by these ugly facts. Nevertheless, Margaria Edwards and Dill granted Hill and Meyerhof the honor of arms. They stated elegantly that their concept defined an oxygen debt in the sense meant by Hill, and they admitted that the lactacid oxygen debt, as originally proposed by Hill and Meyerhof, accounts for the energy of resynthesis of glycogen from lactic acid. Unfortunately, this latter part concerns only a minority of conditions encountered in actual life: those in which the performance of severe exercise is required.

Published in 1933 in the American Journal of Physiology, this study generated a cascade of consequences, among which, incidentally, also the School of Milano. Above all, however, we must recognize that it is the cornerstone of the theory that, after nearly 90 years, is still considered the most correct view of the energetic processes underlying muscle contraction: the hydrolysis of phosphagen as primary energy yielding mechanism. At the steady state of exercise, the hydrolysis of phosphagen and its resynthesis at the expense of oxygen consumption proceed at the same rate. Conversely, in the initial exercise period, because of the inertia of the oxidative machinery, a certain fraction of the phosphagen split to yield the necessary amount of energy cannot be resynthesized; it will be only in the recovery after exercise at the expense of the so-called "alactic" oxygen debt. The net production of lactate is viewed as playing a quantitatively important role in terms of phosphagen resynthesis only above the so-called maximal oxygen consumption.

It should also be noted that, at the time, ATP and PCr were lumped together under the term phosphagen. Only in 1934 Lohmann, who had identified ATP in 1928, suggested it to be the *primum movens* of muscular contraction (Lohmann 1934). This view was further supported when Engelhardt and Lyubimova, discovering the ATPase activity of myosin, bridged the gap between structure and function (Engelhardt and Lyubimova 1939).

No further details as concerns the history of muscle and exercise physiology will be reported here, the interested reader being referred to the magnificent Dorothy Needham's *Machina Carnis* (1971). However, before discussing to a deeper extent Margaria's concept of oxygen debt (see Chap. 3), it seems worthwhile to outline his multifaceted approach to the energetics of muscle and exercise physiology. Indeed, the above conceptual "revolution" that subsumes under one and the same hat data obtained from widely different experimental approaches, such as the identification of phosphagen in vitro and the physiology of exercising men, highlights *ad abundantiam* the extraordinary intellectual creativity of Margaria.

This same general approach will bring Margaria in subsequent years to identify several other relevant aspects of the energetics of muscular exercise. These include the energy equivalent of lactate production, the mechanisms controlling the aerobic energy-yielding processes and the energetics of the exercises at maximal absolute power in humans, to mention just a few (see Chap. 3). Moreover, he expressed his talent in many other areas of human physiology, some of which are outlined below and in the subsequent chapters.

To summarize, the guiding principle that can be evicted from Margaria's multi-faceted scientific production, and that he has transmitted to those of us who had the fortune to work at his side, is precisely the above-mentioned capacity to subsume under coherent and simple theories experimental observations seemingly quite distant.

2.3 The Energetics of Muscular Contraction: A View from Milano

As already mentioned, the revolution generated by Margaria et al. (1933) and the subsequent creation of the concept of the alactic oxygen debt had the remarkable peculiarity of being fully compatible with the general principles of thermodynamics. The successors of Margaria in Milano kept this concept very clear in their mind, and the experiments that they conceived to test Margaria's energetic theory were driven by a thermodynamic vision. This thermodynamic approach is a key point of the enduring success of Margaria's energetic vision of muscular exercise.

The paragraphs that follow are therefore devoted to providing a short summary of the basic principles of thermodynamics, from a historical perspective. Subsequently, we analyze Margaria's vision of thermodynamics as applied to muscular contraction, and summarize data, analyses, and theories resulting from several studies performed in Milano, Göttingen, and Geneva between the 1960s and the 1980s, in which several thermodynamic parameters related to muscle contraction were determined. These studies stemmed from multiple discussions among Margaria himself, Paolo Cerretelli, Johannes Piiper, and many other colleagues directly or indirectly connected with the Milano School. More specifically we will deal with the entangled web existing among (i) intramuscular concentrations of ATP and PCr, (ii) muscle shortening speed, (iii) efficiency of muscle contraction, and (iv) exhaustion time in humans exercising at maximal absolute power.

2.3.1 The Thermodynamics of Muscle Contraction

A crucial tenet of classical thermodynamics is the demonstration that the efficiency of a heat engine is set by the temperature difference between heat source and sink:

$$\varepsilon = (T_h - T_c)/T_h \tag{2.1}$$

where ε is the thermodynamic efficiency, and T_h and T_c are the absolute temperatures (°K) of the heat source and sink, respectively (Klotz 1964).

Equation 2.1 shows that, to attain an efficiency of 1.0, T_c ought to reach absolute zero, which is obviously impossible for any real physical engine. This same Equation shows also that, as already pointed out by Fick (1893), *the muscle cannot operate as a heat engine*. Indeed, were this the case, since T_c must be close to 310 °K (37 °C) and since the efficiency of muscle contraction under optimal isotonic conditions is about 0.25, T_h ought to attain 413–443 °K (140–170 °C), an obvious physiological nonsense.

Indeed, the muscle is a chemical engine that transforms a fraction of the energy made available by ATP hydrolysis into work, whereas the remaining fraction is dissipated as heat. In turn, ATP is reconstituted at the expense of numerous biochemical processes, all in the end depending on the oxygen consumption (VO_2). The details of the complex and refined series of biochemical processes are summarized in Fig. 2.3 and discussed in detail in Chap. 3. Suffice it to say here that, when considering a complete activity cycle: rest – work – recovery – rest, such that the final conditions be the same as the initial ones, the overall energy requirement is covered by the oxidative processes. Hence:

$$VO_2 = \Delta U = h + w \tag{2.2}$$

where h is the dissipated heat, w is the work performed on the surrounding, and ΔU is the overall chemical energy change, proportional to the overall energy expenditure, and hence to the VO_2.

In contracting muscles, as well as in many other physiological processes, a fraction of the work performed is utilized for changes of volume and/or pressure of the muscle itself. Since we are generally interested in the work performed by the muscle on its surroundings, this fraction of the work performed is lost for all practical purposes. It is therefore customary to utilize another quantity, defined enthalpy change (ΔH), which is obtained by subtracting the amount of work done on the muscle itself from the overall chemical energy change (ΔU):

$$\Delta H = \Delta U - \int PdV = w + h \tag{2.3}$$

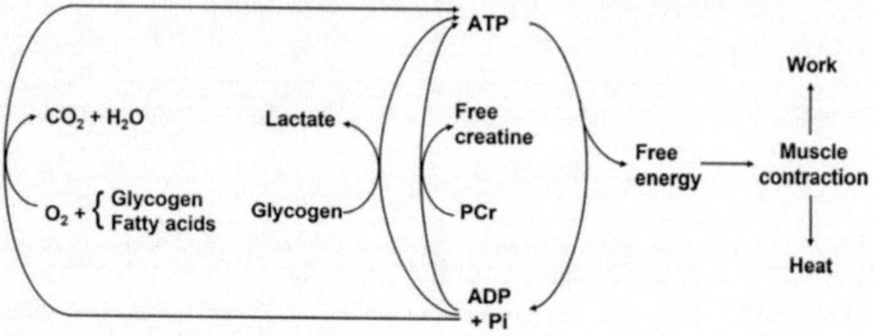

Fig. 2.3 Schematic representation of the energetics of muscle contraction. ATP, adenosine-tri-phosphate; ADP adenosine-di-phosphate; PCr, phosphocreatine; *Pi*, inorganic phosphate. The hydrolysis of ATP yields the energy that the muscle proteins utilize for contraction (work plus heat). In turn, ATP is reconstituted at the expense of the indicated processes. Free energy represents the free energy change of ATP hydrolysis. From Ferretti et al. 2022, modified after di Prampero 1981

where $\int PdV$ is the work done by the muscle on itself, P being the pressure, and dV the volume change. The muscle, however, is substantially incompressible, hence, as a first approximation, we can assume that $\int PdV = 0$. Therefore:

$$\Delta H = \Delta U = w + h \tag{2.4}$$

2.3.2 *Entropy and Efficiency of Biological Processes*

Equations 2.3 and 2.4 would apply also if a muscle, absorbing heat from its surroundings, would produce work and/or store chemical energy. It is noteworthy that in the late 1950s, Hill and Howarth (1959) hypothesized that during eccentric contractions a fraction of the mechanical work absorbed by the muscle could be utilized to resynthesize ATP.

Anyway, apart from these considerations, Eqs. 2.3 and 2.4 cannot specify the direction in which natural events do evolve. More specifically, Eq. 2.3 does not explain the existence of those spontaneous processes, such as the solution of a sugar cube in water, occurring without any difference of chemical energy between reagents and products. Indeed, in this specific instance, since $\Delta H = 0$, there must be another "force" promoting the solution of sugar. The existence of this "force" explains the reason why a sugar cube always dissolves in water, whereas the opposite phenomenon, i.e. the formation of a sugar cube from a dilute solution never occurs spontaneously.

To express quantitatively this set of common sense observations, it is necessary to utilize a quantity defined as entropy. In simplified terms, entropy is a measure of the disorder: it is higher, the greater the disorder of a given system. A lump of bricks is characterized by a greater entropy than the same number of bricks ordered to form a wall, even if the overall chemical energy content of the system is equal. The natural tendency toward a continuous increase of the disorder is expressed quantitatively by the continuous increase of the entropy of the Universe. It necessarily follows that the processes that occur spontaneously are those associated with an increase in entropy: by themselves, the walls fall, but they do not self-construct. The disorder of the system, and hence its entropy, increases as a consequence of the solution of a sugar cube in water. The driving force, allowing the process to proceed spontaneously with a velocity > 0, is proportional to the increase in entropy. Hence, in every real thermodynamic process, be it the solution of a sugar cube in water, or the fall of a wall, the entropy of the Universe increases. The second law of thermodynamics summarizes quantitatively this state of affairs: the inevitable tendency of the entropy of the Universe toward a continuous increase.

When dealing with systems of small dimensions, such as the human body or a muscle, we cannot take the entropy of the whole Universe into account. It is therefore convenient to use a different quantity defined free energy, the symbol of which (G) stems from the name of the American physicist (Josiah Willard Gibbs, 1839–1903) who first proposed it. G, neglecting the remainder of the Universe, is a quantitative measure of the intrinsic tendency of a given system toward spontaneous transformations. Its changes can be expressed as:[3]

$$\Delta G = G\ reagents \ - \ G\ products \tag{2.5}$$

where ΔG is positive in all processes that can proceed spontaneously, in which case the G of the reagents is greater than the G of the products.

ΔG is the sum of two terms: (i) the enthalpy change (ΔH), expressing the sum of w and h (Eq. 2.3), and (ii) the product of the entropy change (ΔS) and the absolute temperature (T):

$$\Delta G = \Delta H + T\Delta S \tag{2.6}$$

Considering again the previous example, the sugar cube can dissolve in water, because, even if under these conditions $\Delta H = 0$, ΔG is positive because of the increase of entropy. Indeed, the disorder of the sugar molecules dissolved in water is greater than that of the same molecules assembled into the sugar crystals of the cube. The reaction can therefore proceed spontaneously. Moreover, with the use of an

[3]The formulation of Eq. 2.5 is not conform to the classical formulation reported in thermodynamics books ($\Delta G = G\ products$ - $G\ reagents$). The aim of the present formulation is to have the energy fluxes positive for exoergonic reactions and negative for endoergonic reactions, as usually found in physiology textbooks. Another consequence of the present notation is that the thermodynamic and the mechanical efficiencies turn out positive in the former case (work performance and heat liberation toward the external environment) and negative in the latter case (work performance and heat liberation toward the internal environment).

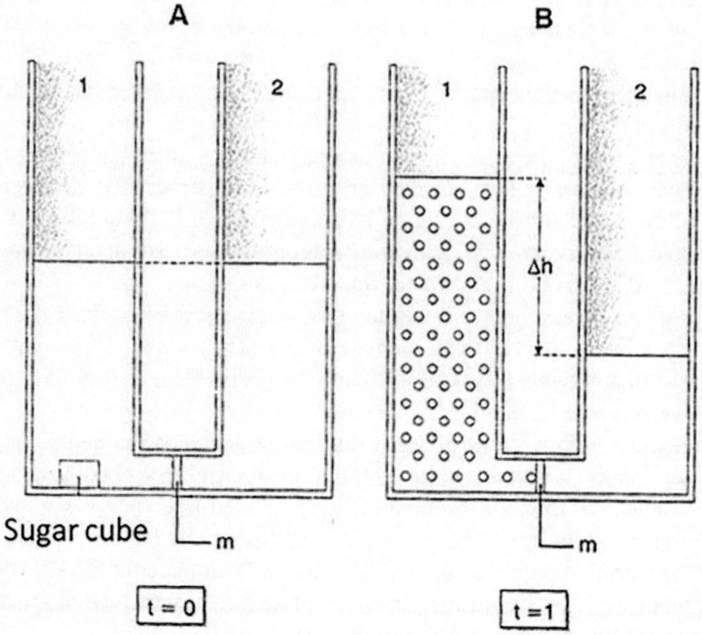

Fig. 2.4 Two water-filled reservoirs separated by a semipermeable membrane (m), can be utilized to perform mechanical work thanks to the free energy changes. In (**a**), the sugar cube is still intact, so that the two reservoirs are in hydrostatic equilibrium. When the sugar cube has dissolved in water (**b**), the resulting increase in osmotic pressure leads to the transfer of a certain amount of water from reservoir 2 to reservoir 1. The increase of the water level in 1 can be harnessed by a simple mechanical tool (e.g. a piston) to produce useful external work. Since under these conditions the enthalpy change is nil, the only driving force is the increased disorder of the system (entropy). It also follows that, to perform external work, as shown by Eq. 2.7, the system must absorb heat from the surroundings, thus leading to a decrease of its temperature. From di Prampero 2015, with the kind permission of Raffaello Grandi, Edi Ermes

appropriate tool, such as a semi-permeable membrane, the system could also perform external work (Fig. 2.4). In this case, combining Eqs. 2.3 and 2.6, and setting $\Delta H = 0$, we obtain:

$$\Delta G - T\Delta S = \Delta H = w + h = 0 \tag{2.7}$$

In such a system, w can be performed only at the expense of h, the value of which must therefore decrease. This necessarily implies a decrease of T: the system becomes colder.

It should be added here that, even if the entropy of the Universe keeps always increasing, when considering tiny portions of the Universe such as the human body, local decreases of entropy occur systematically whenever complex structures (such as proteins) are synthesized from simpler constituents (e.g., amino acids). In this specific case, the disorder (and hence the entropy) of the products (molecular proteins) is smaller than that of the reagents wherefrom they have been synthesized (free amino acids). This can happen since in a coupled reaction (or a series thereof) the increase of entropy is greater than its local fall, thus necessarily leading to a net increase of the overall entropy of the Universe. The same occurs paradigmatically in the development of a human being from a fertilized egg: life requires energy income from outside the body. This capacity of streamlining the fluxes of energy through the system, to obtain a local decrease of entropy, is one of the fundamental characteristics of living systems. All of them are bound by the first and second laws of thermodynamics, in spite of their essentially limitless variety (the number of living species and forms on Earth is only a tiny fraction of the existing possibilities). The two laws of thermodynamics are beautifully summarized by the famous aphorism of the German physicist Rudolf Clausius (1822–1888): "*Die Energie der Welt ist constant. Die Entropie der Welt strebt einem Maximum zu*". (The Energy of the Universe is constant; the Entropy of the Universe tends toward a maximum).

The maximal work that a muscle could perform theoretically under ideal reversible conditions, i.e., at infinitely low speed and with no energy losses, is equal to ΔG. The ratio between w and ΔG is defined as thermodynamic efficiency (ε):

$$\varepsilon = \frac{w}{\Delta G} \tag{2.8}$$

The fraction of ΔG that is not transformed into work is dissipated as heat. This, however, is not the only heat source during muscular contraction. Indeed, assuming hypothetically a thermodynamic efficiency = 1.0, under which conditions the amount of work performed is equal to ΔG, we can rewrite Eq. 2.7 as follows:

$$h = -T\Delta S = h_{rev} \tag{2.9}$$

where h_{rev} is defined as reversible heat. h_{rev} is always exchanged, also under these ideal conditions. The sign of h_{rev} changes with the direction of the reaction: indeed, this is the reason why it is defined "reversible."

On the contrary, when $\varepsilon < 1.0$, as is normally the case, the amount of heat dissipated is equal to the difference between ΔG and w. This additional amount of heat is defined irreversible heat (h_{irr}), because it could not be recovered, even if the direction of the reaction were inverted. The amount of h_{irr} depends on ε, which, in turn, is smaller, the greater the speed of the reaction. Thus, the total amount of heat dissipated during muscular contraction (h_{tot}) is greater, the higher the speed of the process (see Sect. 2.3.4):

$$h_{tot} = h_{rev} + h_{irr} = \Delta G^{*}\ (1\ -\ \varepsilon)\ -\ T\Delta S \tag{2.10}$$

ΔG cannot be determined easily in vivo, since it depends on numerous factors, such as the concentration of products and reagents of all the reactions involved, as well as on the pH and on the ionic force of the medium in which the reactions occur.

For this reason, the efficiency of muscular contraction is conventionally assessed on the basis of the so-called mechanical efficiency (η):

$$\eta = \frac{w}{\Delta H} = \frac{w}{w + h} \tag{2.11}$$

Combining Eqs. 2.4, 2.8, and 2.11 one obtains:

$$\eta = \frac{\varepsilon\, \Delta G}{\Delta H} \tag{2.12}$$

This Equation shows that, for exoergonic reactions, whereas the limits of ε are 0 and 1, those of η are 0 and $\Delta G/\Delta H$. Since the ratio $\Delta G/\Delta H$ can be substantially different from 1, any given experimental measurement of η has a rather limited interest, unless the ratio $\Delta G/\Delta H$ is also known. In practice, however, since the oxidative processes generally occur without large entropy changes, in this specific case we can assume $\Delta H \approx \Delta G$. Thus, the efficiency values assessed from the ratio between mechanical work performed and the corresponding metabolic energy expenditure from aerobic sources are rather close to ε.

Margaria was perfectly aware of the above principles and recognized the need that a theory of the energetics of muscular exercise be necessarily compatible with the thermodynamic constraints. He made use of η in the study of human locomotion and he was definitely convinced that a measure of it informs on the thermodynamic efficiency of muscular contraction. The concepts of oxygen debt and of its payment imply a series of metabolic reactions that are overall exoergonic, with a precise thermodynamic efficiency. The η of concentric muscle contractions (0.20–0.25) corresponds to the η that Margaria computed for uphill walking and running at high positive slope (Margaria 1938), and that Dickinson (1929) first obtained for pedalling on a cycle ergometer. These values result from the themodynamic efficiencies of all reactions involved in resynthesizing ATP and in the transformation of the chemical energy stored in ATP into mechanical work and heat during muscular contraction.

2.3.3 *Partial and Global Efficiencies: The ATP Cycle*

In physiological conditions (Margaria 1958), we deal with numerous metabolic cycles in series, in which the products of a given reaction are the reagents of the subsequent one. Under these conditions, the global efficiency of the process is the product of the individual efficiencies of all the intervening reactions. Let us consider, for instance, an aerobic exercise at the steady state, wherein hydrolysis and resynthesis of ATP proceed simultaneously at the same rate, the resynthesis of ATP occurring at the expense of the oxidative processes. If this is so, the oxidation of one glycosidic unit (162 g of glycogen) yields 2840 kJ and leads to the resynthesis

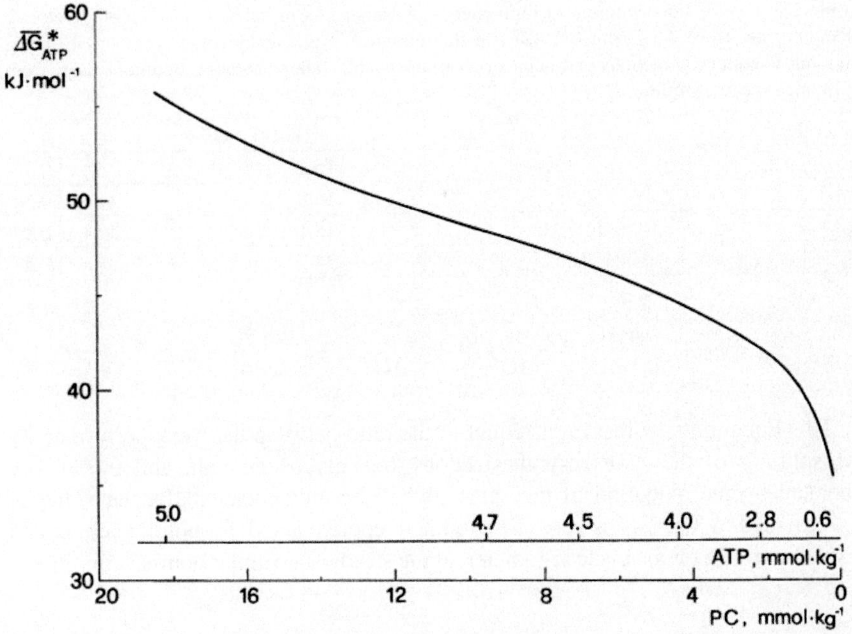

Fig. 2.5 Free energy change per mole ATP (ΔG^*_{ATP}) as a function of the intramuscular concentrations of PCr and ATP. From di Prampero 1981

of 37 moles of ATP, the molar free energy of which (ΔG^*_{ATP}) under physiological conditions is about 50 kJ mol^{-1} (see Fig. 2.5). Hence, the thermodynamic efficiency of the oxidative resynthesis of ATP (ε_R) amounts to:

$$\varepsilon_R = \frac{n\,\Delta G^*_{ATP}}{\Delta G^*_{gl}} = \frac{37 \cdot 50}{2840} = 0.65 \tag{2.13}$$

where ΔG^*_{gl} is the free energy liberated by the oxidation of one glycosidic unit and n (= 37) is the number of ATP moles resynthesized per glycosidic unit oxidized.

In turn, the efficiency of work performance at the expense of ATP splitting (ε_S) is the ratio between the mechanical work produced per mole of ATP split (w^*_{ATP}) and ΔG^*_{ATP}. Hence, since under optimal conditions, in dog gastrocnemius, $w^*_{ATP} \approx$ 20 kJ mol^{-1} (Table 2.1), we obtain:

$$\varepsilon_S = \frac{w^*_{ATP}}{\Delta G^*_{ATP}} = \frac{20}{50} = 0.40 \tag{2.14}$$

In this condition, the overall efficiency of the process (ε_{tot}), i.e., the product of ε_R (Eq. 2.13) times ε_S (Eq. 2.14), can be subdivided into its two components, ATP splitting and resynthesis:

Table 2.1 Overall concentration of high-energy phosphates in muscle ($\tilde{}P_O$), shortening velocity (v_s), work per mole ATP split (w^*_{ATP}) and thermodynamic efficiency (ε) at work onset during rhythmic isotonic contractions of dog gastrocnemius in situ, under anaerobic conditions. Data from di Prampero et al. 1981

$\tilde{}P_O$ (mmol kg^{-1})	v_s (mm s^{-1})	w^*_{ATP} (kJ mol^{-1})	ε
18.7	135.0	15.1	0.30
14.4	99.7	16.8	0.34
12.0	78.0	18.1	0.39
4.7	58.6	24.2	0.57

$$\varepsilon_{tot} = \varepsilon_R\ \varepsilon_S = \frac{n\,\Delta G^*_{ATP}}{\Delta G^*_{gl}}\ \frac{w^*_{ATP}}{\Delta G^*_{ATP}} = \frac{n\,w^*_{ATP}}{\Delta G^*_{gl}} = \frac{37 \cdot 20}{2840} = 0.26 \tag{2.15}$$

This Equation tells that ε_{tot} is equal to the ratio between the work performed by the splitting of the ATP resynthesized by one glycosidic unit, and the energy liberated by the oxidation of the same glycosidic unit. Incidentally, the value of ε_{tot} provided by Eq. 2.15 is very close to the η reported above for uphill walking and aerobic exercise on the cycle ergometer, at the steady state under optimal conditions.

2.3.4 *Contraction Efficiency, Speed of Shortening, and ATP Concentration*

The theoretical analysis performed by Margaria (1958) along the lines reported above is based on the implicit assumption that work performance and free energy liberation per mole of ATP are essentially constant. Margaria's pupils tested this assumption experimentally in the 1960s and 1980s. The results that they obtained on the isolated-perfused muscle preparation demonstrated that things were not so straightforward. Indeed, during rhythmic isotonic contractions of dog gastrocnemius under anaerobic conditions, the muscle shortening velocity decreases, whereas the w^*_{ATP} and the ε of contraction increase with decreasing the high-energy phosphate concentration in muscle (di Prampero et al. 1981). This data are shown in Table 2.1.

The w^*_{ATP} reported in Table 2.1 was calculated from the ratio of the actual work performed and the number of moles of high energy phosphates (sum of ATP and PCr) hydrolyzed, corrected for the amount of lactate produced. In turn, ε was obtained from the ratio of w^*_{ATP} and ΔG^*_{ATP}, obtained as indicated in Fig. 2.5.

Table 2.1 and Fig. 2.5 show that the increase of ε with decreasing concentration of high energy phosphates in muscle is due, on the one side, to the increase of w^*_{ATP}, and on the other, to the decrease of the corresponding ΔG^*_{ATP} brought about by the decrease of the intramuscular PCr concentration. Figure 2.5 shows also that, thanks to the relatively high value of the equilibrium constant of the Lohmann reaction, catalyzed by creatine-kinase ($K = 20$), the ATP concentration is kept to rather high values, until the PCr concentration is substantially reduced.

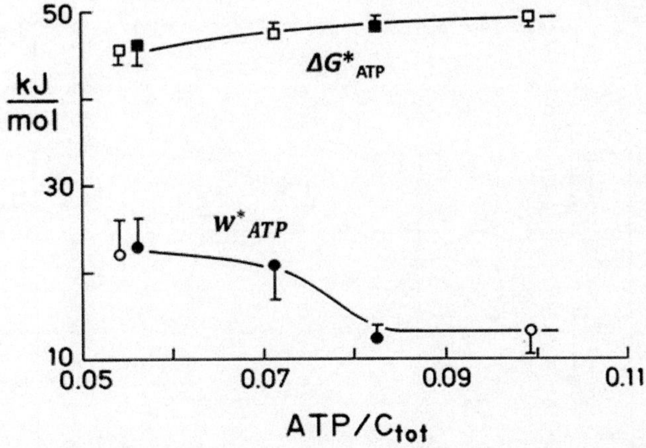

Fig. 2.6 Work performed *(w^*_{ATP})* and free energy output *($\Delta G^*_{\rm ATP}$)* per mole ATP, as a function of the intramuscular ATP concentration, expressed as a ratio of the total creatine concentration (C_{tot}), as given by the sum of PCr and free creatine. Frog Sartorius at 10 °C in nitrogen, poisoned with monoiodoacetic acid. From di Prampero et al. 1988

In the late 1980s, the data and conclusions reported above were further supported by a series of experiments performed on frog sartorius at 10 °C poisoned with monoiodoacetic acid in nitrogen (di Prampero et al. 1988). Under these conditions, data analysis and interpretation are simplified, because, thanks to the inhibition of lactate formation by monoiodoacetic acid poisoning, the only energy source available for work performance is the splitting of ATP and PCr.

The so obtained data, reported in Figs. 2.6 and 2.7, support the view that w^*_{ATP} and ε increase with decreasing the intramuscular concentration of ATP. However, the "bill" to be paid for this increase of ε is a reduction of the shortening speed, because the difference between "driving force" (ΔG^*_{ATP}) and "work output" (w^*_{ATP}) decreases progressively with the fall of intramuscular ATP concentration.

The data reported in Figs. 2.6 and 2.7 are coherent with those obtained more recently by Bruno Grassi and coworkers (Grassi et al. 2011). These authors measured the $\dot{V}O_2$ and the force generated by dog gastrocnemius in situ during rhythmic isometric tetanic contractions after acute inhibition of creatine-kinase by means of iodo-acetamide. At steady state (after 3 min of contractions), after iodo-acetamide poisoning, the time integral of the tension-time curve (tension time index, TTI), the $\dot{V}O_2$ and the ATP concentration were substantially reduced, whereas the ratio TTI/$\dot{V}O_2$ was greater than observed under control conditions (Table 2.2).

It should be remembered that the inhibition of the creatine-kinase activity by iodo-acetamide and hence of ATP resynthesis from PCr and ADP, leads to a rapid fall of ATP without substantial changes of PCr concentration.

As a first approximation, the ratio TTI/$\dot{V}O_2$ can be considered an index of the thermodynamic efficiency of contraction; hence, these data support the view that the efficiency in question increases with decreasing ATP concentration.

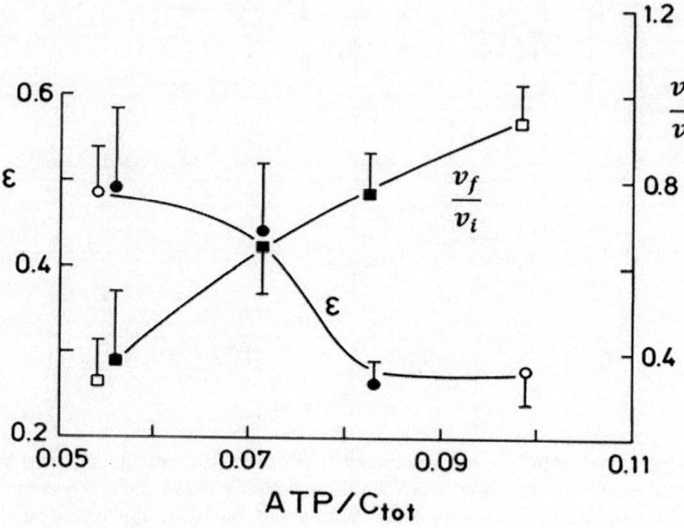

Fig. 2.7 Thermodynamic efficiency (ε), left ordinate, and shortening speed, right ordinate, as a function of the ratio between ATP and total creatine concentrations (C_{tot}). The speed of shortening is expressed as the ratio between the speed of the last (v_f), to the speed of the first (v_i), contraction. Frog Sartorius at 10 °C in nitrogen, poisoned with monoiodoacetic acid. Modified after di Prampero et al. 1988

Table 2.2 Rate of O_2 consumption ($\dot{V}O_2$), tension time index (TTI), ratio TTI/$\dot{V}O_2$ and intramuscular ATP concentration during isometric tetanic contractions in dog gastrocnemius in control conditions and after iodo-acetamide (IA) poisoning. Data from Grassi et al. (2011)

	Control	IA	IA/Control
$\dot{V}O_2$ (ml g^{-1} min^{-1})	0.160	0.046	0.29
TTI (arbitrary units)	1.00	0.60	0.60
TTI/$\dot{V}O_2$ (arbitrary units)	6.25	13.00	2.08
ATP (mmol kg^{-1} dry weight)	18.2	4.8	0.26

2.4 Exercise at Maximal Power in Humans

The considerations summarized in the preceding sections should be taken *cum grano salis* because of the numerous underlying simplifications and approximations; nevertheless, they explain in a rather convincing way other observations on the same topic in humans. Margaria et al. (1971) investigated the maximal vertical velocity over a long flight of stairs in young healthy male subjects. The maximal running bouts were performed either from rest or, without interruption, from the steady state of moderate-intensity aerobic exercises. The so obtained data did show that the stores of anaerobic alactic energy, as given by the sum of ATP and PCr, available at the onset of the maximal intensity running bout, cannot be entirely utilized at maximal power. Ferretti et al. (1987), who repeated the experiment during all-out standing high jumps on a force platform, attained the same conclusion.

Under these conditions, because of the short duration of the all-out efforts (6 s during running, 0.5 s during jumping, 20 ms at maximal instantaneous power), the only energy sources, above the net $\dot{V}O_2$ prevailing at the very onset of the bout, were the splitting of ATP and PCr. This allowed an estimate of the overall amount of energy derived from ATP and PCr splitting, after subtracting the amount derived from $\dot{V}O_2$from the overall energy spent during the bout. The energy equivalent of ATP splitting at rest (19 kJ mol^{-1}) obtained by Ferretti et al. (1987) fits well with the data reported in Table 2.1 for isolated muscle. When the all-out effort was superimposed to a steady state $\dot{V}O_2$ higher than resting, the maximal power decreased by the same amount and in the same manner in both studies, regardless of the duration of the all-out effort. The fall of explosive power was accentuated as the $\dot{V}O_2$ of the preceding exercise increased. The pattern followed by the power fall reflects the characteristics of the ATP changes in contracting muscles, as imposed by the chemical equilibria of the Lohmann reaction.

The explosive effort used by Margaria et al. (1971) lasted about 6 s. In this period, the muscle concentration of PCr decreased substantially. In the initial phase of the explosive bout, the fall of PCr and the resulting (smaller) fall of ATP intramuscular concentrations led to an increase of ε, sufficient to compensate for the concomitant decrease of the shortening velocity, so that the external mechanical power remained essentially unchanged. With increasing the intensity of the preceding exercise, the speed of shortening is reduced to such an extent, that the power must be reduced, even if a substantial fraction of the alactic oxygen stores (ATP + PCr) is still available.

From a general point of view, these considerations suggest that, as far as mammalian skeletal muscle is concerned, biological evolution has given the priority to speed, rather than to economy, thus explaining also why in resting mammalian skeletal muscles the PCr concentration is so much higher than the ATP concentration. Indeed, thanks to creatine kinase, and to its relatively high equilibrium constant, under normal operational conditions, the ATP concentration is always sufficiently high to avoid a substantial reduction of the shortening speed, which would inevitably compromise survival in a prey–predator system.

2.5 Conclusions: Of Animals and Cars

Every self-moving object, whether live or inanimate, is propulsed by an energy – transforming engine. In animals, including humans, the engine is biological and is called muscle. In most types of cars, it is an internal combustion engine. Both require oxidation of an organic substrate (glycogen or fatty acids in animals, gasoline in cars) and consumption of oxygen. Substrate combustion provides the chemical energy that is transformed into mechanical work in internal combustion engines. Substrate oxidation provides the chemical energy that is necessary to resynthesize

ATP, the hydrolysis of which provides the energy that is transformed into mechanical work during the cross-bridge power stroke. As clearly demonstrated by Margaria and his pupils, since ATP can be resynthesized by anaerobic metabolic pathways, there is a remarkable difference between animals and cars: the former can rely on different metabolic pathways, with different inertias and performances, the latter cannot. Nature has created much more versatile motors than human genius has.

Only a fraction of the chemical energy, however, can be transformed into mechanical energy, the remainder being lost as heat: the greater this fraction, the larger the mechanical efficiency of the engine. The maximal mechanical power is proportional to the maximal rate at which the chemical energy transformation occurs. The maximal capacity of the energy transformation is the maximal amount of energy available to the engine. For the car, it is equal to the volume of the fuel tank. For humans, it is variable: very high for aerobic metabolism, which relies on a large amount of glycogen and an even larger amount of fatty acids; very low for anaerobic alactic metabolism, which relies on muscle PCr concentration only.

The maximal mechanical power is also proportional to the maximal speed of the car. The constant relating power to speed has the dimension of a force (F) and includes all forces opposing the movement of an object. When the object moves on Earth, these forces include non-aerodynamic forces, id est the friction forces of the ground (the road for a car) and gearing, and the drag generated by air resistance. When the object is not propelled by an engine, these forces progressively reduce the kinetic energy of the object, which after a given time stops. In objects that are propelled by engines, the speed can be maintained by the energy transformations occurring in the engine.

At the speeds attained by cars moving on a flat road, the aerodynamic forces are by far predominant. In walking humans, non-aerodynamic forces predominate, as detailed in Chap. 5. Chemical power is also directly proportional to speed, but through a proportionality constant which is much greater than F, being equal to the ratio of F to η. This larger constant is equal to the fuel consumption per unit of distance and corresponds to the energy cost as defined by Margaria (1938). As discussed in detail in Chap. 5 and 6, the fraction of it related to aerodynamic forces varies with the square of the speed. For any given maximal power, the maximal speed attained is inversely proportional to this constant. Car designers have therefore strived at reducing its value, by improving cars' aerodynamics and engine's efficiency. Nevertheless, thanks mostly to the work of Margaria and the School of Milano, we can say that muscles perform much better than cars' internal combustion engines. Moreover, muscles can also contract eccentrically, which implies energy absorption by the contracting muscle, and a mechanical efficiency above -1.00, and which cars' engines cannot do. This has an impact on mechanical efficiency during running, as demonstrated within the School of Milano, in particular by Giovanni Cavagna.

We finally point out that, as already mentioned, muscles derive chemical energy from three intermingled metabolic pathways, physiologically described as aerobic metabolism, anaerobic lactic metabolism and anaerobic alactic metabolism. They differ among them in terms of power and capacity. Thus, animals (humans) encompass a large spectrum of performances, depending on the situation and on the needs. In modern athletics, this spectrum is represented, on one extreme, by Usain Bolt running the 100 m dash in 9.58 s, on the other extreme, by Eliud Kipchoge running the marathon in approximately 2 h. In nature, such a flexible muscle energetic system is particularly well adapted to a prey–predator system in which animals move continuously within their vital space.

Most of the activity carried out within the School of Milano and within all the other Schools active in the fields of the energetics of muscular exercise and the energetics and biomechanics of human locomotion since the years 1930s, started with Margaria's falsification of Hill and Meyerhof's theory and the subsequent revolution of knowledge in the field. The ensemble of these activities has produced a deep change and a continuous development of our understanding, not only of exercise physiology, but also of the environmental and ecological conditions of animals' life.

Appendix: Margaria's Tale of the "Revolution" in Muscle Physiology in the 1930s

Some Historical Notes

It has long been known that the amount of energy that is transformed during muscular exercise is proportional to the rate of oxygen consumption: this indicates that the ultimate energy source is in all cases coming from combustions. The muscle, however, is not a combustion engine, and oxidations are not the chemical reactions providing energy for the production of mechanical work.

It is well-known that a muscle can perform a long series of contractions in absence of oxygen, and that the mechanical characteristics of contraction, as well as heat production, the electrical modifications (action potentials) and so on, are exactly the same in presence as in absence of oxygen: the only difference is due to the fact that in absence of oxygen a muscle is unable to perform prolonged work.

Clearly enough, the fundamental chemical reactions that provide energy to carry out mechanical work are anaerobic. In the years 1920s–1930s, people thought that these reactions consisted of lactic acid formation from glycogen (glycolysis). This theory was promoted especially by A.V. Hill and O. Meyerhof, and it seemed to provide a satisfactory explanation both of the energetic transformations occurring in isolated muscles during contraction, accurately analyzed by the former by measuring heat production, and of the chemical alterations (lactic acid formation, glycogen disappearance, oxygen consumption) occurring in muscle during exercise. More precisely, these authors identified the production of lactic acid from glycogen as the fundamental reaction of muscular contraction: without lactic acid, no muscle contraction would have been possible. Oxidations would have taken place at a second time, to provide energy for glycogen resynthesis and lactic acid removal, since the glycolytic reaction is reversible, in order to sustain a continuous, prolonged muscular activity.

This theory, which was called Hill and Meyerhof's theory for muscular contraction, was almost universally accepted by the physiologists of those times. A "revolution" in muscular physiology, to use A.V. Hill's terminology (1932), occurred in 1930, when a young Danish physiologist, E. Lundsgaard, found that a muscle, after poisoning with monoiodoacetic acid, a substance blocking lactic acid formation from glycogen, is still able to carry on multiple contractions, which do not differ at all from the normal ones: only, the total number of contraction is very limited, and at the end the muscle is in a typical contraction state. In these conditions, the muscle, instead of becoming acid, as usual, becomes alkaline. The energy for work production by the muscle seemed to come from the splitting of a substance just discovered in muscles, creatine phosphate, into its two components (creatine and phosphate), a strongly exergonic reaction indeed.

Hill and Meyerhof's theory had to be modified to account for this new discovery. Creatine phosphate splitting was then considered as the reaction that was directly involved in energy supply for mechanical work production, and this reaction was supposed to be located upstream of the reactions of glycolysis.

This last was moved from the first to the second row, but quantitatively speaking maintained its full meaning, as far as the chemical and energetic processes that occur during muscular contraction are concerned. In particular, lactic acid formation from glycogen was still considered an obligatory and necessary step in the chain of reactions involved in muscular contraction; and lactic acid was still considered and obligatory intermediate product of glycogen oxidative metabolism. So, glycolysis was still considered the main, although indirect, anaerobic energy source for muscular contraction.

Lactic acid formation from glycogen still remained the only mechanism for the oxygen debt contraction, in agreement with the original scheme prospected by A.V. Hill, and glycogen resynthesis from lactic acid was still the mechanism for the payment of the same oxygen debt.

According to this hypothesis, we should have found a proportionality between the quantity of lactic acid formed as a consequence of exercise and the amount of the oxygen debt; and the disappearance of lactic acid from blood should have gone on in parallel to the process of oxygen debt payment.

Experiments carried out (1933) on man by R. Margaria, R.H.T. Edwards, and D.B. Dill did not confirm this hypothesis. First of all, they found that lactic acid does not increase in blood at all as a consequence of light or moderate exercise, although also in these cases there is a contraction of a remarkable oxygen debt: an increase of lactic acid in blood occurred only after strenuous muscular exercise, the intensity of which was close to or higher than the maximum oxygen consumption.

In these conditions, the increase of the oxygen debt appeared to proceed at the same rate as, and linearly to, the increase of lactic acid in blood.

On the other side, the kinetics of lactic acid disappearance from blood was definitely different from the kinetics of oxygen debt payment. Lactic acid removal from blood during recovery appeared as a simple exponential process with a half-time of 15 minutes, whereas the time course of oxygen consumption during recovery looked like a more complex process, in which at least two exponential functions were involved: the first being very fast, with a half-time of 0.5 min, and the second very slow, with a half-time of 15 min, id est of the same order of lactic acid removal from blood.

It is evident from this data that lactic acid metabolism is so slow, that it cannot account for the fast and intense oxidative processes that take place in the organism during intense muscular work: clearly enough, glycogen and other substrates, which are burned in the muscle during exercise, cannot take the lactic acid path, because the metabolism of this substance would slow down the entire chain of oxidative reactions to a much slower rate than that actually observed and deducible from oxygen consumption.

In conclusion, the splitting of glycogen to lactic acid did not seem to be a process compatible with the chemical and energetic events that occur during normal muscular exercise, except as an emergency mechanism during strenuous exercise.

In these last conditions, lactic acid formation actually has the meaning of the contraction of an oxygen debt, as suggested by Hill.

Notwithstanding, the oxygen debt contracted during moderate-intensity exercise cannot be due to glycolysis, but to other anaerobic reactions, which Margaria, Edwards, and Dill thought to identify with creatine phosphate splitting. For this reason, they made a distinction between the fraction of the oxygen debt due to creatine phosphate splitting, which for this reason was called alactacid, and is paid very quickly during the recovery period, and the other fraction due to glycolysis (lactacid oxygen debt), which is paid much more slowly, in parallel to lactic acid disappearance from blood.

The quantity of these two fractions of the oxygen debt was plotted as a function of the intensity of exercise (or of the oxygen consumption) in the diagram reported in Fig. 3 (actually Fig. 2.8 in this appendix).

From these experiments it was clear that the combustion coefficient of lactic acid, id est the fraction of lactic acid that is burned, as compared to the total amount of lactic acid that disappears, is not ¼, as observed by Meyerhof on isolated muscle, but sensibly less. The oxygen consumption that is necessary to remove a given quantity of lactic acid from blood in 1 min can be calculated for any given pre-assigned value of the combustion coefficient: if we give to this coefficient a value of ¼, the oxygen consumption involved in the process appeared to be higher than the actually observed one. With this approach, it was remarked that the combustion coefficient of lactic acid could not possibly be higher than about 1/8–1/10.

A few years later, it was found that creatine phosphate splitting can occur only in presence of adenosine-tri-phosphate (ATP), a compound that K. Lohmann isolated in muscle in 1928, and that can be split in adenosine-di-phosphate and inorganic phosphate: this reaction is strongly exergonic too. This substance, or others of the same type, is nowadays considered as the fundamental exergonic reaction, not only in muscular contraction but in all vital processes requiring the production of different forms of energy.

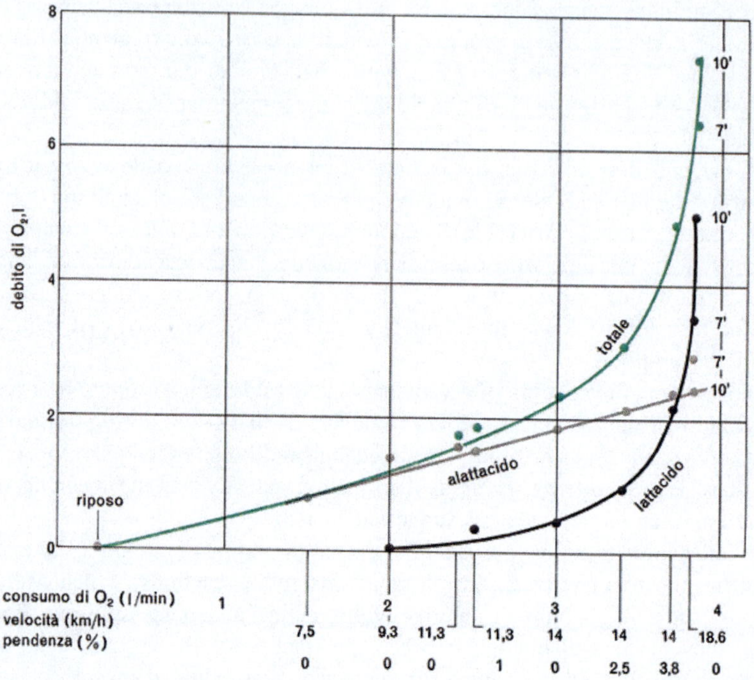

Fig. 2.8 Total alactacid and lactacid debt as a function of oxygen consumption. The type of exercise consisted in running on a treadmill at the speed and incline indicated on the abscissa. The exercise lasted 10 min, except when differently indicated (on the right) at the maximal exercise intensities. From Margaria et al. 1933. This Figure is Fig. 3 of Margaria's book

For this reason, this reaction has been inserted upstream of creatine phosphate splitting in the chain of reactions that occur during muscular contraction.

By so doing, the role of creatine phosphate has progressively evolved to get the meaning of a provider of phosphate and energy for the resynthesis of adenosine-tri-phosphate from adenosine-di-phosphate.

From: Rodolfo Margaria (1975). **Fisiologia muscolare e meccanica del movimento. Biblioteca della EST, Mondadori, Milano, pp. 10–16. Translation from Italian by Guido Ferretti. Figures omitted, except Fig. 3.**

References

Andersson B (1933) Über Co-Zymaseaktievirung einiger Dehydrogenase. Z Physiol Chem 217: 186–190

Atwater WO (1904) Neue Versuche über Stoff- und Kraftwechsel im menschlichen Körper. Ergeb Physiol 3:497–622

Barcroft J, Margaria R (1931) Some effects of carbonic acid on the character of human respiration. J Physiol Lond 72:175–185

Barcroft J, Margaria R (1932) Some effects of carbonic acid in high concentration on respiration. J Physiol Lond 74:156–162

Cerretelli P, Radovani P (1960) The maximum consumption of oxygen in Olympic athletes of various specialties. Boll Soc Ital Biol Sper 36:1871–1872

Cerretelli P, Rossi L, Rovelli E, Marchi S (1960) Cardio-circulatory and metabolic characteristics of a group of Olympic athletes. Boll Soc Ital Biol Sper 36:1874–1875

Chauveau M, Kaufmann M (1887) Expériences pour la determination du coefficient de l'activité nutritive et respiratoire des muscles en repos et en travail. CR Acad Sci Paris D 104:1126–1132

Chauveau M, Tissot J (1896) L'énergie dépensée par le muscle en contraction statique pour le soutien d'une charge d'après les échanges respiratoires. CR Acad Sci Paris D 123:1236–1241

Danilewski A (1880) Thermodynamische Untersuchungen der Muskeln. Pflügers Arch 21:109–152

di Prampero PE (2015) La locomozione umana su terra, in acqua, in aria: fatti e teorie. Edi Ermes, Milano

Dickinson S (1929) The efficiency of bicycle pedalling as affected by speed and load. J Physiol Lond 67:243–255

Eggleton P, Eggleton GP (1927a) The inorganic phosphate and a labile form of organic phosphate in the gastrocnemius muscle of the dog. Biochem J 21:190–195

Eggleton P, Eggleton GP (1927b) The physiological significance of phosphate. J Physiol Lond 63: 155–161

Embden G, Deuticke HJ, Kraft G (1933) Über die intermediären Vorgänge bei der Glykolyse in der Muskulatur. Klein Wochenschr 12:213–215

Embden G, Lawaczeck H (1922) Über die Bildung anorganischer Phosphorsäure bei del Kontraktion des Froschmuskels. Biochem Z 127:181–199

Engelhardt VA, Lyubimova MN (1939) Myosin and adenosine-triphosphatase. Nature 144:668–669

Engelmann TW (1895) On the nature of muscular contraction. Proc Roy Soc B 57:411–435

Ferretti G, Fagoni N, Taboni A, Vinetti G, di Prampero PE (2022) A century of exercise physiology: key concepts on coupling respiratory oxygen flow to muscle energy demand during exercise. Eur J Appl Physiol 122:1317–1365

Ferretti G, Gussoni M, di Prampero PE, Cerretelli P (1987) Effects of exercise on maximal instantaneous muscular power of humans. J Appl Physiol 62:2288–2294

Fick A (1893) Einige Bemerkungen zu Engelmann's Abhandlung über den Ursprung der Muskelkraft. Pflügers Arch 53:606–615

Fiske CH, Subbarow Y (1927) The nature of inorganic phosphate in the voluntary muscle. Science 65:401–403

Fiske CH, Subbarow Y (1928) The isolation and function of phosphocreatine. Science 67:169–171

Fletcher WM, Hopkins FG (1907) Lactic acid in amphibian muscle. J Physiol Lond 35:247–309

Fletcher WM, Hopkins FG (1917) Croonian lecture of 1915: the respiratory processes in muscle and the nature of muscular motion. Proc Roy Soc B 89:444–467

Frank O (1904) Thermodynamik des Muskels. Ergeb Physiol 3:348–513

Grassi B, Rossiter HR, Hogan HC, Howlett RA, Harris JE, Goodwin ML, Dobson JL, Gladden LB (2011) Faster O_2 uptake kinetics in canine skeletal muscle in situ after acute creatine kinase inhibition. J Physiol Lond 589:221–233

Hasselbalch KA (1916) Die Berechnung der Wasserstoffzahl des Blutes aus der freien und gebundenen Kohlensaure desselben, und die Sauerstoffbindung des Blutes als Funktion der Wasserstoffzahl. Biochem Z 78:112–144

Heidenhain R (1864) Mechanische Leistung, Wärmeentwicklung und Stoffumsatz bei der Muskeltätigkeit. Breitkopf und Härtel, Leipzig

Heinemann HN (1901) Experimentelle Untersuchungen am Menschen über den Einfluß der Muskelarbeit auf den Stoffverbrauch und die Bedeutung der einzelnen Nährstoffe als Quelle der Muskelkraft. Pflügers Arch 83:441–476

Helmholtz H (1847) Über die Erhaltung der Kraft. G. Reimer, Berlin
Henderson LJ (1908) The theory of neutrality regulation in the animal organism. Am J Phys 21: 427–448
Hill AV (1913) The energy degraded in the recovery processes of stimulated muscles. J Physiol Lond 46:28–80
Hill AV (1916) Die Beziehungen zwischen der Wärmebildung und dem im Muskel stattfindenden chemischen Prozessen. Ergeb Physiol 15:340–479
Hill AV (1922) The maximum work and mechanical efficiency of human muscles and their most economical speed. J Physiol Lond 56:19–41
Hill AV (1932) The revolution in muscle physiology. Physiol Rev 12:56–67
Hill AV, Howarth JV (1959) The reversal of chemical reactions in contracting muscle during an applied stretch. Proc Roy Soc B 151:169–193
Hill AV, Long CNH, Lupton H (1924) Muscular exercise, lactic acid and the supply and utilization of oxygen. Parts IV – VI. Proc Roy Soc B 97:84–138
Kardel T (1990) Niels Stensen's geometrical theory of muscle contraction (1667): a reappraisal. J Biomech 23:953–965
Klotz IL (1964) Chemical thermodynamics. WA Benjamin Inc., New York, pp 96–141
Lohmann K (1928) Über die isolierung verschiedener natürlicher Phosphorsäure-verbindungen und die Frage ihrer Einheitlichkeit. Biochem Z 194:306–327
Lohmann K (1934) Über die enzymatische Aufspaltung der Kreatin-phosporsäure, zugleich en Beitrag zur Muskelkontraktion. Biochem Z 271:264–277
Lundsgaard E (1930a) Untersuchungen über Muskelkontraktionen ohne Milchsäurebildung. Biochem Z 217:162–177
Lundsgaard E (1930b) Untersuchungen über Muskelkontraktionen ohne Milchsäurebildung. Biochem Z 227:51–82
Margaria R (1928) La resistenza degli animali alla depressione barometrica con varie miscele di ossigeno e anidride carbonica. Arch Sci Biol 11:425–453
Margaria R (1929) Die Arbeitfähigkeit des Menschen bei verminderten Luftdruck. Arbeit 2:261–272
Margaria R (1931) On the state of CO_2 in blood and haemoglobin solutions, with an appendix on some osmotic properties of glycine in solution. J Physiol Lond 73:311–330
Margaria R (1938) Sulla fisiologia e specialmente sul consume energetico della Marcia e della corsa a varia velocità ed inclinazione del terreno. Atti Reale Accad Lincei 7:299–368
Margaria R (1958) Principii di biochimica e fisico-chimica fisiologica. Ambrosiana, Milano
Margaria R (1975) Fisiologia muscolare e meccanica del movimento. Edizioni Scientifiche e Tecniche, Mondadori, Milano
Margaria R, di Prampero PE, Aghemo P, Derevenco P, Mariani M (1971) Effect of a steady state exercise on maximal anaerobic power in man. J Appl Physiol 30:885–889
Margaria R, Edwards HT, Dill DB (1933) The possible mechanism of contracting and paying the oxygen debt and the role of lactic acid in muscular contraction. Am J Phys 106:689–714
Margaria R, von Muralt A (1934) Photoelektrische Messung der pH-Änderung im Muskel während der Kontraction. Naturwissenschaften 22:634
Mayer JR (1845) Die organische Bewegung in ihrem Zusammenhang mit der Stoffwechsel. Dreholerchen, Heilbronn
Meyerhof O (1920) Die Energieumwandlungen im Muskel. I. Über die Beziehungen der Milchsäure zur Wärmebildung und Arbeitsleistung des Muskels in der Anaerobiose. Pflügers Arch 182:232–283
Meyerhof O (1921) Die Energieumwandlungen im Muskel. V. Milchsäurbildung und mechanische Arbeit Pflügers Arch 191:128–183
Meyerhof O (1922) Die Energieumwandlungen im Muskel. VI. Über den Ursprung der Kontraktionswärme. Pflügers Arch 195:22–74
Meyerhof O (1924) Die Energieumwandlungen im Muskel. VII. Weitere Unterschinen über den Ursprung der Kontraktionswärme. Pflügers Arch 204:295–331

Meyerhof O, Geliazkowa N (1947) The rate of anaerobic glycolysis of various hexoses in mammalian tissues. Arch Biochem 12:405–434

Meyerhof O, Lohmann K (1927) Über die enzymatische Milchsäurebildung in Muskelextract. IV. Mitteilung: die Spaltung der Hexosemonophosphorsäuren. Biochem Z 185:113–164

Nachmanson D (1928) Über den Zerfall der Kreatinphosphorsäure in Zusammenhang mit den Tätigkeit des Muskels. Biochem Z 196:73–97

Needham DM (1971) Machina Carnis. Cambridge University Press, Cambridge

Pettenkofer M, Voigt C (1866) Untersuchungen über den Stoffverbrauch des normalen Menschen. Biochem Z 2:459–573

Poole DC, White M, Whipp BJ (2015) The discovery of oxygen. Hektoen International 7: ISSN 2155–3017

di Prampero PE (1981) Energetics of muscular exercise. Rev Physiol Biochem Pharmacol 89:143–222

di Prampero PE, Boutellier U, Marguerat A (1988) Efficiency of work performance and contraction velocity in isotonic tetani of frog sartorius. Pflügers Arch 412:455–461

di Prampero PE, Meyer M, Cerretelli P, Piiper J (1981) Energy sources and mechanical efficiency of anaerobic work in dog gastrocnemius. Pflügers Arch 389:257–262

di Prampero PE, Piñera-Limas F, Sassi G (1970) Maximal muscular power (aerobic and anaerobic) in 116 athletes performing at the XIX Olympic games in Mexico. Ergonomics 13:665–674

Ranvier L (1873) Propriété et structure différente des muscles rouges et des muscles blancs chez les lapins et chez les raies. CR Acad Sci Paris D 77:1030–1034

Rubner M (1894) Die Quelle der tierischen Wärme. Z Biol 30:73–142

Séguin A, Lavoisier A (1789) Premier mémoire sur la respiration des animaux. Mém Acad Sci Paris:566–584

Stella G (1928) The concentration and diffusion of inorganic phosphate in living muscle. J Physiol Lond 66:19–31

Wilson LG (1961) William Croone's theory of muscular contraction. Notes Rec Roy Soc Lond 16: 158–178

Zuntz N (1901) Über die Bedeutung der verschiedenen Nährstoffe als Erzeuger der Muskelkraft. Pflügers Arch 83:557–571

Chapter 3
Margaria's Concept of Oxygen Debt

Pietro Enrico di Prampero and Guido Ferretti

Abstract This chapter is mainly devoted to an analysis of the exercise transients, along the paths traced by Margaria in 1933. The oxygen deficit contracted during light exercise at the expense of anaerobic alactic metabolism (obligatory component of the oxygen deficit), the role of oxygen stores and early lactate in the oxygen deficit, especially at higher work loads, the metabolic control of muscle oxygen consumption during exercise and the slow component of its kinetics at the onset of exercise are discussed. A more detailed description of anaerobic metabolisms is carried out in the second part. Blood lactate accumulation in submaximal exercise, the energy equivalent of lactate and the maximal lactate power are analyzed. The maximal explosive power and the power and capacity of anaerobic alactic power are then discussed. In the appendix, a detailed analysis of the concept of anaerobic threshold is proposed along the energetic way of thinking of the School of Milano.

P. E. di Prampero
Emeritus Professor of Physiology, University of Udine, Udine, Italy
e-mail: Pietro.Prampero@uniud.it

G. Ferretti (✉)
Department of Molecular and Translational Medicine, University of Brescia, Brescia, Italy
e-mail: Guido.Ferretti@unibs.it

G. Ferretti (ed.), *Exercise, Respiratory and Environmental Physiology*, Perspectives in Physiology, https://doi.org/10.1007/978-3-031-19197-8_3

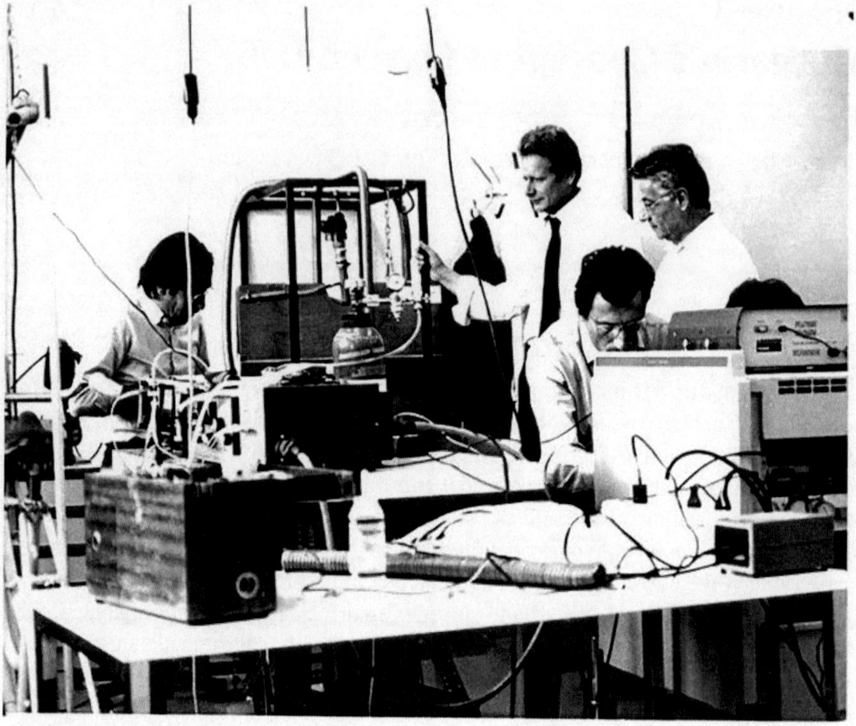

Laboratory work in Milano in the 1970s. From left to right, Pietro Enrico di Prampero, Paolo Cerretelli, Arsenio Veicsteinas (seated) and Giovanni Sassi. With the kind permission of Arsenio Veicsteinas family

3.1 Introduction

In the preceding chapter, we discussed the origin, the meaning, and some consequences of the revolutionary study by Margaria et al. (1933) that, falsifying Hill and Meyerhof's theory of muscular contraction, opened the way to the present view of muscle energetics. We also analyzed several subsequent studies that emerged therefrom within the school of Milano, which shaped and refined that view. We finally discussed how Margaria's concepts were compatible with a thermodynamic vision of energy exchanges of living animals. We will now address a strictly related question: how the concept of oxygen debt evolved from the early 1960s onward within the School of Milano.

3.2 Metabolic Transients and Oxygen Debt "Contraction" and "Payment"

During muscular exercise, the rate of ATP splitting is dictated by the work intensity. At the onset of a constant load square-wave exercise, it increases in an essentially instantaneous way to attain a level that is higher, the greater the work intensity. This level remains subsequently unchanged until the end of exercise, to fall to the pre-exercise resting level, again in an essentially instantaneous way, at the very instant at which the exercise ceases. On the contrary, at variance with ATP splitting, the rate of muscle oxygen consumption ($\dot{V}_mO_2$) lags behind the mechanical events with a certain inertia: for instance, at the onset of an aerobic square-wave exercise of moderate intensity, it increases to reach a steady state in 2–4 min (Fig. 3.1). It follows that, in the first minutes of exercise, the energy yield from the oxidative processes is not sufficient to resynthesize the overall amount of ATP utilized for muscular work. However, in all cases, the rates of ATP splitting and resynthesis must be equal (or very nearly so), since even a small and transitory fall of ATP concentration leads to serious consequences in terms of thermodynamic efficiency and speed of shortening (see Sect. 2.3.4). It follows that a fraction of the ATP

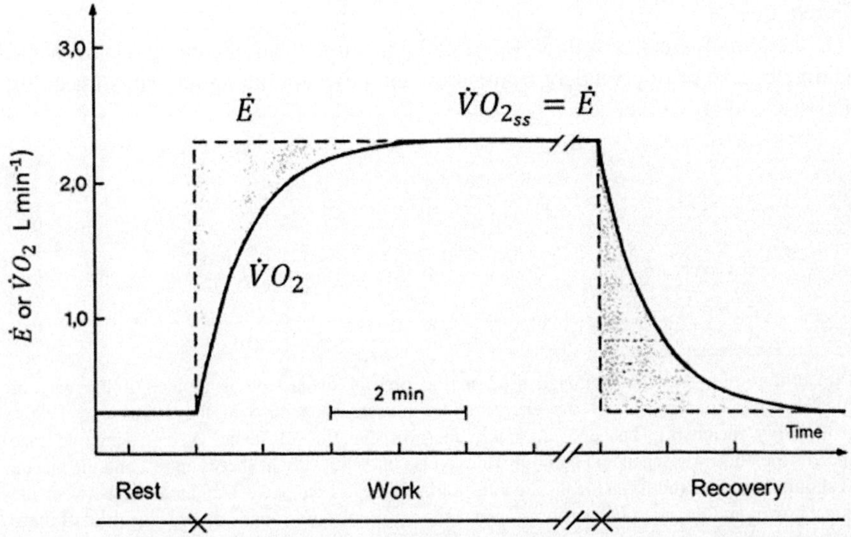

Fig. 3.1 Energy requirement per unit of time for ATP resynthesis in equivalent oxygen units ($\dot{E}$, broken line) and oxygen uptake through the upper airways (continuous line) ($\dot{V}O_2$, L min^{-1}, continuous line), as a function of time, during moderate square wave exercise in humans (schematic). At work onset, $\dot{E}$ exceeds $\dot{V}O_2$; in the recovery, the opposite is true: hence, an oxygen debt is "contracted" at work onset and "paid" in recovery (gray areas). At steady state, requirement and uptake coincide ($\dot{V}O_2 = \dot{E}$). It is noteworthy that, when this type of representation started to be used, $\dot{V}O_2$ was considered equivalent to the rate of muscle oxygen consumption also in an exercise transient. Modified after di Prampero 2015

necessary for work performance must be resynthesized at the expense of energy sources different from aerobic metabolism. This amount of energy was called by Margaria et al. (1933) oxygen debt contraction (Fig. 3.1).

At work offset, $\dot{V}_mO_2$ progressively returns to the resting value, following a time course essentially similar, but opposite, to that observed at work onset; on the contrary, as mentioned above, the rate of ATP splitting falls to the pre-exercise level, again in an essentially instantaneous way. Thus, the energy sources from which the oxygen debt was borrowed at work onset, can be reconstituted at the expense of the extra oxygen utilized in the recovery after exercise, above the resting muscle energy requirement (Fig. 3.1). Margaria defined the latter situation as "payment" of the oxygen debt, to distinguish it from the "contraction" occurring at work onset. Of common usage nowadays is also the terminology "deficit" and "debt" to indicate the contraction and payment, respectively.

Hill et al. (1924) attributed the oxygen debt to the oxidative removal of the extra lactate produced at work onset. However, as discussed in detail in Chap. 2, Margaria et al. (1933) demonstrated that, in exercising humans, regardless of the work intensity, the contraction of the oxygen debt, which they defined "alactacid" (alactic), is independent of lactate removal. They attributed it to the resynthesis of the amount of phosphagen split at work onset, thus paving the way to the theory of the energetics of muscle contraction that is still considered essentially correct (Ferretti 2015).

In the paragraphs that follow, we describe in some detail the energy sources and the time course of oxygen debt contraction and payment, along the paths traced by Margaria and co-workers.[1]

[1]The terminology and the language used in the present discussion is typical of the way of expressing the concepts about the energetics of muscular exercise in the 1970s and 1980s. Terminology meanwhile has evolved, more in form than in substance. A representative case concerns the currently accepted taxonomy of exercise intensities: light (between rest and the lactate threshold), moderate (between the lactate threshold and the critical power), and intense (between the critical power and the maximal aerobic power) (see e.g. Poole and Jones 2012). If the third of these taxonomic categories is sound, the separation of a light and a moderate exercise domain by the lactate threshold looks conceptually weak. Notwithstanding the doubtful meaning of the lactate threshold concept, a steady state for oxygen uptake and for blood lactate concentration, implying no lactate accumulation after completion of the exercise transient, and thus no incurring anaerobic lactic metabolism, means that in both cases, after attainment of the steady state, the metabolism is entirely aerobic. The only difference is that in the light exercise domain, the steady blood lactate concentration corresponds closely to that at rest, whereas in the moderate exercise intensity, the steady blood lactate concentration is higher than at rest, because of "early" lactate accumulation during the exercise transient. We nevertheless acknowledge that the terminology used in more ancient times was somewhat lax.

3.2.1 Aerobic Exercise

Aerobic exercise encompasses the range of work intensities wherein, after the initial transient phase, during which the oxygen debt is contracted, a steady state is attained, i.e., a condition in which no net lactic acid accumulation, neither net ATP nor phosphocreatine (PCr) splitting do occur. Under these conditions, at the onset of a "square wave" exercise of moderate intensity, $\dot{V}_m O_2$ was assumed to increase, as a first approximation, as a mono-exponential function of time (t):

$$\dot{V}_m O_{2t} = \dot{V}_m O_{2ss}(1 - e^{-t/\tau}) \tag{3.1}$$

where $\dot{V}_m O_{2t}$ and $\dot{V}_m O_{2ss}$ indicate the net (above resting) rates of muscle oxygen consumption at time t and at steady state (ss), respectively, and τ is the time constant of the exponential equation describing the phenomenon. In a wide range of metabolic intensities, τ is about 23 s; it follows that, after about 92 s (i.e., for $t \approx 4 * \tau$), $\dot{V}_m O_2$ has essentially attained a steady state ($\dot{V}_m O_{2t} \approx 0.98\ \dot{V}_m O_2$), as shown schematically in Fig. 3.2 (Piiper et al. 1968; Binzoni et al. 1992).

Indeed, at work onset, the intramuscular PCr concentration decreases to attain, after about 3 min, a constant level. The energy liberated by this initial fall of PCr is

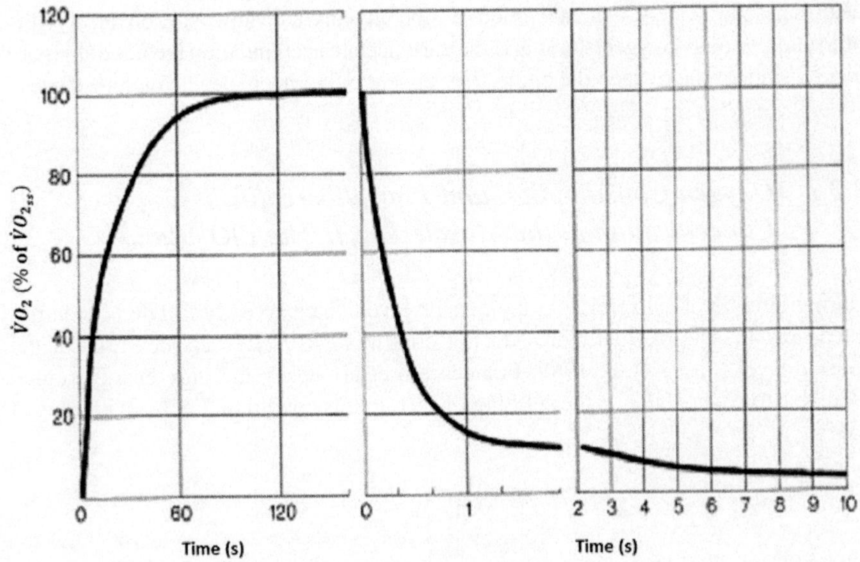

Fig. 3.2 Net oxygen consumption ($\dot{V}O_2$, percentage of the value at steady state, $\dot{V}O_{2ss}$) as a function of time, during aerobic work and in the following recovery, in dog gastrocnemius (schematic). At work onset, the oxygen consumption at the muscle level increases mono-exponentially with a time constant of about 22 s, whereas in recovery after exercise a slow component of the oxygen consumption decrease is also appreciable. Modified after Piiper et al. 1968

utilized for the resynthesis of a fraction of the ATP necessary for work performance, the intramuscular concentration of which is essentially unchanged. Furthermore, at work onset, in addition to the amount of oxygen taken up through the lungs, also a decrease in the body oxygen stores (id est of the amount of oxygen bound to hemoglobin in venous blood, and to muscle myoglobin) contributes to the energy requirements. It seems worthwhile to point out here that the contribution of the body oxygen stores to the overall oxygen consumption, due to the decrease of the amount of oxygen already present in the body before the onset of exercise, is not detected at the lung level (di Prampero et al. 1970a). Finally, during aerobic exercises of medium to high intensity, at work onset a certain amount of lactate can also be produced, thus contributing to the muscle energy requirement (Cerretelli et al. 1979, see also the Appendix to this Chapter). Hence, the oxygen deficit ($DefO_2$) can be expressed as the sum of three terms:

$$DefO_2 = VO_2{}^{PCr} + VO_2{}^{eLa} + \Delta VO_2{}^{st} \tag{3.2}$$

where $VO_2{}^{PCr}$ and $VO_2{}^{eLa}$ ("early lactate") represent the oxygen equivalent of net PCr hydrolysis and of the amount of lactate produced (in fact accumulated) before the attainment of the steady state, and $\Delta VO_2{}^{st}$ is the change of the oxygen stores at work onset.

The hydrolysis of PCr is a necessary component of the oxygen deficit, since it determines the kinetics of activation of the glycolytic pathway. Conversely, the utilization of body oxygen stores and the early lactate accumulation are not obligatory components of the oxygen deficit, as they appear only under specific circumstances.

3.2.2 *Oxygen Consumption and Phosphocreatine Concentration at the Muscle Level: The P/O$_2$ Ratio*

During light aerobic exercise, at the muscle level, since no early lactate is accumulated, and muscle oxygen stores do not contribute to $DefO_2$, as measured at the muscle level (Piiper et al. 1968; Francescato et al. 2003), the only energy source available to cover $DefO_2$ is the splitting of PCr. In this condition, Eq. (3.2) is reduced to

$$DefO_2 = VO_2{}^{PCr} = \dot{V}_mO_{2ss}\,\tau^{al} \tag{3.3}$$

where τ^{al} ($\approx$ 23 s) is the time constant of the exponential decrease of PCr concentration, which corresponds to the time constant of the exponential increase of $\dot{V}_mO_2$[2]

[2] At steady state, the oxygen consumption (above resting), as measured at the upper airways level, is equal to that of the active muscles. However, during the metabolic transients at the onset and offset of the exercise, the muscle oxygen consumption is not necessarily equal to that measured at the upper airways, because of the "buffering" effect of the body oxygen stores.

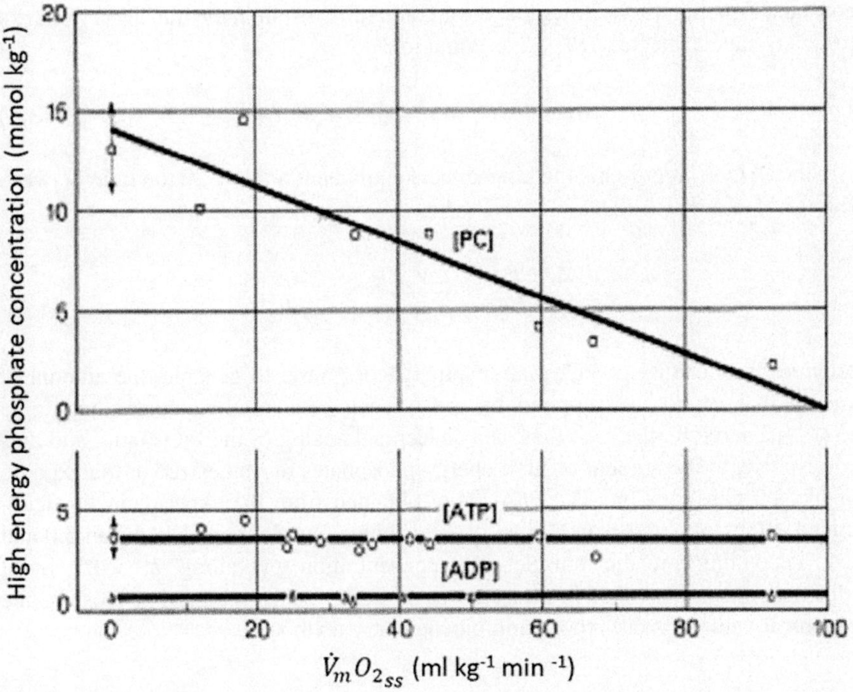

Fig. 3.3 Concentration of high energy phosphates in dog gastrocnemius as a function of the net oxygen consumption per kg of muscle during aerobic work at steady state ($\dot{V}_m O_{2ss}$). ATP and ADP are essentially unchanged, whereas PCr concentration decreases linearly with increasing $\dot{V}_m O_{2ss}$. From Piiper et al. 1968

(see Figs. 3.1 and 3.2). Equation (3.3) reports in quantitative terms the relationship between the intramuscular PCr concentration and the exercise intensity at steady state, both expressed in terms of equivalent oxygen consumption. VO_2^{PCr} is traditionally defined "net alactic oxygen deficit" and expresses the "obligatory" fraction of the total deficit Eq. (3.2), in so far as it cannot be modulated by simple experimental manipulations (di Prampero and Margaria 1968). A direct consequence of Eq. (3.3) is that, during light aerobic exercise, VO_2^{PCr} is a linear increasing function of the exercise intensity. Hence, since in this case, the kinetics of muscle PCr concentration is a mirror image of that of $\dot{V}_m O_2$ (di Prampero and Margaria 1968), muscle PCr concentration must be a linear decreasing function of the net steady state $\dot{V}_m O_{2ss}$ (Binzoni et al. 1992; Piiper et al. 1968), as reported in Fig. 3.3 for isolated dog gastrocnemius in situ:

$$PCr_{ss} = PCr_0 - a\ \dot{V}_m O_{2ss} \tag{3.4}$$

where PCr_{ss} and PCr_0 are the muscle PCr concentrations at the exercise steady state and at rest before the exercise onset, respectively, and a is the appropriate correlation constant. Equation (3.4) shows that the amount of PCr split in the transition from rest to steady state exercise (ΔPCr_{ss}), is equal to:

$$PCr_0 - PCr_{ss} = \Delta PCr_{ss} = a\ \dot{V}_m O_{2ss} \tag{3.5}$$

Since ΔPCr_{ss} represents the biochemical equivalent of VO_2^{PCr}, the ratio between Eqs. (3.5) and (3.3):

$$\frac{\Delta PCr_{ss}}{VO_2^{PCr}} = \frac{a\ \dot{V}_m O_{2ss}}{\tau^{al}\ \dot{V}_m O_{2ss}} = \frac{a}{\tau^{al}} \tag{3.6}$$

provides the quantity of PCr that is split at work onset to generate the amount of energy that cannot be supplied by the oxygen consumption, corresponding to VO_2^{PCr}. Therefore, the ratio a/τ^{al} is a conceptual analog of the P/O_2 ratio, and thus corresponds to the amount of high-energy phosphates resynthesized at the expense of oxygen consumption. The P/O_2 ratio, obtained from the experimentally determined values of a and τ on human gastrocnemius muscle, ranges between 5.0 and 6.4, depending on the muscle PCr concentration prevailing at work onset (di Prampero et al. 2003; Francescato et al. 2003, 2008), i.e., rather close to the canonical value of ≈6.0 reported in biochemistry textbooks.

3.2.3 Oxygen Stores, Early Lactate and the Overall Oxygen Deficit

The contribution of ΔVO_2^{st} to the overall oxygen deficit increases to a lesser extent than the work intensity (di Prampero et al. 1970a); on the contrary, VO_2^{eLa} becomes progressively larger the higher the work intensity (Cerretelli et al. 1979).

The relative role of the three terms of Eq. (3.2) is reported in Fig. 3.4 for exercises of increasing intensity during treadmill walking and running. Whereas VO_2^{PCr} is dictated by $\dot{V}_m O_{2ss}$, the relative contribution of VO_2^{eLa} and to a lesser extent of ΔVO_2^{st}, depends also on the type of the exercising muscles, on the subject's training level and on the body position during exercise. For instance, ceteris paribus, VO_2^{eLa} is greater, as compared to the data reported in Fig. 3.4, during cycloergometric or arm exercise. We can also predict it to be greater in untrained than in trained subjects, and in supine than in upright position.

We analyze now the kinetics of whole-body oxygen uptake, as measured at the upper airways ($\dot{V}_L O_2$), at the onset of square wave aerobic exercise of various intensities. Since the $\dot{V}_L O_2$ increases to attain a constant steady-state value after about 3 min, and considering the energetics of the oxygen debt as defined by Margaria, this kinetics was described by mono-exponential Equations of the following type:

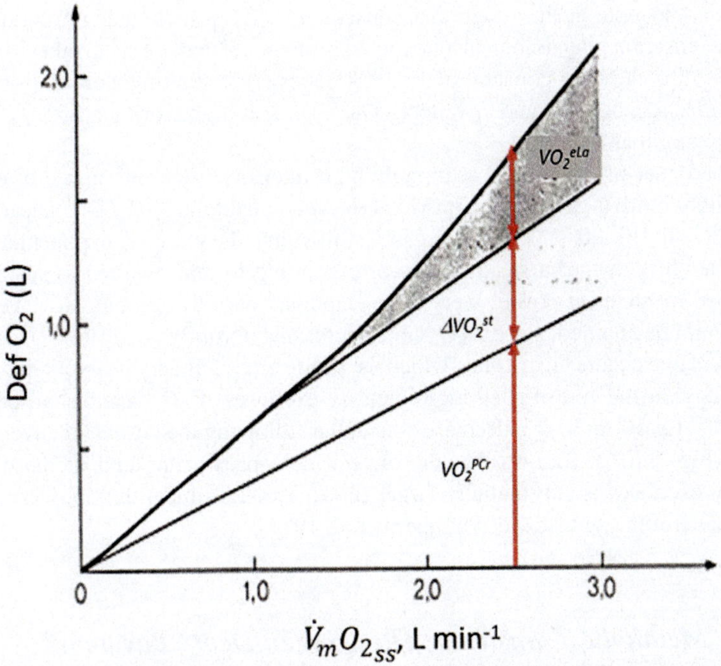

Fig. 3.4 Oxygen deficit ($DefO_2$) as a function of the net oxygen consumption at the steady state ($\dot{V}_mO_{2ss}$) in humans during treadmill walking or running. The overall deficit is indicated by the uppermost function; its three components (hydrolysis of phosphocreatine, $VO_2{}^{PCr}$, oxygen stores utilization, $\Delta VO_2{}^{st}$, early lactate accumulation, $VO_2{}^{eLa}$) correspond to the vertical distance between the appropriate functions (the red arrows highlight their values at $\dot{V}_mO_{2ss}$= 2.5 L min^{-1}). The indicated values refer specifically to walking and running in active non-athletic subjects. Modified after di Prampero 2015

$$\dot{V}_LO_{2t} = \dot{V}_LO_{2ss}\left(1 - e^{t/\tau_L}\right) \tag{3.7}$$

where $\dot{V}_LO_{2t}$ and $\dot{V}_LO_{2ss}$ are the net lung oxygen uptake at time t and at the steady state, respectively, and τ_L, which ranges between 30 and 120 s, is the time constant of the process as measured at the mouth (di Prampero 1981). In exercises with a relatively large muscle mass and of light to moderate intensity, τ_L is generally about 40–45 s. For higher exercise intensities, under which conditions $VO_2{}^{eLa}$ is proportionately larger, the kinetics of $\dot{V}_LO_2$ becomes slower (Cerretelli et al. 1979). In addition, during arm exercise, the $\dot{V}_LO_2$ kinetics, other things being equal, is slower than during leg exercise (Cerretelli et al. 1977). These different responses were attributed to the varying behavior of $VO_2{}^{eLa}$ and of $\Delta VO_2{}^{st}$ at work onset: indeed, a large contribution of one, or both these terms, leads to a larger overall $DefO_2$, and hence to a slower $\dot{V}_LO_2$ kinetics at work onset.

We should note at this stage that constant τ^{al} of Eq. (3.1) and τ_L of Eq. (3.7) describe different physiological phenomena, albeit related to each other. Indeed, whereas τ^{al} is the time constant of $\dot{V}_mO_2$ kinetics, τ_L is the time constant of $\dot{V}_LO_2$ kinetics. Hence, both τ^{al} and τ_L reflect VO_2^{PCr} and, at high work loads, VO_2^{eLa}, but only τ_L encompasses also ΔVO_2^{st}.

At the onset of exercise of extremely high intensity, requiring a rate of energy expenditure more than twice the maximal oxygen consumption ($\dot{V}O_2^{max}$), leading to exhaustion in 15–60 s, $\dot{V}_LO_2$ increases, following the "usual" exponential time course, tending toward a steady state corresponding to that required to sustain the developed mechanical power, were the metabolism entirely aerobic. However, this level cannot be attained, since $\dot{V}_LO_2$ stops increasing abruptly once $\dot{V}O_2^{max}$ has been reached. This occurs in a time, which is shorter, the higher the work intensity. Conversely, at the end of such high-intensity exercises, $\dot{V}_LO_2$ remains at, or close to, $\dot{V}O_2^{max}$ for 10-to-30 s; it decreases thereafter following the "usual" recovery time course (Fig. 3.5). It follows that the oxygen debt paid at the end of these high-intensity exercises is substantially larger (up to 1.6–2.0 times) than observed after maximal aerobic exercises (di Prampero et al. 1973).

3.2.4 Metabolic Transients: The Oxygen Debt "Payment"

In the recovery after exercise, $\dot{V}_LO_2$ tends to the pre-exercise value following a complex function that, as originally described by Margaria et al. (1933), can be dissected into two major components. The first includes a fast exponential $\dot{V}_LO_2$ decrease with a time constant of about 42–45 s. The time integral of this exponential function represents the so-called payment of the alactic oxygen debt. It comprises PCr resynthesis and oxygen stores replenishment. So it is, at least, equal to the sum of VO_2^{PCr} plus VO_2^{eLa}.

The second component is described by an exponential function with a much longer time constant, of about 12–17 min. This time constant corresponds to that of lactate removal at the end of exercise (Margaria et al. 1933). It includes an "increase of the resting metabolism" of the muscle due to the preceding exercise, the physiological bases of which are rather difficult to pinpoint. However, it is tempting to attribute this slow component of the time course of muscle metabolism during recovery after exercise to the changes of muscle pH (pH_m) due to the preceding exercise. Indeed, during intense exercise, pH_m decreases, to resume the value prevailing at rest ($pH_m \approx 7.0$) in the subsequent recovery period, following once more an exponential function with a time constant similar to that reported above for lactate washout ($\approx$ 14 min) (Harris et al. 1976). This state of affairs leads to small changes of the equilibrium constant of Lohmann's reaction, which depends on the prevailing pH_m:

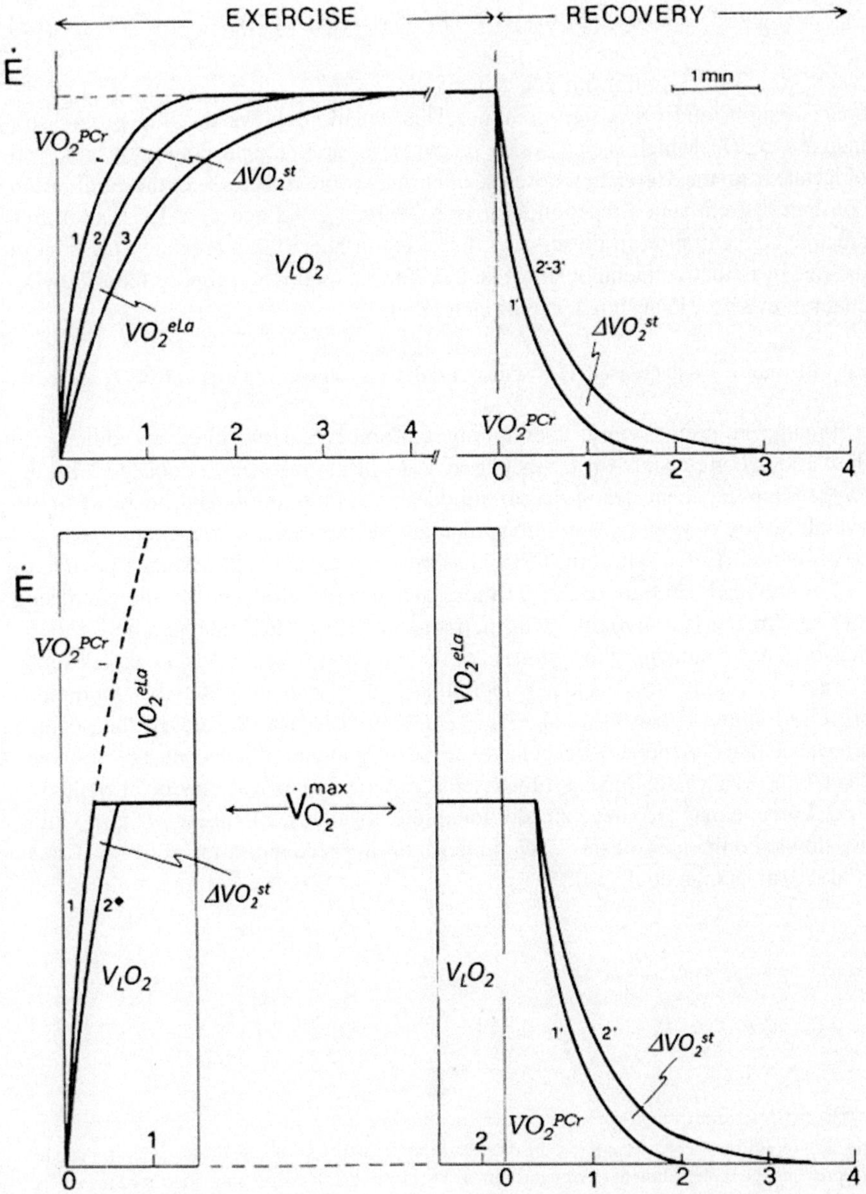

Fig. 3.5 Time course of energetic processes in humans (schematic). **Upper panel:** Metabolic power ($\dot{E}$) is below maximal oxygen consumption ($\dot{V}O_2^{max}$). Curve 1 accounts for (i) the oxygen uptake through the airways (curve 3), (ii) early lactate accumulation (curve 2) and (iii) oxygen stores depletion. The energetic contribution of the last two processes is given by areas VO_2^{eLa} (between curves 2 and 3) and ΔVO_2^{st} (between curves 1 and 2); the area V_LO_2 indicates the oxygen taken up through the lungs. The difference between Ė (horizontal upper dotted line) and curve 1 is covered by PCr splitting (VO_2^{PCr}). Thus, the area VO_2^{PCr} yields the net alactic oxygen deficit. In the recovery phase, area VO_2^{PCr} indicates the payment of the oxygen debt, and curve 1 describes the

$$\log K_{eq} = \log \{([\text{ATP}]^* [\text{Cr}])/([\text{ADP}]^* [\text{PCr}]\} \quad (3.8)$$

where K_{eq} is the equilibrium constant of the reaction, square parentheses indicate concentration, and Cr is free creatine. This equation shows that the progressive increase in pH_m, which occurs during recovery, is parallel to the continuous removal of lactate from the exercising muscles, and brings about a decrease of the equilibrium constant of Lohmann's reaction (Harris et al. 1977), and hence, a decrease in free creatine concentration in muscle. As discussed in Sect. 3.2.6, creatine may control the rate of oxidative metabolism. It is, therefore, tempting to propose the following chain of events (H^+ designates hydrogen ions):

Lactate removal → decreased $[H^+]$ → decreased [Cr] → slow component of $\dot{V}_LO_2$ decrease.

Finally, it seems worth mentioning that moderate exercise, in addition to the various components of the increased $\dot{V}_LO_2$ in the recovery associated with the oxygen debt payment, leads also to a modest, but more prolonged, increase of the overall resting oxygen consumption, generally defined excess post-exercise oxygen consumption (EPOC). In turn, EPOC has been associated with irisine, a peptide of 112 amino-acids and about 20 kD molecular weight, produced by the exercising muscles. Irisine is a myokine with thermogenic effects. It increases adipose tissue metabolism by inducing the synthesis of uncoupling protein 1. Irisine expression is elevated in highly active elderly individuals. It is also increased by continuous exercise training (Boström et al. 2012; Zhang et al. 2016). Several of the positive effects of daily moderate muscular exercise in patients affected by the so-called "metabolic syndrome" (obesity, hypertension, type 2 diabetes) may be attributed to EPOC. Irisine may also have an additional role, together with that of pH_m, in setting the slower component of the $\dot{V}_LO_2$ kinetics during recovery after exercise (Larsen et al. 2014; Tjønna et al. 2008).

Fig. 3.5 (continued) corresponding time course of the process. The area between curve 1 and curve 2 represents the replenishment of oxygen stores. Thus, curve 2 describes the kinetics of the oxygen uptake through the airways during recovery. Lactate is not represented, because the kinetics of its recovery is much slower. **Lower panel.** $\dot{E}$ is 1.75 $\dot{V}O_2{}^{max}$, the exercise time to exhaustion is 2.25 min (see the broken time axis). The areas $VO_2{}^{PCr}$, V_LO_2 and $\Delta VO_2{}^{st}$ have the same meaning as in the upper panel. Curve 1 shows the kinetics of muscle oxidative processes. The onset of anaerobic lactic energy production (dotted line, theoretical) starts when $\dot{V}O_2{}^{max}$ is attained (horizontal arrows). The energy provided by lactate accumulation, in oxygen equivalents, is given by area $VO_2{}^{eLa}$. At the exercise end, $VO_2{}^{PCr}$ is paid by oxidative processes, which remain at the $\dot{V}O_2{}^{max}$ level for about 20 s. From di Prampero 1981

3.2.5 The Control of Muscle Oxygen Consumption

The state of affairs described above can be interpreted univocally, assuming that the $\dot{V}_m O_2$ is in some way controlled by the intramuscular concentration of PCr, or, rather, by its mirror image, free creatine. In other words, VO_2^{PCr} can be considered as the controller of muscle metabolism (di Prampero and Margaria 1968; di Prampero et al. 2003; Mader 2003; Mahler 1985). It follows that the fraction of the oxygen deficit represented by VO_2^{PCr} must necessarily occur, to "bring" the muscle to the metabolic level required by the ongoing exercise intensity.

This simplified view of the role of creatine and/or PCr in regulating $\dot{V}_m O_2$ should be taken with caution. The work by Grassi et al. (2011) supports a dual role of PCr breakdown at exercise onset: an almost instantaneous increase in energy yield for muscle contraction, coupled with an important role in $\dot{V}_m O_2$ regulation (di Prampero 1981; Whipp and Mahler 1980; Meyer et al. 1984; Meyer 1988; Kaasik et al. 1999; Greenhaff 2001; Rossiter et al. 2005). Again, according to Grassi et al. (2011): "... *the bioenergetic mechanism which is fast enough to meet sudden increases in metabolic demand, that is PCr breakdown, is functionally related, through the level of some of its metabolites to the regulation of oxidative phosphorylation, the most important mechanism for ATP resynthesis*." For detail and further references on this and related matters, the reader is referred to Grassi et al. (1996, 2002, 2005).

These considerations further highlight the reasons why the School of Milano considered that an alactic oxygen deficit must occur at work onset, nor can it be "paid" before the end of the exercise, regardless of its intensity. Indeed, were the alactic oxygen deficit not incurred, the muscle could not increase its metabolism; on the other side, were the oxygen debt paid before the end of the exercise, the muscle metabolism would necessarily fall below the level required. The alactic oxygen deficit is a physiological necessity.

3.2.6 The Slow Component

In more recent years, a third phase of the $\dot{V}_L O_2$ kinetics at work onset, defined "slow component," has been identified. It consists of a moderate and slow $\dot{V}_L O_2$ increase, which often, in medium-to-high intensity exercises, persists beyond the time necessary and sufficient to attain the steady state (2.5–3 min). The slow component has been analyzed in some detail, for both theoretical and practical reasons. Indeed if, on the one side, the identification of the underlying biochemical and physiological mechanisms is of substantial theoretical interest, on the other, the $\dot{V}_L O_2$ increase due to the slow component may severely compromise aerobic performances of long terms events, e.g., leading to a substantial increase in the energy cost of running during a marathon or semi-marathon.

The slow component has been described mathematically by means of either exponential equations with a rather long time constants ($\tau \geq \approx 90$ s), or of linear increasing functions (Barstow and Molé, 1991; Camus et al. 1988; Capelli et al. 1993; Poole et al. 1988, 1994). None of these studies mentions a theoretical reason for choosing an exponential rather than a linear treatment of it. In the time domain considered by many authors (generally ≈ 10 min, or thereabout), both models fit the experimental data rather well. The choice, however, is not without consequences. On the one side, a linear function, after a sufficiently long time, will inevitably lead to the attainment of $\dot{V}O_2^{max}$, and hence to exhaustion; on the other side, an exponential is consistent with a steady state, even after a longer time and to a higher level than observed under "canonical" conditions, i.e., in the absence of any slow component.

The slow component becomes evident during high-intensity exercise. Nevertheless, Jones et al. (2010) reported that a slow component may appear also during moderate exercise, around or slightly below the critical power: in this case a $\dot{V}_LO_2$, were the slow component conveniently described by an exponential equation, a steady state in $\dot{V}_LO_2$ may still be attained at a metabolic power ($\dot{E}$) lower than $\dot{V}O_2$ ^{max}, but higher than that expected at steady state from the classical $\dot{V}_LO_2$ versus power relationship. If this was the case, an exponential model of the slow component may indeed be retained. This option is also supported by further energetic arguments, as pointed out hereafter.

We have previously seen that the progressive increase of pH_m, occurring in recovery after heavy exercise, might be responsible for the slow component of the $\dot{V}_LO_2$ kinetics observed in this condition. We have hypothesized that this is due to the effects of pH on the equilibrium constant of Lohmann's reaction, and hence on creatine concentration Eq. (3.8). Therefore, it would be logical to hypothesize that something similar may occur also during exercise, whenever work intensity leads to an increase of intramuscular lactate concentration and to a decrease of pH_m. If the exercise is not of exceedingly high intensity, the resulting increase in creatine concentration ceases as soon as an equilibrium between lactate production and removal is attained (see Appendix).

This hypothesis is supported by a study by Capelli et al. (1993), who observed that, during prolonged (20–40 min) exercise on a cycle ergometer, the intensity of which ranged from 60 to 80% of the subjects' $\dot{V}O_2^{max}$: (1) $\dot{V}_LO_2$ kept increasing also after the first 4–5 min of exercise and (2) the rate of increase of $\dot{V}_LO_2$ after the fifth minute of exercise was directly proportional to that of lactate accumulation in blood. On the contrary, for exercise intensities <60% $\dot{V}O_2^{max}$, the rates of both $\dot{V}_LO_2$ and blood lactate accumulation did indeed attain a constant value after a few minutes of exercise. In similar metabolic conditions, Rossiter et al. (2005) observed a slow component in the kinetics of the changes in PCr concentration at exercise onset, investigated by ^{31}P-NMR spectroscopy. A possible role for pH_m changes in intense exercise was suggested also by Jones et al. (2011).

In contrast with this hypothesis, Poole et al. (1994) discussed other potential mechanisms explaining the slow component, namely (1) a possible metabolic effect of increased blood catecholamine concentration, (2) an increased work of non-locomotor muscles, especially respiratory muscles, above the so-called ventilatory threshold; (3) potential differences in contraction efficiency between type I and type II muscle fibers. Overall, none of these mechanisms interferes with the energetic processes of muscle, as long as (1) blood catecholamine concentration increases also in light exercise, when no slow component appears (Leuenberger et al. 1993; Miyamoto et al. 2006); (2) non-locomotor muscles, if any, increase internal work, and (3) no differences in the delta-efficiency of cycling exercise was found between athletes and non-athletes (Moseley et al. 2004).

The slow component challenges the concept of a constant delta-efficiency of cycling exercise, as long as it implies the notion that the relationship between $\dot{V}_L O_2$ and mechanical power may not be linear along the entire $\dot{E}$ range from rest to $\dot{V}O_2^{max}$. Depending on what the causes of the slow component are, this may impose a significant revision of the concept of maximal aerobic power and of its meaning within the general theory of the energetics of muscular exercise.

3.3 The Lactate Mechanism

When the work intensity exceeds that corresponding to $\dot{V}O_2^{max}$, there is a substantial and continuous accumulation of lactate in blood. Under these conditions, the subject reaches exhaustion when blood lactate concentration ($[La]_b$) attains 15–17 mM. In addition, the time necessary to reach this critical $[La]_b$ is shorter the higher the work intensity. The reason for this is that a fraction of the ATP yielding the energy for muscle work performance is resynthesized at the expense of the anaerobic formation of lactate. Indeed, under these conditions, the rate of pyruvate formation from glycogen is greater than that of its oxidation to carbon dioxide and water via the Krebs cycle. As a consequence, the excess pyruvate is transformed anaerobically into lactate, and the subject can sustain exercise intensities greater than $\dot{V}O_2^{max}$, albeit for relatively short times (at most a few minutes). However, because lactic acid is a strong acid that at physiological pH is almost completely dissociated, lactate formation leads to a fall of pH_m that, when a given threshold is attained, blocks the enzymatic formation of lactate itself, thus preventing an excessive and dangerous hydrogen ion accumulation in the body fluids. In addition, lactate production is rather uneconomical energetically, as far as it yields only three moles of ATP per glycosidic unit, as compared to the 37 obtained from the complete oxidation of the same amount of glycogen (Table 3.1).

Neglecting the initial exercise period in which $\dot{V}_L O_2$ has not yet attained its maximal value, the energetics of muscular exercise under these conditions can be described by:

$$\overleftarrow{\dot{\text{ATP}}} = b\,\dot{La} + c\,\dot{V}O_2^{max} \tag{3.9}$$

where $\overleftarrow{\dot{\text{ATP}}}$ and $\dot{La}$ are the rates of ATP resynthesis and of lactate accumulation, respectively (moles per unit of time); b and c are proportionality constants

Table 3.1 ATP production per mole of substrate, or of oxygen, consumed along various metabolic pathways

Reaction	ATP/mole		RQ
	Substrate	Oxygen	
Glycogen* ↔ lactate	3		
Glucose ↔ lactate	3		
Lactate ↔ $CO_2 + H_2O$	17	5.2	1.00
Glycogen* ↔ $CO_2 + H_2O$	37	6.2	1.00
Glucose ↔ $CO_2 + H_2O$	36	6.0	1.00
FFA ↔ $CO_2 + H_2O$	138	5.6	0.71
Acetate ↔ $CO_2 + H_2O$	23	5.7	0.73
β-Hydroxibutirrate↔ $CO_2 + H_2O$	26	5.8	0.80

The corresponding respiratory quotient (RQ) is also reported. The asterisk indicates that the corresponding data are per glycosidic unit. FFA, free fatty acids. Modified after di Prampero 1981

corresponding, respectively, to the moles of ATP resynthesized per mole of lactate accumulated, or of oxygen consumed (Table 3.1).

The aim of the following paragraphs is dual. On one side, we demonstrate that, with the exception of the initial exercise period, the transition between the conditions (partially anaerobic) described by Eq. (3.9) and the completely aerobic ones, occurs in the vicinity of $\dot{V}O_2^{max}$, and in any case above the critical power. On the other side, we argue that assessment of $[La]_b$ can be safely utilized for determining the overall energy expenditure during exercise, provided that the underlying limits and assumptions are correctly interpreted (di Prampero and Ferretti 1999).

3.3.1 *Blood Lactate in Submaximal Exercises: The Anaerobic Threshold*

In a large range of low-to-moderate intensity exercises, $[La]_b$ does not increase above the value prevailing at rest (≈ 1.0–2.0 mM). At higher intensities, but still below $\dot{V}O_2^{max}$, $[La]_b$ increases during the first minutes of exercise, to attain a constant level, or very nearly so, within 5–10 min (Fig. 3.6).

When these $[La]_b$ values are plotted as a function of the exercise intensity, expressed as a fraction of $\dot{V}O_2{}^{max}$, one obtains graphs as the one reported in Fig. 3.7a. The so obtained curves depend on many variables such as: type of exercise, training level, subjects' body position, time after work onset of blood sampling. They are generally interpreted as an index of the fact that, above a work intensity corresponding to about 65% $\dot{V}O_2{}^{max}$, the exercise becomes partially anaerobic, since, above this work intensity, $[La]_b$ becomes substantially higher than at rest. As such, the exercise level, above which $[La]_b$ becomes greater than 4 mM, is generally defined "anaerobic threshold" (Fig. 3.7b).

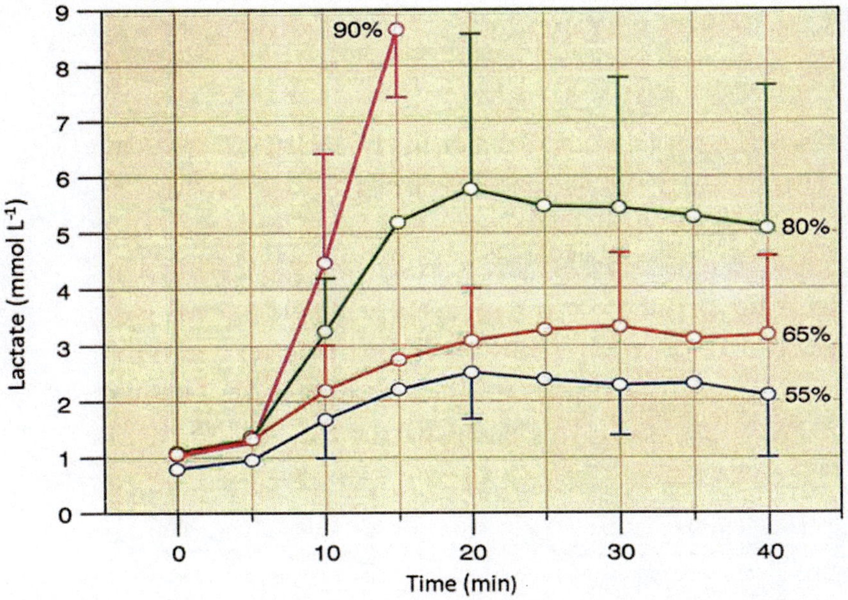

Fig. 3.6 Blood lactate concentration as a function of time during cycloergometric exercise at the indicated metabolic intensities (% of maximal oxygen consumption). From Ribeiro et al. 1986, modified as in di Prampero 2015, with the kind permission of Raffaello Grandi, Edi Ermes

The increase in $[La]_b$ occurring whenever the anaerobic threshold is attained leads to a corresponding increase in pulmonary ventilation (measured as expired ventilation, $\dot{V}_E$), and of the ratio between $\dot{V}_E$ and carbon dioxide output ($\dot{V}_E/\dot{V}CO_2$) (Fig. 3.7C). On this basis, Wasserman et al. (1973) proposed a respiratory method to assess the anaerobic threshold. In addition, at the anaerobic threshold, or close to it, the relationship between heart rate and work intensity ceases to be linear; indeed, Conconi et al. (1982) suggested defining the anaerobic threshold as the exercise intensity at which this relationship deviates from linearity (Fig. 3.7D). The numerous methods for determining the anaerobic threshold yield similar results, the interested reader being referred to Lindinger and Whipp (2008) for further references and data.

However, in spite of its practical relevance, within the context of the energetic view of the School of Milano, the anaerobic threshold cannot be considered as an index of the workload above which the exercise becomes partially anaerobic, because of the reasons outlined herewith.

The ordinate of Fig. 3.7A is proportional to the amount of energy, not to the power, derived from anaerobic lactic metabolism, whereas the abscissa of the same figure is proportional to the overall $\dot{E}$, id est to the energy expenditure per unit of time. Thus, unless the exercise duration is also reported, any given point of this figure cannot express quantitatively the amount of $\dot{E}$ derived from anaerobic sources. Indeed, any given lactate concentration value has a widely different significance according to whether it is reached, e.g., after 30 s or 15 min of exercise.

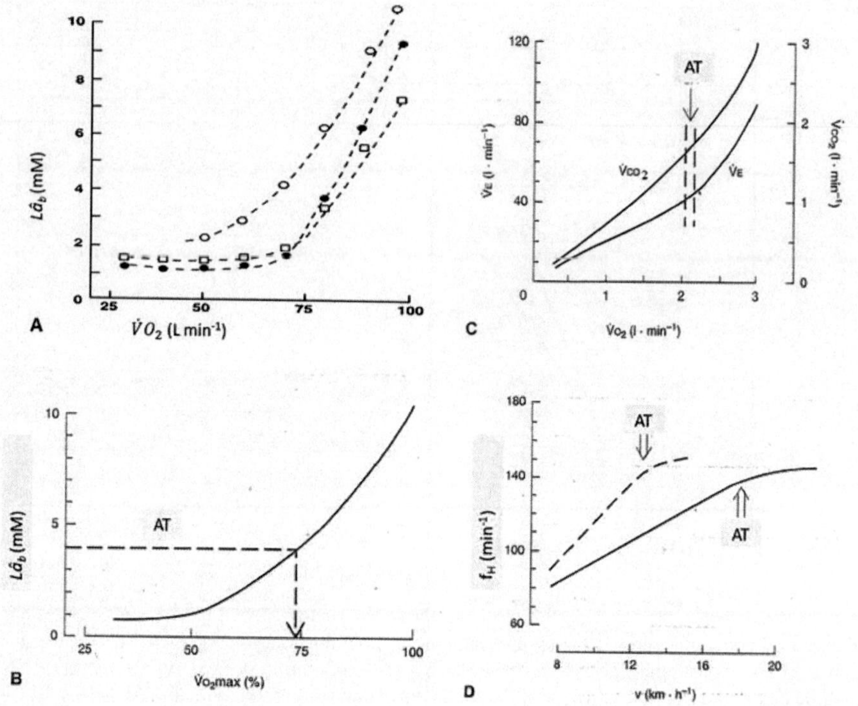

Fig. 3.7 (**A**) Peak blood lactate concentration ($Lâ_b$) as a function of the exercise intensity, expressed as a percentage of $\dot{V}O_2^{max}$; average values in cyclists (open dots), sprinters (full dots), and endurance runners (open squares). (**B**) The anaerobic threshold (AT) is generally defined as the exercise intensity (as a percentage of $\dot{V}O_2^{max}$) at which blood lactate concentration reaches 4 mM. (**C**) The AT can also be determined on the basis of the metabolic intensity at which the pulmonary ventilation ($\dot{V}_E$), or the carbon dioxide output ($\dot{V}CO_2$), increase in excess as compared to the corresponding oxygen consumption, or (**D**) at the running speed at which the relationship between heart rate (f_H) and the speed (v) ceases to be linear (Conconi test). Modified after di Prampero 2015

Moreover, a constant $[La]_b$, regardless of its absolute value, represents an entirely aerobic condition for the organism as a whole, since it results from an equilibrium between lactate production and removal. Indeed, as long as the amount of lactate eliminated as such by the kidney or by the sweat glands is negligible, lactate removal occurs in part by complete oxidation to carbon dioxide and water, following the reconversion of lactate into pyruvate, and in part via resynthesis to glycogen. Both processes require a continuous consumption of oxygen. The oxygen utilized and the energy liberated by the complete oxidation of glycogen are the same, regardless of the reversible transformation pyruvate – lactate – pyruvate. It necessarily follows that a condition in which $[La]_b$ is constant in time, regardless of its absolute value, is, from the whole body point of view, entirely aerobic, and thus the rate of oxygen consumption is a quantitatively correct measure of the whole body energy expenditure, even if $[La]_b$ is higher than 4 mM.

On the contrary, under partially anaerobic conditions, the amount of pyruvate transformed into lactate is greater than that undergoing the opposite transformation, thus leading to a continuous increase of $[La]_b$ in time. In this case, a correct assessment of the whole body energy expenditure can be obtained only by adding the energy derived from the lactate accumulated during the period in question to the amount of oxygen consumed (di Prampero and Ferretti 1999).

Beneke (2003) defined "Maximal Lactate Steady State" (MLSS) as the maximal La_b concentration value that can be maintained at a constant level for at least 10–15 min. This definition is closer to the concept discussed above, in so far as the exercise intensity at which MLSS is attained, represents indeed the highest power, above which lactate production exceeds its oxidative removal. Thus, we insinuate that MLSS is related in someway to the concept of critical power (Ferretti 2015; Jones et al. 2010; Leo et al. 2022).

Nevertheless, and independently of these conceptual problems, above the so-called anaerobic threshold, the relevant modifications occurring in terms of ventilation, gas exchange ratio, and acid-base equilibrium do have a substantial role in reducing the exercise duration, even if this last does indeed remain entirely aerobic from the energetic point of view.

Furthermore, above the anaerobic threshold and after the first few minutes of exercise, $[La]_b$ can be maintained at a constant level only if there is an equal increase in lactate production and removal. Thus, the muscle fibers that produce lactate (type II fibers) deplete their glycogen stores at a rate 13 times faster than those oxidizing their glycogen stores completely (type I fibers). It follows that the lactate-producing muscle fibers may reach the minimal glycogen level compatible with the required exercise intensity in a shorter time than those that oxidize glycogen completely. If this is so, type II fibers may become the factor limiting the exercise duration and/or leading to a decrease in the exercise intensity that can be sustained by a subject. This does indeed explain why, in long-distance events such as the marathon, semi-marathon, cross-country skiing, and so on, individual performances are better correlated with the anaerobic threshold, rather than with the $\dot{V}O_2^{max}$ of the subject. It should be noted, however, that $\dot{V}O_2^{max}$ and anaerobic threshold are not completely independent, in so far as a high $\dot{V}O_2^{max}$ is inevitably correlated with a high anaerobic threshold and vice-versa.

For a more detailed discussion of these problems, the reader is referred to the Appendix.

3.3.2 *The Energy Equivalent of Lactate*

When the work intensity exceeds $\dot{V}O_2^{max}$, $[La]_b$ keeps increasing beyond the first minutes of exercise, thus indicating a continuing anaerobic lactate contribution to the overall energy requirement of the body. At the end of this type of exercises, $[La]_b$ increases further to attain a peak between the fifth and the seventh minute in recovery

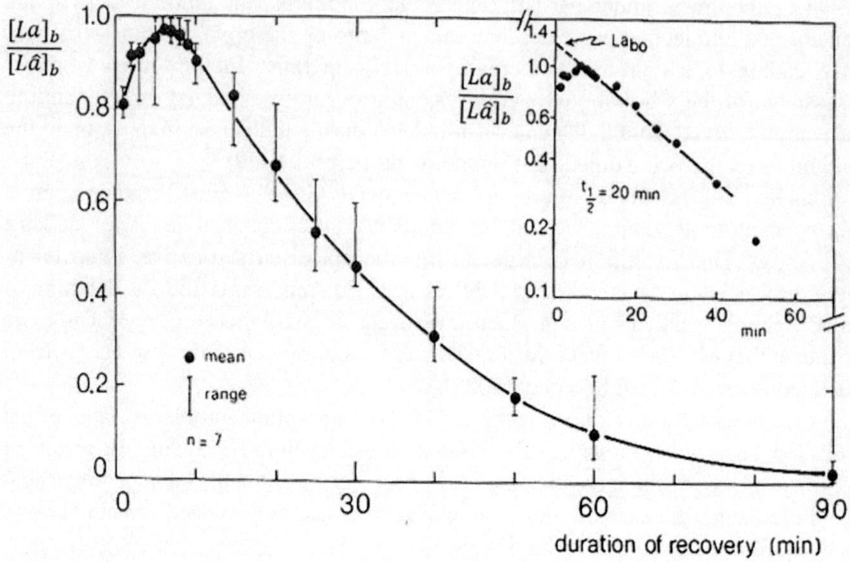

Fig. 3.8 Average blood lactate concentration (*[La$_b$]*, dots), expressed as the ratio to the peak attained in the recovery (*Lâb*), and the corresponding range (vertical bars), as a function of time in the recovery after supra-maximal exercise. In the inset, the average values are reported on a semi-log scale. The time constant of the simple exponential function calculated between the 6th and 40th minute of recovery is ≈ 14 min (half-time ~ 20 min). Back extrapolation at $t = 0$ s of the so obtained function (La_{b0}) represents the hypothetical blood lactate concentration that would be attained at the very end of the exercise, where the diffusion and distribution of lactate in the water phase of the body are instantaneous. Modified after di Prampero 1981

(Fig. 3.8). In a rather large range of work intensities and durations, the peak La_b concentration attained in recovery ($L\hat{a}_b$) is proportional to the overall amount of lactate produced per kg body mass during the preceding exercise period. For a detailed analysis and references on this matter, the reader is referred to di Prampero 1981; di Prampero and Ferretti 1999.

$[La]_b$ is greater the longer the exercise (Fig. 3.9) and, for any given duration, the greater the exercise intensity (Fig. 3.10, **inset**). If the initial exercise period is neglected, the rate of lactate accumulation in blood, ($\dot{La}_b$) and $\dot{E}$ are linearly correlated, as described by:

$$\dot{E} = 1.01\ \dot{V}O_2^{max} + 2.96\ \dot{La}_b \tag{3.10}$$

$$\dot{E} = 0.85\ \dot{V}O_2^{max} + 2.72\ \dot{La}_b \tag{3.11}$$

$$\dot{E} = 0.78\ \dot{V}O_2^{max} + 2.83\ \dot{La}_b \tag{3.12}$$

in running, front crawl swimming and cycling, respectively, where $\dot{E}$ and $\dot{V}O_2{}^{max}$ are expressed in ml kg^{-1} min^{-1} and $\dot{La}_b$ in mM min^{-1}, both quantities referring to the

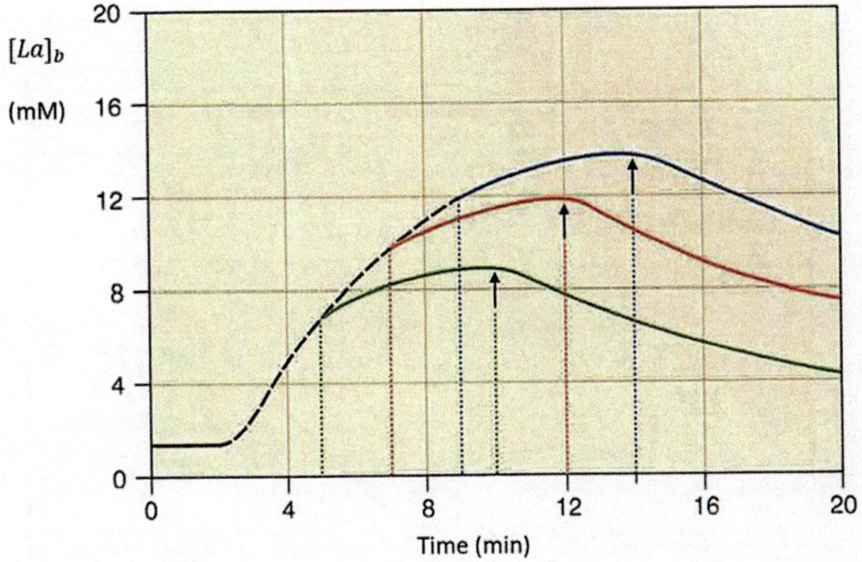

Fig. 3.9 Blood lactate concentration (La_b) as a function of time during (broken line) and after (continuous lines) three supra-maximal exercises of equal intensity but of increasing duration, indicated by the vertical dotted lines (schematic). The maximal concentration (arrows) is attained about 5 min after the end of the exercise. After di Prampero 2015, with the kind permission of Raffaello Grandi, Edi Ermes

appropriate net value (above resting) (Margaria et al. 1963b; di Prampero et al. 1978; di Prampero and Ferretti 1999).

The intercepts on the y axis of the three functions of Fig. 3.10, i.e. the metabolic intensities corresponding to a null $\dot{La}_b$, neglecting the initial exercise period, indicate the relative power, at which lactate starts accumulating continuously in blood, and hence the exercise becomes partially anaerobic. These powers are ≈ 100% $\dot{V}O_2{}^{max}$ in running, ≈ 85% in swimming and ≈ 78% in cycling. This supports the hypothesis that lactate production becomes energetically relevant only at metabolic intensities close to $\dot{V}O_2{}^{max}$, above the critical power. On the other side, the differences reported above among the three forms of locomotion considered Eqs. (3.10, 3.11, and 3.12) show that the data obtained in one specific exercise type may not be transposed, as such, to another exercise type.

The slope of the three regressions of Fig. 3.10 ($\Delta\dot{E}/\Delta\dot{La}_b$) is similar, and, as demonstrated by the set of Eqs. (3.10, 3.11, and 3.12), ranges between 2.7 and 3.0 $mlO_2\ kg^{-1}\ mM^{-1}$. Thus, in these three exercise types, the energy made available whenever $[La]_b$ increases by 1 millimole per liter of blood, is equivalent to that made available by the consumption of ≈ 2.85 $mlO_2\ kg^{-1}$ body mass: this is the energy equivalent of blood lactate accumulation. This value is extremely useful in practice; indeed it allows an estimate of the overall energy expenditure simply determining $\dot{V}O_2$ and $[La]_b$, even if the overall energy requirement exceeds $\dot{V}O_2{}^{max}$.

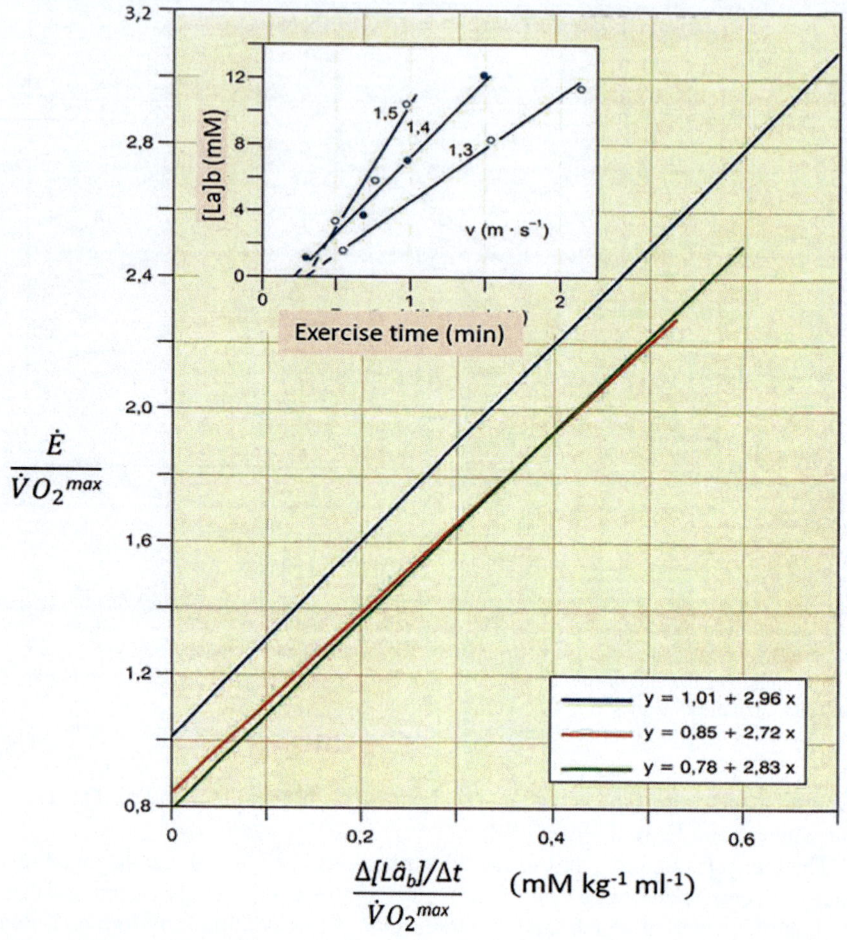

Fig. 3.10 Energy requirement per unit of time ($\dot{E}$) in running (blue), front crawl swimming (red) and cycling (green) as a function of the rate of blood lactate accumulation ($\dot{La}_b$). To compare subjects of varying athletic capacity, $\dot{E}$ and $\dot{La}_b$ are expressed relative to the individual $\dot{V}O_2^{max}$. $\dot{La}_b$ was obtained from the peak (above pre-exercise resting) attained in the recovery divided by the duration of the preceding exercise. The inset shows the rate of blood lactate accumulation during front crawl swimming at three supra-maximal work intensities corresponding to a net oxygen requirement of 70, 80, and 90 ml kg^{-1} min^{-1}: the slopes of the so obtained regression lines yield $\dot{La}_b$. From di Prampero 1981, modified as in di Prampero, 2015, with the kind permission of Raffaello Grandi, Edi Ermes

It is also noteworthy that the energy equivalent of blood lactate accumulation does not depend on any predefined assumption concerning the distribution of lactate in the body fluids, provided that, once $L\hat{a}_b$ is attained during recovery, the lactate distribution among the different compartments of the body has attained an equilibrium, regardless of the absolute $L\hat{a}_b$ value. This assertion is supported by the observation that, if the lactate concentration is expressed as a ratio of the corresponding peak value, rather than in absolute values, the kinetics of lactate disappearance from blood is independent of the specific peak values, being described by an exponential with a time constant of ≈ 14 min (see. Fig. 3.8).

3.3.3 *The Maximal Lactic Power*

$\dot{La}_b$ increases with the exercise intensity above $\dot{V}O_2^{max}$ (see Fig. 3.10, **inset**). However, when the exercise intensity exceeds a given level, $\dot{La}_b$ does not increase any further, regardless of the workload (Fig. 3.11). In other words, as was the case for $\dot{V}O_2$, also $\dot{La}_b$ attains a maximal value that cannot be overcome, itself a measure of the maximal lactic power.

This state of affairs, originally suggested by Margaria and Edwards (1934), was demonstrated experimentally 30 years later by Margaria et al. (1964), who showed that, in a group of well-trained young non-athletic subjects, whose net $\dot{V}O_2^{max}$ amounted to ≈ 50 ml kg^{-1} min^{-1}, the maximal lactic power during treadmill running was ≈ 0.4 mM s^{-1}. So, on the basis of an energy equivalent of 3 mlO_2 kg^{-1} mM^{-1}, as from Eq. (3.10) for running, the corresponding $\dot{E}$, expressed in oxygen equivalents, amounted to 72 ml kg^{-1} min^{-1}, i.e., about 1.5 times the net $\dot{V}O_2^{max}$ of the investigated subjects.

In spite of the obvious practical and theoretical relevance of these data, as far as we know, in the more than 50 years elapsed since the paper by Margaria et al. (1964)

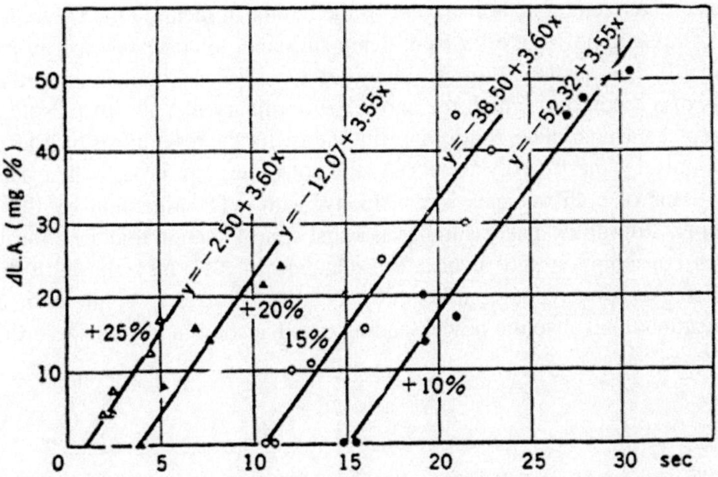

Fig. 3.11 Peak lactate concentration above resting attained in recovery (La_b in the text, ΔL.A. in the original Figure), as a function of the duration of the preceding exercise. Treadmill running at 18 km h^{-1}, at the indicated inclines (10, 15, 20, and 25%). The corresponding metabolic power is 92, 108, 135, and 159 mlO_2 kg^{-1} min^{-1}, respectively. Exhaustion occurs after about 30, 22, 11 e 5 s, respectively, i.e. before the attainment of the maximal lactate concentration, because of the complete utilization of the anaerobic alactic stores (mainly PCr hydrolysis). Note that the La_b can be easily expressed in mM, considering that the molecular weight of lactate is 90 dalton. From Margaria et al. 1964

was published, the maximal lactic power was rarely investigated. We know that it is higher in power athletes than in non-athletes (Hermansen 1971) and that it decreases at altitude (Grassi et al. 1995). We can nevertheless predict, or hypothesize, that it should (1) be related to muscle fiber composition, (2) be low in aerobic athletes and high in power athletes; (3) decrease with age and generally speaking whenever muscle atrophy occurs; (4) increase with power training (sports practice indeed suggest this to be the case).

3.4 The Anaerobic Alactic Energy Sources

The whole body $\dot{E}$ can increase by more than 50-fold in a fraction of a second: from the resting value ($\approx 1\ \mathrm{W \cdot kg^{-1}}$) to $\approx 55\ \mathrm{W\ kg^{-1}}$ in exercises at the maximal absolute power of about 5–7 s duration in non-athletic subjects (Margaria et al. 1966). This increase occurs exclusively at the level of the active muscles that, in this type of exercise, do not exceed $\approx$30% of the muscle mass, corresponding to $\approx$20 kg of muscle for a 70 kg subject. It follows that, inside the active muscles, the increase of $\dot{E}$ amounts to $\approx 55/0.3 \approx 180$ fold. Furthermore, in "explosive" exercises, such as a maximal standing high jump off both feet, the peak power developed is about 4 times larger than that observed in maximal exercise of 5–7 s duration (Ferretti et al. 1987). If this is correct, we may speculate that human muscles can increase their $\dot{E}$ by more than 700-fold in a fraction of a second.

This extraordinary increase of muscle $\dot{E}$ can be due neither to aerobic metabolism, nor to anaerobic lactic metabolism, nor to their sum. In fact: (1) the corresponding values of maximal power are far from being sufficient to cover the energy requirement and (2) their response times are by far too slow, as compared to the time (fractions of a second) in which the above extraordinary increase of muscle power does take place. Under these conditions, the energy for the resynthesis of ATP can be provided only by the hydrolysis of PCr, the sole energetic process that can keep pace, in terms of both intensity and velocity, with ATP utilization by the active muscle fibers. It follows that, assuming as a first approximation that the contribution of oxygen consumption and of lactate production to ATP resynthesis in the first seconds of exercise are negligible, the energetics of muscle contraction during exercises at maximal absolute power leading to exhaustion in 5–7 s is described by:

$$\overleftarrow{A\dot{T}P} = \rightarrow P\dot{C}r \tag{3.13}$$

where $\overleftarrow{A\dot{T}P}$ is he rate of ATP resynthesis, and $\rightarrow P\dot{C}r$ indicates the rate of PCr splitting. These conditions are traditionally defined as anaerobic alactic and represent formally the energetics of all the exercises of very short duration (< 7 s) at maximal absolute intensity.

Equation (3.13) represents formally a situation in which no steady state can be attained; indeed the intramuscular PCr concentration decreases continuously until

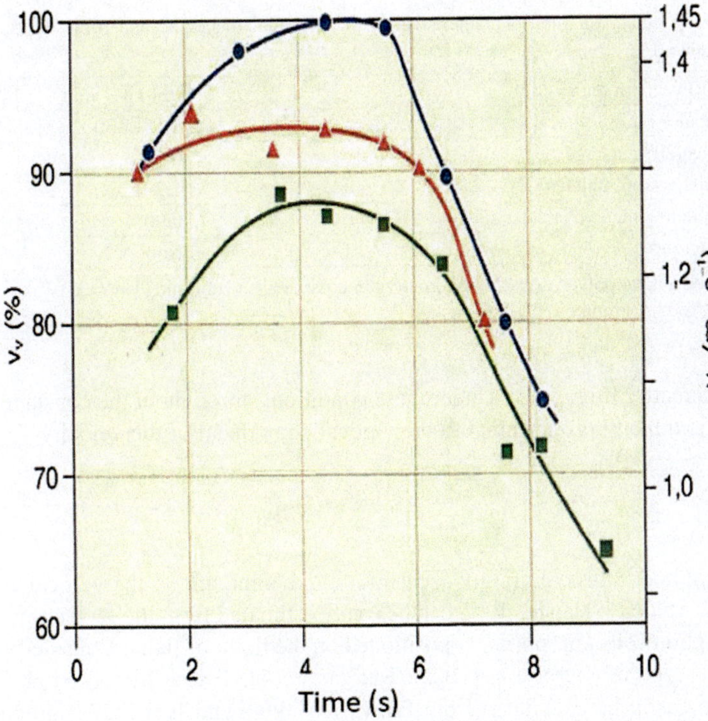

Fig. 3.12 Vertical velocity v_v as a function of time after the onset of an all-out running bout on a flight of stairs. The all-out run was performed from rest (blue dots) or, without interruption, from a steady-state aerobic exercise, the net oxygen consumption of which amounted to 18 (red triangles) or 35 (green squares) ml kg^{-1} min^{-1}. The corresponding external mechanical power can be calculated from Eq. (3.15). From Margaria et al. 1971, modified as in di Prampero, 2015, with the kind permission of Raffaello Grandi, Edi Ermes

the exercise must be interrupted or its intensity reduced. The mechanical power ($\dot{w}$) developed in these conditions can be expressed as:

$$\dot{w} = \eta \, \dot{A}l \tag{3.14}$$

where $\dot{A}l$ indicates the net rate of utilization of the anaerobic alactic energy sources (essentially PCr splitting), expressed in oxygen equivalent, for ATP resynthesis, and η is the mechanical efficiency of the exercise.

Equation (3.14) allows an assessment of the maximal anaerobic alactic power ($\dot{A}l^{max}$) in man, on the basis of the maximal mechanical power ($\dot{w}^{max}$) during very short all-out exercises (of a few seconds duration), the mechanical efficiency of which is known. This elegant approach was proposed and applied for the first time by Margaria et al. (1966), who assessed $\dot{A}l^{max}$ on subjects climbing a normal flight of stairs at maximal speed (Margaria's test). In such a trial, the speed attains a maximal value in about 3 ≈ s, it remains at this level for an additional ≈ 3 s and falls

Table 3.2 Summary of main characteristics of the maximal metabolic anaerobic alactic power (Margaria's test)

Value at 20 years in males	(ml kg^{-1} min^{-1}) 140–160
Males/Females ratio	1.15–1.30
60 years age/20 years age	0.60
Power athletes/Non-athletes	1.2–1.5
Fraction due to genetic characteristics	0.90
Ethnic differences	None

The maximal alactic power is expressed in oxygen equivalents (1 ml min^{-1} = 0.35 W kg^{-1} body mass, data from di Prampero 2015).

rapidly thereafter (Fig. 3.12). Under these conditions, throughout the constant speed phase, $\dot{w}$ is directly proportional to the vertical component of the velocity:

$$\dot{w}^{max} = \frac{Mgh}{t} = Mgv_v \tag{3.15}$$

M being body mass, g gravity acceleration, h height, and v_v the vertical component of the velocity. Hence, $\dot{w}^{max}$ can be easily estimated from the v_v determined by means of photocells appropriately positioned on the flight of stairs. Considering also that in this type of exercise $\eta = 0.25$ (Margaria et al. 1963a; Minetti et al. 2002), $\dot{Al}^{max}$ can easily be calculated from Eq. (3.15). Ikuta and Ikai (1972) proposed a similar approach to assess $\dot{Al}^{max}$ during all-out cycloergometric exercises.

Davies and Rennie (1968) proposed an alternative method, allowing a measure of the maximal absolute mechanical power ($\widehat{w}$) during standing high jumps off both feet on a force platform. Bosco et al. (1983) simplified this method and made it available to a wider audience. For a detailed analysis of the numerous methods available for assessing $\widehat{w}$, the reader is referred to Cerretelli (2001), Dal Monte (1983), di Prampero (2015); Ferretti (2015).

$\dot{w}^{max}$ is higher in power athletes and lower in endurance athletes than in non-athletic individuals, decreases with age and is lower in women than in men, in proportion to the size of the active muscle mass (Bonnefoy et al. 1998; Chamari et al. 1995; di Prampero et al. 1970b; Ferretti et al. 1994; Grassi et al. 1991; Murphy et al. 1986). In chronic hypoxia, $\dot{w}^{max}$ is reduced because of the reduction in muscle cross-sectional area (Ferretti et al. 1990). Lastly, since the rate of chemical processes is temperature-dependent, $\dot{w}^{max}$ is lower, the colder is the contracting muscle mass (Ferretti et al. 1992). The average values of $\dot{Al}^{max}$ in exercises of about 6 s duration, together with the main physiological characteristics thereof, such as the effects of age, sex and training, are summarized in Table 3.2.

In "explosive" exercises, such as a standing high jump off both feet, the mean mechanical power during the push phase is about twice that observed during the Margarita's test. This depends on the fact that, in the former case, both lower limbs are active simultaneously, instead of one at a time. The mean mechanical power during the jump ranges between 20 and 40 W kg^{-1}. During the push phase of a standing high jump off both feet, the $\widehat{w}$ can attain 50 to 70 W kg^{-1}. This is of the same order as the mechanical power of a medium size car (80 HP = 58.7 kW, 800 kg mass: 73.4 W kg^{-1}). The obvious difference is that this mechanical power can be sustained for only a fraction of a second by an exercising human, as compared to quite long periods of time for a car: this is the only advantage of a car engine over muscles.

On the practical side, the measurement of $\dot{w}^{max}$ and $\widehat{w}$ during explosive exercises yields useful assessment of the "sprint" characteristics of an individual. It should be noted, however, that, as long as $\widehat{w}$ is concerned, the data depend not only on the physiological characteristics of the subjects but also on their neuromuscular coordination and determination to express the maximum during the test. In fact, the implicit assumption is that at $\widehat{w}$ all the fibers of the active muscle mass are activated simultaneously. However, this may not be the case in some special conditions, such as long-term microgravity exposure, in which a disruption of the motor control of the lower limbs occurs (see Chap. 11).

3.5 Conclusion: The General Equation of the Energetics of Muscular Exercise

To sum up, starting from the revolutionary work of Margaria et al. (1933), the School of Milano has collected a remarkable bunch of data on the energetics of muscular exercise. Margaria and his pupils, as detailed in the previous paragraphs, defined the quantitative characteristics and the thermodynamic constraints of the main three energy metabolisms (aerobic, anaerobic lactic, and anaerobic alactic). The evidence collected and the theoretical developments up to the end of the 1970s were put together and condensed in a coherent framework by di Prampero (1981). In that review, he summarized the basic principles of the energetics of muscular exercise in the following simple equation:

$$\dot{E} \propto \overrightarrow{\mathrm{A\dot{T}P}} = \overleftarrow{\mathrm{A\dot{T}}P} = c\,\dot{V}O_2 + b\,\dot{La} + a\,\overrightarrow{\mathrm{P\dot{C}r}} \tag{3.16}$$

where $\overrightarrow{\mathrm{A\dot{T}P}}$ and $\overleftarrow{\mathrm{A\dot{T}P}}$ are the rates of ATP hydrolysis and resynthesis, respectively. Equation (3.16) has been defined as the general equation of the energetics of muscular exercise. In this Equation, the three constants *a, b,* and *c* indicate the moles of ATP resynthesized, respectively, by a mole of PCr hydrolyzed, a mole of lactate accumulated, and a mole of oxygen consumed. Lohmann's reaction tells that *a* is equal to 1; *c* (otherwise P/O_2 ratio) takes a mean value of 6.17 for the oxidation of a glycosidic unit into glycolysis. For constant *b*, Fig. 3.10 demonstrates that it is equivalent to 1.5, yielding an energy equivalent of ~2.8–3.0 ml of oxygen per kg body mass per mM.

All experimental conditions analyzed in the previous paragraphs represent a particular case of Eq. (3.16). Light aerobic exercise, exercise transients, intense exercise above the anaerobic threshold, supramaximal exercise, all-out explosive efforts, they can all be represented by Eq. (3.16), or by simplified forms of it, if one or more terms become equal to zero. The three terms of the right side of Eq. (3.16) are characterized by a maximal power and a maximal capacity. The maximal power is the $\dot{V}O_2^{max}$ for $c\ \dot{V}O_2$, the maximal lactic power for $b\ \dot{L}a$, and $\dot{A}l^{max}$ for $\overrightarrow{aP\dot{C}r}$. Concerning capacity, that of aerobic exercise depends on exercise intensity and on substrate availability, whereas the maximal lactic capacity is related to the maximal blood lactate concentration and the maximal alactic capacity to muscle PCr concentration.

Equation (3.16), which was formulated 40 years ago (di Prampero 1981), has resisted the further evolution of knowledge in the field of exercise physiology remarkably well (Ferretti 2015). Nowadays, it is still a pillar of our knowledge of the energetics of muscular exercise, and the best summary of the conjectures on the energetics of muscular exercise that, after Margaria's revolution of 1933, still drive experiments and thinking on this subject.

Appendix

The Anaerobic Threshold

This appendix is devoted to a critical analysis of the anaerobic threshold concept, to show that it cannot be seen as an index of the metabolic intensity above which the exercise becomes partially anaerobic.

To this aim, let us first consider a group of muscle fibers, all "aerobically" active in a homogeneous way. Under these conditions, graphically represented in Fig. 3.13A, each muscle fiber oxidizes completely the pyruvate produced by glycolysis, and the number of moles of ATP resynthesized per glycosidic unit (162 grams of glycogen) amounts to 3 + 34 = 37. In turn, this requires the consumption of 6 moles of oxygen, thus leading to a P/O_2 ratio of 37/6 = 6.17.[3]

Let us now consider a group of muscle fibers all active "anaerobically" in a homogeneous way, as represented graphically in Fig. 3.13B. Assume further that under these conditions, which should be more appropriately defined "hypo-aerobic," each muscle fiber oxidizes only half the amount of pyruvate derived from glycolysis. Of course, the remaining pyruvate that cannot be oxidized at the mitochondrial level via the Krebs' cycle, is anaerobically transformed into lactate in the cytosol, thus leading to an overall resynthesis of 37 + 3 = 40 moles of ATP. As a consequence,

[3]The symbol P/O_2, as utilised in this section, refers to the overall ratio between ATP resynthesized and oxygen utilised. As such, the biochemical details concerning ATP resynthesis at the subcellular level will not be dealt with.

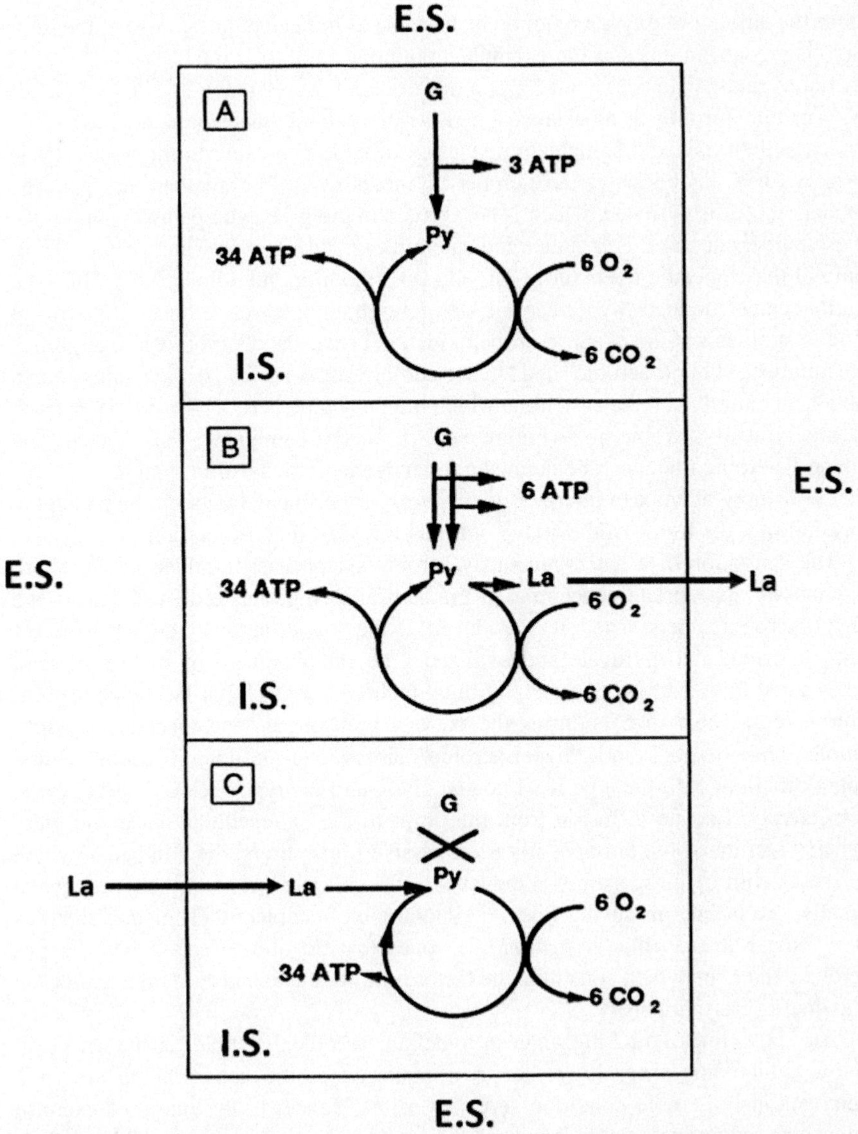

Fig. 3.13 (**A**) Muscle fiber in "aerobic" conditions: all the pyruvate (Py) produced from glycogen (G) is oxidized to carbon dioxide and water within the fiber itself (I.S. = intracellular space; E.S. = extracellular space). The oxidation of 1 glycosidic unit produces 2 moles of pyruvate, requires the consumption of 6 moles of oxygen and leads to the production of 6 moles of carbon dioxide and to the resynthesis of 37 moles of ATP (3 in the glycolytic phase and 34 in Krebs' cycle and oxidative phosphorylation). Hence the overall P/O_2 ratio is: (34 + 3)/6 = 6.17. (**B**) Muscle fiber in "hypo-aerobic" conditions: the amount of pyruvate produced from glycogen in the first (glycolytic) phase of the process is twice as much as it can be oxidized within the fiber itself. Hence, out of four moles of pyruvate produced, two are transformed anaerobically into lactate (La) and leave the fiber, and the other two are oxidized within the mitochondria. As a consequence, the overall P/O_2 ratio is: (34 + 6)/6 = 6.67. (**C**) Muscle fiber in "hyper-aerobic" conditions: the lactate that diffuses from E.S. to I.S is transformed to pyruvate and completely oxidized to carbon dioxide and water. Hence,

since the amount of oxygen required is the same as indicated in Fig. 3.13A, the P/O_2 ratio is greater, in this specific example amounting to 40/6 = 6.67.

Let us finally consider a third group of muscle fibers all active in a homogeneous way in transforming lactate taken up from the extracellular space into pyruvate, which is then oxidized completely to carbon dioxide and water in the Krebs' cycle (Fig. 3.13C). In this case, the number of moles of ATP resynthesized from the oxidation of this pyruvate, which is not formed in the glycolytic pathway within the muscle fibers at stake, is less than indicated in Fig. 3.13A or B. Thus, also the P/O_2 ratio of these specific fibers turns out reduced, amounting to 34/6 = 5.67. This type of fiber can be defined "hyper-aerobic" to highlight their lower "efficacy" in terms of ATP resynthesis at the expense of the oxidative processes. Combining an appropriate number of "hypo-aerobic" and "hyper-aerobic" fibers (Fig. 3.14), it is possible to obtain an entirely aerobic system, in which the P/O_2 ratio is (6 + 34 + 34)/12 = 6.17, as in a typically aerobic fiber (Figure 3.13A). Such a combination of "hyper-" and "hypo-" aerobic fibers can be defined a **heterogeneously aerobic system**.

Let us now attempt to interpret qualitatively experimental data on the anaerobic threshold, in the frame of the above-outlined theoretical considerations.

The activation of a heterogeneously aerobic system, as described in Fig. 3.14, requires an appropriate concentration gradient allowing lactate diffusion from the "hypo-aerobic" fibers, where it is produced, to the "hyper-aerobic" ones, where it is re-transformed into pyruvate and oxidized. The establishment of such a gradient takes some time from work onset, during which the extracellular lactate concentration increases: this appears during the exercise transient as "early lactate" accumulation. The "hypo-" and "hyper-aerobic" fibers can be identified, as a first approximation, with the type II white glycolytic and the type I red oxidative fibers, respectively. Lactate diffusion from the intra- to the extracellular space and vice-versa is facilitated by, but does not need passive transporters, the number of which increases with training, mainly at the level of the cell membrane of the type I fibers. Finally, the lactate-producing fibers ("hypo-aerobic") deplete their glycogen stores at a faster rate than the "hyper-aerobic" ones that do utilize lactate. The "hypo-aerobic" fibers may then constitute the factor limiting the exhaustion time and/or the maximal exercise intensity.

The above line of thinking suggests to define "aerobic threshold" as the workload below which all, or at least the great majority of the active fibers are in a homogenously aerobic condition (P/O_2 = 6.17). Hence, in the range of exercise intensities below the aerobic threshold the blood lactate concentration ($[La]_b$) does not exceed the values prevailing at rest: $[La]_b \leq 2$ mM. In a large range of exercise intensities, the $[La]_b$ does not change substantially, as compared to the resting values. When muscle fiber metabolism is homogeneously aerobic, each muscle

---

Fig. 3.13 (continued) the glycogen stores of the fiber are untouched and the overall P/O_2 ratio is: 34/6 = 5.67. Modified after Antonutto and di Prampero 1995

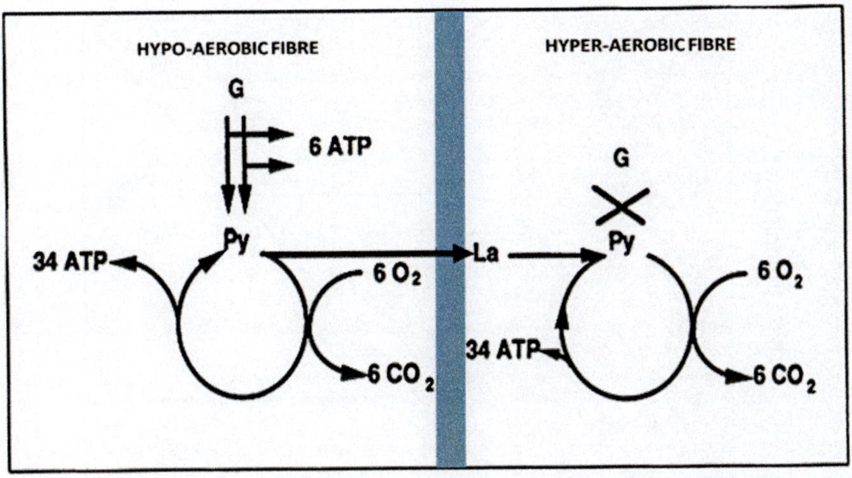

Fig. 3.14 An appropriate combination of "hypo-" and "hyper-" aerobic fibers can lead to a completely aerobic system, characterized by a typical "aerobic" P/O_2 ratio: $(34 + 34 + 6)/12 = 6.17$. In order for the system to be operational, it is necessary for lactate to diffuse from the intracellular space of the "hypo-aerobic" fibers to the extracellular space (vertical blue band) and finally to the intracellular space of the "hyper-aerobic" fibers, thus requiring the implementation (by early lactate production) of an appropriate concentration gradient, as indicated by the horizontal arrow La. Modified after Antonutto and di Prampero 1995

fiber (or group thereof) oxidizes all, and only, the pyruvate that it produces from glycogen (Fig. 3.13A).

At exercise intensities greater than the aerobic threshold, as defined here above, an aerobic status can still be achieved thanks to the appropriately combined activity of hypo- and hyper-aerobic fibers, to yield a "*heterogeneously aerobic*" condition, in which hyper-aerobic fibers oxidize the excess lactate produced by the hypo-aerobic ones (Fig. 3.14). This condition requires an initial increase in La_b (early lactate) to prime the system. All work intensities in which, after the initial ≈ 5 min, $[La]_b$ attains a constant level, higher than resting, belong to this exercise category, regardless of the absolute lactate level. It is interesting to note that this analysis bears an analogy with Brooks' theory of the lactate shuttle (Brooks 1985, 1986, 2000, 2009).

The constant value that $[La]_b$ attains after work onset in a heterogeneously aerobic condition, is greater the higher the exercise intensity. The highest exercise intensity compatible with a constant $[La]_b$, MLSS (Beneke 2003), represents the maximal power that can be sustained by the whole body oxidative processes, albeit heterogeneously distributed. Above the work intensity corresponding to the MLSS, lactate keeps increasing with time, thus leading to a corresponding continuous increase of $[La]_b$. The MLSS is attained at a power, which is probably higher than the critical power. Throughout the whole range of the aerobic work intensities, regardless of the fact that the aerobic metabolism sustaining them be homogeneously or heterogeneously distributed, after the initial 5 min of exercise (classical duration of the exercises transient, during which early lactate may be accumulated, Cerretelli

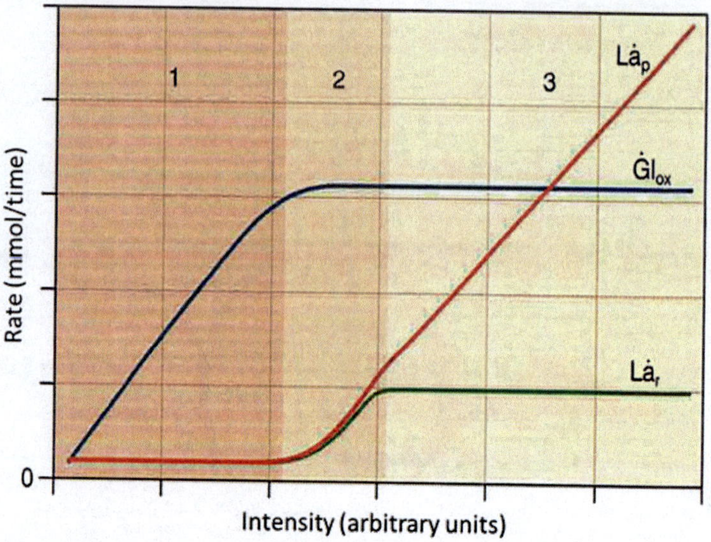

Fig. 3.15 Rates of: glycogen oxidation ($\dot{Gl}_{ox}$); production ($\dot{La}_p$) and oxidative removal ($\dot{La}_r$) of lactate (mmoles per unit of time) as a function of the exercise intensity (arbitrary units). In zone "1" each and every fiber oxidizes all the pyruvate that it produces. Throughout zone "2," production and oxidative removal of lactate proceed in parallel. In zone "3" lactate production exceeds lactate oxidation. Hence, after the initial exercise period, in zones "1" and "2" no net lactate accumulation occurs, whereas in "3," the difference between the two functions $\dot{La}_p$ and $\dot{La}_r$ leads to a constant accumulation of lactate. Zone "1" can be defined "homogeneously" aerobic; zone "2" "heterogenously" aerobic; zone 3 "anaerobic." From Antonutto and di Prampero 1995, as modified in di Prampero 2015, with the kind permission of Raffaello Grandi, Edi Ermes

et al. 1979), whole body oxygen consumption is a correct quantitative measure of the overall rate of energy expenditure, as long as the terms $b\,\dot{La}$ and $a \rightharpoonup \dot{PCr}$ of the general equation of the energetics of muscular exercise (Eq. 3.16) are nil.

From an energetic viewpoint, it seems legitimate to ask whether the maximal aerobic power, at which $\dot{E} = \dot{V}O_2{}^{max}$, does, or does not, coincide with the power corresponding to the MLSS. According to Busso and Chatagnon (2006), there exists a range of workloads close to, but lower than, $\dot{V}O_2{}^{max}$, in which $[La]_b$ keeps increasing with time. This being so, the maximal aerobic power would be higher than that corresponding to the MLSS. This is coherent with the data reported in Fig. 3.10, showing that, at least in cycling and swimming, $[La]_b$ starts increasing continuously at intensities below $\dot{V}O_2{}^{max}$, specifically at 85 and 78% thereof, respectively (Eqs. 3.10, 3.11, and 3.12).

It should finally be noted that the power at which $[La]_b$ keeps increasing continuously even after the initial phase, is crucially dependent not only on the intensity and type of exercise but also on the subject's training level. At even higher exercise intensity, the activity and/or number of the "hypo-aerobic" fibers exceeds the capacity of the "hyper-aerobic" ones to oxidize lactate. Thus, $[La]_b$ keeps increasing also beyond the initial minutes of exercise. This exercise intensity level constitutes the true whole-body anaerobic threshold. It represents the exercise intensity above which the oxygen consumption is not sufficient to cover the total energy requirement and is generally attained at workloads closer to $\dot{V}O_2^{max}$: it definitely does not correspond to the traditional anaerobic threshold set at a conventional $[La]_b$ of 4 mM. For a schematic representation of these three levels of exercise intensities, see Fig. 3.15. Concerning the relation of early lactate to oxygen delivery, see Chap. 4.

After this analysis, we are confident to conclude that the conventional anaerobic threshold is neither a "threshold," in so far as it does not represent a net transition between two different conditions, nor "anaerobic" (at least if the whole body is considered), in so far as, under constant blood lactate concentration, regardless of its absolute value, the whole organism is always in aerobic conditions. It follows from this conclusion that the widespread use of the conventional anaerobic threshold concept has introduced a strong bias in the interpretation of the energetics of muscular exercise, especially as far as the meaning of anaerobic lactic metabolism is concerned. Only the School of Milano has been fighting fiercely against the shadow cast by this bias, unfortunately with less success than it would have deserved.

References

Antonutto G, di Prampero PE (1995) The concept of lactate threshold: a short review. J Sports Med Phys Fitness 35:6–12

Beneke R (2003) Maximal lactate steady state concentration (MLSS): experimental and modelling approaches. Eur J Appl Physiol 88:361–369

Binzoni T, Ferretti G, Schenker K, Cerretelli P (1992) Phosphocreatine hydrolysis by ^{31}P-NMR at the onset of constant-load exercise in humans. J Appl Physiol 73:1644–1649

Bonnefoy M, Kostka T, Arsac LM, Berthouze SE, Lacour JR (1998) Peak anaerobic power in elderly men. Eur J Appl Physiol 77:182–188

Bosco C, Luhtanen P, Komi PV (1983) A simple method for measurement of mechanical power in jumping. Eur J Appl Physiol 50:273–282

Boström P, Wu J, Jedrychowski MP, Korde A, Ye L, Lo JC, Rasbach KA, Almer Boström E, Choi JH, Long JZ, Kajimura S, Zingaretti MC, Vind BF, Tu H, Cinti S, Højlund K, Gygi SP, Spiegelman BM (2012) A PGC1-α-dependent myokine that drives brown-fat-like development of white fat and thermogenesis. Nature 481:463–468

Brooks G (1985) Anaerobic threshold: review of the concept and directions for future research. Med Sci Sports Exerc 17:22–31

Brooks GA (1986) Lactate production under fully aerobic conditions: the lactate shuttle during rest and exercise. Fed Proc 45:2924–2929

Brooks GA (2000) Intra- and extra-cellular lactate shuttles. Med Sci Sports Exerc 32:790–799

Brooks GA (2009) Cell-cell and intracellular lactate shuttles. J Physiol Lond 587:5591–5600

Busso T, Chatagnon M (2006) Modelling of aerobic and anaerobic energy production in middle distance running. Eur J Appl Physiol 97:745–754

Camus G, Atchou G, Bruckner JC, Giezendanner D, di Prampero PE (1988). Slow upward drift of $\dot{V}O_2$ during constant-load cycling in untrained subjects. Eur J Appl Physiol 58: 197–202

Capelli C, Antonutto G, Zamparo P, Girardis M, di Prampero PE (1993) Effects of prolonged cycloergometer exercise on maximal power and oxygen uptake in humans. Eur J Appl Physiol 66:189–195

Cerretelli P (2001) Fisiologia dell'Esercizio. U., Roma, Sport, Ambiente, Età, Sesso, S.E

Cerretelli P, Pendergast D, Paganelli WC, Rennie DW (1979). Effects of specific muscle training on $\dot{V}O_2$ response and early blood lactate. J Appl Physiol 47: 761–769

Cerretelli P, Shindell D, Pendergast DP, di Prampero PE, Rennie DW (1977) Oxygen uptake transients at the onset and offset of arm and leg work. Respir Physiol 30:81–97

Chamari K, Ahmaidi S, Fabre C, Massé-Biron J, Préfaut C (1995) Anaerobic and aerobic peak power output and the force-velocity relationship in endurance-trained athletes: effects of aging. Eur J Appl Physiol 71:230–234

Conconi F, Ferrari M, Ziglio PG, Droghetti P, Codeca L (1982) Determination of the anaerobic threshold by a non invasive field test in runners. J Appl Physiol 52:869–873

Dal Monte A (1983) La Valutazione Funzionale dell'Atleta. Sansoni Nuova, Firenze

Davies CTM, Rennie E (1968) Human power output. Nature 217:770–771

di Prampero PE (1981) Energetics of muscular exercise. Rev Physiol Biochem Pharmacol 89:143–222

di Prampero PE (2015) La Locomozione Umana su Terra, in Acqua, in Aria. Fatti e Teorie, Edi-Ermes, Milano

di Prampero PE, Davies CTM, Cerretelli P, Margaria R (1970a) An analysis of O_2 debt contracted in submaximal exercise. J Appl Physiol 29:547–551

di Prampero PE, Piñera-Limas F, Sassi G (1970b) Maximal muscular power (aerobic and anaerobic) in 116 athletes performing at the XIX Olympic games in Mexico. Ergonomics 13:665–674

di Prampero PE, Ferretti G (1999) The energetics of anaerobic muscle metabolism: a reappraisal of older and recent concepts. Respir Physiol 118:103–115

di Prampero PE, Margaria R (1968) Relationship between O_2 consumption, high-energy phosphates and the kinetics of the O_2 debt in exercise. Pflügers Arch 304:11–19

di Prampero PE, Peeters L, Margaria R (1973) Alactic O_2 debt and lactic acid production after exhausting exercise in man. J Appl Physiol 34:628–633

di Prampero PE, Pendergast DR, Wilson DW, Rennie DW (1978) Blood lactic acid concentration in high velocity swimming. In: Eriksson B, Furberg B (eds) Swimming medicine IV. University Park Press, Baltimore MD, USA, pp 249–261

di Prampero PE, Francescato MP, Cettolo V (2003) Energetics of muscular exercise at work onset: the steady state approach. Pflügers Arch 445:741–746

Ferretti G (2015) Energetics of muscular exercise. Springer, Heidelberg

Ferretti G, Gussoni M, di Prampero PE, Cerretelli P (1987) Effects of exercise on maximal instantaneous muscular power of humans. J Appl Physiol 62:2288–2294

Ferretti G, Hauser H, di Prampero PE (1990) Muscular exercise at high altitude. VII. Maximal muscular power before and after exposure to chronic hypoxia. Int J Sports Med 11, suppl 1:S31–S34

Ferretti G, Ishii M, Moia C, Cerretelli P (1992) Effects of temperature on the maximal instantaneous muscle power of humans. Eur J Appl Physiol 64:112–116

Ferretti G, Narici MV, Binzoni T, Gariod L, Le Bas JF, Reutenauer H, Cerretelli P (1994) Determinants of peak muscle power: effects of age and physical conditioning. Eur J Appl Physiol 68:111–115

Francescato MP, Cettolo V, di Prampero PE (2003) Relationship between mechanical power, O_2 consumption, O_2 deficit and high-energy phosphates during calf exercise in humans. Pflügers Arch 445:622–628

Francescato MP, Cettolo V, di Prampero PE (2008) Influence of phosphagen concentration on phosphocreatine breakdown kinetics. Data from human gastrocnemius mucle. J Appl Physiol 105:158–164

Grassi B, Cerretelli P, Narici MV, Marconi C (1991) Peak anaerobic power in master athletes. Eur J Appl Physiol 62:394–399

Grassi B, Ferretti G, Kayser B, Marzorati M, Colombini A, Marconi C, Cerretelli P (1995) Maximal rate of blood lactate accumulation during exercise at altitude in humans. J Appl Physiol 79:331–339

Grassi B, Hogan MC, Greenhaff PL, Hamann JJ, Kelley KM, Aschenbach WG, Constantin-Teodosiu D, Gladden LB (2002). $\dot{V}O_2$ on-kinetics in dog gastrocnemius *in situ* following activation of pyruvate dehydrogenase by dichloroacetate. J Physiol Lond 538: 195–207

Grassi B, Hogan MC, Kelley KM, Howlett RA, Gladden LB (2005) Effect of NOS inhibition by I-NAME on oxygen uptake kinetics in isolated canine muscle *in situ*. J Physiol Lond 568:1021–1033

Grassi B, Poole DC, Richardson RS, Knight DR, Eriksson BK, Wagner PD (1996) Muscle O_2 uptake kinetics in humans: implications for metabolic control. J Appl Physiol 80:988–998

Grassi B, Rossiter HR, Hogan MC, Howlett RA, Harris JE, Goodwin ML, Dobson JL, Gladden LB (2011) Faster O_2 uptake kinetics in canine skeletal muscle in situ after acute creatine kinase inhibition. J Physiol Lond 589:221–233

Greenhaff PL (2001) The creatine-phsphocreatine system: there's more than one song in its repertoire. J Physiol Lond 537:657

Harris RC, Edwards RH, Hultman E, Nordesjö LO, Nylind B, Sahlin K (1976) The time course of phosphorylcreatine resynthesis during recovery of the quadriceps muscle in man. Pflügers Arch 367:137–142

Harris RC, Sahlin K, Hultman E (1977) Phosphagen and lactate contents of m. quadriceps femoris of man after exercise. J Appl Physiol 43:852–857

Hermansen L (1971) Lactate production during exercise. In: Pernow B, Saltin B (eds) Muscle metabolism during exercise. Plenum Press, New York, pp 401–407

Hill AV, Long CNH, Lupton H (1924) Muscular exercise, lactic acid, and the supply and utilisation of oxygen. Parts IV – VI. Proc Roy Soc B 97:84–138

Ikuta K, Ikai M (1972) Study on the development of maximum anaerobic power in man with bicycle ergometer. Res J Physiol Ed Japan 17:151–157

Jones AM, Grassi B, Christensen PM, Krustrup P, Bangsbo J, Poole DC (2011). Slow component of $\dot{V}O_2$ kinetics: mechanistic basis and practical applications. Med Sci Sports Exerc 43: 2046–2062

Jones AM, Vanhatalo A, Burnley M, Morton RH, Poole DC (2010). Critical power: implications for determination of $\dot{V}O_2{}^{max}$ and exercise tolerance. Med Sci Sports Exerc 42: 1876–1890

Kaasik A, Minajeva A, De Sousa E, Ventura-Clapier R, Veksler V (1999) Nitric oxide inhibits cardiac energy production via inhibition of mitochondrial creatine kinase. FEBS Lett 444:75–77

Larsen I, Welde B, Martins C, Tjønna AE (2014) High- and moderate-intensity aerobic exercise and excess post-exercise oxygen consumption in men with metabolic syndrome. Scand J Med Sci Sports 24:e174–e179

Leo P, Spragg G, Podlogar T, Lawley JS, Mujika I (2022) Power profiling and the power-duration relationship in cycling: a narrative review. Eur J Appl Physiol 122:301–316

Leuenberger U, Sinoway L, Gubin S, Gaul L, Davis D, Zelis R (1993) Effects of exercise intensity and duration on norepinephrine spillover and clearance in humans. J Appl Physiol 75:668–674

Lindinger MI, Whipp BJ (2008) The anerobic threshold: fact or misinterpretation? In: Taylor N, Groeller H (eds) Physiological bases of human performance during work and exercise. Churchill Livingstone Elsevier, Edinburgh, pp 191–199

Mader A (2003) Glycolysis and oxidative phosphorylation as a function of cytosolic phosphorylation state and power output of the muscle cell. Eur J Appl Physiol 88:317–338

Mahler M (1985). First-order kinetics of muscle oxygen consumption, and equivalent proportionality between $\dot{V}O_2$ and phophorylcreatine level. Implications for the control of respiration. J Gen Physiol 86: 135–165

Margaria R, Aghemo P, Rovelli E (1966) Measurement of maximal power (anaerobic) in man. J Appl Physiol 21:1662–1664

Margaria R, Cerretelli P, Aghemo P, Sassi G (1963a) Energy cost of running. J Appl Physiol 18: 367–370

Margaria R, Cerretelli P, di Prampero PE, Massari C, Torelli G (1963b) Kinetics and mechanism of oxygen debt contraction in man. J Appl Physiol 18:371–377

Margaria R, Cerretelli P, Mangili F (1964) Balance and kinetics of anaerobic energy release during strenuous exercise in man. J Appl Physiol 19:623–628

Margaria R, di Prampero PE, Aghemo P, Derevenco P, Mariani M (1971) Effect of a steady state exercise on maximal anaerobic power in man. J Appl Physiol 30:885–889

Margaria R, Edwards HT (1934) The sources of energy in muscular work performed in anaerobic conditions. Am J Phys 108:341–348

Margaria R, Edwards HT, Dill DB (1933) The possible mechanism of contracting and paying the oxygen debt and the role of lactic acid in muscular contraction. Am J Phys 106:689–714

Meyer RA (1988) A linear model of muscle respiration explains monoexponential phosphocreatine changes. Am J Physiol Cell Physiol 254:C548–C553

Meyer RA, Sweeney HL, Kushmerick MJ (1984) A simple analysis of the "phosphocreatine shuttle". Am J Physiol Cell Physiol 246:C365–C377

Minetti AE, Moia C., Roi G.S, Susta D, Ferretti G. (2002). Energy cost of walking and running at extreme uphill and downhill slopes. J Appl Physiol 93: 1039–1046

Miyamoto T, Oshima Y, Ikuta K, Kinoshita H (2006) The heart rate increase at the onset of high-work intensity exercise is accelerated by central blood volume loading. Eur J Appl Physiol 96: 86–96

Moseley L, Achten J, Martin JC, Jeukendrup AE (2004) No differences in cycling efficiency between world-class and recreational cyclists. Int J Sports Med 25:374–379

Murphy MM, Patton JF, Frederick FA (1986) Comparative anaerobic power of men and women. Aviat Space Environ Med 57:636–641

Piiper J, di Prampero PE, Cerretelli P (1968) Oxygen debt and high-energy phosphates in gastrocnemius muscle of the dog. Am J Phys 215:523–531

Poole DC, Barstow TJ, Gaesser GA, Willis WT, Whipp BJ (1994) O_2 slow component: physiological and functional significance. Med Sci Sports Exerc 26:1354–1358

Poole DC, Jones AM (2012) Oxygen uptake kinetics Compr Physiol 2:933–966

Poole DC, Ward SA, Gardner GW, Whipp BJ (1988) Metabolic and respiratory profile of the upper limit for prolonged exercise in man. Ergonomics 31:1265–1279

Ribeiro JP, Hughes V, Fielding RA, Holden W, Evans W, Knuttgen HG (1986) Metabolic and ventilatory responses to steady state exercise relative to lactate thresholds. Eur J Appl Physiol 55:215–221

Rossiter HB, Howe FA, Ward SA (2005). Intramuscular phosphate and pulmonary $\dot{V}O_2$ kinetics during exercise: Implications for control of skeletal muscle O_2 consumption. In: "Oxygen uptake kinetics in Sport, Exercise and Medicine, Jones AM, Poole DC (eds.). Routledge, London, pp 154–184

Tjønna AE, Lee SJ, Rognmo Ø, Stølen TO, Bye A, Haram PM, Loennechen JP, Al Share QY, Skogvoll E, Slørdahl SA, Kemi OJ, Najjar SM, Wisløff U (2008) Aerobic interval training versus continuous moderate exercise as a treatment for the metabolic syndrome: a pilot study. Circulation 118:346–354

Wasserman K, Whipp BJ, Koyal SN, Beaver WL (1973) Anaerobic threshold and respiratory gas exchange during exercise. J Appl Physiol 35:236–243

Whipp BJ, Mahler M (1980) Dynamics of pulmonary gas exchange during exercise. In: West JB (ed) Pulmonary gas exchange, vol II. Academic Press, New York, pp 33–96

Zhang Y, Xie C, Wang H, Foss RM, Clare M, George EV, Li S, Katz A, Cheng H, Ding Y, Tang D, Reeves WH, Yang LJ (2016) Irisin exerts dual effects on browning and adipogenesis of human white adipocytes. Am J Physiol Endocrinol Metab Physiol 311:E530–E541

Chapter 4
Further Developments on Exercise Transients: Los Angeles Versus Milano

Guido Ferretti

Abstract This chapter deals with the alternate view of exercise transients that stemmed from the School of Los Angeles, set by Karlman Wasserman and Brian Whipp. The pathway that led to the creation of the double exponential model of the oxygen uptake kinetics upon exercise start is traced, and discussed along lines different than those along which some of these concepts were dealt with by the School of Milano. Specific aspects like the exercise transients in hypoxia, the oxygen uptake kinetics limitation, the kinetics of cardiac output upon exercise start and the effects of priming exercise are discussed. The correspondence between the kinetics of muscle oxygen consumption and that of lung oxygen uptake, a key issue in Los Angeles vision, is critically examined. A reappraisal of the slow component under this perspective is carried out. In this context, the pathways are highlighted, along which the two Schools of Milano and Los Angeles arrived at some shared conclusions, namely that muscle oxygen consumption kinetics is controlled by the metabolic changes initiated by muscle contractions at exercise onset, at least during exercise with large muscle masses in normoxia.

G. Ferretti (✉)
Department of Molecular and Translational Medicine, University of Brescia, Brescia, Italy
e-mail: Guido.Ferretti@unibs.it

G. Ferretti (ed.), *Exercise, Respiratory and Environmental Physiology*, Perspectives in Physiology, https://doi.org/10.1007/978-3-031-19197-8_4

Karlman Wasserman (left) and Brian Whipp (right), portrayed at a scientific meeting in Japan. With the kind permission of Susan Ward

4.1 The Birth of a New Vision of Exercise Transients

In Chap. 3, we analyzed how the School of Milano developed a metabolic vision of the oxygen deficit. All started in 1933, when Margaria et al. (1933) created the concept of alactic oxygen debt, thus falsifying Hill and Meyerhof's theory of the energetics of muscular contraction (see Chap. 2). That notion of oxygen deficit was refined through different steps: the role of changes in oxygen stores in the oxygen deficit was identified, the concept of an obligatory (alactic oxygen deficit) and a facultative (early lactate, changes in oxygen stores) component of the oxygen deficit was created, the biochemical meaning of the obligatory component of the oxygen deficit was clarified. Pietro Enrico di Prampero and Paolo Cerretelli made a very clear synthesis of the metabolic vision of the oxygen deficit (Cerretelli and di Prampero 1987; di Prampero 1981), which we report in detail in Chap. 3. The model behind the metabolic vision of the oxygen deficit is the mono-exponential increase of oxygen consumption in the working muscles ($\dot{V}_mO_2$) and of oxygen uptake at the mouth ($\dot{V}_LO_2$). The School of Milano always considered the latter a direct consequence of the former, with some distortions, because between the mouth and the mitochondria there is an entire respiratory system, which carries and stores oxygen, and which consists of numerous resistances in-series.

In fact the kinetics of $\dot{V}_mO_2$ cannot be directly measured inside the active muscle mass in humans. The collaboration between Milano and Göttingen made it possible to assess the kinetics of $\dot{V}_mO_2$ upon the start of induced isotonic contractions of an isolated-perfused muscle preparation (Cerretelli et al. 1969; Piiper et al. 1968). The energy balance of the oxygen deficit resulting from these studies was compatible with the conclusion arrived at by previous studies on humans by the School of Milano. The Göttingen studies were interpreted by di Prampero and Margaria (1968), who understood that the alactic oxygen deficit is a physiological necessity, since its biochemical counterparts, i.e., muscle phosphocreatine (PCr) and free creatine, contribute to the regulation of glycolysis, and thus of the metabolic pathways controlling energy flows and $\dot{V}_mO_2$.

However, the gap between $\dot{V}_LO_2$ and $\dot{V}_mO_2$ is so huge that some authors were reluctant to accept a metabolic origin of the oxygen deficit, despite the collected evidence (Dejours 1964; Wasserman et al. 1967; West and Jones 1965). Indeed, one of the main reasons for the perplexities about Margaria's concept of oxygen deficit was the demonstration of a very rapid ventilatory response at exercise onset, which was well known since long (Krogh and Lindhard 1913, see also below). Thus, the cultural climate was pushing toward the exploration of new tracks, which were investigated especially within the Los Angeles (LA) School, founded by Karlman Wasserman (1927–2020), then led by Brian Whipp (1937–2011). Whipp, a Welshman who trained in physical education, was a former post-doctoral fellow of Wasserman, who kept collaborating with him for a long time at the University of California at Los Angeles.

Gilbert et al. (1967) were the first to point out a rapid $\dot{V}_LO_2$ response at exercise start. This finding was soon confirmed and became a pole of attraction for the LA School (Whipp and Wasserman 1972). A schematic representation of the transient suggested by this view is reported in Fig. 4.1.

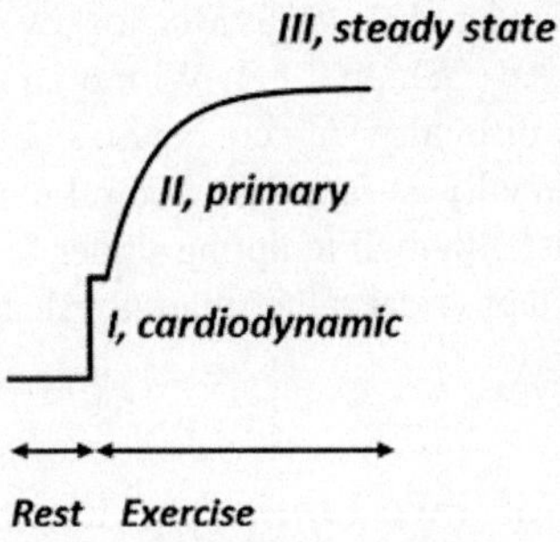

Fig. 4.1 Schematic representation of the kinetics of respiratory variables, which includes the cardiodynamic increase postulated by Whipp and Wasserman (1972). This scheme applies to oxygen uptake ($\dot{V}_LO_2$). It applies also to carbon dioxide output ($\dot{V}_LCO_2$) and expired ventilation ($\dot{V}_E$), albeit with different time constants. From Poole and Jones (2012)

Wasserman et al. (1974) attributed the occurrence of the rapid component of the exercise transient, which appears in the first seconds of exercise, to a sudden increase in cardiac output ($\dot{Q}$). These findings suggested that a mono-exponential model might be insufficient to describe the $\dot{V}_LO_2$ response during a square-wave exercise transient. The rapid component of the $\dot{V}_LO_2$ response became a preferential object of research in the field ever since, particularly by the LA School. The results led to the subsequent double-exponential model with time delay of the $\dot{V}_LO_2$ kinetics upon exercise start, by Barstow and Molé (1987). These authors also restricted the domain of application of the metabolic vision of the oxygen deficit to the second exponential only, which some authors later defined as the primary component of the $\dot{V}_LO_2$ response during the exercise transient (see, e.g., Poole and Jones 2012). In this view, the time delay represents the physiological delay between the increase in $\dot{V}_mO_2$ and $\dot{V}_LO_2$, corresponding approximately to half circulation time (Lindholm et al. 2006).

It is interesting to note that, when Wasserman et al. (1974) attributed the rapid phase of the $\dot{V}_LO_2$ kinetics at exercise onset to a sudden increase in $\dot{Q}$, several authors, besides Krogh and Lindhard (1913), had already recognized a fast component in the dynamics of pulmonary ventilation (Asmussen 1973; Asmussen and Nielsen 1948; Comroe and Schmidt 1943; D'Angelo and Torelli 1971; Dejours 1964). This observation generated the conjecture that ventilatory control in the exercise transient is dissociated from the metabolic control of the $\dot{V}_LO_2$ kinetics, and that the increase in ventilation was to precede the increase in $\dot{V}_LO_2$, although the mechanisms controlling $\dot{V}_LO_2$ were yet to be elucidated. The latter notion was accepted even within the School of Milano (Cerretelli and di Prampero 1987) and everybody agreed that neurogenic mechanisms were behind the rapid ventilatory response at exercise onset.

Notwithstanding this acknowledgment, a widespread and long-lasting debate took place on the origin of the rapid ventilatory response. Two opposing fields faced each other, one supporting a peripheral view (reflex mechanism originating in the contracting limbs), the other a central view (a feed-forward mechanism originated in the cerebral cortex, often called central command) (Comroe 1944; Dejours 1964; Eldridge et al. 1981; Favier et al. 1983; McCloskey and Mitchell 1972; Torelli and Brandi 1961; Turner 1991; Whipp 1994; Whipp and Ward 1982). The debate was complicated by the difficulty of conceiving coherent neural pathways supporting either view, especially as far as the central command is concerned. The central command hypothesis, although tempting under several respects, is unfortunately still confined in a metaphysical rather than physical domain.

4.2 The Coexistence of Two Opposing Visions

David Poole and Andrew Jones, in their impressive review on exercise transients (Poole and Jones 2012), qualified Brian Whipp as one of the foremost pioneers in the field of metabolic control and $\dot{V}_LO_2$ kinetics. This is right enough, another being

Rodolfo Margaria. Yet there is a bias in their historical development. They set the origin of studies on exercise transients in the work of August Krogh (1874–1949) and Archibald Vivian Hill (Hill and Lupton 1923; Hill et al. 1924; Krogh and Lindhard 1913, 1920). It is a matter of fact that these authors and their collaborators described the dynamics of heart rate and of ventilation at the onset of a square wave exercise and that Hill understood that the $\dot{V}_LO_2$ kinetics was related to the changes in muscle metabolic rate. Hill's intuition of the exponential nature of the exercise transient set the basis for many of the further developments by the LA and the Milano Schools, including Margaria's refutation of Hill and Meyerhof's theory.

However, Poole and Jones (2012) set an ideal continuity between Krogh and Hill, on one side, and Wasserman and Whipp on the other, whereas they downgrade the contribution of Margaria and pupils to a mere energetic analysis of the alactic and lactic components of the oxygen deficit and of its payment, an aspect which they seem to consider of lesser importance. They forgot, or did not realize, that Hill's thinking on the oxygen transient was confined within a theory of the energetics of muscular exercise, which Margaria et al. (1933) falsified (see Chap. 2 for details on this issue), meanwhile setting the fundaments for all future work in the field of the energetics of muscular exercise. Notwithstanding this misunderstanding, it is a matter of fact that much of the significant work on the exercise transients and the $\dot{V}_LO_2$ kinetics was carried out in Milano and in Los Angeles, although we should not forget the thoughtful contributions of the Canadian School created by Richard Hughson (Hughson and Morrissey 1983; Hughson 1990, 2009; Hughson et al. 2001).

Within the energetic view of the $\dot{V}_LO_2$ kinetics of the School of Milano, one of the most crucial questions was: where does the missing energy during the exercise transient come from? The answer was: from alactic and lactic anaerobic metabolisms, and from changes in body oxygen stores. As already pointed out in Chap. 3, the alactic component of the oxygen deficit was considered to be a physiological necessity (di Prampero and Margaria 1968). The changes in oxygen stores covered a gap between the kinetics of $\dot{V}_LO_2$ and of $\dot{V}_mO_2$; in the moderate and intense exercise domains, the anaerobic lactic component covered a gap between speed of activation of glycolysis, on the one side, and of aerobic metabolism on the other (early lactate, Cerretelli et al. 1979).

The second fundamental question formulated by the School of Milano concerned the mechanism of the oxygen deficit. To them, the activation of glycolysis depends on the dynamics of activation of phospho-fructo-kinase (PFK), a key regulatory enzyme of the glycolytic pathway. PCr, free creatine, and inorganic phosphate (*Pi*) modulate PFK activity. Hence, the priming of aerobic metabolism requires an initial decrease of muscle PCr. This carries along an increase in muscle *Pi*, thus reducing the so-called phosphorylation potential (ATP/(ADP + *Pi*)), which, besides controlling PFK, is a key regulator of oxidative phosphorylation in the mitochondria (Korzeniewski 2003; Mader 2003; Meyer 1988; Meyer and Foley 1996).

As soon as an exercise starts, ATP is immediately used in muscle contraction. The first, rapid source of ATP resynthesis is Lohmann's reaction: PCr falls, *Pi* goes up, the phosphorylation potential is reduced, and the ensemble of these chemical

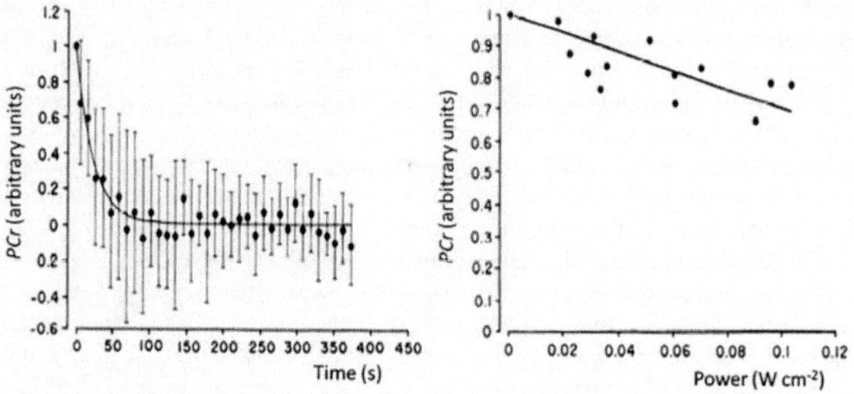

Fig. 4.2 **Left:** Time course of phosphocreatine concentration (PCr) decreases upon exercise onset, as determined by ^{31}P nuclear magnetic resonance. The exponential function has a time constant of 23.3 s. Bars indicate standard deviation. **Right:** Linear relationship between steady-state PCr and applied power, from the same experiment. Modified after Binzoni et al. 1992

changes accelerates PFK activity. As long as *Pi* falls, the energy flux along the glycolytic pathway increases and prompts the aerobic metabolism overall. This chain of events provides the biochemical basis to the concept that the exercise transient has a metabolic origin (di Prampero 1981; di Prampero and Ferretti 1999; Ferretti 2015; Grassi 2000). The energetic vision of the oxygen deficit implies that, at the exercise steady state, there is a linear relationship with negative slope between $\dot{V}_mO_2$ (or $\dot{V}_LO_2$: at steady state $\dot{V}_mO_2 = \dot{V}_LO_2$) and muscle PCr concentration (Binzoni and Cerretelli 1991; Binzoni et al. 1992; Cerretelli et al. 1969). Moreover, in absence of early lactate accumulation, the kinetics of muscle PCr concentration at exercise onset must have the same time constant as that of $\dot{V}_mO_2$ (Binzoni and Cerretelli 1991). Several studies, in which this time constant was calculated by ^{31}P-NMR spectroscopy (Fig. 4.2), demonstrated the correctness of both these predictions even in humans (Binzoni et al. 1992; di Prampero et al. 2003; Francescato et al. 2013; Rossiter et al. 1999, 2002; Whipp et al. 1999). Finally, experiments on the isolated-perfused muscle preparation, carried out in San Diego and realized by Bruno Grassi, demonstrated that, when muscle blood flow is artificially increased at rest or rapidly accelerated at exercise onset, no reduction in the time constant of the kinetics of $\dot{V}_mO_2$ does occur (Grassi et al. 1998a).

The LA School tried to develop a more holistic interpretation of the $\dot{V}O_2$ kinetics. To their mind, the energetic vision of Margaria and his pupils was limitative. They strived to set an integrated vision of the $\dot{V}_LO_2$ kinetics in the context of an integrated response of the respiratory system. This led them to the alternative concept that the delayed response of the respiratory system dictates the oxygen deficit and the $\dot{V}_LO_2$ kinetics as determined at the mouth (Whipp and Ward 1982, 1990), even if the viewpoint of the LA School later evolved in a different direction (Poole and Jones 2012). Richard Hughson also shared this view (Hughson and Morrissey 1983; Hughson et al. 2001).

A key point to them is that the $\dot{V}_L O_2$ kinetics in humans is at best determined at the mouth on a breath-by-breath basis. An assumption behind their view is that the $\dot{V}_L O_2$ kinetics at the mouth reflects what happens in the working muscles (Poole and Jones 2012). Yet the wide gap between what happens at the mouth, which we can assess breath by breath, and what happens in a contracting muscle, is a matter of fact, and this gap encompasses the entire respiratory system. To the Milano people, accounting for changes in blood oxygen stores was enough to fully explain the discrepancies between the kinetics of $\dot{V}_m O_2$ and that of $\dot{V}_L O_2$ (di Prampero et al. 1970; Davies et al. 1972), since it was sufficient to explain the energetics of the oxygen deficit. The LA colleagues instead considered these discrepancies as a challenge, and thus as an opportunity to investigate the interrelations between $\dot{V}_L O_2$, ventilation and $\dot{Q}$, and so came the identification of what they called the "cardiodynamic phase" of the $\dot{V}_L O_2$ kinetics (Wasserman et al. 1974). They further reinforced the concept when they observed a sudden $\dot{V}_L O_2$ increase at exercise start under controlled ventilation at resting level (Weissman et al. 1982). They deeply investigated the interrelationships between the rapid responses of $\dot{V}_L O_2$ and of pulmonary ventilation (Casaburi et al. 1978, 1989a; Ward and Whipp 1980; Wasserman et al. 1986). The latter turned out synchronous with the instantaneous increase in pulmonary blood flow (Gilbert et al. 1967; Wasserman et al. 1974; Weissman et al. 1982; Whipp and Ward 1982).

Admitting a rapid component in the $\dot{V}_L O_2$ kinetics, clearly unrelated to $\dot{V}_m O_2$ kinetics, had a profound impact on the theoretical approach to the meaning of the exercise transient. These consequences were clearly recognized firstly by Whipp et al. (1982), who preconized a possible subdivision of the $\dot{V}_L O_2$ kinetics into two-phases, one fast, the other slow. Their thinking opened the way to a major step forward, the construction of the double exponential model of the $\dot{V}_L O_2$ kinetics (Barstow and Molé 1987). This model relies on the notion that an exponential increase in $\dot{V}_L O_2$ (phase II), which Whipp et al. (1982) considered related to metabolic adaptation in skeletal muscle, is preceded by a faster $\dot{V}_L O_2$ increase in the first seconds of exercise (phase I), which the LA school attributed to an immediate cardiovascular response upon exercise onset (see e.g. Whipp and Ward 1982). Barstow and Molé (1987) treated phase I also as an exponential, and so they considered the system as consisting of two capacitances of different sizes and meanings. According to this model, we thus have:

$$\dot{V}_L O_{2t} = A_1\left(1 - e^{-k_1 t}\right) + A_2\left(1 - e^{-k_2(t-d)}\right) \tag{4.1}$$

where $\dot{V}_L O_{2t}$ is the net $\dot{V}_L O_2$ at time t, k_1 and k_2 are the rate constants in phase I and II, respectively (their reciprocals, τ_1 and τ_2, are the corresponding time constants), d is the time delay, and A_1 and A_2 are the amplitudes of the $\dot{V}_L O_2$ that increase during phase I and phase II, respectively. We underline that d, reflecting the time taken by venous blood to reach the pulmonary capillaries from the active muscle mass, cannot assume negative values. From time 0 (exercise start) to time d, the mixed venous oxygen concentration ($C_{\bar{v}}O_2$) remains unchanged after exercise onset (Barstow and

Molé 1987). In fact, the $C_{\bar{v}}O_2$ changes upon exercise start that Casaburi et al. (1989b) called abrupt have a slower kinetics than reported for phase I, so that the changes actually occurring during d are rather small.

During light exercise, τ_1 is extremely short, functionally instantaneous, indicating a sudden, immediate upward shift of $\dot{V}_LO_2$ since the first breath; τ_2 is slower than τ_1, and similar to, perhaps slightly faster than, the τ provided by the single exponential model. This slight difference is a mere mathematical consequence of having introduced a time delay. According to Whipp and Ward (1990), τ_2 during light exercise is around 30 s. This value is somewhat higher than the time constant of the PCr decrease in the exercise transient reported above.

4.3 The Kinetics of Cardiac Output at Exercise Onset

The creation of the double exponential model of the $\dot{V}_LO_2$ kinetics is one of the most elegant and remarkable achievements in exercise physiology in the 1980s. Yet a brick was still missing in this construction. If phase I has a cardiovascular origin, cardiovascular changes must explain the amplitude of $\dot{V}_LO_2$ changes in phase I. This was yet to be demonstrated on a quantitative basis at exercise onset. Wasserman et al. (1974) did not demonstrate a correspondence between a phase I kinetics of $\dot{Q}$ and a phase I kinetics of $\dot{V}_LO_2$, inasmuch as they obtained measurements of aortic blood flow in catheterized resting anesthetized dogs, on which they merely demonstrated a temporal correspondence between the sudden increase in $\dot{Q}$ after isoproterenol injection and the subsequent sudden increase in ventilation and in $\dot{V}_LO_2$. Their allegation of the cardiodynamic origin of the early increase in ventilation and $\dot{V}_LO_2$ at exercise start did not derive from experiments on exercising humans.

Paolo Cerretelli in Buffalo investigated the kinetics of $\dot{Q}$ for the first time, but not on a beat-by-beat basis (Cerretelli et al. 1966). For this reason, the kinetics of $\dot{Q}$ appeared relatively slow in that study. Even so, Cerretelli et al. (1966) already realized that it was faster than that of $\dot{V}_LO_2$ and admitted the possibility of two components, since they wrote that initially it increases rapidly, to slow down successively. They concluded that the discrepancy between the kinetics of $\dot{Q}$ and of $\dot{V}_LO_2$ demonstrates that the former is under neurogenic control. This was perfectly coherent with the ideas of the School of Milano on the metabolic origin of the oxygen deficit: the kinetics of $\dot{Q}$ was not dictating the kinetics of $\dot{V}_LO_2$. Davies et al. (1972) obtained similar results a few years later.

Until the early 1980s, only the kinetics of heart rate could be studied on a beat-by-beat basis in humans. The first heart rate recordings at exercise, obtained on rabbits, date back to 1895 (Johansson 1895). Pulse pressure recordings through a cannula placed in the carotid artery were used. Krogh and Lindhard (1913) had been the first to describe a heart rate kinetics at exercise onset in humans. The subsequent continuous developments of ECG techniques made heart rate recordings a customary measure in exercise physiology. Reckoning hundredths of studies is a huge task of little importance in the present context. The interested reader can refer to any of the numerous reviews and books on the subject (see, e.g., Cerretelli and di Prampero 1987; Poole and Jones 2012; Rowell 1993).

Of great importance, however, appears a paper by Fagraeus and Linnarsson (1976), for its impact on further thinking. These authors investigated the effects of parasympathetic blockade with atropine on the heart rate kinetics at exercise onset in humans. They identified a phase I response of heart rate in control condition, which however disappeared under atropine. So they provided a very strong evidence supporting their concept of vagal tone withdrawal at exercise start. Similar conclusions were attained, 10 years later, by Maciel et al. (1986), who studied also beta-blockade with propranolol and proposed that the heart rate increase during dynamic exercise is mediated by a biphasic mechanism, initially depending on rapid vagal release, the intensity of which is greater the higher the exerted power, whereas an increased sympathetic activity appears with some delay, especially at the higher levels of activity. If Maciel et al. (1986) were correct, then the withdrawal of vagal tone at exercise start would not be a switch-off mechanism. It would rather reflect a sudden down-modulation of vagal activity.

The appearance of beat-by-beat techniques for the determination of $\dot{Q}$ in the mid-1980s led to the demonstration that, not only the heart rate kinetics, but also the kinetics of $\dot{Q}$ at exercise onset is very fast, much faster than stated by Cerretelli et al. (1966) and faster than that of $\dot{V}_LO_2$ (Cummin et al. 1986; De Cort et al. 1991; Eriksen et al. 1990; Yoshida and Whipp 1994). These findings were the strongest piece of evidence supporting, at least on a qualitative basis, the "cardiodynamic" origin of phase I. Moreover, Cummin et al. (1986) highlighted a close correspondence between the rapid increase of $\dot{Q}$ and ventilation at exercise onset, suggesting the hypothesis of an overall integrated response of the entire respiratory system in the exercise transient. To then, however, no modeling of phase I for $\dot{Q}$ was proposed.

The School of Milano was the first to extend the application of the dual exponential model of Barstow and Molé (1987) to the analysis of the kinetics of $\dot{Q}$ and of systemic oxygen delivery ($\dot{Q}_aO_2$) (Lador et al. 2006). During light aerobic exercise, these authors obtained very low τ_1 values for both variables, as a consequence of the extremely rapid kinetics of the two determinants of $\dot{Q}$, heart rate and stroke volume. Lador et al. (2006) attributed the rapid changes of the former to vagal withdrawal (Fagraeus and Linnarsson 1976), and of the latter to an increase in heart contraction force of mechanical origin. This hypothesis relied essentially on animal studies (Sheriff et al. 1993), but direct evidence in healthy humans was still missing in those days. It came only recently when Fagoni et al. (2020) demonstrated the increased amplitude of the stroke volume response in phase I under progressively increasing levels of lower body negative pressure.

The extremely low τ_1 for $\dot{V}_LO_2$ and $\dot{Q}$ suggested a strong link between the two variables in the exercise transient and supported the "cardiodynamic" origin of phase I, but it was not a demonstration yet of the dependence of $\dot{V}_LO_2$ kinetics on $\dot{Q}$ kinetics. Application of the Fick principle helps understand the issue. Because of a delay between $\dot{V}_mO_2$ and $\dot{V}_LO_2$, we can reasonably assume that during the first seconds of exercise the composition of mixed venous blood, and hence the arterial-venous oxygen difference ($CaO_2 - C\overline{v}O_2$), remain constant and equal to those at rest

(Barstow and Molé 1987; Weissman et al. 1982). If this is so, any increase in $\dot{V}_LO_2$ during phase I would be due only to an increase in $\dot{Q}$. Lador et al. (2006) found that the amplitude of phase I (A_1) for $\dot{Q}$ was on average 4.3 L min^{-1}. For an average resting $CaO_2 - C\overline{v}O_2$ of 87 ml L^{-1}, this would carry along a corresponding immediate $\dot{V}_LO_2$ increase of 374 ml min^{-1}. This figure is very close to the observed A_1 for $\dot{V}_LO_2$, reported in the same study (355 ± 148 ml min^{-1}). So, they concluded that the A_1 for $\dot{V}_LO_2$ is indeed entirely accounted for by the $\dot{Q}$ increase during phase I, supporting the hypothesis put forward by Wasserman et al. (1974) and Whipp et al. (1982). The LA school was right under this respect, but there was more. Lador et al. (2008) found faster τ_2 of $\dot{Q}$ and $\dot{Q}ao2$ than of $\dot{V}_LO_2$ in normoxia, with equal time delay (Table 4.1). Therefore, they hypothesized that the kinetics of $\dot{Q}$ is under different control from that of $\dot{V}_LO_2$ even in phase II. They attributed the phase II increase in $\dot{Q}$ to the progressive activation of sympathetic flow to the heart.

This is coherent with the widely accepted notion that the sympathetic system predominates in controlling circulation in steady-state exercise (Cerretelli and di Prampero 1987; Nyberg 1981; Robinson et al. 1966; Rowell and O'Leary 1990; Strange 1999; Toska and Eriksen 1994; Turner 1991). Moreover, the amount of oxygen made available to the contracting muscles in phase II overrides the oxygen demand, determining a transient decrease of $CaO_2 - C\overline{v}O_2$. Since Lador et al. (2006) observed no CaO_2 changes in the exercise transient in normoxia, they postulated that the decrease of $CaO_2 - C\overline{v}O_2$ was to produce a transient increase of venous oxygen stores.

4.4 The Oxygen Uptake Kinetics in Hypoxia

The time constant of the primary component of the $\dot{V}_LO_2$ kinetics (τ_2 of the double exponential model) is slower in hypoxia than in normoxia, independently of the applied model (Bowen et al. 2013; Cleuziou et al. 2005; Engelen et al. 1996; Hughson and Kowalchuk 1995; Lador et al. 2008, 2013; Springer et al. 1991). However, only one study, issued from the School of Milano, reported at the same time the values of early lactate accumulation (Lador et al. 2013), thereby allowing the computation of the energy balance of the oxygen deficit. So, we refer to this paper for the analysis reported here below.

The volume of oxygen that is taken up through the lungs (V_LO_2) during a constant-load exercise is given by:

$$V_LO_2 = \int_0^t d\dot{V}_LO_2\, dt \tag{4.2}$$

where t is the exercise time. V_LO_2 is less than $t \cdot V_LO_{2ss}$ (suffix ss indicates steady state) by an amount that is equal to the oxygen deficit ($DefO_2$) determined at the

Table 4.1 Parameters of the double exponential equations describing the kinetics of cardiac output ($\dot{Q}$), oxygen delivery ($\dot{Q}aO_2$), heart rate (f_H) and lung oxygen uptake ($\dot{V}_LO_2$) upon light exercise onset in normoxia (N) and acute hypoxia (H)

	Condition	A_1	d_1	τ_1	A_2	d_2	τ_2
$\dot{Q}$	**N**	5.0 ± 1.0	1.1 + 0.8	2.9 + 1.3	2.6 + 0.6	16.3 + 4.1	7.7 + 4.4
	H	3.6* ± 1.8	0.6 + 0.7	4.6 + 3.6	4.0* + 1.1	16.9 + 4.2	29.8* + 12.7
$\dot{Q}aO_2$	**N**	0.80 ± 0.23	1.1 + 1.0	2.1 + 1.1	0.47 + 0.15	14.4 + 4.8	8.3 + 6.8
	H	0.45* ± 0.16	0.5 + 0.7	3.0 + 1.5	0.64 + 0.34	15.6 + 2.3	31.5* + 21.7
f_H	**N**	22 ± 8	0.7 + 0.6	2.7 + 1.6	12 + 7	14.5 + 8.4	10.2 + 8.1
	H	21 ± 11	0.5 + 0.6	3.9 + 2.1	29 + 5	19.1 + 12.1	28.2* + 10.1
$\dot{V}_LO_2$	**N**	0.54 + 0.15	0.2 + 0.4	1.3 + 1.5	0.83 + 0.19	15.2 + 1.9	17.8 + 3.1
	H	0.49 + 0.15	0.0 + 0.1	1.9 + 1.2	0.90 + 0.15	16.4 + 4.5	28.4* + 5.4

A, amplitude; *d*, time delay; τ, time constant. Suffixes 1 and 2 refer to phase I and phase II. Asteriscs on hypoxia data indicate significant differences with respect to the corresponding normoxia values. Data from Lador et al. 2008 mouth. $DefO_2$ includes also the changes in venous blood oxygen stores ($\Delta VO_2{}^{st}$) during the transient, which are equal to (Barstow et al. 1990):

$$\Delta VO_2^{st} = \left[\left(\frac{\dot{V}o_2}{\dot{Q}}\right)_r - \left(\frac{\dot{V}O_2}{\dot{Q}}\right)_{ss}\right] V_{vb} \tag{4.3}$$

where suffixes r and ss refer to the pre-exercise resting and to the steady state conditions, respectively, and V_{vb} is the venous blood volume, that Barstow et al. (1990) assumed equal to 3 L and invariant within the exercise transient. Application of Eq. (4.3) to the data of Lador et al. (2013), after admitting the correctness of Barstow's assumption, yielded average $\Delta VO_2{}^{st}$ equal to 260 and to 230 ml in normoxia and hypoxia, respectively. This allows the computation of a muscle oxygen deficit (DO_{2M}) as:

$$DO_{2M} = DefO_2 - \Delta VO_2^{st} = VO_2^{pcr} + VO_2^{eLa} \tag{4.4}$$

Equation (4.4) shows that DO_{2M} consists of the alactic oxygen deficit ($VO_2{}^{PCr}$), i.e. the obligatory component of the oxygen deficit, plus the energy derived from early lactate accumulation ($VO_2{}^{eLa}$). $VO_2{}^{PCr}$ results from an exponential $\dot{V}_mO_2$ kinetics, the time constant of which, in homogeneously aerobic conditions, i.e. when $VO_2{}^{eLa}$ is nil, is equal to the time constant of the mono-exponential kinetics of muscle PCr decrease at exercise onset (Binzoni and Cerretelli 1991). As already pointed out, this time constant ranges between 20 and 25 s, as demonstrated by ^{31}P-NMR spectroscopy experiments, and is invariant and independent of the mechanical power imposed (Binzoni et al. 1992, 1997; di Prampero et al. 2003; Francescato et al. 2008; Rossiter et al. 1999, 2002). As a consequence, when $VO_2{}^{eLa}$ is nil, $VO_2{}^{PCr}$ turns out directly proportional to the steady state $\dot{V}O_2$, and $DO_{2M} = VO_2{}^{PC}$.

Lador et al. (2013) found no increase in blood lactate concentration during light exercise in normoxia. Their subjects exercised in homogeneously aerobic conditions (see the appendix of Chap. 3). The estimated time constant of $\dot{V}_mO_2$ kinetics, after accounting for $\Delta VO_2{}^{st}$, corresponded well to that of muscle PCr kinetics. This was not so in hypoxia, in which case DO_{2M} was larger than in normoxia. Since the mechanical power in hypoxia was the same as in normoxia, and so was $\dot{V}_mO_{2ss}$, $VO_2{}^{PCr}$ was necessarily equal in both conditions. Thus, the DO_{2M} increase in hypoxia was exclusively due to an increase in $VO_2{}^{eLa.}$. Hence, the energy provided by $VO_2{}^{eLa.}$ in hypoxia must be equal to the difference in DO_{2M} between hypoxia and normoxia. Ferretti (2015) calculated the ratio between $VO_2{}^{eLa}$ and the net average lactate accumulation after the data of Lador et al. (2013). Once expressed per unit of body mass, this ratio resulted equal to 2.5 $mlO_2\ mM^{-1}\ kg^{-1}$, a value fairly close to the range normally admitted for the energy equivalent of blood lactate accumulation (di Prampero 1981; di Prampero and Ferretti 1999). This definitely demonstrates that the energy balance of DO_{2M} is compatible with the energetic principles exposed in Chap. 3.

Lador et al. (2008) found that, in normoxia, the τ_2 of $\dot{Q}$ and $\dot{Q}aO_2$ were smaller than that of $\dot{V}_LO_2$ (Table 4.1). The oxygen delivery to the working muscle was sufficient to sustain the rate of increase of $\dot{V}_mO_2$. In contrast, the τ_2 of $\dot{Q}$ and $\dot{Q}a$o2 increased (slower kinetics) in hypoxia, to become even higher than that of $\dot{V}_LO_2$ in normoxia. This entailed a slowing down of the kinetics of $\dot{V}_LO_2$ as well. The oxygen delivery to the working muscle became insufficient to sustain the expected rate of increase of $\dot{V}_mO_2$, so that the kinetics of $\dot{V}_mO_2$ became slower than that in normoxia, and slower than what would be necessary to sustain the increased rate of energy flux along the glycolytic pathway. Pyruvate was accumulated in muscles and transformed into lactate (early lactate). The conditions for heterogeneous aerobic metabolism at steady state were set.

The above-mentioned studies have the advantage of providing data of both $\dot{V}_LO_2$ and $\dot{Q}$ kinetics. Yet there are several other conditions, in which the $\dot{V}_LO_2$ kinetics is slower than during light exercise in normoxia. Considering only physiological conditions, we may cite supine exercise, arm exercise, induced ischemia, beta-adrenergic blockade (Faisal et al. 2010; Hughson 1984; Hughson and Kowalchuk 1991; Hughson et al. 1991, 1993, 1996; Koga et al. 1996; McDonald et al. 1998; Paganelli et al. 1989). Richard Hughson from Canada was particularly active in these fields, a work that made him the herald of the theory of oxygen delivery regulation of $\dot{V}_mO_2$ kinetics, which we partially disagree with, as pointed out below.

4.5 The Limits of Muscle Oxygen Consumption Kinetics

In contracting muscles, the rate of energy flow increases with the same time constant as that describing the muscle PCr drop. If this occurs in a condition in which the phase II kinetics of $\dot{Q}$ and $\dot{Q}aO_2$ is faster than that of the PCr drop, sufficient oxygen is made available to contracting muscles to sustain the higher metabolic energy flux. In this condition, the rate of increase of energy flow dictates the rate at which $\dot{V}_mO_2$ increases (Bowtell et al. 2007; Connett et al. 1985; Korzeniewski and Zoladz 2002).

However, any slowing of $\dot{Q}$ and $\dot{Q}aO_2$ kinetics entails a slowing of both the kinetics of $\dot{V}_LO_2$, and of $\dot{V}_mO_2$. Moreover, as soon as the kinetics of $\dot{Q}$ and $\dot{Q}aO_2$ become slower than the rate at which the energy flow at the muscle fiber level increases, pyruvate accumulates in contracting muscle fibers, thus promoting muscle, then blood lactate accumulation (early lactate, Cerretelli et al. 1979). In this condition, $\dot{Q}aO_2$, rather than the rate of energy flow, limits the kinetics of $\dot{V}_mO_2$. Early lactate provides the extra energy that is required to build up the oxygen deficit.

This assertion can take a more general form, as follows: whenever the rate of glucose degradation along the glycolytic pathway exceeds the rate of $\dot{V}_mO_2$, and thus the rate at which ATP resynthesis takes place along the Krebs' cycle and the oxidative phosphorylation pathways, pyruvate is accumulated in the working muscles. The amount of pyruvate that cannot be fed into the Krebs' cycle is reduced to lactate, whence the appearance of early lactate ($VO_2{}^{eLa}$ becomes positive).

Two are the possible causes of such disequilibrium between glycolysis and the downward components of the oxidative metabolic chain: (i) the rate of glycolysis exceeds the maximal rate of aerobic metabolism, or (ii) oxygen delivery is insufficient to sustain aerobic metabolism. The former case, which is discussed in Chap. 3, concerns the intense exercise domain, and thus all conditions in which the maximal $\dot{V}_mO_2$ of type II fast muscle fibers is exceeded. The latter case, which is discussed in the previous paragraphs for acute hypoxia, concerns those conditions in which the kinetics of $\dot{Q}$ and $\dot{Q}aO_2$ become slower than the rate at which muscle energy flow can increase in the aerobic pathways.

This is a classical reasoning reflecting the way of thinking of the School of Milano. Yet, similar reasoning arose also within the LA School, although with a different flavor. As Poole and Jones (2012) put it, *there is no question that muscle oxygen availability has the potential to influence metabolic control* (Erecinska and Wilson 1982; Marcinek et al. 2003*). If mitochondrial oxygen supply is truly insufficient, the rate of oxidative metabolism will be restricted and this will be manifest as slower* $\dot{V}O_2$ *kinetics across a metabolic transient.* This is not far from what we have just stated. But there is more, as long as they continue as follows. *However, the contention in this review is that muscle oxygen delivery does not limit* $\dot{V}O_2$ *kinetics during most forms of exercise (those in which the heart is positioned above the bulk of the working muscle mass such as in running, cycling, and rowing) in most subjects (*i.e., *apparently healthy, physically active people below the age of approximately 50 years), and even when the exercise intensity is high (*i.e., *severe).* These statements close 40 years of discussions, placing the Milano and the LA Schools on the same wavelength.

Just one point still separates the two Schools: early lactate, as long as Whipp and his successors do not seem to accept the concept of energy equivalent of blood lactate accumulation yet, thereby partially denying the energetic meaning of DO_{2M}. To us, early lactate is a keystone of the entire story, to the point that we dare to end this paragraph by the following hazardous statement. When there is no early lactate accumulation in the exercise transient, muscle oxygen delivery does not limit $\dot{V}_mO_2$ or $\dot{V}_LO_2$ kinetics; on the contrary, it does indeed, when there is early lactate accumulation in the exercise transient, with the exception of intense exercise, when the glycolytic flux becomes very high.

4.6 On the Correspondence Between Muscle Oxygen Consumption and Lung Oxygen Uptake Kinetics

Historically, the implicit assumption behind the investigation of the $\dot{V}_LO_2$ kinetics is that its primary component reflects the time course of $\dot{V}_mO_2$. The double exponential model of the $\dot{V}_LO_2$ kinetics challenges this concept, at least concerning phase I. It challenges the same concept also for the kinetics of leg oxygen flow, which Grassi et al. (1996) analyzed with a single exponential model. This challenge was not dangerous to the vision of the school of Milano. In fact, as already pointed out, this problem was accounted for by admitting a buffer allowing transient variations in blood oxygen stores (di Prampero 1981), and by accounting for early lactate accumulation as an energy source for the enlargement of the oxygen deficit at high-intensity exercise (di Prampero and Ferretti 1999).

On the contrary, this challenge was felt dangerous to the LA vision. Wasserman and Whipp, although they generated the cultural milieu allowing the creation of the double exponential model, were strongly attached to a strict relation between the phase II $\dot{V}_LO_2$ kinetics and the $\dot{V}_mO_2$ kinetics. To them, this was an essential element, a pillar of the construction of the double exponential model. Contrary to the School of Milano, they transformed it from a conjecture into an axiom.

Yet the double exponential model challenged this axiom. Poole and Jones (2012) were particularly sensitive to this challenge and discussed it widely. Their discussion started from the theoretical analysis of Barstow et al. (1990), who stated that the size of venous volume determines the time delay between $\dot{V}_LO_2$ and $\dot{V}_mO_2$, without altering the respective time constants. Moreover, Barstow et al. (1990) claimed that altering the $\dot{Q}$ response affects primarily the $\dot{V}_LO_2$ response in phase I. Of course, these authors were not aware that the τ_2 of $\dot{Q}$ and $\dot{Q}aO_2$ are faster than that of $\dot{V}_LO_2$ (Lador et al. 2006, 2008). Then, Poole and Jones (2012) criticized di Prampero's considerations on the role of changes in oxygen stores on the $\dot{V}_LO_2$ kinetics and argued that changes in oxygen stores occurring in phase I only would not dissociate $\dot{V}_LO_2$ kinetics from $\dot{V}_mO_2$ kinetics.

To support their point, they summarize numerous studies in a thoughtful paragraph. We draw attention to a study involving also the School of Milano: Grassi et al. (1996) determined the time course of leg blood flow and $\dot{V}_mO_2$ in humans, and compared them with simultaneous determinations of $\dot{V}_LO_2$ kinetics at the mouth. In condition of normal leg blood flow, the kinetics of $\dot{V}_mO_2$ and that of the primary component (phase II) of $\dot{V}_LO_2$ had the same time constants. For Poole and Jones (2012), this is a major finding supporting the concept that the $\dot{V}_LO_2$ kinetics is representative of the $\dot{V}_mO_2$ kinetics. However, they had to admit that this was not the case for the phase I increase of $\dot{V}_LO_2$, which appeared unchanged even when $\dot{V}_mO_2$ increased very slowly, a finding confirmed also by further studies (Bangsbo et al. 2000; Krustrup et al. 2009). Moreover, this discrepancy was observed also in the high-intensity domain (Bangsbo et al. 2000; Krustrup et al. 2009; Nyberg et al. 2010). Indeed, phase I is a phenomenon that has nothing to do with $\dot{V}_mO_2$, but responds to other functional imperatives (Ferretti 2015).

Readers interested in the complex reasoning, by which Poole and Jones (2012) tried to reconcile their vision to experimental observations, are referred to the original publication. We just point out that part of their astonishment came from their reticence to think in terms of energetics of muscle contraction, as taught by Rodolfo Margaria and the School of Milano (see Chap. 3). Nevertheless, they were ready to conclude that, once allowance is made for phase I, which is the real great contribution of the LA School to our knowledge of exercise transients, phase II kinetics is indeed a close analog of $\dot{V}_mO_2$ kinetics.

However, this is not the crucial point, to our eyes, as we try to explain in the next lines. Bruno Grassi's findings have a deeper value than that of mere support of the correspondence between a measure at the mouth and a measure in the muscles. The most important concept behind his experiments is that the $\dot{V}_L O_2$ kinetics and the $\dot{V}_m O_2$ kinetics, although closely matched, are under different control systems and respond to different homeostatic needs. This concept appears not only from the study on the single-leg exercise model (Grassi et al. 1996), but even more neatly in the studies on the isolated-perfused muscle preparation (Grassi et al. 1998a, 1998b) (Fig. 4.3). In the latter studies, the kinetics of the oxygen delivery through the femoral artery was found to be only slightly faster than that of the $\dot{V}_m O_2$, but much slower than that reported by others of the local muscle blood flow at exercise onset, which is known to be extremely rapid (Rådegran and Saltin 1998; Toska and Eriksen 1994; Walloe and Wesche 1988). Clearly, leg oxygen delivery through the femoral artery was unable to respond as fast as local muscle blood flow, suggesting that systemic oxygen delivery, muscle oxygen diffusion and perfusion, and oxygen consumption may be the consequence of different, independent, though optimized phenomena.

Grassi et al. (1998a) also demonstrated that the time constant of $\dot{V}_m O_2$ kinetics was the same in conditions of normal and forced oxygen delivery, and equal to that calculated for the kinetics of muscle PCr decrease, further reinforcing the concept of an independent metabolic control of the kinetics of $\dot{V}_m O_2$.

It is a matter of fact that the $\dot{V}_L O_2$ kinetics corresponds closely to the $\dot{V}_m O_2$ kinetics, once allowance is made for changes in blood oxygen stores. This concept is partially challenged by the double exponential model, which assumes a cardiovascular origin of phase I. However, stating that the kinetics of $\dot{V}_L O_2$ and $\dot{V}_m O_2$ are characterized by the same time constants is not tantamount to state that the $\dot{V}_L O_2$ kinetics and the $\dot{V}_m O_2$ kinetics are under the same control system. They are coupled, and they are tightly matched (except for phase I), but this does not mean that the $\dot{V}_L O_2$ kinetics is under the same metabolic control as the $\dot{V}_m O_2$ kinetics. Metabolic changes dictate exclusively the kinetics of PCr decrease and the obligatory component of the oxygen deficit.

4.7 The Effect of Priming Exercise

The kinetics of $\dot{Q}aO_2$ may be accelerated if the blood flow to the contracting muscle mass is artificially elevated before the start of a square-wave exercise. In humans, this can be done, for instance, by performing an intense exercise a few minutes before the start of the "experimental" exercise. During recovery after intense exercise, the decrease of blood flow is slower than the corresponding increase at exercise onset. Thus, the second exercise starts when muscle blood flow has not yet returned to its resting value. In this case, a smaller (faster) τ_2 describes the $\dot{V}_L O_2$ kinetics, than that observed when the same exercise is started from a resting blood flow (see e.g. Burnley et al. 2011; Gerbino et al. 1996). This was called the priming exercise

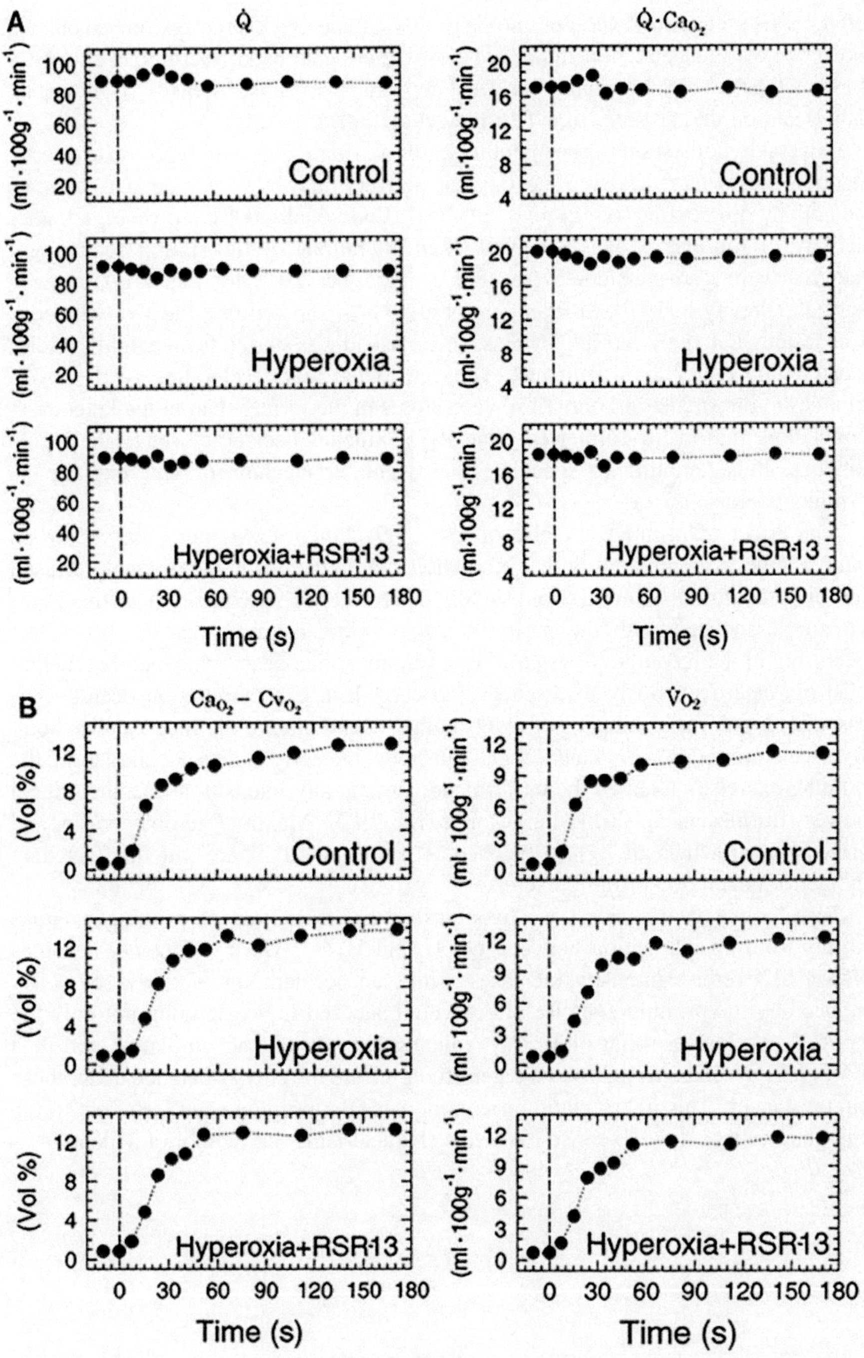

Fig. 4.3 **Panel A.** Average values of muscle blood flow (left) and muscle oxygen delivery (right). **Panel B.** Average values of artero-venous oxygen difference across muscle (left) and muscle oxygen uptake (right) at rest and during contractions in three experimental conditions, control,

effect. This effect, however, does not appear when the first exercise is carried out at a power below the lactate threshold, or below the critical power (Gerbino et al. 1996). In the intense exercise domain, the effect depends on exercise intensity (Bailey et al. 2009) and on the recovery time (Burnley et al. 2001).

Several hypotheses have been put forward to explain the priming exercise effects. Among these, increased muscle oxygen delivery may play a crucial role in its potentially direct effect (DeLorey et al. 2007; Endo et al. 2005). However, a higher activity of muscle oxidative enzymes and alterations of motor unit recruitment patterns were also considered (see Poole and Jones 2012 for details). Curiously enough, already in 1970, di Prampero et al. (1970) showed that the $\dot{V}_LO_2$ kinetics was faster when the same step increase in workload was started from a light exercise steady state rather than from rest. They attributed this acceleration of the $\dot{V}_LO_2$ kinetics to the smaller amount of oxygen stores in the former than in the latter case, concluding that in the former case the $\dot{V}_LO_2$ kinetics became closer to the $\dot{V}_mO_2$ kinetics. It is tempting to speculate that a similar mechanism may explain the priming exercise effect.

The effect of priming exercise on the $\dot{V}_LO_2$ kinetics was not observed when studying the $\dot{V}_mO_2$ kinetics during blood flow manipulations in the isolated-perfused muscle preparation (Grassi et al. 1998a). Therefore, the acceleration of the $\dot{V}_LO_2$ kinetics associated with the performance of priming exercise is due to events occurring in the respiratory system. The phenomenon appears when aerobic metabolism is heterogeneously distributed, and early lactate accumulation occurs. It is associated with faster kinetics of $\dot{Q}aO_2$ in the second than in the first exercise bout. By accelerating the $\dot{V}_LO_2$ kinetics, and thus reducing early lactate accumulation, the priming exercise increases the performance time at any intensity above the critical power (Burnley et al. 2011; Ferguson et al. 2010). Maximal aerobic power and critical power, which are tightly matched (Adami et al. 2013), are both unaffected by the performance of a priming exercise.

Based on what precedes, we hypothesize that the effect of priming exercise results from an interaction between $\dot{Q}aO_2$ and $\dot{V}_mO_2$. The τ_2 of $\dot{Q}aO_2$ becomes slower as exercise intensity increases. Thus, an acceleration of the τ_2 of $\dot{Q}aO_2$ induced by the priming exercise effect would succeed in accelerating not only the τ_2 of $\dot{V}_LO_2$, but also that of $\dot{V}_mO_2$, reducing early lactate accumulation and thus VO_2^{eLa}. Thus, changes in blood oxygen stores ensure the energy balance under these circumstances. This interpretation accommodates the priming exercise effect in the energetic vision of the exercise transients characterizing the School of Milano.

Fig. 4.3 (continued) hyperoxia and hyperoxia plus RSR-13 (a drug that reduces hemoglobin oxygen affinity). Vertical hatched lines, contraction onset. From Grassi et al. 1998a

4.8 Further Reflexions on the Slow Component

In the intense exercise domain, beside the appearance of early lactate, a new component of the $\dot{V}_LO_2$ kinetics becomes evident, the so-called slow component of the $\dot{V}_LO_2$ response to exercise (Camus et al. 1988; Capelli et al. 1993; Paterson and Whipp 1991; Poole et al. 1988; Whipp and Wasserman 1972). The slow component prevents the attainment of a steady state during constant-load exercise: $\dot{V}_LO_2$ increases until it reaches its maximum and exercise is terminated (Poole et al. 1988). Barstow and Molé (1991) treated the slow component as a third exponential of the $\dot{V}_LO_2$ kinetics (phase III), the amplitude of which would be nil at light exercise, and would be positive above the maximal lactate steady state (MLSS), eventually higher, the higher the mechanical power. These authors reported time constants higher than 100 s for phase III $\dot{V}O_2$ kinetics.

The view of the slow component by the School of Milano has been discussed in Chap. 3. In short, based on the chain of biochemical events described by Harris et al. (1976, 1977) during recovery after exercise, it was hypothesized (Capelli et al. 1993) that the progressive decrease of muscle pH, occurring in high-intensity exercise, is responsible for the slow component of the $\dot{V}_LO_2$ response. In fact, muscle pH affects the equilibrium constant of Lohmann's reaction, and hence the muscle PCr and the free creatine concentrations, thus increasing muscle Pi and reducing the phosphorylation potential (ATP/(ADP + Pi)). This, as is well known, is a key controller of PFK activity.

However, and thanks essentially to the work by the LA School, other interpretations were also considered. These include a mixture of fiber type functional characteristics, fatigue, increase in muscle temperature, and differences in efficiency between type I and type II fibers (Jones et al. 2011; Poole et al. 1994).

The slow component represents a challenge to the classical notion of a linear $\dot{V}_LO_2$ versus power relationship from rest to maximal exercise (Ferretti 2015). In fact, the appearance of the slow component suggests a non-linear relationship between $\dot{V}_LO_2$ and mechanical power and the lack of a clearly visible steady state for $\dot{V}_LO_2$ during intense exercise. Several scientists took this as a refutation of the concept that maximal aerobic power is the power requiring a $\dot{V}_LO_2$ equivalent to the maximum (Poole and Richardson 1997; Zoladz et al. 1995).

Let us first discuss the effects on the mechanical efficiency of exercise (η). If, on one side, the extra-oxygen consumption due to the slow component would tend to decrease η at a given constant mechanical power, on the other side it is also true that the mechanical work done per mole of ATP split increases (di Prampero et al. 1988; Ferretti et al. 1987), which increases η. These two phenomena tend to compensate each other, thus attenuating, if not canceling out, the distortive effects on the linear $\dot{V}O_2$ versus power relationship, and thus maintaining η essentially independent of power, and also of exercise duration.

Concerning maximal aerobic power, we can speculate as follows. In Chap. 3, we have highlighted the role of free creatine in determining the slow component of the $\dot{V}_LO_2$ kinetics during recovery. A mirror speculation can be proposed for the interpretation of the slow component of the $\dot{V}_LO_2$ kinetics during intense exercise, in which $\dot{La}_b > 0$:

$$\textbf{Lactate formation} \rightarrow \textbf{increased } [\mathrm{H}^+] \rightarrow \textbf{increased } [\mathrm{Cr}]$$
$$\rightarrow \textbf{slow component of } \dot{V}_LO_2 \textbf{ increase.}$$

If this is so, (i) the slow component could indeed be described by an exponential equation; (ii) the $\dot{V}_LO_2$ increase due to the slow component does not occur in the contracting muscle fibers; (iii) the $\dot{V}_LO_2$ increase due to the slow component is an attempt to remove lactate accumulated in active muscle, and thus to reduce $\dot{La}_b$; (iv) a three-exponential model of the $\dot{V}_LO_2$ kinetics becomes admissible (Borrani et al. 2001; Burnley et al. 2000), as described by the following equation:

$$\dot{V}_LO_{2t} = A_1\left(1 - e^{-k_1 t}\right) + A_2\left(1 - e^{-k_2(t-d)}\right) + A_3\left(1 - e^{-k_3(t-d_s)}\right) \quad (4.5)$$

where d_s is the time delay of the slow component, which the third exponential refers to. The amplitude of the third, slowest exponential would be nil at powers lower than the MLSS or the critical power. Therefore, below the MLSS, the three-exponential model would reduce to a simpler double exponential. Under these conditions, Eq. (4.1) becomes a particular case of Eq. (4.5), being and remaining a good descriptor of the $\dot{V}_LO_2$ kinetics in the light and moderate power ranges.

If this line of reasoning were correct, the maximal aerobic power would still be the lowest power at which the maximal oxygen consumption is attained, considering however that, above the MLSS, the fraction of $\dot{V}_LO_2$ related to the slow component (A_3) is not used to sustain mechanical power. Therefore, an apparent reduction of exercise efficiency appears, as reported by some studies (Borrani et al. 2003; Zoladz et al. 1995), although the mechanical efficiency (η) of the transformation of metabolic power into mechanical power ($\dot{w}$) in muscular contraction remains unchanged. In this context, η can be redefined as follows:

$$\eta = \frac{\dot{w}}{\dot{V}_LO_2 - \dot{V}_LO_{2r}} = \frac{\dot{w}}{\dot{V}_mO_2 - \dot{V}_mO_{2r}} = \frac{\dot{w}}{A_1 + A_2} \quad (4.6)$$

where $\dot{V}_LO_{2r}$ and $\dot{V}_mO_{2r}$ are the lung and muscle oxygen flows at rest, respectively, expressed in watts, as well as A_1 and A_2. The equality established by Eq. (4.6) applies at powers up to the MLSS or up to the critical power. At higher powers, in the intense exercise domain, since $\dot{V}_LO_2 - \dot{V}_LO_{2r} > A_1 + A_2$, because the former term includes also A_3, we have:

$$\frac{\dot{w}}{\dot{V}_mO_2 - \dot{V}_mO_{2r}} = \frac{\dot{w}}{A_1 + A_2} = \eta > \frac{\dot{w}}{\dot{V}_LO_2 - \dot{V}_LO_{2r}} \quad (4.7)$$

In this case, the term $\frac{\dot{w}}{\dot{V}_L O_2 - \dot{V}_L O_{2r}}$ would provide the gross efficiency of exercise, accounting for the apparent decrease of efficiency in the severe exercise domain, whereas the actual mechanical efficiency of power generation in the contracting muscle mass remains equal to $\frac{\dot{w}}{A_1 + A_2}$.

The appearance of the slow component has an impact also on the characteristics of the $\dot{V}_L O_2$ response during sinusoidal exercise. This can be defined as a type of exercise in which the workload varies continuously between a minimum and a maximum following a sinusoidal pattern of given wavelength. The LA School typically investigated this type of exercise. In the light exercise domain, $\dot{V}_L O_2$ follows a symmetric sinusoidal pattern, with a time lag corresponding to the time constant of the primary component of the $\dot{V}_L O_2$ kinetics (Casaburi et al. 1977; Haouzi et al. 1993; Whipp et al. 1982). This is because in this type of exercise there is neither the rapid phase I of the $\dot{V}_L O_2$ kinetics, nor the slow component. Therefore, the respiratory system behaves as a first-order system, providing essentially linear responses. This is not so in the intense exercise domain, at powers implying the appearance of the slow component. Because of a progressive increase of the time constant of the primary component, implying early lactate accumulation, and of the superposition of the slow component, the sinusoidal $\dot{V}_L O_2$ response is distorted. On one side, the time lag between the power and the $\dot{V}_L O_2$ sinusoidal patterns expands as $\dot{La}_b$ goes up (Haouzi et al. 1993). On the other, an asymmetry between the increasing and decreasing parts of the $\dot{V}_L O_2$ response is expected, due to the divergent characteristics between the on- and off- phases of the $\dot{V}_L O_2$ pattern. The latter prediction requires a closer experimental scrutiny.

4.9 The Problem of Gas Flow Analysis on a Breath-by-Breath Basis

The measurement and computation techniques used to investigate the $\dot{V}_L O_2$ kinetics at exercise onset deeply affected our understanding of the exercise transient. Initially, no breath-by-breath recording was possible, and ingenious techniques of serial sampling were adopted (Cerretelli and Brambilla 1958; Henry 1951). The time resolution of the alveolar oxygen transfer attained the single breath level when Auchincloss et al. (1966) published their original algorithm for $\dot{V}_L O_2$ computation. This was a great achievement, which started the modern history of the $\dot{V}_L O_2$ kinetics.

However, Auchincloss' algorithm requires a correct determination of the changes in lung gas stores within each breath. This quantity cannot be measured. Therefore, Auchincloss et al. (1966) assumed a fixed pre-defined value of it, which they set equal to the functional residual capacity. Several authors introduced variants to the Auchincloss algorithm (Busso and Robbins 1997; Swanson and Sherrill 1983; Wessel et al. 1979), in an attempt at identifying more correct values of end-expiratory lung volume. However, di Prampero and Lafortuna (1989) demonstrated that the end-expiratory lung volume was not equal to the functional residual capacity, was varying continuously among breaths, and was unpredictable. This demonstration of the impossibility of attaining a correct estimate of end-expiratory lung volume at each breath was a fundamental theoretical result. It established that it was impossible to reduce the experimental error of single-breath alveolar oxygen transfer determination by refining the Auchincloss algorithm, the theoretical value of which was slain by an elusive variable.

Although many studies still rely on the Auchincloss algorithm, which undermines their value, and although several metabolic carts of common use in clinical environment as well as in physiological laboratories have implemented a computation software based on the Auchincloss algorithm, other tracks had to be taken to reduce the signal-to-noise ratio. A remarkable computational improvement was achieved after Capelli et al. (2001) had resumed an alternative algorithm (Grønlund 1984), which did not require an independent a-priori estimate of end-expiratory lung volume. Using Grønlund's algorithm, Capelli et al. (2001) demonstrated a two-times improvement of the signal-to-noise ratio in breath-by-breath determination of alveolar gas transfer as compared with Auchincloss-like algorithms. Moreover, Cautero et al. (2002) demonstrated that Grønlund's algorithm was able to provide lower τ_2 values than any variant of Auchincloss' algorithm, no matter what value was attributed to the end-expiratory lung volume. This was a crucial achievement, suggesting that the primary component of the $\dot{V}_LO_2$ kinetics might indeed be faster than previously admitted, even faster than the kinetics of $\dot{V}_mO_2$ (Adami et al. 2011; De Roia et al. 2012; Lador et al. 2006). A further evolution of Grønlund's algorithm was recently proposed (Cettolo and Francescato 2018), with similar results (Francescato and Cettolo 2019).

Such algorithm improvements, however, did not act on the time resolution, the second fundamental limitation in the study of gas exchange dynamics. Since the single breath is a physiological barrier, further improvements in the time resolution could rely only on computational manipulations, such as superposition of several trials and interpolation procedures. This is a need if we wish to use the double exponential model of Barstow and Molé (1987). This model in fact yields very low τ_1 values, so low (so fast), that τ_1 was systematically undersampled, because it was resolved within one, at most two breaths, from exercise start.

The first attempts in this direction relied on interpolation methods on the 1 s basis (Beaver et al. 1981; Hughson et al. 1993; Lamarra et al. 1987). All these methods introduced filters distorting the physiological signal. So they introduced a further source of error, which tends to overestimate calculated time constants. Lamarra et al. (1987) were aware of this problem. They honestly admitted that their step interpolation generated "*mean physiological responses that are smoothed by a filter whose time constant is the mean breath duration (typically 3-4 s for these studies)*". For this reason, they restricted application of their procedure to the analysis of phase II, the time constant of which was at least one order of magnitude greater than the interpolation interval. Hughson et al. (1993) circumvented the limits of stepwise interpolation by introducing a linear interpolation procedure. They assumed a continuous, linear $\dot{V}O_2$ increase within a given breath. Yet a filter effect is still present, although of smaller impact, especially in the determination of τ_1. The interpolation methods remain intrinsically inaccurate.

More recently, at least in the light exercise domain, mere stacking of multiple repetitions was proposed and treated as if the data were from the same rest-to-exercise transient (Bringard et al. 2014; Francescato et al. 2014a, 2014b). This procedure prevented possible signal distortion by superposing raw data obtained during several identical exercise transients, on the assumption that each repetition on a given subject is representative of the same physiological situation. The number of observations encompassed by a fitting procedure is greater. Thus, the fitting is reinforced by mere addition of data. The stacking procedure provided equivalent τ_2 values, but lower τ_1 values than the two interpolation procedures (Bringard et al. 2014), for the same $\dot{V}_L O_2$ computational algorithm (Grønlund's). It is not unlikely that the $\dot{V}_L O_2$ kinetics upon exercise onset be faster than usually reported, because of insufficient or inadequate data treatment tools. This would reinforce the concept of the independence of $\dot{V}_L O_2$ kinetics from $\dot{V}_m O_2$ kinetics and metabolic control.

4.10 Conclusions

We owe much of the present knowledge on the exercise transient to the work done by the Milano and the LA Schools. They competed fairly for some 40 years, starting from different points and cultural visions. They pursued separate tracks, had divergent theoretical backgrounds, and therefore performed different experiments. But they finally converged on a shared conclusion, namely that $\dot{V}_m O_2$ kinetics is controlled by the metabolic changes initiated by muscle contractions at exercise onset, at least during exercise with big muscle masses in normoxia. There are, however, conditions in which the kinetics of muscle oxygen delivery becomes slower than that of activation of the muscle metabolic machinery. When this is the case, then the time constant of the $\dot{V}_m O_2$ kinetics increases and the process becomes limited by cardiovascular oxygen transport, for the glory of Richard Hughson and the Canadian School. In all conditions, the primary component of the $\dot{V}_L O_2$ kinetics has the same time constant as the $\dot{V}_m O_2$ kinetics, once allowance is made for a time delay between working muscles and the lungs and for the changes in blood oxygen stores. It is therefore justified to infer on muscle energetics from gas exchange kinetics. However, when the $\dot{V}_m O_2$ kinetics becomes slower than that of muscle PCr decrease, and thus early lactate is accumulated in muscle and in blood, such inference becomes possible only if early lactate is determined and an energy equivalent for lactate accumulation is admitted. In spite of this, and despite the collected evidence (see Chap. 3), this concept is not accepted by all scientists in the field even nowadays.

References

Adami A, Pogliaghi S, De Roia G, Capelli C (2011) O_2 uptake, cardiac output and muscle deoxygenation at the onset of moderate and supramaximal exercise in humans. Eur J Appl Physiol 111:1517–1527

Adami A, Sivieri A, Moia C, Perini R, Ferretti G (2013) Effects of step duration in incremental ramp protocols on peak power and maximal oxygen consumption. Eur J Appl Physiol 113:2647–2653

Asmussen E (1973) Ventilation at transition from rest to exercise. Acta Physiol Scand 89:68–78

Asmussen E, Nielsen M (1948) Studies on the initial changes in respiration at the transition from rest to work and from work to rest. Acta Physiol Scand 16:270–285

Auchincloss JH Jr, Gilbert R, Baule GH (1966) Effect of ventilation on O_2 transfer during early exercise. J Appl Physiol 21:810–818

Bailey SJ, Vanhatalo A, Wilkerson DP, DiMenna FJ, Jones AM (2009) Optimizing the "priming" effect: influence of prior exercise intensity and recovery duration on O_2 uptake kinetics and severe-intensity exercise tolerance. J Appl Physiol 107:1743–1756

Bangsbo J, Krustrup P, Gonzalez-Alonso J, Boushel R, Saltin B (2000) Muscle O_2 kinetics at onset of intense dynamic exercise in humans. Am J Phys 279:R899–R906

Barstow TJ, Lamarra N, Whipp BJ (1990) Modulation of muscle and pulmonary O_2 uptakes by circulatory dynamics during exercise. J Appl Physiol 68:979–989

Barstow TJ, Molé PA (1987) Simulation of pulmonary O_2 uptake during exercise transients in humans. J Appl Physiol 63:2253–2261

Barstow TJ, Molé PA (1991) Linear and non-linear characteristics of O_2 uptake kinetics during heavy exercise. J Appl Physiol 71:2099–2106

Beaver WL, Lamarra N, Wasserman K (1981) Breath-by-breath measurement of true alveolar gas exchange. J Appl Physiol 51:1662–1675

Binzoni T, Cerretelli P (1991) Muscle ^{31}P-NMR in humans: estimate of bias and qualitative assessment of ATPase activity. J Appl Physiol 71:1700–1704

Binzoni T, Ferretti G, Schenker K, Cerretelli P (1992) Phosphocreatine hydrolysis by ^{31}P-NMR at the onset of constant-load exercise in humans. J Appl Physiol 73:1643–1649

Binzoni T, Hiltbrand E, Yano T, Cerretelli P (1997) Step vs. progressive exercise: the kinetics of phosphocreatine hydrolysis in human muscle. Acta Physiol Scand 159:209–215

Borrani F, Candau R, Millet GY, Perrey S, Fuchslocher J, Rouillon JD (2001). Is the $\dot{V}O_2$ slow component dependent on progressive recruitment of fast-twitch fibers in trained runners? J Appl Physiol 90: 2212–2220

Borrani F, Candau R, Perrey S, Millet GY, Millet GP, Rouillon JD (2003). Does the mechanical work in running change during the $\dot{V}O_2$ slow component? Med Sci Sports Exerc 35: 50–57

Bowen TS, Rossiter HB, Benson AP, Amano T, Kondo N, Kowalchuk JM, Koga S (2013) Slowed O_2 uptake kinetics in hypoxia correlate with the transient peak and reduced spatial distribution of absolute skeletal muscle deoxygenation. Exp Physiol 98:1585–1596

Bowtell JL, Marwood S, Bruce M, Constantin-Teodosiu D, Greenhaff DL (2007) Tricarboxylic acid cycle intermediate pool size: functional importance for oxidative metabolism in exercising human skeletal muscle. Sport Med 37:1071–1088

Bringard A, Adami A, Moia C, Ferretti G (2014) A new interpolation-free procedure for breath-by-breath analysis of O_2 uptake in exercise transients. Eur J Appl Physiol 114:1983–1994

Burnley M, Davison G, Baker JR (2011). Effects of priming exercise on $\dot{V}O_2$ kinetics and the power-duration relationship. Med Sci Sports Exerc 43: 2171–2179

Burnley M, Doust JH, Carter H, Jones AM (2001) Effects of prior exercise and recovery duration on oxygen uptake kinetics during heavy exercise in humans. Exp Physiol 86:417–425

Burnley M, Jones AM, Carter H, Doust JH (2000) Effects of prior heavy exercise on phase II pulmonary oxygen uptake kinetics during heavy exercise. J Appl Physiol 89:1387–1396

Busso T, Robbins PA (1997) Evaluation of estimates of alveolar gas exchange by using a tidally ventilated nonhomogenous lung model. J Appl Physiol 82:1963–1971

Camus G, Atchou G, Bruckner JC, Giezendanner D, di Prampero PE (1988). Slow upward drift of $\dot{V}O_2$ during constant-load cycling in untrained subjects. Eur J Appl Physiol 58: 197–202

Capelli C, Antonutto G, Zamparo P, Girardis M, di Prampero PE (1993) Effects of prolonged cycoergometer exercise on maximal power and oxygen uptake in humans. Eur J Appl Physiol 66:189–195

Capelli C, Cautero M, di Prampero PE (2001) New perspectives in breath-by-breath determination of alveolar gas exchange in humans. Pflügers Arch 441:566–577

Casaburi R, Barstow TJ, Robinson T, Wasserman K (1989a) Influence of work rate on ventilatory and gas exchange kinetics. J Appl Physiol 67:547–555

Casaburi R, Daly J, Hansen JE, Effros RM (1989b) Abrupt changes in mixed venous blood gas composition after the onset of exercise. J Appl Physiol 67:1106–1112

Casaburi R, Whipp BJ, Wasserman K, Beaver WL, Koyal SN (1977) Ventilatory and gas exchange dynamics in response to sinusoidal work. J Appl Physiol 42(300–311):1977

Casaburi R, Whipp BJ, Wasserman K, Stremel RW (1978) Ventilatory control characteristics of the exercise hyperpnea as discerned from dynamic forcing techniques. Chest 73(Suppl 2):280–283

Cautero M, Beltrami AP, di Prampero PE, Capelli C (2002) Breath-by-breath alveolar O_2 transfer at the onset of step exercise in humans. Eur J Appl Physiol 88(203–213):2002

Cerretelli P, Brambilla I (1958) Cinetica della contrazione di un debito di O_2 nell'uomo. Boll Soc Ital Biol Sper 34:679–682

Cerretelli P, di Prampero PE (1987) Gas exchange at exercise. In: Farhi LE, Tenney SM (eds) Handbook of physiology, the respiratory system IV. Bethesda, MD, Am Physiol Soc, pp 555–632

Cerretelli P, di Prampero PE, Piiper J (1969) Energy balance of anaerobic work in the dog gastrocnemius muscle. Am J Phys 217:581–585

Cerretelli P, Pendergast DR, Paganelli WC, Rennie DW (1979) Effects of specific muscle training on $\dot{V}O_2$ on-response and early blood lactate. J Appl Physiol 47:761–769

Cerretelli P, Sikand R, Farhi LE (1966) Readjustments in cardiac output and gas exchange during onset of exercise and recovery. J Appl Physiol 21:1345–1350

Cettolo V, Francescato MP (2018) Assessing breath-by-breath alveolar gas exchange: is the contiguity in time of breaths mandatory? Eur J Appl Physiol 118:1119–1130

Cleuziou C, Perrey S, Lecoq AM, Candau R, Courteix D, Obert P (2005) O_2 uptake kinetics during moderate and heavy intensity exercise in humans: the influence of hypoxia and training status. Int J Sport Med 26:356–362

Comroe JH (1944) The hyperpnea of muscular exercise. Physiol Rev 24:319–339

Comroe JH, Schmidt CF (1943) Reflexes of the limbs as a factor in the hyperpnoea of muscular exercise. Am J Phys 138:536–547

Connett RJ, Gayeski TE, Honig CR (1985) Energy sources in fully aerobic rest-work transition: a new role for glycolysis. Am J Phys 248:H922–H929

Cummin AR, Iyawe VI, Mehta N, Saunders KB (1986) Ventilation and cardiac output during the onset of exercise, and during voluntary hyperventilation, in humans. J Physiol Lond 370:567–583

D'Angelo E, Torelli G (1971) Neural stimuli increasing respiration during different types of exercise. J Appl Physiol 30:116–121

Davies CTM, di Prampero PE, Cerretelli P (1972) Kinetics of cardiac output and respiratory gas exchange during exercise and recovery. J Appl Physiol 32:618–625

De Cort SC, Innes JA, Barstow TJ, Guz A (1991) Cardiac output, O_2 consumption and arterio-venous O_2 difference following a sudden rise in exercise level in humans. J Physiol 441:501–512

Dejours P (1964) Control of respiration in muscular exercise. In: Fenn WO, Rahn H (eds) Handbook of physiology, respiration. Washington DC, American Physiological Society, pp 631–648

DeLorey DS, Kowalchuk JM, Heenan AP, duManoir GR, Paterson DH (2007) Prior exercise speeds pulmonary O_2 uptake kinetics by increases in both local muscle O_2 availability and O_2 utilization. J Appl Physiol 103:771–778

De Roia G, Pogliaghi S, Adami A, Papadopoulou C, Capelli C (2012) Effects of priming exercise on the speed of adjustment of muscle oxidative metabolism at the onset of moderate-intensity step transitions in older adults. Am J Phys 302:R1158–R1166

di Prampero PE (1981) Energetics of muscular exercise. Rev Physiol Biochem Pharmacol 89:143–222

di Prampero PE, Boutellier U, Marguerat A (1988) Efficiency of work performance and contraction velocity in isotonic tetani of frog sartorius. Pflügers Arch 412:455–461

di Prampero PE, Davies CTM, Cerretelli P, Margaria R (1970) An analysis of O_2 debt contracted in submaximal exercise. J Appl Physiol 29:547–551

di Prampero PE, Ferretti G (1999) The energetics of anaerobic muscle metabolism: a reappraisal of older and recent concepts. Respir Physiol 118:103–115

di Prampero PE, Francescato MP, Cettolo V (2003) Energetics of muscle exercise at work onset: the steady-state approach. Pflügers Arch 445:741–746

di Prampero PE, Lafortuna CL (1989) Breath-by-breath estimate of alveolar gas transfer variability in man at rest and during exercise. J Physiol Lond 415:459–475

di Prampero PE, Margaria R (1968) Relationship between O_2 consumption, high energy phosphates and the kinetics of O_2 debt in exercise. Pflügers Arch 304:11–19

Eldridge FL, Millhorn DE, Waldrop TG (1981) Exercise hyperpnea and locomotion: parallel activation from the hypothalamus. Science 211:844–846

Endo M, Okada Y, Rossiter HB, Ooue A, Miura A, Koga S, Fukuba Y (2005) Kinetics of pulmonary $\dot{V}O_2$ and femoral artery blood flow and their relationship during repeated bouts of heavy exercise. Eur J Appl Physiol 95:418–430

Engelen M, Porszasz J, Riley M, Wasserman K, Maehara K, Barstow TJ (1996) Effects of hypoxic hypoxia on O_2 uptake and heart rate kinetics during heavy exercise. J Appl Physiol 81:2500–2508

Erecinska M, Wilson DF (1982) Regulation of cellular energy metabolism. J Mem Biol 70:1–14

Eriksen M, Waaler BA, Walloe L, Wesche J (1990) Dynamics and dimensions of cardiac output changes in humans at the onset and the end of moderate rhythmic exercise. J Physiol Lond 426: 423–437

Fagoni N, Bruseghini P, Adami A, Capelli C, Lador F, Moia C, Tam E, Bringard A, Ferretti G (2020) Effect of lower body negative pressure on phase I cardiovascular responses at exercise onset. Int J Sports Med 41:209–218

Fagraeus L, Linnarsson D (1976) Autonomic origin of heart rate fluctuations at the onset of muscular exercise. J Appl Physiol 40:679–682

Faisal A, Beavers KR, Hughson RL (2010) O_2 uptake and blood pressure regulation at the onset of exercise: interaction of circadian rhythm and priming exercise. Am J Phys 299:H1832–H1842

Favier R, Desplanches D, Frutoso J, Grandmontagne M, Flandrois R (1983) Ventilatory transients during exercise: peripheral or central control? Pflügers Arch 396:269–276

Ferguson C, Rossiter HB, Whipp BJ, Cathcart AJ, Murgatroyd SR, Ward SA (2010) Effect of recovery duration from prior exhaustive exercise on the parameters of the power-duration relationship. J Appl Physiol 108:866–874

Ferretti G (2015) Energetics of muscular exercise. Springer, Heidelberg

Ferretti G, Gussoni M, di Prampero PE, Cerretelli P (1987) Effects of exercise on maximal instantaneous muscular power of humans. J Appl Physiol 62:2288–2294

Francescato MP, Cettolo V (2019) The "independent breath" algorithm: assessment of O_2 uptake during exercise. Eur J Appl Physiol 119:495–508

Francescato MP, Cettolo V, Bellio R (2014a) Confidence intervals for the parameters estimated from simulated O_2 uptake kinetics: effects of different data treatments. Exp Physiol 99:187–195

Francescato MP, Cettolo V, Bellio R (2014b) Assembling more O_2 uptake responses: is it possible to merely stack the repeated transitions? Respir Physiol Neurobiol 200:46–49

Francescato MP, Cettolo V, di Prampero PE (2008) Influence of phosphagen concentration on phosphocreatine breakdown kinetics. Data from human gastrocnemius muscle. J Appl Physiol 105:158–164

Francescato MP, Cettolo V, di Prampero PE (2013) O_2 uptake kinetics at work onset: role of cardiac output and of phosphocreatine breakdown. Respir Physiol Neurobiol 185:287–295

Gerbino A, Ward SA, Whipp BJ (1996) Effects of prior exercise on pulmonary gas-exchange kinetics during high-intensity exercise in humans. J Appl Physiol 80:99–107

Gilbert R, Auchincloss JH Jr, Baule GH (1967) Metabolic and circulatory adjustments to unsteady-state exercise. J Appl Physiol 22:905–912

Grassi B (2000) Skeletal muscle $\dot{V}O_2$ on-kinetics: set by O_2 delivery or by O_2 utilization? New insights into an old issue. Med Sci Sports Exerc 32:108–116

Grassi B, Gladden LB, Samaja M, Stary CM, Hogan MC (1998a) Faster adjustment of O_2 delivery does not affect $\dot{V}O_2$-on kinetics in isolated in situ canine muscle. J Appl Physiol 85:1394–1403

Grassi B, Gladden LB, Stary CM, Wagner PD, Hogan MC (1998b) Peripheral O_2 diffusion does not affect $\dot{V}O_2$ on-kinetics in isolated in situ canine muscle. J Appl Physiol 85:1404–1412

Grassi B, Poole DC, Richardson RS, Knight DR, Erickson BK, Wagner PD (1996) Muscle O_2 uptake kinetics in humans: implications for metabolic control. J Appl Physiol 80:988–998

Grønlund J (1984) A new method for breath-to-breath determination of O_2 flux across the alveolar membrane. Eur J Appl Physiol 52:167–172

Haouzi P, Fukuba Y, Casaburi R, Stringer W, Wasserman K (1993) O_2 uptake kinetics above and below the lactic acidosis threshold during sinusoidal exercise. J Appl Physiol 75:1683–1690

Harris RC, Edwards RH, Hultman E, Nordesjö LO, Nylind B, Sahlin K (1976) The time course of phosphorylcreatine resynthesis during recovery of the quadriceps muscle in man. Pflügers Arch 367:137–142

Harris RC, Sahlin K, Hultman E (1977) Phosphagen and lactate contents of m. quadriceps femoris of man after exercise. J Appl Physiol 43:852–857

Henry FM (1951) Aerobic O_2 consumption and alactic debt in muscular work. J Appl Physiol 2: 481–487

Hill AV, Long CNH, Lupton H (1924) Muscular exercise, lactic acid and the supply and utilization of oxygen. Parts IV – VI. Proc R Soc London (Biol) 97:84–138

Hill AV, Lupton H (1923) Muscular exercise, lactic acid, and the supply and utilization of O_2. Q J Med 16:135–171

Hughson RL (1984) Alterations in the O_2 deficit-O_2 debt relationships with beta-adrenergic receptor blockade in man. J Physiol 349:375–387

Hughson RL (1990) Exploring cardiorespiratory control mechanisms through gas exchange dynamics. Med Sci Sports Exerc 22:72–79

Hughson RL (2009) Oxygen uptake kinetics: historical perspective and future directions. Appl Physiol Nutr Metab 34:840–850

Hughson RL, Cochrane JE, Butler GC (1993) Faster O_2 uptake kinetics at onset of supine exercise with than without lower body negative pressure. J Appl Physiol 75:1962–1967

Hughson RL, Kowalchuk JM (1991) Beta-blockade and O_2 delivery to muscle during exercise. Can J Physiol Pharmacol 69:285–289

Hughson RL, Kowalchuk JM (1995) Kinetics of O_2 uptake for submaximal exercise in hyperoxia, normoxia, and hypoxia. Can J Appl Physiol 20:198–210

Hughson RL, Morrissey MA (1983) Delayed kinetics of $\dot{V}O_2$ in the transition from prior exercise. Evidence for O_2 transport limitation of $\dot{V}O_2$ kinetics: a review. Int J Sports Med 4:31–39

Hughson RL, Shoemaker JK, Tschakovsky ME, Kowalchuk JM (1996) Dependence of muscle $\dot{V}O_2$ on blood flow dynamics at onset of forearm exercise. J Appl Physiol 81:1619–1626

Hughson RL, Tschakovsky ME, Houston ME (2001) Regulation of O2 consumption at the onset of exercise. Exerc Sport Sci Rev 29:129–133

Hughson RL, Xing HC, Borkhoff C, Butler GC (1991) Kinetics of ventilation and gas exchange during supine and upright cycle exercise. Eur J Appl Physiol 63:300–307

Johansson JE (1895) Über die Einwirkung der Muskelthätigkeit auf die Atmung und die Hertzthätigkeit. Skand Arch Physiol 5:20–66

Jones AM, Grassi B, Christensen PM, Krustrup P, Bangsbo J, Poole DC (2011) Slow component of kinetics: mechanistic bases and practical applications. Med Sci Sports Exerc 43:2046–2062

Koga S, Shiojiri T, Shibasaki M, Fukuba Y, Fukuoka Y, Kondo N (1996) Kinetics of O_2 uptake and cardiac output at onset of arm exercise. Respir Physiol 103:195–202

Korzeniewski B (2003) Regulation of oxidative phosphorylation in different muscles and various experimental conditions. Biochem J 375:799–804

Korzeniewski B, Zoladz JA (2002) Influence of rapid changes in cytosolic pH on oxidative phosphorylation in skeletal muscle: theoretical studies. Biochem J 365:249–258

Krogh A, Lindhard J (1913) The regulation of respiration and circulation during the initial stages of muscular work. J Physiol Lond 47:112–136

Krogh A, Lindhard J (1920) The changes in respiration at the transition from work to rest. J Physiol Lond 53:431–439

Krustrup P, Jones AM, Wilkerson DP, Calbet JA, Bangsbo J (2009) Muscular and pulmonary O_2 uptake kinetics during moderate and high-intensity sub-maximal knee-extensor exercise in humans. J Physiol 587:1843–1856

Lador F, Azabji-Kenfack M, Moia C, Cautero M, Morel DR, Capelli C, Ferretti G (2006) Simultaneous determination of the kinetics of cardiac output, systemic O_2 delivery and lung O_2 uptake at exercise onset in man. Am J Phys 290:R1071–R1079

Lador F, Tam E, Adami A, Azabji-Kenfack M, Bringard A, Cautero M, Moia C, Morel DR, Capelli C, Ferretti G (2013) Cardiac output, O_2 delivery and VO_2 kinetics during step exercise in acute normobaric hypoxia. Respir Physiol Neurobiol 186:206–213

Lador F, Tam E, Azabji-Kenfack M, Cautero M, Moia C, Morel DR, Capelli C, Ferretti G (2008) Phase I dynamics of cardiac output, systemic O_2 delivery and lung O_2 uptake at exercise onset in men in acute normobaric hypoxia. Am J Phys 295:R624–R632

Lamarra N, Whipp BJ, Ward SA, Wasserman K (1987) Effect of interbreath fluctuations on characterizing exercise gas exchange kinetics. J Appl Physiol 62:2003–2012

Lindholm P, Karlsson L, Gill H, Wigertz O, Linnarsson D (2006) Time components of circulatory transport from the lungs to a peripheral artery in humans. Eur J Appl Physiol 97:96–102

Maciel BC, Gallo L Jr, Marin Neto JA, Lima Filho EC, Martins LE (1986) Autonomic nervous control of the heart rate during dynamic exercise in normal man. Clin Sci 71:457–460

Mader A (2003) Glycolysis and oxidative phosphorylation as a function of cytosolic phosphorylation state and power output of the muscle cell. Eur J Appl Physiol 88:317–338

Marcinek DJ, Ciesielski WA, Conley KE, Schenkman KA (2003) O_2 regulation and limitation to cellular respiration in mouse skeletal muscle in vivo. Am J Phys 285:H1900–H1908

Margaria R, Edwards HT, Dill DB (1933) The possible mechanism of contracting and paying the O_2 debt and the role of lactic acid in muscular contraction. Am J Phys 106:689–714

McCloskey DI, Mitchell JH (1972) Reflex cardiovascular and respiratory responses originating in exercising muscle. J Physiol 224:173–186

McDonald MJ, Shoemaker JK, Tschakovsky ME, Hughson RL (1998) Alveolar O_2 uptake and femoral artery blood flow dynamics in upright and supine leg exercise in humans. J Appl Physiol 85:1622–1628

Meyer RA (1988) A linear model of muscle respiration explains monoexponential phosphocreatine changes. Am J Phys 254:C548–C553

Meyer RA, Foley JM (1996) Cellular processes integrating the metabolic response to exercise. In: handbook of physiology. Exercise: regulation and integration of multiple systems. Bethesda, MD: am Physiol Soc, section 12. Chapter 18:841–869

Nyberg G (1981) Vagal and sympathetic contributions to the heart rate at rest and during isometric and dynamic exercise in young healthy men. J Cardiovasc Pharmacol 3:1243–1250

Nyberg M, Mortensen SP, Saltin B, Hellsten Y, Bangsbo J (2010) Low blood flow at onset of moderate-intensity exercise does not limit muscle O2 uptake. Am J Phys 298:R843–R848

Paganelli W, Pendergast DR, Koness J, Cerretelli P (1989) The effect of decreased muscle energy stores on the $\dot{V}O_2$ kinetics at the onset of exercise. Eur J Appl Physiol 59:321–326
Paterson DH, Whipp BJ (1991) Asymmetries of O_2 uptake transients at the on- and offset of heavy exercise in humans. J Physiol 443:575–586
Piiper J, di Prampero PE, Cerretelli P (1968) O_2 debt and high-energy phosphates in the gastrocnemius muscle of the dog. Am J Phys 215:523–531
Poole DC, Barstow TJ, Gaesser GA, Willis WT, Whipp BJ (1994) O_2 slow component: physiological and functional significance. Med Sci Sports Exerc 26:1354–1358
Poole DC, Jones AM (2012) O_2 uptake kinetics. Compr Physiol 2:933–996
Poole DC, Richardson RS (1997) Determinants of oxygen uptake. Implications for exercise testing Sports Med 24:308–320
Poole DC, Ward SA, Gardner GW, Whipp BJ (1988) Metabolic and respiratory profile of the upper limit for prolonged exercise in man. Ergonomics 31:1265–1279
Rådegran G, Saltin B (1998) Muscle blood flow at onset of dynamic exercise in man. Am J Physiol Heart Circ Physiol 274:H314–H322
Robinson BF, Epstein SE, Beiser GD, Braunwald E (1966) Control of heart rate by the autonomic nervous system. Studies in man on the interrelation between baroreceptor mechanisms and exercise. Circ Res 19:400–411
Rossiter HB, Ward SA, Doyle VL, Howe FA, Griffiths JA, Whipp BJ (1999) Inferences from pulmonary O_2 uptake with respect to intramuscular [phosphocreatine] kinetics during moderate exercise in humans. J Physiol Lond 518:921–932
Rossiter HB, Ward SA, Kowalchuk JM, Howe FA, Griffiths JR, Whipp BJ (2002) Dynamic asymmetry of phosphocreatine concentration and O_2 uptake between the on- and off-transients of moderate- and high-intensity exercise in humans. J Physiol Lond 541:991–1002
Rowell LB (1993) Human cardiovascular control, vol 25. Oxford University Press, New York, p 1432
Rowell LB, O'Leary DS (1990) Reflex control of the circulation during exercise: chemoreflexes and mechanoreflexes. J Appl Physiol 69:407–418
Sheriff DD, Rowell LB, Scher AM (1993) Is rapid rise in vascular conductance at onset of dynamic exercise due to muscle pump? Am J Phys 265:H1227–H1234
Springer C, Barstow TJ, Wasserman K, Cooper DM (1991) O_2 uptake and heart rate responses during hypoxic exercise in children and adults. Med Sci Sports Exerc 23:71–79
Strange S (1999) Cardiovascular control during concomitant dynamic leg exercise and static arm exercise in humans. J Physiol Lond 514:283–291
Swanson GD, Sherrill DL (1983) A model evaluation of estimates of breath-to-breath alveolar gas exchange. J Appl Physiol 55:1936–1941
Torelli G, Brandi G (1961) Regulation of the ventilation at the beginning of muscular exercise. Int Z Angew Physiol 19:134–142
Toska K, Eriksen M (1994) Peripheral vasoconstriction shortly after onset of moderate exercise in humans. J Appl Physiol 77:1519–1525
Turner DL (1991) Cardiovascular and respiratory control mechanisms during exercise: an integrated view. J Exp Biol 160:309–640
Walloe L, Wesche J (1988) Time course and magnitude of blood flow changes in the human quadriceps muscle during and following rhythmic exercise. J Physiol 405(257–273):1988
Ward SA, Whipp BJ (1980) Ventilatory control during exercise with increased external dead space. J Appl Physiol 48:225–231
Wasserman K, Van Kessel AL, Burton GG (1967) Interaction of physiological mechanisms during exercise. J Appl Physiol 22:71–85
Wasserman K, Whipp BJ, Casaburi R (1986) Respiratory control during exercise. In: Widdicombe G (ed) Handbook of physiology: the respiratory system. Control of breathing. Cherniack NS. The American Physiological Society, Bethesda, MD, pp 595–619

Wasserman K, Whipp BJ, Castagna J (1974) Cardiodynamic hyperpnea: hyperpnea secondary to cardiac output increase. J Appl Physiol 36:457–464

Weissman ML, Jones PW, Oren A, Lamarra N, Whipp BJ, Wasserman K (1982) Cardiac output increase and gas exchange at start of exercise. J Appl Physiol 52:236–244

Wessel HU, Stout RL, Bastanier CK, Paul MH (1979) Breath-by-breath variation of FRC: effect on O_2 and CO_2 measured at the mouth. J Appl Physiol 46:1122–1126

West JB, Jones NL (1965) Effects of changes in topographical distribution of lung blood flow on gas exchange. J Appl Physiol 20:825–835

Whipp BJ (1994) Peripheral chemoreceptor control of exercise hyperpnea in humans. Med Sci Sports Exerc 26:337–347

Whipp BJ, Rossiter HB, Ward SA, Avery D, Doyle VL, Howe FA, Griffiths JR (1999) Simultaneous determination of muscle ^{31}P and O_2 uptake kinetics during whole body NMR spectroscopy. J Appl Physiol 86:742–747

Whipp BJ, Ward SA (1982) Cardiopulmonary coupling during exercise. J Exp Biol 100:175–193

Whipp BJ, Ward SA (1990) Physiological determinants of pulmonary gas exchange kinetics during exercise. Med Sci Sports Exerc 22:62–71

Whipp BJ, Ward SA, Lamarra N, Davis JA, Wasserman K (1982) Parameters of ventilatory and gas exchange dynamics during exercise. J Appl Physiol 52:1506–1513

Whipp BJ, Wasserman K (1972) O_2 uptake kinetics for various intensities of constant-load work. J Appl Physiol 33:351–356

Yoshida T, Whipp BJ (1994) Dynamic asymmetries of cardiac output transients in response to muscular exercise in man. J Physiol 480:355–359

Zoladz JA, Rademaker AC, Sargeant AJ (1995) Non-linear relationship between O_2 uptake and power output at high intensities of exercise in humans. J Physiol Lond 488:211–217

Chapter 5
The Energetics and Biomechanics of Walking and Running

Pietro Enrico di Prampero and Guido Ferretti

Abstract This chapter is on walking and running, the present knowledge of which is due essentially to the School of Milano. After a short historical synopsis, the energetics of walking and running is discussed. In this context, after having introduced Margaria's concept of energy cost, the effects of incline, of the terrain, of pathological gaits and of body mass and aging are treated. In the second part, the contribution of the School of Milano to the knowledge of the biomechanics of walking and running is analyzed, with special attention to the fundamental role played by Giovanni Cavagna. The pendulum-like mechanism of walking and the elastic mechanism of running are described. Concepts like internal and external work and mechanical efficiency are discussed. Further, the transition between walking and running is discussed along the lines proposed by Alberto Minetti. Finally, the energetics of sprint running and the effects of acceleration are discussed.

P. E. di Prampero
Emeritus Professor of Physiology, University of Udine, Udine, Italy
e-mail: Pietro.Prampero@uniud.it

G. Ferretti (✉)
Department of Molecular and Translational Medicine, University of Brescia, Brescia, Italy
e-mail: Guido.Ferretti@unibs.it

G. Ferretti (ed.), *Exercise, Respiratory and Environmental Physiology*, Perspectives in Physiology, https://doi.org/10.1007/978-3-031-19197-8_5

Three athletes during training for the marathon race of the 1896 Athens Olympic Games, on the road from Marathon, Greece. The athlete in the middle is Charilaos Vasilakos (1877–1964), a student from Athens, who ended at the second place, after the shepherd Spyridon Louis (1873–1940). Property of WikiCommons, in free domain

5.1 Introduction

The aim of the present chapter is to subsume under a coherent picture the vast collection of data on the bioenergetics and biomechanics of walking and running, arising from the numerous papers devoted to these topics by the School of Milano, along the lines of the seminal study by Rodolfo Margaria (1938) on the energy cost of walking and running. More specifically, after a brief historical note, we discuss the basic principles along which an analysis of human locomotion can be performed, and then we describe more in detail the energetics and biomechanics of walking and running. The effects of the incline are analyzed for both. Competitive walking is also considered. The School was active in the study also of other types of locomotion, both on land (speed skating, skiing, but mostly cycling) and in water (swimming, rowing and other forms of assisted locomotion), along the same general principles. These are discussed in Chap. 6.

5.2 The Founders

It is well known since the Hellenistic era that muscles are the site of animal movement. Herophilus (≈325 – ≈255 B.C.), one of the founders of the medical school of Alexandria, was already aware that muscles can shorten and lengthen. Although he is credited to have been one of the first to establish a bridge between hard science and medical science, we are not aware as to whether he created what we would call nowadays a physical theory of animal motion. Most of his work has been lost and present information on him is indirect, from citation in other texts (Von Staden 1989). To set a theory of locomotion, we need an adequate physical system and the Hellenistic one was lost. It was only with the development of modern classical mechanics starting from the seventeenth century, and the creations of concepts like work and power, that we got an instrument for the analysis and interpretation of animal locomotion.

As already pointed out in Chap. 2, the second necessary instrument came in the eighteenth century, with the discovery of oxygen and carbon dioxide. Antoine-Laurent de Lavoisier (1743–1794) was the first to demonstrate that both gases are necessary to life, as humans consume oxygen and produce carbon dioxide; but even more important, he demonstrated, together with Armand Séguin (1767–1835), that oxygen consumption, carbon dioxide elimination and heat production all increase in exercise, in rough proportion between them (Perkins Jr 1964). This observation implies three consequences: (i) chemical energy transformation is an oxidation process of some "organic" fuel, with carbon dioxide as end product; (ii) there must be a tight coupling of oxygen consumption to fuel degradation, and (iii) the chemical energy transformation into mechanical energy occurs during muscular contraction. The basis for an energetic analysis and interpretation of human locomotion was set.

It does not come as a surprise that the earlier attempts at describing the energetics of human locomotion in quantitative terms were devoted to walking. Indeed, in the second half of the nineteenth century, several authors determined the energy expenditure during level walking, wherefrom the energy cost of walking (C_w) on the level at speeds between 3.8 and 4.8 km h^{-1} (1.05–1.33 m s^{-1}) can be calculated (Katzenstein 1891; Smith 1859; Sondén and Tigerstedt 1895). The obtained values were astonishingly close to the currently accepted ones (from 0.32 to 0.51 kcal kg^{-1} km^{-1}, i.e., from 1.34 to 2.13 J kg^{-1} m^{-1}). The mechanical efficiency of uphill walking, as obtained from the ratio of the potential energy (E_p) changes to the corresponding metabolic energy expenditure, was also determined, obtaining values ranging from ≈ 0.20 to ≈ 0.37 (Katzenstein 1891; Löwy et al. 1897; Schumburg and Zuntz 1896). At the beginning of the twentieth century, Brezina and Kolmer (1912) reported that the energy cost of level walking, per unit of transported mass and distance, increases sharply at speeds >1.33 m s^{-1}. Shortly afterward, Barkan et al. (1914) observed that the energy expenditure of level walking at speeds between 0.8 and 1.3 m s^{-1}, at the altitude of the "Laboratorio Angelo Mosso" (2900 m above sea level, see Chap. 1), was essentially equal to that observed by others at sea level

(0.37–0.55 kcal kg^{1} km^{1}). Similar results were reported also by Benedict and Murschhauser (1915).

The energy cost of running (C_r) can be assessed from the studies by Waller (1919) and by Liljestrand and Stenström (1919), whence values ranging between 0.80 and 1.30 kcal kg^{1} km^{1} were obtained, again not far from the presently accepted ones; whereas Hill (1928) was probably the first to investigate the effects of the air resistance in running. However, as mentioned above, the most comprehensive study on the energy cost of walking and running was undertaken by Margaria (1938). Margaria's study was a turning point of the history of the energetics of human locomotion, for the formalization of the concept of energy cost (C), which is still at the core of our interpretation of the energy transformations during walking and running, and actually during any type of locomotion.

The study of the biomechanics of human locomotion took great advantages from the technological revolution that occurred in the second half of the nineteenth century, thanks to the dramatic development of photography. Several attempts at reproducing movement using photographic tools preceded the Lumière brothers' invention of what we call *cinema*. Two men, interested in understanding human movement, had a major role in these developments: Etienne-Jules Marey (1830–1904), a French physiologist, and Eadweard Muybridge (1830–1904), a British photographer who practised in the USA, who, amazingly enough, shared the years of birth and death. The former invented the chronophotographic gun, a tool that allowed taking sequences of 12 frames per second reproduced on the same plate (classical movie cameras filmed at a frequency of 24 images per second); with this technique, Marey (1878) decomposed animal movement to get detailed analysis of it. Similarly, Muybridge placed 12, then 24 cameras at given distances, taking pictures at given predetermined times along the path of a moving animal. He copied his images onto a spinning disc allowing projection at predetermined speed of the sequence of images through a source of light: by doing so, he reproduced motion at a pre-cinematographic speed (Muybridge 1887).

Their techniques opened the way not only to modern cinema, but also to the cinematographic techniques of locomotion analysis, which allowed Wallace Fenn (1893–1971) to perform the first computations of mechanical work during walking and running (Fenn 1930a, 1930b). Pretty much as Margaria's 1938 study for the energetics, Fenn's papers represented a turning point of the history of the biomechanics of human locomotion. Incidentally, Margaria and Fenn were friends. Fenn later created Rochester, which then spread to Buffalo: their friendship is the base of the Buffalo—Milano connection.

5.3 Definitions and General Concepts

The maximal absolute velocities in human locomotion encompass a wide range of values (Fig. 5.1), whereas the maximal muscular power of the athletes competing in the different events, for any given effort duration, is substantially equal. Thus, the

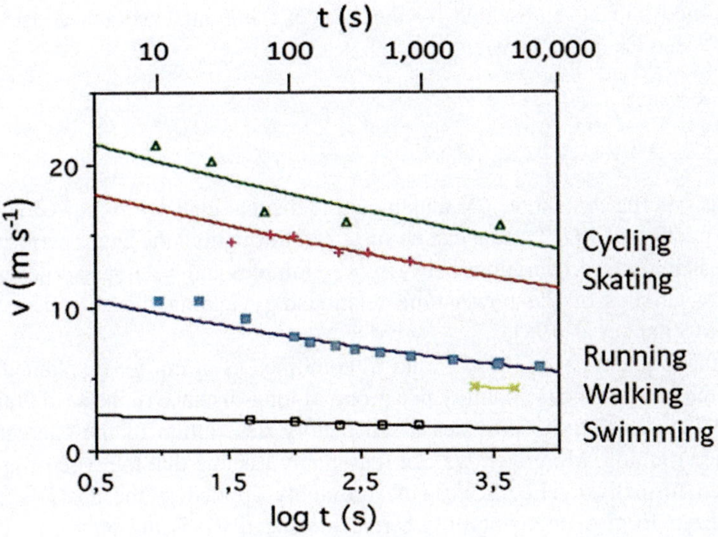

Fig. 5.1 Average speed (v) of some extremely high performances, or 2014 world records, as a function of time in the indicated forms of human locomotion. Since we refer to records, walking is competitive walking. The time is expressed in seconds (upper abscissa, log scale) or as the corresponding logarithm, lower abscissa. Modified after di Prampero 2015

above speed differences cannot be attributed to the athletes' "engine"; on the contrary, they depend on the set of intrinsic characteristics of any given form of locomotion that determine *C*, is usually expressed in kJ km^{-1}. However, to compare subjects of different body size, it is often convenient to express *C* per kg of body mass (J kg^{-1} m^{-1}). It should also be pointed out that 1 kcal = 4.186 kJ, and that the consumption of 1 L of oxygen in the human or animal body yields ≈ 5 kcal (≈ 21 kJ), the precise value ranging from 4.68 to 5.05 kcal L^{-1} depending on the substrates utilized, and hence on the respiratory quotient (Margaria 1975).

The rate of energy expenditure per unit of time (metabolic power, $\dot{E}$) is the product of *C* times the velocity (*v*):

$$\dot{E} = C\, v \tag{5.1}$$

Equation 5.1 tells that, in any form of locomotion, and for any given *C* value, the maximal attainable speed (v_{max}) is directly proportional to the subject's maximal metabolic power ($\dot{E}_{max}$). Solving Eq. 5.1 for v_{max}, we obtain:

$$v_{max} = \frac{\dot{E}_{max}}{C} \tag{5.2}$$

Equation 5.2 applies regardless of the energy sources yielding $\dot{E}_{max}$; however, it becomes particularly useful in long-duration aerobic performances, wherein $\dot{E}_{max}$ is well defined and is represented by the subject's maximal oxygen consumption ($\dot{V}O_2{}^{max}$, see Chap. 7). Hence:

$$v_{max} = \frac{\varphi\, \dot{V}O_2^{max}}{C} \tag{5.3}$$

In this specific context, v_{max} is tantamount to the maximal aerobic speed, whereas φ is the fraction of $\dot{V}O_2{}^{max}$ that can be sustained throughout the entire performance. We note that, for effort durations between ≈ 30 minutes and ≈ 2 hours, φ depends on the characteristics of the power-time relationship yielding the so-called critical power (Moritani et al. 1981).

The three quantities φ, $\dot{V}O_2{}^{max}$ and C, combined as in Eq. 5.3, explained about 70% of the performance variability in a group of long-distance runners (di Prampero et al. 1986). So, Eq. 5.3 provides a satisfactory description of the energetics of endurance running. Moreover, we can reasonably assume that the type of analysis expressed formally by Eq. 5.3 can be profitably applied to the analysis of any standardized form of locomotion, wherein the speed: (i) is the only criterion for evaluating performances, and (ii) depends directly on the subject's metabolic power. Thus, Eq. 5.3, the first formulation of which we owe to Margaria (1938), has been the basis for the analysis of the energetics of human locomotion carried out in the last 80 years within the School of Milano, as discussed in the next paragraphs.

In all forms of terrestrial locomotion, a fraction of C is utilized to overcome non-aerodynamic forces. This is defined as the non-aerodynamic energy cost (C_{n-a}) and includes the fraction of C that is necessary to overcome (i) gravity, to lift and lower, and inertia, to accelerate and decelerate the body center of mass at each stride; (ii) the friction of the point of contact with the terrain, whenever there is motion of one in respect to the other; (iii) the internal work necessary to move the limbs in respect to the subject's center of mass; (iv) the static muscular contractions to maintain the body posture; (v) the work of the respiratory muscles and of the heart to sustain respiration and circulation.

Over and above C_{n-a}, in all forms of terrestrial locomotion, an additional amount of energy is dissipated against the air resistance that increases with the square of v, where v is the velocity in respect to the air, not in respect to the terrain: these two velocities become equal in the absence of wind. In fact, the amount of C used against the air resistance, defined as the aerodynamic energy cost (C_a),[1] is given by:

[1] In fact, C is a force. Dimensionally and conceptually, a force (F) is equal to the work (w) or energy per unit of distance (length, L): $F = w\,L^{-1}$. This depends on the fact that the work is defined as the product of the force and the distance along the direction of motion. Expressing the quantity in question as force opposing motion, or as work, or energy, per unit of distance, is therefore conceptually identical.

Table 5.1 Constant relating the energy cost (expressed per unit of distance and per m^2 of body surface area) to the square of the speed (k_e); non-aerodynamic energy cost per unit of distance and per kg body mass (C_{n-a}, J m^{-1} kg^{-1}); overall energy cost per unit of distance for a man of 70 kg and 1.75 m (C, J m^{-1}) and aerodynamic coefficient (C_x) in the indicated forms of human locomotion on solid compact terrain at constant speed (v, in m s^{-1}) in the absence of wind at sea level (Pressure = 760 mmHg, Temperature = 20 °C). Cycling on a traditional racing bike in dropped posture. Data from di Prampero 1986

	Walking (competitive)	Running	Speed skating (ice)	Cycling
k_e	0.46	0.40	0.44	0.42
C_{n-a}	1,43 + 0,91 v	3,86	1,00	0,17
C	100 + 64 v + 0,83 v^2	270 + 0,72 v^2	70 + 0,79 v^2	13 + 0,77 v^2
C_x	1.10	1.10	0.50	0.75

$$C_a = k_e\, v^2 \tag{5.4}$$

where k_e is the proportionality constant relating the two variables. It depends on (i) the area projected on the frontal plane (A_f), (ii) the body shape, (iii) the air density (ρ), and (iv) the efficiency of transformation of the metabolic energy into mechanical work (η):

$$k_e = \frac{0.5\, C_x A_f\, \rho}{\eta} = \frac{k_w}{\eta} \tag{5.5}$$

where k_w is the proportionality constant relating the force opposing motion to the square of the speed, and C_x is a dimensionless coefficient, which is smaller the more "aerodynamic" is the shape of the moving object. C_x is the ratio of the drags of two objects of equal A_f: one is the body at stake, the other is an object of standard shape with a squared frontal area. Indeed the drag of any given moving body varies as a function of the Reynolds number, and thus, ceteris paribus, it is greater, the higher is the speed of the body with respect to the fluid. When the drag of the body at stake is equal to that of the standard object, the C_x is equal to 1. This is the case for a cube. In the four forms of locomotion discussed in this Chapter and in Chap. 6, C_x ranges from ≈ 0.5 to ≈ 1 (Table 5.1), whereas in old fashion cars is ≈ 0.6 and slightly lower than ≈ 0.3 for contemporary streamlined cars.

To sum up, C is equal to:

$$C = C_a + C_{n-a} \tag{5.6}$$

Combination of Eqs. 5.4 and 5.6 yields:

$$C = k_e\, v^2 + C_{n-a} \tag{5.7}$$

A substantial body of data collected over the years has provided rather accurate values of k_e (Table 5.1). This table shows that C_{n-a} is independent of v in most

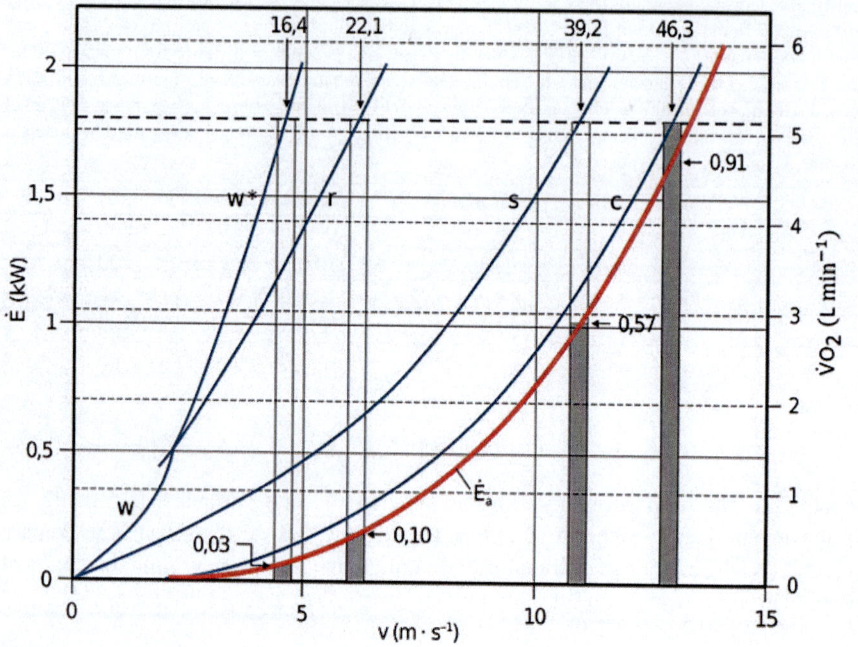

Fig. 5.2 Metabolic power ($\dot{E}$, kW, left ordinate) and corresponding oxygen uptake ($\dot{V}O_2$, L min^{-1}, right ordinate) above resting as a function of speed (v, m s^{-1}) during natural (w) or competitive (w*) walking, running (r), ice speed-skating (s) and cycling (c) in a standard subject of 70 kg body mass, 1.75 m stature, at constant speed on flat compact terrain, at sea level in the absence of wind. The red curve ($\dot{E}_a$) describes the power dissipated against the air resistance, essentially equal in the four forms of locomotion. The vertical distance between the red curve and the blue ones is the power dissipated against non-aerodynamic forces ($\dot{E}_{n-a}$). The upper black broken horizontal line is the net maximal oxygen consumption of an elite athlete (1.8 kW, or 5.17 LO_2 min^{-1}). The intersection between this line and the blue ones is a measure of the maximal theoretical aerobic speed (vertical arrows and corresponding velocities, in km hr^{-1}) of this hypothetical athlete in the four forms of locomotion. The dark columns indicate the fraction of the overall power dissipated against the air resistance at the maximal aerobic speed. Modified after di Prampero 1986

forms of human locomotion on land. In this case, plotting C as a function of v^2 provides straight lines with slope equal to k_e and y-axis intercept equal to C_{n-a}. However, this is not so for competitive walking, because in this case C_{n-a} is a linear function of v. At variance, when we walk or run on a treadmill, C_a is nil, so that C corresponds to $C_{n\text{-}a}$.

The overall $\dot{E}$, as given by the product between C and v in the absence of wind (Eq. 5.6), is reported as a function of v in Fig. 5.2 for natural or competitive walking, running, ice speed-skating and cycling (for details on these last two forms of locomotion, see Chap. 6) (di Prampero 1986). The two components of the overall metabolic power (aerodynamic, $\dot{E}_a$ and non-aerodynamic, $\dot{E}_{n-a}$) can be estimated for each form of locomotion, since the curve depicting the $\dot{E}_a$ versus v relationship, in red on the graph, is essentially the same for all types of locomotion.

This figure shows that, in competitive walking as in running, since $\dot{E}_a$ is but a minor fraction of $\dot{E}$, no substantial advantage can be obtained by ameliorating the subject's aerodynamic profile. On the contrary, as discussed in some detail in Chap. 6, an improvement of the aerodynamics of the bicycle provides astonishing increases of the speed on flat terrain, due to the large predominance of $\dot{E}_a$ over $\dot{E}_{n-a}$.

Figure 5.2 shows also the maximal aerobic speed that a hypothetical elite athlete could achieve in these four forms of locomotion, were he/she able to compete with the same technical ability in all of them. Indeed, the maximal aerobic speed in each form of locomotion is set by the intersection between the horizontal line representing the subject's $\dot{V}O_2\ ^{max}$ (above resting) and the appropriate metabolic power versus speed function.

It should finally be noted that all the functions shown in Fig. 5.2 for the reported locomotion modes had been obtained within the School of Milano, thanks essentially to the work of Rodolfo Margaria, Paolo Cerretelli, Pietro Enrico di Prampero, Gabriele Cortili (1939–2009), Piero Mognoni (1936–2008), and Franco Saibene (1932–2008). Figure 5.2 subsumes one of the most important legacies of the School of Milano to the physiological community: the theoretical and experimental basis of our understanding of the energetics of human locomotion, on which current studies still rely.

Combining Eqs. 5.1 and 5.4, one obtains:

$$\dot{E}_a = C_a\ v = k_e\ v^3 \tag{5.8}$$

It should be noted that Eq. 5.8, as such, applies only in the absence of wind. If this is not the case, the analysis becomes more complicated, as briefly discussed in Chap. 6.

A direct consequence of Eq. 5.8 is that, at speeds close to the maximal aerobic ones in elite athletes, $\dot{E}_a$ varies greatly among different exercise modes. As an example, at the $\dot{V}O_2\ ^{max}$ represented in Fig. 5.2, $\dot{E}_a$ is almost negligible in competitive walking (≈ 0.066 kW or 0.19 L O_2 min^{-1}), rather small in level running (≈ 0.19 kW or 0.54 L O_2 min^{-1}), much higher in speed skating (up to ≈ 1.08 kW or 3.10 L O_2 min^{-1}), to reach about 90% of the total metabolic power in cycling (≈ 1.71 kW or 4.92 LO_2 min^{-1}).

5.4 The Energy Cost of Walking and Running

The relationship between $C_{n\text{-}a}$ and v in natural level walking on the treadmill has a U shape with a minimum at a speed between 1.1 and 1.4 m s^{-1} (Fig. 5.3a). On the contrary, in level running, $C_{n\text{-}a}$ is essentially constant and independent of the speed (Fig. 5.3b), amounting to about 3.8–4.1 J kg^{-1} m^{-1}, a value about twice that observed in level walking at the most economical speed (Margaria et al. 1963). However, above ≈ 2.2 m s^{-1}, the $C_{n\text{-}a}$ of walking on flat terrain becomes higher than that of running. Indeed, it seems that, at speeds around 2.2 m s^{-1}, uninformed

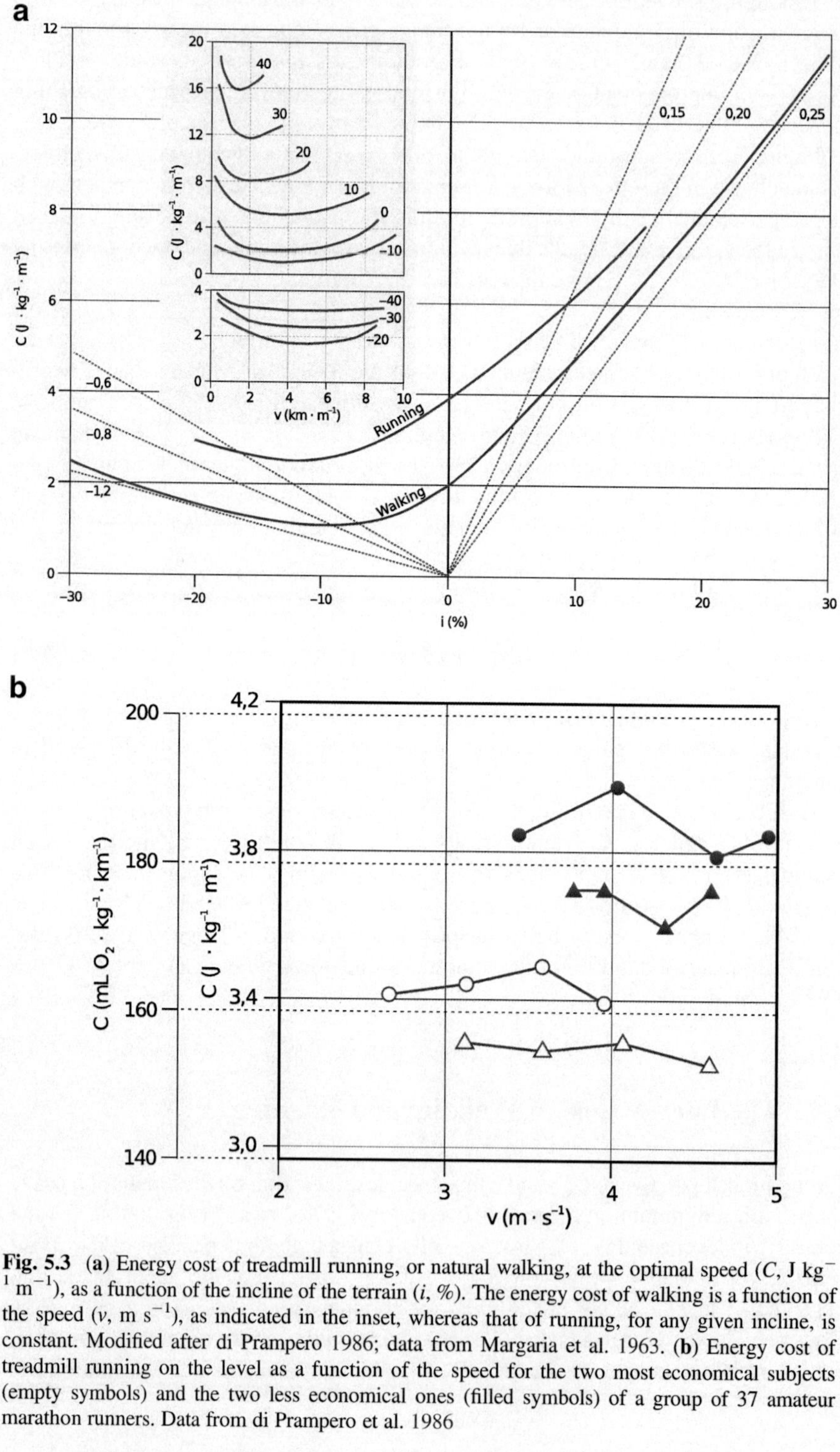

Fig. 5.3 (**a**) Energy cost of treadmill running, or natural walking, at the optimal speed (C, J kg^{-1} m^{-1}), as a function of the incline of the terrain (i, %). The energy cost of walking is a function of the speed (v, m s^{-1}), as indicated in the inset, whereas that of running, for any given incline, is constant. Modified after di Prampero 1986; data from Margaria et al. 1963. (**b**) Energy cost of treadmill running on the level as a function of the speed for the two most economical subjects (empty symbols) and the two less economical ones (filled symbols) of a group of 37 amateur marathon runners. Data from di Prampero et al. 1986

subjects shift spontaneously from walking to running (or vice-versa, if the speed is decreasing). The way this transition occurs is discussed more in detail in Sect. 5.6.

The same applies, in an even more elaborate way, in animals, like dogs and horses, characterized by three gaits: pace, trot, and gallop. The shift from one to the other of these forms of locomotion, as a function of the speed, occurs always in such a way as to select the most economical gait at the speed in question.

However, with a little bit of patience, it is possible to persuade humans or horses to utilize un-economical gaits: trot races for horses, competitive walking for humans. In this latter case, C, even if higher than that of running, is nevertheless substantially lower than could be expected from the extrapolation to high speeds of the C versus v relationship for natural walking. This feat is obtained thanks to the well-known technique of locomotion of competitive walking (Cappozzo et al. 1976; Cavagna and Franzetti 1981; Marchetti et al. 1983), somewhat funny at first glance, but energetically more economical than natural walking at the competition speed, even if not as much as level running.[2]

As far as C_a is concerned, Eq. 5.7 applies to running as well, as demonstrated on top-level marathon runners from Kenya (Tam et al. 2012), who could be investigated up to speeds as high as 5.0–5.5 m s^{-1} (18–20 km h^{-1}) in fully aerobic conditions. Concerning natural walking, C_a is so small, that it can be practically neglected, so that the C versus v relationship on a flat terrain (an athletic field, for instance) is essentially the same as that on the treadmill.

5.4.1 The Effect of the Incline

As mentioned above, in 1938, Margaria investigated in depth the energetics of walking and running at speeds between 0.14 and 2.64 m s^{-1} and slopes between +40 and – 40% in walking and between 1.80 and 4.20 m s^{-1} from +15 to – 30% in running, respectively. His data are reported in Fig. 5.3a. This Figure shows that $C_{n\text{-}a}$, as expected, increases uphill and decreases downhill; in this latter case, however, if the slope becomes steeper than – 15%, $C_{n\text{-}a}$ increases again. It shows also that in walking, the optimal speed, yielding the minimum $C_{n\text{-}a}$, is smaller, the steeper the incline (Fig. 5.3a). In running, at all slopes, the $C_{n\text{-}a}$ is independent of the speed (Fig. 5.3b). Finally, the $C_{n\text{-}a}$ of running turns out always higher than that of walking at the optimal speed; on flat terrain, as mentioned above, the former is about two times higher than the latter.

Minetti et al. (2002) studied C_w or C_r uphill or downhill in a much larger range of slopes than did Margaria, i.e., from – 45% to +45%. The so obtained data, on the one side, confirmed the data reported by Margaria (1938), as did also Minetti et al.

[2] At the 2014 European Championships in Zürich, Yohann Diniz from France established the current world record in the 50 km distance, which he completed in 3 h 32 min and 33 s. His average speed corresponded to 14.11 km h^{-1} or 3.92 m s^{-1}

(1994a), and, on the other side, have shown that the $C_{n\text{-}a}$ of treadmill running is independent of v even at these very steep uphill or downhill slopes. This has allowed condensing these data into one and the same empirical relationship (independent of the speed) between the net $C_{n\text{-}a}$ (in J kg^{-1} m^{-1}) and the incline of the terrain (i):

$$C_{n-a} = 155.5\, i^5 - 30.4\, i^4 - 43.3\, i^3 + 46.3\, i^2 + 19.5\, i + 3.6 \qquad (5.9)$$

where i is the tangent of the angle β between the terrain and the forward direction of the treadmill's movement (positive for uphill and negative for downhill), and 3.6 is the $C_{n\text{-}a}$ value observed on their subjects during level running at constant speed.

Knowledge of $i = \tan\beta$ allows a trigonometric computation of the vertical distance h, whence the work against gravity (Mgh, where M is the body mass and g is the gravity acceleration) can be obtained. In fact h is equal to:

$$h = L_{ho} \tan\beta = L_t \sin\beta = L_{ho}\, i \qquad (5.10)$$

where L_{ho} is the distance covered on the horizontal plane (that of the floor on which the treadmill is placed) and L_t is the distance covered along the treadmill's axis (direction of movement),[3] so that:

$$L_t = L_{ho} \frac{\tan\beta}{\sin\beta} \qquad (5.10a)$$

Equation 5.10 shows also that i is the h per unit L_{ho}, so that the abscissa of Fig. 5.3a is a representation of h, and thus, for any given human on Earth, of Mgh. The ordinate of this same figure reports the corresponding $C_{n\text{-}a}$, so that, when both $C_{n\text{-}a}$ and Mgh are expressed per unit of distance, their ratio is the reciprocal of the apparent net mechanical efficiency of the vertical work (η_v) of walking (at the optimal speed), or of running. Thus, in Fig. 5.3a, we can draw a family of isopleths departing from the origin of the axes (three of which are shown in the Figure), along which η_v is constant. However, h is expressed per unit of L_{ho}, whereas $C_{n\text{-}a}$ is expressed per unit of L_t. For this reason, since, when $\beta > 0$, and thus $L_t > L_{ho}$, the nominal η_v turns out larger than the real η_v, by a factor equal to $\tan\beta/\sin\beta$ (Eq. 5.10a). For inclines ≤ 0.25, corresponding to $\beta \leq 14.04$ deg., $\tan\beta/\sin\beta \leq 1.03$. Hence, within this range of inclines, the above approach does not lead to

[3]Consider as an example a 10% slope: under these conditions the subject, for $L_{ho} = 1$ m, lifts his body mass by 0.1 m; as a consequence, the mechanical work performed against gravity (Mgh), expressed per unit body mass and distance, amounts to: (9.81 * 0.1) = 0,98 J/kg^{-1} m^{-1}. The y axis of Figure 5.3a shows that the energy cost corresponding to $i = 0.10$ amounts to 4,70 and 6,18 J kg^{-1} m^{-1} of L_{ho}, for walking and running, respectively. Hence, since $L_t = L_{ho} \frac{\tan\alpha}{\sin\alpha}$ (Eq. 5.10a), the corresponding efficiencies (η_v) amount to (0,98*1.005)/4,70 = 0,210 for walking at the optimal speed and to (0.98*1.005)/6.18 = 0,159 in running. Neglecting $\frac{\tan\beta}{\sin\beta}$, one would obtain $\eta_v = 0{,}98/4{,}70 = 0{,}208$ for walking at the optimal speed and to 0,98/6,18 = 0,158 in running, a very negligible difference indeed.

any gross over-estimates of η_v; for steeper slopes, this error turns out important, so that it becomes necessary to apply this correction procedure.

When walking or running downhill, Mgh is negative, in so far as it is performed by gravity on the mass of the subject, whose muscles contract eccentrically to apply a brake to the fall of the body, thus absorbing mechanical work: hence the resulting η_v is also negative.

During locomotion on flat terrain at constant speed, both in walking and in running, η_v is nil, because no net vertical work is performed. On the contrary, when moving uphill, η_v does increase, tending to a value of the order of 0.25 for i values close to 0.25 in both walking and running (Fig. 5.3a). The η_v at these slopes corresponds well to the maximal efficiency of isotonic muscle contractions, in which the muscle performs only positive (concentric) work. Similarly, when moving downhill, η_v becomes progressively more negative, tending to a value between -1.0 and -1.2 for i values close to -0.20 (Minetti et al. 2002). This η_v corresponds well to the maximal efficiency of isotonic eccentric muscle contractions, in which the muscle performs only negative work. It is therefore reasonable to assume that, at the inclines of the terrain at which these asymptotic values of η_v are attained, the muscles perform only positive or negative work.

This state of affairs is likely the consequence of the biomechanics of walking and running. Indeed, when walking or running on flat terrain, the subject's center of mass is lifted and lowered in the vertical plane at each stride, over the same distance upward and downward (see Sect. 5.5). On the contrary, when moving uphill, the ascent of the body center of mass is greater than the subsequent descent, the opposite being true during downhill locomotion. It seems therefore reasonable to assume that, at the inclines at which the efficiency attains its asymptotic values, the descent (in uphill walking or running) or the ascent (in downhill walking or running) of the center of mass cease altogether.

5.4.2 Effects of the Terrain

The data reported so far, and summarized in Figs. 5.2 and 5.3, and in Table 5.1, were obtained during natural or competitive walking and in running on the treadmill or on smooth compact terrain. On irregular or soft terrain, such as on sand, grass, or snow, C increases by an amount that depends on the terrain characteristics. An example thereof is reported in Fig. 5.4 for natural walking and running on dry sand. Under these conditions, at variance with what happens on compact terrain, C_w increases monotonically with the speed and is 2 to 3 times greater than on compact terrain. Similarly, also C_r is about 1.5 times larger than on compact terrain (Lejeune et al. 1998; Pinnington and Dawson 2001; Soule and Goldman 1972; Zamparo et al. 1992).

This state of affairs is likely due to the fact that, on soft terrain, the pendulum-like mechanism that leads to the transformation of potential energy into kinetic energy, and vice-versa, in natural walking, and the recovery of elastic energy in running, are

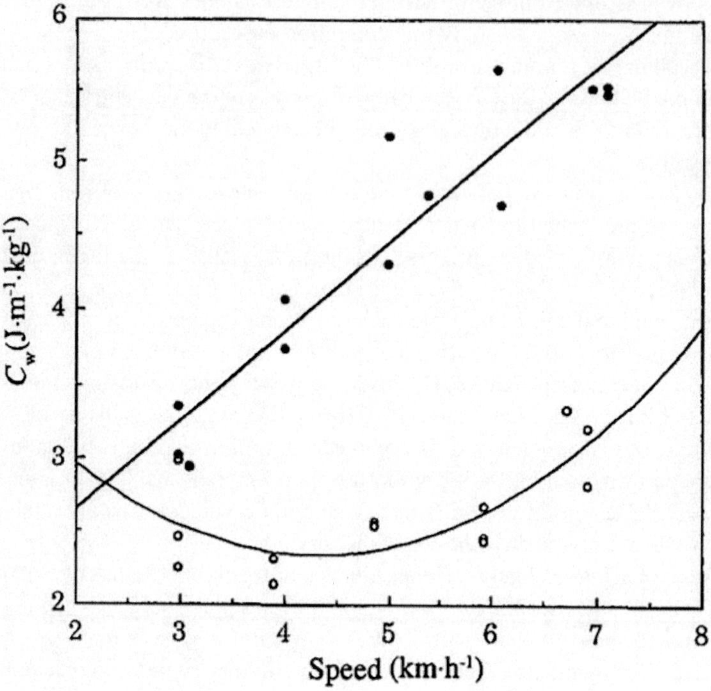

Fig. 5.4 Energy cost of natural walking (C_w, J kg^{-1} m^{-1}) on compact terrain (empty symbols), or on dry sand (filled symbols), as a function of the speed. From Zamparo et al. 1992

less effective than on compact terrain (see Sect. 5.6) (Pinnington and Dawson 2001). These mechanisms act optimally only if there is a firm grip of the foot on the ground. In the absence of this, the deficit of elastic energy recovery in running on sand implies increased muscle activation, associated with greater hip and knee range of motion than on firm terrain (Pinnington et al. 2005). On grass, C_r is about 20% larger than on compact terrain; this lesser increase, as compared to dry sand, is probably because grass is less compliant than dry sand, thus permitting a greater recovery of elastic energy (Pinnington and Dawson 2001).

5.4.3 *Pathologies of Locomotion*

C_w increases substantially also in numerous pathologies, both neurological (multiple sclerosis, Parkinson's disease, hemiplegic paralysis) and orthopedic (hip or knee prosthesis) (Fusi et al. 2002; Maggioni et al. 2012; Olgiati et al. 1986; Zamparo et al. 1995), by similar amounts at all walking speeds.

When comparing the C_w in pathological conditions to the normal physiological values, the effects of v cannot be neglected. To take this into account, Fusi et al.

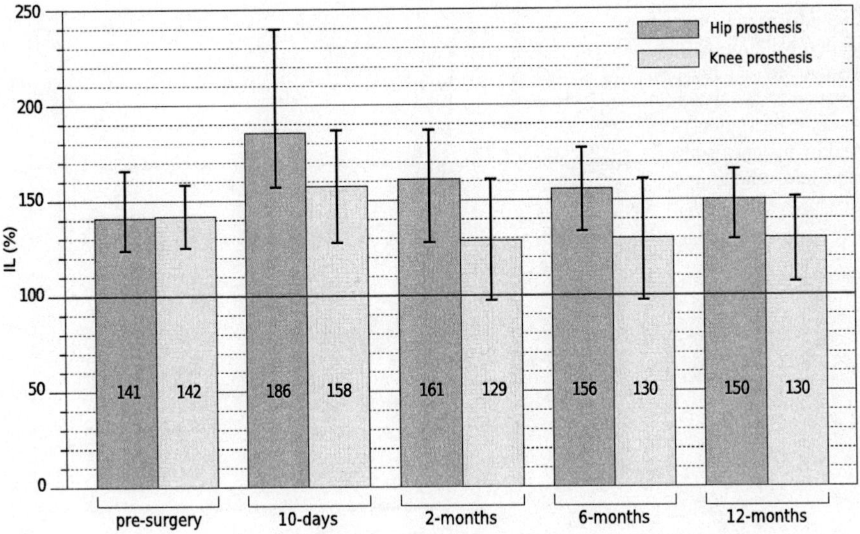

Fig. 5.5 "Index Locomotorius" (IL) before and after hip or knee prosthesis. Vertical bars indicate 1 standard deviation. Red horizontal line: IL in healthy subjects. From Fusi et al. 2002

(2002) proposed to utilize the so-called locomotor index, defined as the ratio between the energy cost in the investigated pathology and that observed in healthy subjects at the same speed. The evolution of the locomotor index as a function of time after surgical intervention is reported in Fig. 5.5 for hip or knee prosthesis carriers. It is noteworthy that, about 6 to 12 months after hip or knee prosthesis, these patients still have a locomotor index of 1.3–1.4, i.e., significantly higher than the physiological value of ≈ 1.0, even if the patient has completely recovered from surgery and is considered clinically "healthy." This is probably due to an abnormal gait pattern that the patient has acquired to minimize pain in the long period preceding the implant of the prosthesis (in the vast majority of instances the implant of the prosthesis is due to long-duration arthrosis processes of the hip or knee joint). This interpretation suggests that, in spite of the clinical recovery, the patient is not able, at least over relatively short periods of time, to resume the normal neuromuscular gait pattern.

On the contrary, in hemiplegic patients, the locomotor index is only marginally increased (≈ 1.1). This can presumably be attributed to the fact that the affected limb, even without contributing energetically to the progression, is nevertheless an effective "pillar" permitting the pendulum-like mechanism, an essential biomechanical mechanism optimizing C during natural walking. However, the muscular insufficiency of the affected limb necessarily leads to a slow speed; hence, the corresponding C moves up the ascending part of the curve relating C and v, thus leading to a larger energy expenditure per unit of distance (Fig. 5.6). As a consequence, in these patients, the overall C is greater than in normal subjects because of

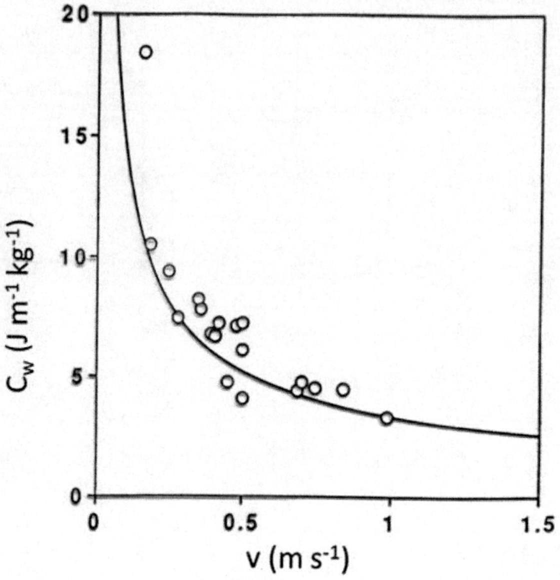

Fig. 5.6 Energy cost of natural walking (C_w) in hemiplegic patients, as a function of the speed (v). Average energy cost of healthy patients is also indicated (continuous curve). From Zamparo et al. 1995

their impossibility to attain the optimal walking speed (v_{opt}), a fact that inevitably leads to a greater sense of fatigue and hence to a sedentary life style.

5.4.4 The Effect of Body Mass and Age

In several animal species, the C_r per unit of body mass decreases with increasing the animal's mass, as indicated in Fig. 5.7 for a mass range from 20 g to 4000 kg. The relationship has an allometric exponent of −0.316, slightly higher than that calculated by Taylor et al. (1970) in the mass range from 10 g to 1000 kg (−0.40). However, within a given species, three of which are represented in the figure: humans, horses, elephants, C is essentially constant and independent of body mass. Coherently, Taboga et al. (2012) observed that, in severely obese subjects, whose body mass ranged from 109 to 172 kg, the C_r on the treadmill at 2.2 m s^{-1}, expressed per kg of body mass, was similar to that observed in non-obese subjects (4,10 ± 0,30 J kg^{-1} m^{-1}, obese, 3,90 ± 0,35 J kg^{-1} m^{-1}, non-obese). This is a common phenomenon when studying the effects of body mass, since allometric cross-species relationships encompass a very wide range of body mass values (in Fig. 5.7, the biggest body mass is 200,000 times larger than the smallest one), whereas within-species relationships cover a by far smaller body mass range (the body mass of the obese subjects was less than 60% higher than that of non-obese subjects in the study by Taboga et al. 2012).

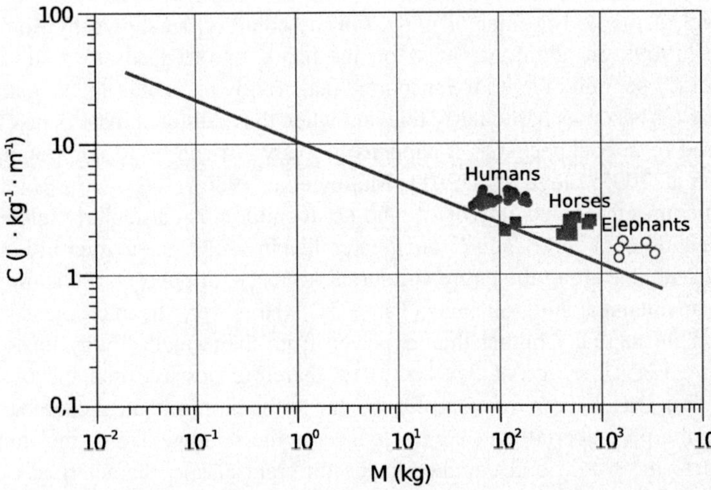

Fig. 5.7 Energy cost (*C*) of running in humans (full dots), horses (filled squares) and elephants (open dots) as a function of the body mass (*M*). Logarithmic coordinates. Straight line, described by $C = 10.7\ M^{-0.316}$, indicates the energy cost of running, as a function of the body mass, in several species of mammals. Modified after di Prampero 2015, data from Griffin et al. 2004; Langman et al. 2012; Taboga et al. 2012

Table 5.2 Energy cost of running and walking at the optimal speed on compact flat terrain at various ages. Mass and optimal walking speed are also indicated. Modified after di Prampero 2008

Age	Cost of running	Mass	Cost of walking at optimal speed	Optimal walking speed
(years)	(ml kg^{-1} m^{-1})	(kg)	(ml kg^{-1} m^{-1})	(m s^{-1})
5	0.390	20	0.09	1.0
7–10	0.245	28	0.09	1.3
11–13	0.212	37	0.09	1.3
14–16	0.195	60	0.10	1.4
Adult	0.182	70	0.10	1.4

At variance with these data, and in apparent contrast with the above considerations, other studies, all from the same group (Bourdin et al. 1993; Lacour et al. 1990; Padilla et al. 1992) observed a substantial reduction of C_r with increasing body mass (from about 30 to about 80 kg) in basketball players and in long-distance runners, similar to, albeit somewhat larger, than reported in Fig. 5.7. However, the high C_r observed by Bourdin et al. (1993) in their smallest subjects may be due, at least in part, to the fact that these subjects were also the youngest of the investigated group. Indeed, since the C_r in children and adolescents is greater than in adults (Table 5.2), we cannot exclude that the differences reported by these authors be due, at least in part, to the physiological variations of C_r itself with age.

Finally, Thorstensson (1986) showed that, both in adults and in children (10 years of age, 30 kg body mass), C_r decreases by 3–7% when the overall mass is artificially increased by 10% by adding an external load, provided that the added mass is placed at the level of the body center of mass. On the contrary, as shown by Soule and Goldman (1969), an additional mass on the hands or feet leads to a substantial increase of C_w per unit of overall transported mass (body mass plus added load). The increase of C_w becomes particularly relevant when the additional mass is positioned on the head or in backpacks, as it happens in many African or Alpine populations (Bastien et al. 2005; Lloyd et al. 2011; Maloiy et al. 1986).

In this connection, the study of C_w and C_r in children is particularly interesting; indeed, in children, both C_w and C_r are greater than in adults, due to their little size; it decreases with age along the entire childhood, to reach, at puberty, the adult value, which is maintained until old age (Table 5.2). However, the increase of C_r in children is substantially higher than expected from their smaller body mass alone (by 70 to 90% at the age of 5 years). It is therefore possible that the observed increase, or at least a large fraction of it, be due to the biomechanical characteristics typical of the pre-pubertal age, e.g., as concerns the relative size of the limbs, as compared to the whole body, and/or to the different elastic characteristics of the muscle tendon complex. A school backpack on children's shoulders increased C_w by about 7% (Merati et al. 2001), which appears in line with the observations on adults who carried additional loads far from the body center of mass (Maloiy et al. 1986; Soule and Goldman 1969).

To conclude this section, we note that Askew et al. (2011) showed that wearing medieval armors, the weight of which was on the order of 30–50 kg, led to an increase of C_w or C_r of about 1.9-to-2.3 times, as compared to "free" locomotion at the same speed. And this, without considering that a tight armor may also generate lung and chest wall restriction, with reduction of total lung capacity, inspiratory flow limitation and acute hypoventilation. It does not seem appropriate to open here a chapter on the "Physiology of History"; nevertheless it is interesting to consider whether and to what extent this enormous metabolic extra-load (for warriors and horses alike!) may have influenced the final outcome of the battles and wars, and hence of the history, of medieval times.

5.5 The Biomechanics of Walking and Running

5.5.1 *Definitions; Efficiency of Locomotion, Internal Work and External Work*

The quantitative assessment of the mechanical work performed, either absolute or per unit of distance, is often rather complicated. In all forms of locomotion, a fraction of the overall mechanical work is not utilized against external forces, nor does it lead to any changes in the position of the body center of mass in the vertical or horizontal plane (e.g., when an upper or lower limb is projected forward and the other symmetrically backwards). This fraction of the overall mechanical work, generally defined internal work (w_i), can only be assessed by means of graphical (video) analyses. Moreover, w_i includes the mechanical work of breathing and of the heart pump. The remaining fraction of mechanical work, called external work (w_e), can by determined not only by means of graphical (video) analyses, but also by means of traditional ergometric approaches and/or by means of force platforms (Cavagna 1975). Obviously enough, the overall mechanical work performed (w_{tot}) by the subject is the sum of internal and external work:

$$w_{tot} = w_i + w_e \tag{5.11}$$

This section is devoted to an analysis of w_i and w_e, and hence of w_{tot}, as a function of the speed during natural walking and running, based essentially on the data reported in a remarkable series of studies by Cavagna and co-workers that followed two seminal papers by Cavagna, Saibene, and Margaria (Cavagna et al. 1963, 1964). These papers initiated a glorious, long-lasting, highly influential scientific line of the School of Milano, the flag of which is still carried by Alberto Minetti, Paola Zamparo and their collaborators.

5.5.2 *Walking and Running on Earth*

In these two forms of locomotion, the center of mass of the body is lifted and lowered in the vertical plane, and accelerated and decelerated in respect to the average speed in the horizontal plane, throughout each stride. In this section, we consider only walking or running on flat terrain and at constant speed, in which case the lifting and lowering of the center of mass, as well as the acceleration and deceleration within each stride are equal in absolute terms.

At any given time instant, the total energy of the center of mass in motion is given by the sum of the changes (Δ) in kinetic energy (E_k) and in potential energy (E_p) (thus, ΔE_k and ΔE_p, respectively), in respect to a pre-defined reference value. If, in any given phase of the stride, the overall energy increases, we state, by definition, that w_e has been performed by the muscles on the center of mass. On the contrary, if the overall energy decreases, the system or parts thereof have absorbed external work; in this case we state, somewhat loosely, that the muscles have done negative w_e. Hence, the overall w_e performed at each stride can be expressed by (Cavagna 1969):

$$w_e = \Delta E_p + \Delta E_k \tag{5.12}$$

As such, w_e takes into account the work performed against gravitational and inertial forces, as well as that performed against the air resistance. Indeed, at each stride, the deceleration in the forward direction is greater, the greater the air resistance; as a consequence, at constant average speed the greater must also be the

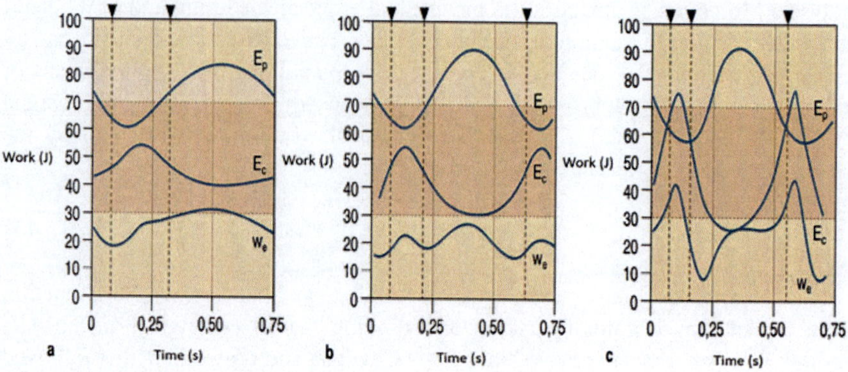

Fig. 5.8 Potential (E_p) and kinetic (E_c in the Figure, E_k in the text) energy (in J) of the body center of mass as a function of time (s) during walking at 0.9, (**a**) 1.4, (**b**) and 2.2, (**c**)m s^{-1}. At any given time during the stride cycle, the external work (w_e) is the sum $\Delta E_p + \Delta E_k$ (Eq. 5.12). The instant at which one foot touches the ground is indicated by the first arrow (and vertical dotted line); that at which the other foot is lifted above the ground is indicated by the second arrow (and vertical dotted line). Modified after Cavagna et al. 1963, Figure taken from di Prampero 2015 with the kind permission of Raffaello Grandi, Edi Ermes

subsequent acceleration, thus leading to an increased value of ΔE_k, and hence of w_e at each stride.

As mentioned above, after a given number of strides on flat terrain, the overall net external work is nil, since positive and negative work are equal in absolute terms. However, this is strictly true only if the air resistance is neglected; nevertheless, at the speeds prevailing in natural walking, the air resistance is so small, that its effects can indeed be neglected.

During natural walking, the center of mass is lifted in the initial phase of the stride, in which its E_p increases at the expense of the simultaneous decrease of E_k due to decreased forward velocity: hence, in this phase, ΔE_p is positive, whereas ΔE_k is negative. The so accumulated E_p is utilized in the immediately following phase to increase the forward velocity of the body: hence, in this second phase, ΔE_p is negative, whereas ΔE_k is positive.

In 1963 Cavagna, Saibene, and Margaria did show that, at v_{opt}, the E_p reaches its maximum almost at the very moment at which E_k attains its minimum and vice-versa (Fig. 5.8). Hence, the biomechanics of walking is characterized by pendulum-like energy exchanges, as in an inverted oscillating pendulum, or in an egg rolling on its major axis. In fact, in an oscillating pendulum, E_p attains its maximum when the sphere is at its highest position and stops (E_k is nil). This occurs at the two extremes of the trajectory of the pendulum, when the sphere is still, or when the major axis of the egg is vertical. On the opposite side, at the lowest point of the trajectory, the translational velocity is maximal (maximal E_k) and the E_p attains its minimum (lowest position of the sphere) (Fig. 5.9).

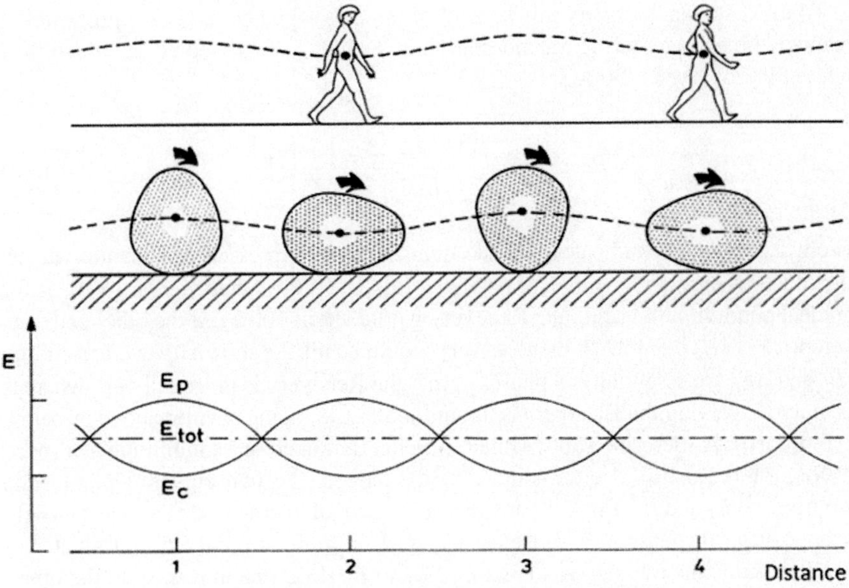

Fig. 5.9 Schematic representation of the potential, kinetic, and total energy (E_p, E_k and E_{tot}, respectively, arbitrary units) of the body center of mass as a function of the distance (arbitrary units) during walking. An egg rolling on its major axis represents a simplified model of walking at the optimal speed. In the rolling egg, as well as in walking, E_p attains a maximum when the center of mass reaches the highest point of its trajectory; at this same instant, E_k attains a minimum. In the absence of any frictional forces, as in an ideal pendulum, E_{tot} (as given by the sum of E_p and E_k) is constant, and the egg, once set in motion, would keep rolling forever thanks to the transformation of one energy form into the other. Redrawn after di Prampero 2015

In an oscillating pendulum, the curves of E_p and E_k are in perfect opposition of phase and of identical amplitude. Thus, in vacuo, were it possible to reduce the friction to zero, the total energy (sum of E_p and E_k) would be constant, so that an ideal pendulum would oscillate *in aeternum*. Obviously enough, this is not the case in natural walking, wherein ΔE_p and ΔE_k do not completely compensate each other, in so far as the respective curves are slightly out of phase and not perfectly symmetrical (Fig. 5.8). Therefore, the overall energy, as given by the sum of E_p and E_k, would decrease at each stride, were it not for the appropriate muscle work, performed on the center of mass, which is equal to the algebraic sum of the two curves. Hence, the w_e in natural walking depends on the absolute values of ΔE_p and ΔE_k, indicating the total w_e, as well as on the respective phase of the two curves (Fig. 5.8), indicating how much of the total w_e comes from the transformation of E_k into E_p and vice-versa. The difference between these two values is equal to the amount of energy that muscle contraction must add to keep w_e, and thus the average forward speed, constant. We define this extra-energy as w_m.

To express quantitatively this state of affairs, Cavagna et al. (1976) proposed to evaluate the efficacy of the transformation of E_k into E_p and vice-versa, which they defined Recovery, as follows:

$$Recovery = \frac{|\Delta E_p| + |\Delta E_k| - |w_m|}{|\Delta E_p| + |\Delta E_k|} \tag{5.13}$$

where the symbol | indicates that the quantities are expressed in absolute values, regardless of the sign, positive, or negative. Eq. 5.13 shows that, were $w_m = 0$, as for an ideal pendulum in vacuo, the Recovery would attain 100%. At the other extreme, were $w_m = |\Delta E_p| + |\Delta E_k|$, the Recovery would be nil. As shown below, in running, ΔE_p and ΔE_k are essentially in phase, so that the Recovery is practically nil, whereas for natural walking the Recovery is about 0.65 at v_{opt}, being smaller above or below it (Fig. 5.10). As described above, the C_w on flat terrain attains a minimum at a speed of about 1.1–1.4 m s^{-1} (v_{opt}, Figure 5.3a). Indeed, the two curves depicting the evolution of E_k and E_p (Fig. 5.8) are very nearly equal in amplitude and most closely in opposition of phase at v_{opt}. Hence w_m is at its minimum, whereas at greater or smaller speeds the two curves depicting E_k and E_p shift, one in respect to the other, and the more so the farther from v_{opt} (Fig. 5.8), thus leading to a progressively higher w_m and hence C_w than at v_{opt}.The relationship between Recovery and speed reported in Fig. 5.10 was obtained on young adults of average body size. However, in order to eliminate the effects of inter-individual differences, it can be normalized, expressing the speed itself as a function of the subjects' size. To this aim, we can use the theory of dynamic similarity, allowing a comparison of the behavior of objects of different size, the motion of which depends on the cyclic transformation of E_p into E_k and vice-versa, as typically happens in natural walking (Alexander 1989; McMahon and Tyler Bonner 1983). According to this theory, two geometrically similar objects of different size behave similarly when they move at the same Froude number (Fr):

$$Fr = \frac{v^2}{L\,g} \tag{5.14}$$

where L is the length of a relevant body segment (the lower limb in this specific instance). The Froude number, so defined in honor of the naval engineer William Froude (1810–1879), who utilized it to infer the characteristics of actual ships as obtained from small-size models, is a non-dimensional quantity, corresponding to the ratio of E_k to E_p.

When the Recovery is expressed as a function of Fr, the resulting relationship is substantially equal for children (aged 1 to 12 years), normal-size adults, hypophysarian dwarves and pygmies (Cavagna et al. 1983; Minetti et al. 1994c, 2000). In all these subjects, the Recovery attains a maximum on the order of 0.70 for $Fr \approx 0.25$ (Saibene and Minetti 2003). In addition, since v_{opt} corresponds to the speed at which the Recovery attains its maximum (Fig. 5.10), and, as mentioned above, this occurs at $Fr \approx 0.25$, we can write:

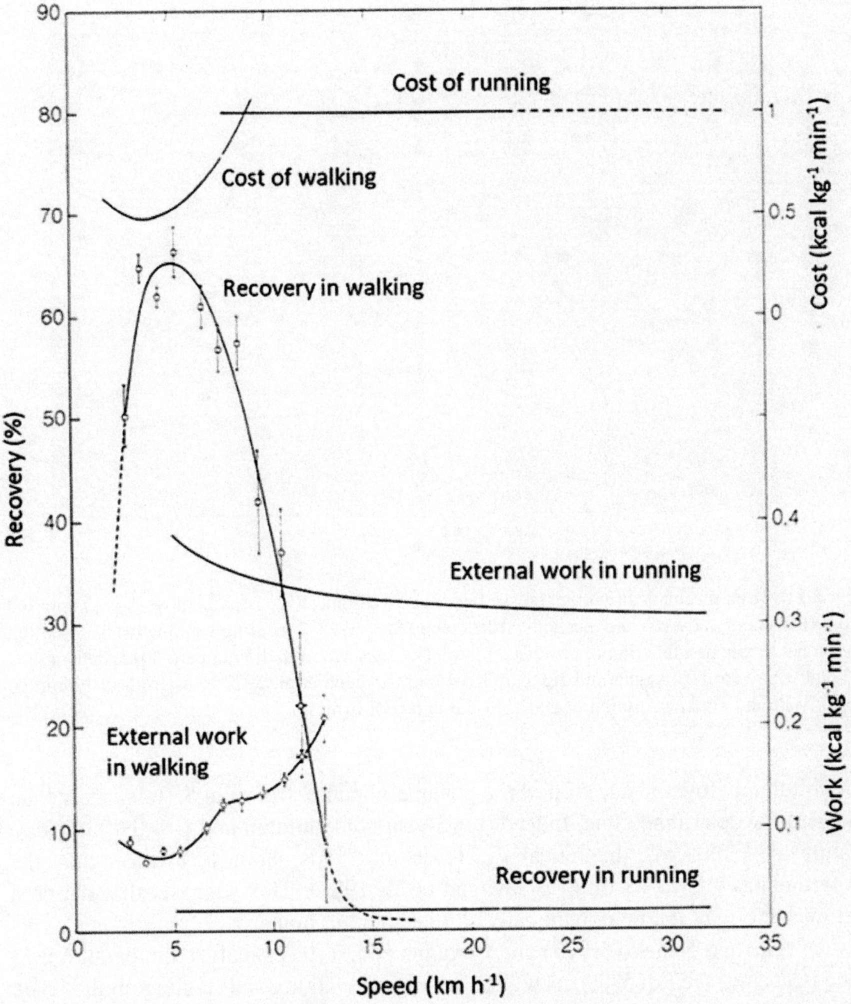

Fig. 5.10 Energy Recovery (see Eq. 5.13), external work and energy cost as a function of the speed in walking and running. From Cavagna 1988

$$\frac{{v_{opt}}^2}{L\,g} = 0.25 \tag{5.15}$$

whence

$$v_{opt} = 0.5\,\sqrt{L\,g} \tag{5.16}$$

Assuming $L = 0.9$ m (average lower limb length of adult males), Eq. 5.16 yields $v_{opt} = 1.49$ m s^{-1} (5,36 km h^{-1}), close to the value reported in Figure 5.3a. However,

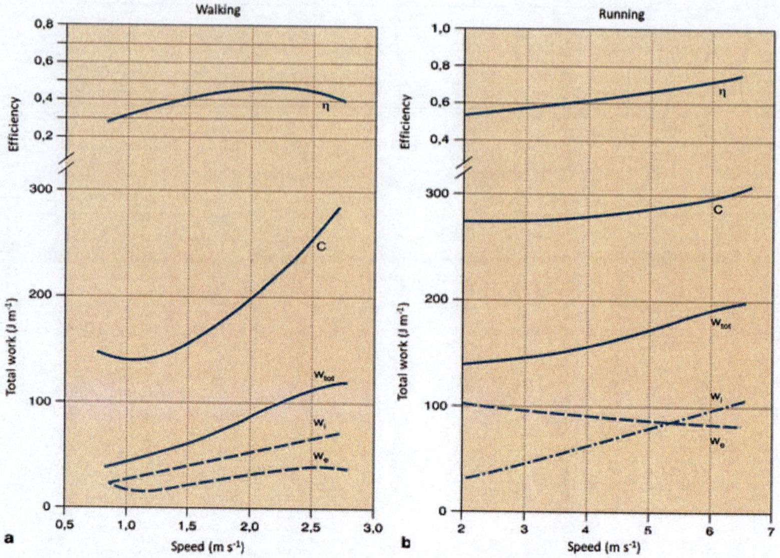

Fig. 5.11 Total mechanical work expressed per unit of distance (w_{tot}), external work (w_e), internal work (w_i), energy cost (C) and mechanical efficiency ($\eta = w_{tot}\, C^{-1}$) during walking (**a**) and running (**b**) on flat terrain as a function of the speed for a 70 kg subject. From di Prampero 2015; data of w_{tot}, w_e and w_i are from Cavagna and Kaneko 1977 and Cavagna et al. 1977; C is from di Prampero 1986. With the kind permission of Raffaello Grandi, Edi Ermes

for children below the age of 6, the v_{opt} value obtained from Eq. 5.16 is larger than the actually determined one. Indeed, for 3-years old children and $L \approx 0.45$ m (for a stature of $\approx$ 0.95 m), theoretical v_{opt} (1.06 m s^{-1}) is about 40% larger than the experimental one (0.75 m s^{-1}, Cavagna et al. 1983). This suggests that children below age 6 may not be dynamically similar to adult humans.

Cavagna and Kaneko (1977) and Cavagna et al. (1977) analyzed in detail E_p, E_k (and hence w_e, see Eq. 5.11), as well as w_i, during walking and running on the level. The so obtained data, summarized in Fig. 5.11, show that, in walking, w_{tot} increases with v:

$$w_{tot} = w_e + w_i = 0.07 + 0.6\, v \tag{5.17}$$

where w_{tot}, w_e and w_i are expressed in J kg^{-1} m^{-1} and v in m s^{-1}.

Since C_w as a function of v is known, Eq. 5.17 allows a calculation of η, as the ratio between w_{tot} and C_w (see Sect. 5.5.1). As shown in Figure 5.11a, η attains a maximal value of about 0.48 in a range of speeds between 2 and 2.5 m s^{-1}. This value is greater than that observed during isotonic muscular contractions ($\approx$0.25). As v is increased, the Recovery diminishes: at the speed of spontaneous transition between walking and running, it is about 0.50, and therefore, η is slightly lower than at v_{opt} (0.40 instead of 0.48).

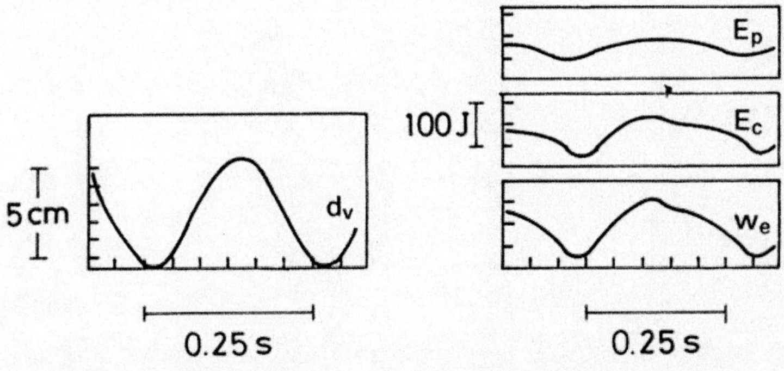

Fig. 5.12 Potential (E_p) and kinetic (E_c in the Figure, E_k in the text) energy, external work (w_e, J) and vertical displacement of the body center of mass (d_v, *cm*) as a function of time during running at 5.56 m s^{-1}. From Cavagna et al. 1964

When running on the level, at the onset of each stride the body center of mass is projected upwards and forwards by the muscle activity of the "pushing" limb. In the subsequent phase, after a brief time interval in which both feet are off the ground, the center of mass is lowered and decelerated, thanks to the muscle action of the "braking" limb. At the onset of the subsequent stride, the "braking" limb becomes the "pushing" limb. It follows that ΔE_p and ΔE_k are always in phase (Fig. 5.12). Hence, at variance with walking, the E_p accumulated in the initial phase of the stride cannot be used to accelerate the body forward in the subsequent phase: no transformation of one type of energy into the other is possible, and the Recovery, as described by Eq. 5.13, is always nil (see Fig. 5.10).

Also in running, as in walking, w_e is the sum of ΔE_p and ΔE_k (Eq. 5.12). However, since in running ΔE_p and ΔE_k are always in phase, at the onset of the stride, w_e has a large positive value, to reach an equally large negative value at the end of the same stride. Therefore, the active muscles must consume energy in the initial phase of the stride (push phase) to perform positive work, as well as in the final phase (braking phase) to absorb negative work. In the latter phase, a fraction of the negative work is stored in the series elastic elements of the contracting muscles.[4] The so accumulated elastic energy is utilized in the subsequent rebounding phase, when the absorbing limb becomes the pushing one. This leads to a substantial economy of metabolic energy, as shown by the high values of η observed in running (see Figure 5.11b). Hence, the biomechanics of running is similar to the bouncing of a rubber ball on the terrain. When the ball hits the ground, it is initially compressed, because of the simultaneous decrease of E_p and E_k, leading to a large negative value of w_e; in this phase, however, the decrease of w_e is partly counterbalanced by an increase of the energy stored in the elastic structures of the ball itself. In the following phase, this elastic energy is released with a simultaneous increase of both E_p and E_k (Fig. 5.13), allowing the ball to rebound. Cavagna et al. (1968) demonstrated a similar

[4]About 30% of the overall elastic energy is stored and released by the elastic structures of the arch of the foot (Ker et al. 1987).

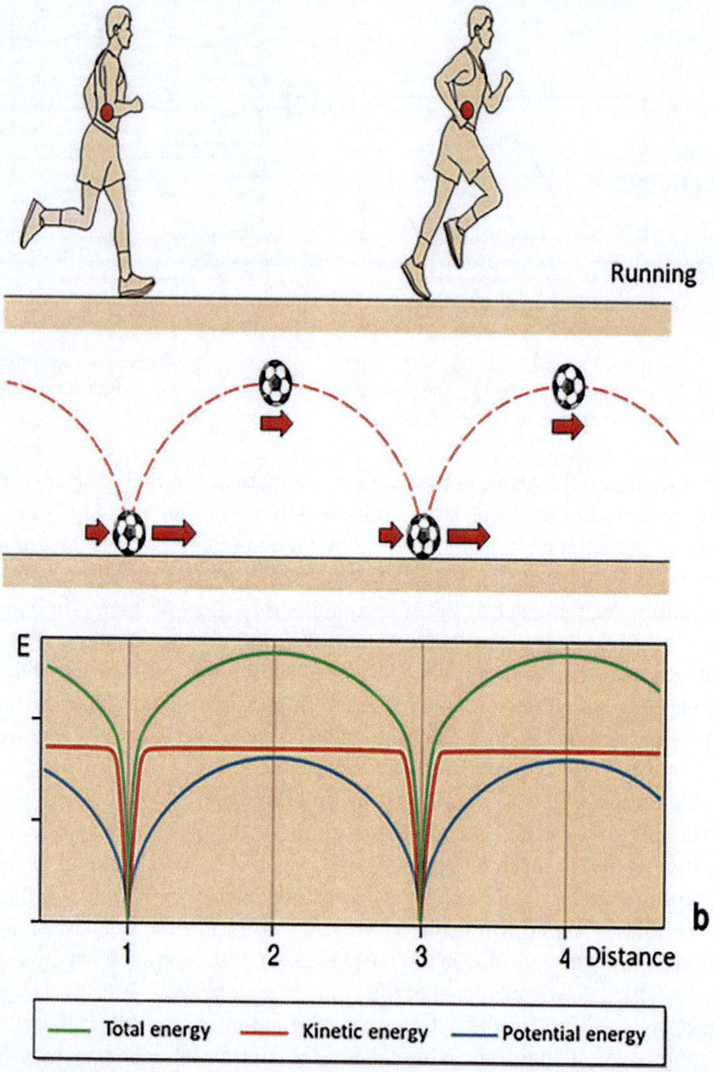

Fig. 5.13 A ball bouncing on the terrain represents a simplified model of running. When the ball hits the ground, the potential and kinetic energy attain simultaneously a minimum value and, for a very short time, the ball is motionless at the lowest point of its trajectory. In this phase, a fraction of the total energy is stored in the elastic structure of the ball itself; it is recovered in the immediately subsequent rebound phase, with a simultaneous increase of both potential and kinetic energy. In running, as well as in a bouncing ball, there cannot occur any transformation of potential into kinetic energy or vice-versa; however, the recovery of elastic energy in the rebound phase is a substantial "energy-sparing" mechanism. From di Prampero 2015, with the kind permission of Raffaello Grandi, Edi Ermes

mechanism in isolated muscles, which perform a greater work when they contract immediately after stretching: the elastic elements of muscle release the stored elastic energy at the onset of contraction.

A recovery of elastic energy analogous to that described above for running does occur also in other movements, in which the shortening of the muscle is immediately preceded by its stretching (eccentric contraction). For instance, a series of flexions and extensions of the knees performed without interruption, other things being equal, requires a lesser metabolic energy consumption than the same series of movements performed with a brief pause between flexion and extension. In this latter case, in fact, the elastic energy accumulated in the extensor muscles that during knee bending are forcefully stretched, is dissipated during the pause; hence, the concentric action that follows must be performed entirely at the expense of metabolic energy. As a consequence, the overall mechanical efficiency of the exercise is lower (η = 0,19) in this latter case, as compared to the same series of movements performed without interruption between flexion and extension (η = 0,26) (Thys et al. 1972). This shows that the recovery of elastic energy is possible only if the eccentric contraction (stretching of the muscle) is immediately followed by its concentric counterpart (shortening); if this is not the case, the elastic energy is dissipated as heat before the subsequent concentric contraction starts.

w_e and w_i are reported as a function of v during level running in Figure 5.11b (Cavagna and Kaneko 1977). Their sum (w_{tot}) is described by the following empirical equation:

$$w_{tot} = 0.66\, v^{-1} + 1.19 + 0.24\, v \tag{5.18}$$

where w_{tot} is in J kg^{-1} m^{-1} and v in m s^{-1}. The sum of the first two terms of Eq. 5.18 is a measure of w_e, which is essentially independent of speed in physiological human running conditions, since in the velocity range (2.8–8.3 m s^{-1}) investigated by Cavagna and Kaneko (1977), it decreases only from 1.43 to 1.27 J kg^{-1} m^{-1}. On the contrary, w_i, represented by the sum of the second and third terms of Eq. 5.18, increases linearly with v.

Knowledge of w_{tot} and of C_r allows calculation of η for level running: as shown in Figure 5.11b, η increases with v from ≈ 0.5 at $v \approx 2$ m s^{-1} to ≈ 0.7 at $v \approx 6.5$ m s^{-1}. These high values of η, substantially larger than the maximal ones observed during purely isotonic muscle contractions, demonstrate that the recovery of elastic energy occurring at each stride in running is larger, the higher the running speed.

The recovery of elastic energy is an extremely useful mechanism for reducing C_r: it occurs to a various extent in several mammalian species (Cavagna et al. 1977), but it reaches an extraordinary extent in kangaroos, which can easily reach speeds of 40–50 km h^{-1} (11–14 m s^{-1}) that, however, they maintain for short time periods only (Dawson 1977). Indeed, at each bout, the thick tendon of the tail of the kangaroo becomes an effective spring (Alexander and Vernon 1975). Thanks to this "tail effect," the recovery of elastic energy increases with the speed, so much so, that the C_{n-a} of "running" (hopping) in this animal species decreases from ≈

7.5 J kg^{-1} m^{-1} at 2.8 m s^{-1} to ≈ 2.8 J kg^{-1} m^{-1} at 6.4 m s^{-1}. It follows that, in this speed range, the $\dot{E}$ of "running" in the kangaroo is constant, amounting to about 20 W kg^{-1} (60 mlO_2 kg^{-1} min^{-1}).

The $\dot{V}O_2\ ^{max}$ of contemporary top-level long-distance runners is ≈ 70 ml kg^{-1} min^{-1} above resting (Tam et al. 2012). Assuming that no recovery of elastic energy occurs during running, and thus $\eta = 0.25$, Eq. 5.18 shows that a hypothetical athlete with these characteristics would attain at $\dot{V}O_2\ ^{max}$ a speed of 2.9 m s^{-1}, and thus would cover the 5.000 m distance in 28 min 44.14 s, as compared to the actual world record of 12 min 35.36 s. This back of the envelope calculation stresses dramatically the great importance of elastic energy recovery at each stride in running humans.

Recognizing the importance of the role of the recovery of elastic energy in setting C_r further highlights, in terms of athletic performances, the roles of training regimes leading to appropriate increases of the maximal muscular power. Indeed, muscle fiber composition and specificity of training do affect muscle performance as well, in so far as training may increase the number of active muscle fibers and lead to the metabolic potentiation of specific fiber types. Hence, since the mechanical characteristics of the various fiber types are diverse, especially in terms of maximal contraction velocity (Bottinelli 2001), it appears reasonable to suggest that the effects on the recovery of elastic energy provided by a training regime based on long-term high-intensity aerobic exercise, may be different from those provided by a training regime based on explosive exercise and force development.

The effects of the recovery of elastic energy on the running economy, and hence on performance, have also been discussed in terms of the potential advantage that artificial metallic prostheses, to be utilized only in competitions, may offer to athletes having undergone mono- or bi-lateral limb amputations. Indeed these prostheses may ameliorate artificially the athlete's running economy thanks to an increased recovery of elastic energy (as compared to natural running), thus constituting a kind of "mechanical doping." This line of reasoning led to the exclusion of athletes suffering from this type of handicap from competitions, as was the case of Oskar Pistorius being excluded from the 2008 Olympics in Peking.

5.6 The Spontaneous Transition between Walking and Running

In Sect. 5.5, we underlined that at speeds around 2.2 m s^{-1}, uninformed subjects shift spontaneously from walking to running (or vice-versa, if the speed is decreasing). This speed was defined as the speed of spontaneous transition between walking and running. The traditional interpretation assumes that humans always choose the most economical gait. On this basis, Margaria et al. (1963) hypothesized that at speeds higher than the spontaneous transition speed, C_r started to be lower than C_w, the opposite being the case at speeds below the transition speed.

The spontaneous transition between walking and running does not occur at a fix speed, but in a speed range between 1.8 and 2.5 m s^{-1} (Saibene and Minetti 2003). This may depend, at least in part, on the geometric characteristics of the subjects, since the spontaneous transition seems to occur at a speed that is lower, the shorter the limb (Thorstensson and Roberthson 1987).

Minetti et al. (1994b) carried out a systematic analysis of the transition speed during treadmill running. According to them, the speed of spontaneous transition is slightly lower than that at which walking and running have the same $C_{n\text{-}a}$. On this basis, they suggested that the transition may not have an energetic origin only. When passing from walking to running, there is a sudden significant, though small, increase of speed, and the subject suddenly moves at a higher step frequency. So, they proposed a hypothesis involving both mechanical and energetic factors. They plotted C as a function of v after expressing C in Joules per kg per *step* instead of J kg^{-1} m^{-1}, as customary. Because of the differences in step frequency in the transition zone, it turned out that the speed of transition was the speed at which both walking and running had the same C per step (see Fig. 5.14). This observation is still partly compatible, although on a slightly different basis, with Margaria's hypothesis.

The question remains as to how a step can be "sensed" such as to determine a gait transition. More favorable patterns of neuromuscular activation during running at the transition speed and a possible mechanical reflex arising from lower limb proprioceptors have been hypothetically evoked (Prilutsky and Gregor 2001; Segers et al. 2013).

5.7 Sprint Running and the Role of Acceleration

The massive utilization of anaerobic sources and the short duration of sprint running events make any direct measurements of $\dot{E}$ under these conditions rather problematic. So far, the energetics of sprint running has been mainly estimated indirectly from biomechanical analyses (Cavagna et al. 1971; Fenn 1930a, 1930b; Mero et al. 1992; Plamondon and Roy 1984). Otherwise, it was assessed by means of rather indirect procedures (Arsac 2002; Arsac and Locatelli 2002; di Prampero et al. 1993; van Ingen Schenau et al. 1991, 1994; Ward-Smith and Radford 2000).

An alternative approach is to assume that, during the acceleration phase, sprint running on flat terrain is biomechanically equivalent to uphill running at constant speed, the slope being dictated by the forward acceleration, and that, conversely, during the deceleration phase it is biomechanically equivalent to running downhill. Since the energy cost of uphill (downhill) running at constant speed over a fairly large range of inclines is rather well know (Minetti et al. 2002), the energy cost of accelerated (decelerated) running, can be easily estimated once the acceleration (deceleration) is known (di Prampero et al. 2005).

Figure 5.15 (left panel) shows that a runner accelerating on flat terrain leans forward, so that the angle α between the runner's mean body sagittal axis and the

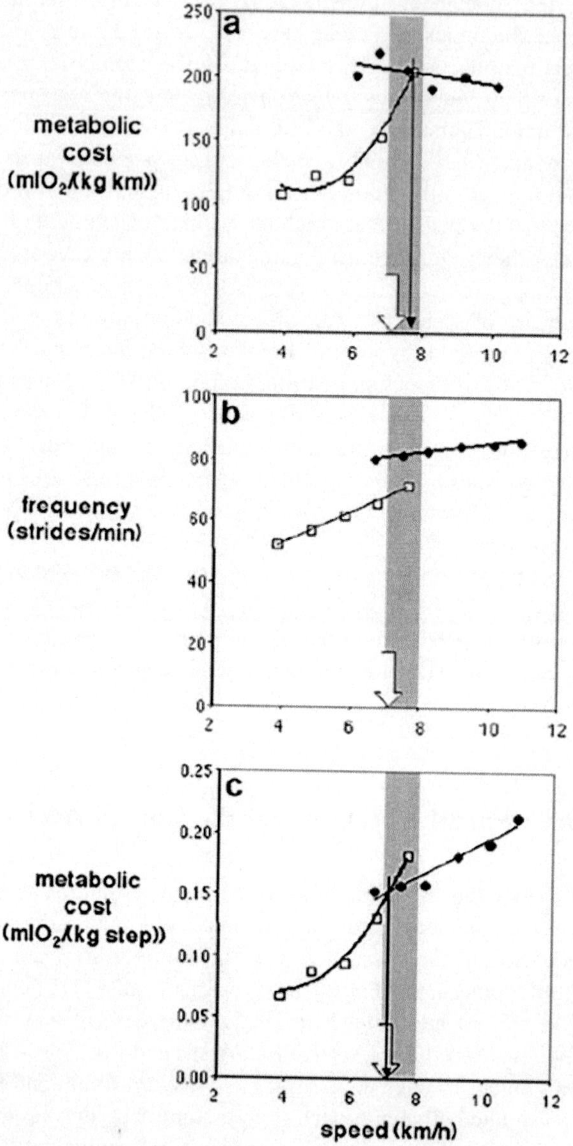

Fig. 5.14 An analysis of the spontaneous transition speed (thick white arrow) between walking (unfilled squares) and running (filled circles). (**a**) The regression curves of energy cost per unit distance, intersect at a higher speed (thin arrow) than the spontaneous one. (**b**) Differences in stride frequency between walking and running are shown. (**c**) The energy cost is expressed per step, instead of per meter. In this case, the intersection between regression curves occurs at the same speed as for spontaneous transition. The grey area depicts the speed range at which humans may either walk or run. From Saibene and Minetti 2003; data from Minetti et al. 1994b

terrain (average throughout a whole stride) is smaller the greater the forward acceleration (a_f). Thus, this state of affairs is analogous to running uphill at constant speed, provided that α is unchanged (Fig. 5.15, right panel). It necessarily follows that the complement of α, i.e., the angle between the terrain and the horizontal plane ($90 - \alpha$), increases with a_f when running on flat terrain, or with the incline of the terrain when running uphill at constant speed.

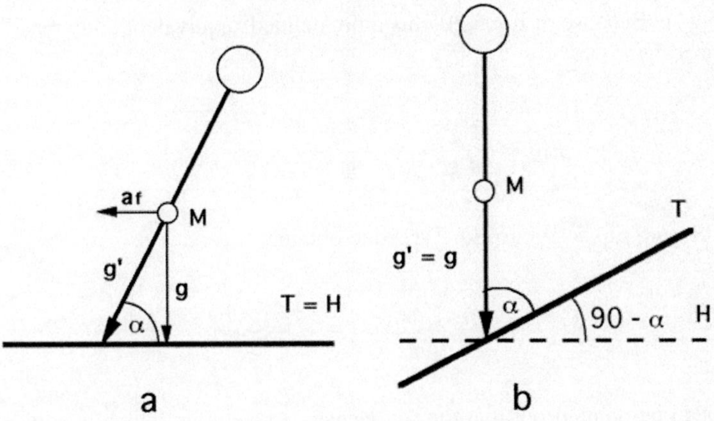

Fig. 5.15 The subject is accelerating forward while running on flat terrain (**a**), or is running uphill at constant speed (**b**). *M*, subject's body mass; a_f, forward acceleration; *g*, acceleration of gravity; $g' = \sqrt{a_f^2 + g^2}$, vectorial sum of a_f plus *g*; T, terrain; H, horizontal; *α*, angle between the runner's mean body axis throughout the stride and T; 90 – *α*, angle between T and H. Note that, in this representation, the angle 90-*α* corresponds to the angle *β* of Eqs. 5.9 and 5.10, and in the footnote 3. See text for details. Modified after di Prampero et al. 2015

The incline of the terrain (*i*, see also Eq. 5.9) is generally expressed as the tangent of 90 – *α* (*β* in Eq. 5.9). During accelerated running on flat terrain, *i* is equivalent to the ratio between a_f and *g* (Fig. 5.15, left panel):

$$i = \tan(90 - \alpha) = \frac{a_f}{g} = \text{ES} \tag{5.19}$$

Hence, the ratio a_f/g yields the tangent 90 – *α*, which makes accelerated running on flat terrain biomechanically equivalent to running at constant speed up to a corresponding slope (Fig. 5.15, right panel), defined as equivalent slope (ES).

Inspection of Fig. 5.15 also shows that accelerated running is characterized by yet another difference, as compared to constant speed running. Indeed, the acceleration, which the runner undergoes along the vertical axis of his/her body (*g'*), is greater than *g*, corresponding to the vectorial sum of a_f and *g*, which is equal to:

$$g' = \sqrt{a_f^2 + g^2} \tag{5.20}$$

It follows that the force that the runner must develop (average throughout a whole stride), as given by the product of the body mass and the acceleration, is greater in accelerated running (= *M g'*) as compared to constant speed running (= *M g*), because *g'* > *g* (Fig. 5.15, left panel). Thus, accelerated running is equivalent to uphill running wherein, however, the body weight is increased in direct proportion to the ratio g'/g. Because of Eq. 5.20, this ratio, defined "equivalent body mass" (EM), is described by:

$$\mathrm{EM} = \frac{M\,g'}{M\,g} = \frac{\sqrt{a_f{}^2 + g^2}}{g} = \sqrt{\frac{a_f{}^2}{g^2} + 1} \tag{5.21}$$

Substituting Eq. 5.19 into Eq. 5.21, one obtains:

$$\mathrm{EM} = \sqrt{\frac{a_f{}^2}{g^2} + 1} = \sqrt{\mathrm{ES}^2 + 1} \tag{5.22}$$

It must also be pointed out that during decelerated running, which is equivalent to downhill running, and in which case ES is negative, EM assumes nevertheless a positive value, because ES in Eq. 5.22 is raised to the square.

In conclusion, if the time course of v during accelerated/decelerated running is determined, and the corresponding instantaneous accelerations/decelerations are calculated, Eqs. 5.19 and 5.22 allow computation of the appropriate ES and EM values, thus converting accelerated/decelerated running on flat terrain into the equivalent constant speed uphill/downhill running. Hence, if C_r at constant speed at any incline is known, the corresponding energy cost of accelerated/decelerated running can be easily obtained.

On these premises, the empirical equation described by Minetti et al. (2002) (current Eq. 5.9) can be used to compute the energy cost of running during acceleration/deceleration (C_{acc}), by substituting ES for i, denoting the energy cost of constant speed level running as C_r, and multiplying then the entire equation by EM:

$$C_{acc} = \mathrm{EM}\left(155.4\,\mathrm{ES}^5 - 30.4\,\mathrm{ES}^4 - 43.3\,\mathrm{ES}^3 + 46.3\,\mathrm{ES}^2 + 19.5\,\mathrm{ES} + C_r\right) \tag{5.23}$$

In Eq. 5.19, only the forward acceleration is accounted for, whereas the effect of the air resistance is neglected. Therefore, only the non-aerodynamic component of ES ($ES_{n\text{-}a}$) is considered, as would be the case in treadmill running, so that ES $= ES_{n\text{-}a}$. To overcome the air resistance, the runner must lean forward more than indicated in Fig. 5.15 during running on a track, thus reducing angle α and increasing the ES by an amount, due to the air drag (aerodynamic ES, ES_a), which is equal to:

$$ES_a = \frac{k_w\,v^2}{g} \tag{5.24}$$

where v is the instantaneous velocity and k_w is the same constant as in Eqs. 5.5, however expressed per unit of body mass. Hence, the overall ES, as set by both a_f and the air velocity, i.e., by the sum of $ES_{n\text{-}a}$ and ES_a, is given by:

$$\mathrm{ES} = ES_{n-a} + ES_a = \frac{a_f}{g} + \frac{k_w\, v^2}{g} \tag{5.25}$$

Equation 5.25 tells that during accelerated running, the air resistance leads to an increase of ES, whereas the opposite is true in decelerated running, in which case a_f is negative.

The effects of the air resistance are rather minor, as compared to those due to the forward acceleration: indeed, assuming k_w (per unit of body mass)≈ 0.0037 m^{-1}, in the range of speeds between 2 and 10 m·s^{-1}, ES_a ranges from 0.15 to 3.8%. This is coherent with the usual practice of simulating the air resistance, when running on the treadmill at speeds of 5.55 m·s^{-1}, by inclining it upwards by about 1%.

In conclusion, if the time course of velocity during accelerated/decelerated running is determined, and the corresponding instantaneous accelerations/decelerations are calculated, Eqs. 5.25 and 5.22 allow computation of the appropriate ES and EM values. Accelerated or decelerated running can then be easily converted into equivalent constant speed uphill or downhill running. Hence, the corresponding energy cost of accelerated/decelerated running can be obtained.

It should be noted here that the effects of the acceleration on C_r can be astonishingly high. Indeed, at the very onset of a 100 m dash, in medium-level sprinters, C_r attains about 50 J kg^{-1} m^{-1}, as compared to about 4 J kg^{-1} m^{-1} for constant speed running on flat terrain (di Prampero et al. 2005). Furthermore, knowledge of C_r allows calculation of the corresponding metabolic power, as given by the instantaneous product of C_r and v. In medium-level sprinters, this attains a peak of about 80 W kg^{-1}, equivalent to an oxygen consumption of 230 ml kg^{-1} min^{-1} above resting, i.e., about four times larger than the $\dot{V}O_2\,^{max}$ of the investigated subjects.

Similar calculations can be made also on Usain Bolt during his 100 meters dash world record (9.58 s). However, because of his astonishingly high initial a_f, his ES falls largely outside the range of application of Eq. 5.9. Depending on how extrapolation of Eq. 5.9 is performed, the peak metabolic power, attained about 0.8 s after the start, may result as high as 160 W kg^{-1} (di Prampero et al. 2015).

5.8 Conclusions

Walking and running are natural forms of bipedal locomotion, as trotting galloping and ambling are natural forms of quadrupedal locomotion. There are more locomotion paradigms in quadrupedal than in bipedal locomotion, and their analogies were investigated by Alberto Minetti (Minetti 1998). Walking is characterized by the exchanges between E_k and E_p at each step in both cases. Running is an analogue of trotting, both being characterized by storage and release of elastic energy at each step, whereas E_k and E_p vary in phase, so that no exchanges between these two forms of energy are possible indeed. Minetti (1998) proposed that skipping gaits are analogues of galloping and represent the third paradigm of bipedal locomotion. In fact, the simultaneous presence of pendulum-like and elastic mechanisms in skipping differentiates this gait from running. Gravity may matter in determining the predominance of skipping on other gaits, as long as skipping becomes the "natural" form of locomotion in low-gravity environments such as the Moon (Pavei et al. 2015), as originally suggested by Margaria and Cavagna (1964), well before the first manned flight to the Moon in 1969 (see Chap. 11).

The role of the School of Milano in the history of the physiology of locomotion is central. It sounds logical that scientists interested in exercise be attracted by the energetics and biomechanics of human locomotion. Yet few combined the two issues, exercise and locomotion, within the same cultural framework, and none in a systematic, programmatic manner as Margaria did. This is a further feature characterizing the exceptional, farseeing vision of Margaria. Along the groove traced initially by Marey and Muybridge, then by Fenn, Margaria started the study of the energetics and promoted the study of the biomechanics of human locomotion. Cavagna and Saibene in the latter, Cerretelli and di Prampero in the former took the relay. Cavagna generated a vassal kingdom at Leuven, Belgium, with Norman Heglund, Peter Willems and their successors; Saibene trained Minetti, who carries the flag of locomotion in Milano. Cerretelli and di Prampero left a patrol of scientific heirs (Carlo Capelli, Guido Ferretti, Bruno Grassi, Paola Zamparo) who ensures scientific continuity to the study of the energetics of muscular exercise and of human locomotion within the cultural framework of the School of Milano. No other School can compete in terms of impact, continuity and longevity with the School of Milano, as far as the study of the energetics and biomechanics of human locomotion is concerned.

References

Alexander RMN (1989) Optimization and gaits in the locomotion of vertebrates. Physiol Rev 69: 1199–1227

Alexander RMN, Vernon A (1975) The mechanics of hopping by kangaroos (*Macropodidae*). J Zool Lond 177:265–303

Arsac LM (2002) Effects of altitude on the energetics of human best performances in 100-m running: a theoretical analysis. Eur J Appl Physiol 87:78–84

Arsac LM, Locatelli E (2002) Modeling the energetics of 100-m running by using speed curves of world champions. J Appl Physiol 92:1781–1788

Askew GN, Formenti F, Minetti AE (2011) Limitations imposed by wearing Armour on medieval soldiers' locomotor performance. Proc Roy Soc B 279:640–644

Barkan O, Giuliani F, Higgins HL, Signorelli E, Viale G (1914) Gli effetti dell'alcool sulla fatica in montagna. Arch Fisiol 13:277–295

Bastien GJ, Schepens B, Willems PA, Heglund NC (2005) Energetics of load carrying in Nepalese porters. Science 308:1755

Benedict FG, Murschhauser H (1915) Energy transformations during horizontal walking, vol 1. Carnegie Institution of Washington, Washington DC, p 597
Bottinelli R (2001) Functional heterogeneity of mammalian single muscle fibres: do myosin isoforms tell the whole story? Pflügers Arch 443:6–17
Bourdin M, Pastene J, Germain M, Lacour J-R (1993) Influence of training, sex, age and body mass on the energy cost of running. Eur J Appl Physiol 66:439–444
Brezina E, Kolmer W (1912) Über den Energieverbrauch bei der Geharbeit unter dem Einflussverschiedener Geschwindigkeiten und Verschiedener Belastungen. Biochem Z 38: 129–153
Cappozzo A, Figura F, Marchetti M, Pedotti A (1976) The interplay of muscular and external forces in human ambulation. J Biomech 9:35–43
Cavagna GA (1969) Travail mécanique dans la marche et la course. J Physiol Paris, Suppl 61:3–42
Cavagna GA (1975) Force platforms as ergometers. J Appl Physiol 39:174–179
Cavagna GA (1988) Muscolo e Locomozione. Raffaello Cortina, Milano
Cavagna GA, Dusman B, Margaria R (1968) Positive work done by a previously stretched muscle. J Appl Physiol 24:21–32
Cavagna GA, Franzetti P (1981) Mechanics of competition walking. J Physiol London 315:243–251
Cavagna GA, Franzetti P, Fuchimoto T (1983) The mechanics of walking in children. J Physiol Lond 343:232–239
Cavagna GA, Heglund NC, Taylor CD (1977) Mechanical work in terrestrial locomotion: two basic mechanisms for minimizing energy expenditure. Am J Phys 233:R242–R261
Cavagna GA, Kaneko M (1977) Mechanical work and efficiency in level walking and running. J Physiol Lond 268:467–481
Cavagna GA, Komarek L, Mazzoleni S (1971) The mechanics of sprint running. J Physiol Lond 217:709–721
Cavagna GA, Saibene FP, Margaria R (1963) External work in walking. J Appl Physiol 18:1–9
Cavagna GA, Saibene FP, Margaria R (1964) Mechanical work in running. J Appl Physiol 19:249–256
Cavagna GA, Thys H, Zamboni A (1976) The sources of external work in level walking and running. J Physiol Lond 262:639–657
Dawson TJ (1977) Kangaroos. Sci Am 237:78–89
di Prampero PE (1986) The energy cost of human locomotion on land and in water. Int J Sports Med 7:55–72
di Prampero PE (2008) Physical activity in the 21st century: challenges for young and old. In: Taylor NAS, Groeller H (eds) Physiological bases of human performance during work and exercise. Churchill Livingstone Elsevier, Edinburgh, pp 267–273
di Prampero PE (2015) La Locomozione Umana su Terra, in Acqua, in Aria. Fatti e Teorie, Edi-Ermes, Milano
di Prampero PE, Atchou G, Brueckner J-C, Moia C (1986) The energetics of endurance running. Eur J Appl Physiol 55:259–266
di Prampero PE, Botter A, Osgnach C (2015) The energy cost of sprint running and the role of metabolic power in setting top performances. Eur J Appl Physiol 115:451–469
di Prampero PE, Capelli C, Pagliaro P, Antonutto G, Girardis M, Zamparo P, Soule RG (1993) Energetics of best performances in middle-distance running. J Appl Physiol 74:2318–2324
di Prampero PE, Fusi S, Sepulcri L, Morin JB, Belli A, Antonutto G (2005) Sprint running: a new energetic approach. J Exp Biol 208:2809–2816
Fenn WO (1930a) Frictional and kinetic factors in the work of sprint running. Am J Phys 92:583–611
Fenn WO (1930b) Work against gravity and work due to velocity changes in running. Am J Phys 93:433–462
Fusi S, Campailla E, Causero A, di Prampero PE (2002) The locomotory index: a new proposal for evaluating walking impairments. Int J Sports Med 23:105–111

Griffin TM, Kram R, Wickler SJ, Hoyt DF (2004) Biomechanical and energetic determinants of the walk-trot transition in horses. J Exp Biol 207:4215–4223

Hill AV (1928) The air resistance to a runner. Proc Roy Soc B 102:380–385

Katzenstein G (1891) Ueber die Einwirkung der Muskeltätigkeit auf den Stoffverbrauch des Menschen. Pflügers Arch 49:330–404

Ker RF, Bennett MB, Bibby SR, Kester RC, Alexander RMN (1987) The spring in the arch of the human foot. Nature 325:147–149

Lacour JR, Padilla S, Barthélémy JC, Dormois D (1990) The energetics of middle distance running. Eur J Appl Physiol 60:38–43

Langman VA, Rowe MF, Roberts TJ, Langman NV, Taylor CR (2012) Minimum cost of transport in Asian elephants: do we really need a bigger elephant? J Exp Biol 215:1509–1514

Lejeune TM, Willems PA, Heglund NC (1998) Mechanics and energetics of human locomotion on sand. J Exp Biol 201:2071–2080

Liljestrand G, Stenström M (1919) Respirationsversuche beim Gehen, Laufen, Ski- und Schlittschuhlaufen. Skand Arch Physiol 39:167–206

Lloyd R, Parr B, Davies S, Cooke C (2011) A kinetic comparison of back-loading and head-loading in Xhosa women. Ergonomics 54:380–391

Löwy A, Löwy J, Zuntz L (1897) Über den Einfluss der verdünnten Luft und des Höhenklimas auf den Menschen. Pflügers Arch 66:477–538

Maggioni MA, Veicsteinas A, Rampichini S, Cè E, Nemni R, Riboldazzi G, Merati G (2012) Energy cost of spontaneous walking in Parkinson's disease patients. Neurol Sci 33:779–784

Maloiy GM, Heglund NC, Prager LM, Cavagna GA, Taylor CR (1986) Energetic cost of carrying loads: have African women discovered an economic way? Nature 319:668–669

Marchetti M, Cappozzo A, Figura F, Felici F (1983) Race walking versus ambulation and running. In: Matsui H, Kobayashi K (eds) Biomechanics VIII-B. Human Kinetics, Champaign IL, USA, pp 669–675

Marey EJ (1878) La machine animale. Locomotion terrestre et aérienne, Baillière, Paris

Margaria R (1938) Sulla fisiologia e specialmente sul consumo energetico della marcia e della corsa a varia velocità ed inclinazione del terreno. Atti R Accad Lincei 7:299–368

Margaria R (1975) Fisiologia muscolare e meccanica del movimento. Edizioni Scientifiche e Tecniche, Mondadori, Milano

Margaria R, Cavagna GA (1964) Human locomotion in subgravity. Aerosp Med 35:1140–1146

Margaria R, Cerretelli P, Aghemo P, Sassi G (1963) Energy cost of running. J Appl Physiol 18:367–370

McMahon TA, Tyler Bonner J (1983) On size and life. Scientific American Library, WH Freeman Co., New York

Merati G, Negrini S, Sarchi P, Mauro F, Veicsteinas A (2001) Cardio-respiratory adjustments and cost of locomotion in school children during backpack walking (the Italian backpack study). Eur J Appl Physiol 85:41–48

Mero A, Komi PV, Gregor RJ (1992) Biomechanics of sprint running. A review. Sports Med 13: 376–392

Minetti AE (1998) The biomechanics of skipping gaits: a third locomotion paradigm? Proc Roy Soc B 265:1227–1235

Minetti AE, Ardigò LP, Saibene F (1994a) Mechanical determinants of the minimum energy cost of gradient running in humans. J Exp Biol 195:211–225

Minetti AE, Ardigò LP, Saibene F (1994b) The transition between walking and running in humans: metabolic and mechanical aspects at different gradients. Acta Physiol Scand 150:315–323

Minetti AE, Ardigò LP, Saibene F, Ferrero S, Sartorio A (2000) Mechanical and metabolic profile of locomotion in adults with childhood-onset GH deficiency. Eur J Endocrinol 142:35–41

Minetti AE, Moia C, Roi GS, Susta D, Ferretti G (2002) Energy cost of walking and running at extreme uphill and downhill slopes. J Appl Physiol 93:1039–1046

Minetti AE, Saibene F, Ardigò LP, Atchou G, Schena F, Ferretti G (1994c) Pygmy locomotion. Eur J Appl Physiol 68:285–290

Moritani T, Nagata A, de Vries HA, Muro M (1981) Critical power as a measure of physical work capacity and anaerobic threshold. Ergonomics 24:339–350

Muybridge E (1887) Animal locomotion. University of Pennsylvania, The Philadelphia Print Shop

Olgiati R, Jacquet R, di Prampero PE (1986) Energy cost of walking and exertional dyspnea in multiple sclerosis. Am Rev Resp Dis 134:1005–1010

Padilla S, Bourdin M, Barthélémy J-C, Lacour J-R (1992) Physiological correlates of middle distance running performance. A comparative study between men and women. J Appl Physiol 65:561–566

Pavei G, Biancardi CM, Minetti AE (2015) Skipping vs. running as the bipedal gait of choice in hypogravity. J Appl Physiol 119:93–100

Perkins JF Jr (1964) Historical development of respiratory physiology. In: Fenn WO, Rahn H (eds) Handbook of physiology. Respiration. Section 3, vol I. American Physiological Society, Washington, DC, pp 1–62

Pinnington HC, Dawson B (2001) Running economy of elite surf iron men and male runners, on soft dry beach sand and grass. Eur J Appl Physiol 86:62–70

Pinnington HC, Lloyd DG, Besier TF, Dawson B (2005) Kinematic and electromyography analysis of submaximal differences running on a firm surface compared with soft, dry sand. Eur J Appl Physiol 94:242–253

Plamondon A, Roy B (1984) Cinématique et cinétique de l'accélération dans le sprint. Can J Sport Med 9:42–52

Prilutsky BI, Gregor RJ (2001) Swing- and support-related muscle actions differentially trigger human walk-run and run-walk transitions. J Exp Biol 204:2277–2287

Saibene F, Minetti AE (2003) Biomechanical and physiological aspects of legged locomotion. Eur J Appl Physiol 88:297–316

Schumburg, Zuntz N (1896) Zur Kenntniss der Einwirkungen des Hochgebirges auf den menschlichen Organismus. Pflügers Arch 63:461–494

Segers V, Desmet K, Van Caekenberghe I, Aerts P, De Clercq D (2013) Biomechanics of spontaneous overground walk-to-run transition. J Exp Biol 216:3047–3054

Smith E (1859) Experimental inquiries into the chemical and other phenomena of respiration, and their modifications by various physical agencies. Phil Trans R Soc London 149:681–714

Sondén K, Tigerstedt R (1895) Untersuchungen über die Respiration und den Gesammtstoffwechsel des Menschen. Skand Arch Physiol 6:1–244

Soule RG, Goldman RF (1969) Energy cost of loads carried on the head, hands, or feet. J Appl Physiol 27:687–690

Soule RG, Goldman RF (1972) Terrain coefficients for energy cost prediction. J Appl Physiol 32: 706–708

Taboga P, Lazzer S, Fessehatsion R, Agosti F, Sartorio A, di Prampero PE (2012) Energetics and mechanics of running men: the influence of body mass. Eur J Appl Physiol 112:4027–4033

Tam E, Rossi H, Moia C, Berardelli C, Rosa G, Capelli C, Ferretti G (2012) Energetics of running in top-level marathon runners from Kenya. Eur J Appl Physiol 112:3797–3896

Taylor CR, Schmidt-Nielsen K, Raab JL (1970) Scaling of energetic cost of running to body size in mammals. Am J Phys 219:1104–1107

Thorstensson A (1986) Effects of moderate external loading on the aerobic demand of submaximal running in men and in 10 year-old boys. Eur J Appl Physiol 55:569–574

Thorstensson A, Roberthson H (1987) Adaptations to changing speed in human locomotion. Acta Physiol Scand 131:211–214

Thys H, Faraggiana T, Margaria R (1972) Utilisation of muscle elasticity in exercise. J Appl Physiol 32:491–494

van Ingen Schenau GJ, de Koning JJ, de Groot G (1994) Optimisation of sprinting performance in running, cycling and speed skating. Sports Med 17:259–275

van Ingen Schenau GJ, Jacobs R, de Koning JJ (1991) Can cycle power predict sprint running performance? Eur J Appl Physiol 63:255–260

Von Staden H (1989) Herophilus. The art of medicine in early Alexandria. Cambridge University Press, Cambridge
Waller AD (1919) The cost of mechanical work in terms of CO_2 expired. J Physiol Lond 52:ii–iii
Ward-Smith AJ, Radford PF (2000) Energy conversion rates during sprinting with an emphasis on the performance of female athletes. J Sports Sci 18:835–843
Zamparo P, Francescato MP, De Luca G, Lovati L, di Prampero PE (1995) The energy cost of level walking in patients with hemiplegia. Scand J Med Sci Sports 5:348–352
Zamparo P, Perini R, Orizio C, Sacher R, Ferretti G (1992) The energy cost of walking or running on sand. Eur J Appl Physiol 65:183–187

Chapter 6
Cycling, Swimming and Other Forms of Locomotion on Land and in Water

Pietro Enrico di Prampero and Guido Ferretti

Abstract This chapter is on other forms of human locomotion. In the first part, assisted locomotion on land in treated. Most attention is paid to cycling, the energetics and biomechanics of which is described in details. The effects of rolling resistance and of aerodynamic resistance are analyzed. The effects of shape and size, of altitude, of cycling uphill and downhill are discussed. Cross-country skiing is also accounted for. In the second part, after a short account on the effects of water density and buoyancy, locomotion in water is treated, with special reference to swimming. The effects of skill, swimming styles and body fat distribution are analyzed. Then, the biomechanics of swimming is discussed along with the effects of drag and efficiency. In the third part, assisted locomotion in water is analyzed, with special attention to rowing and sculling. Whereas the study of cycling was essentially a matter of Milano, the study of swimming was the result of a deep and long-lasting collaboration between Milano and Buffalo, a tribute to which is heartly acknowledged.

P. E. di Prampero
Emeritus Professor of Physiology, University of Udine, Udine, Italy
e-mail: Pietro.Prampero@uniud.it

G. Ferretti (✉)
Department of Molecular and Translational Medicine, University of Brescia, Brescia, Italy
e-mail: Guido.Ferretti@unibs.it

G. Ferretti (ed.), *Exercise, Respiratory and Environmental Physiology*, Perspectives in Physiology, https://doi.org/10.1007/978-3-031-19197-8_6

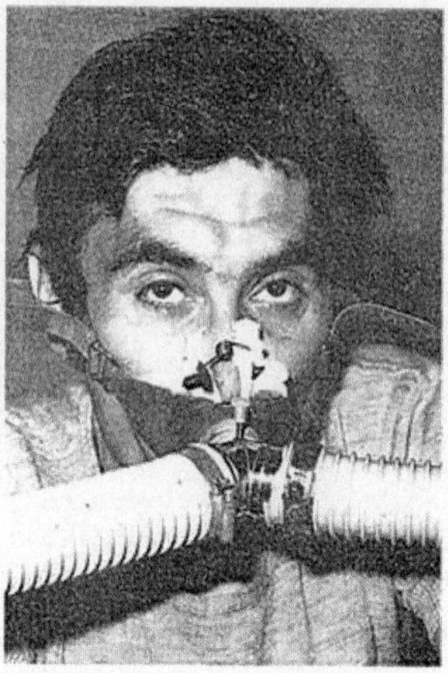

Left: Eddy Merckx during his successful one-hour world record trial at Mexico City on October 25th, 1972. *Right:* Eddy Merckx one month earlier, during exercise testing at the Istituto di Fisiologia, University of Milano. With the kind permission of the University of Milano

6.1 Introduction

This is the second chapter of this book that we devote to human locomotion and the contribution of the School of Milano to the field. In Chap. 5, we discussed natural locomotion, i.e., walking and running. In this Chapter, we analyze locomotion in water, whether unassisted (swimming) or assisted (fin swimming, rowing etc.), and assisted locomotion on land (chiefly cycling). The term assisted locomotion includes all locomotion modes, in which, although the power sustaining the displacement of the body center of mass along the direction of movement is generated by muscle contraction, specific mechanical tools, invented by human creativity and intuition, lead to a reduction of the energy expenditure per unit of distance, thereby allowing an increment of the forward speed at any given metabolic power. The most typical and most successful example of such a tool is the bicycle, wherein muscle power causes the pedals to rotate, thus moving a chain that transmits rotation to the rear axle through the pin of the chain. This is inserted on the axle on which the rear wheel fits, making it moving. The speed amplification effect is because the diameter of the wheel is by far larger than that of the rear pinion.

In Chap. 5, we initially defined the general principles on which the analysis of human locomotion relies. We introduced the concept of energy cost (C) and we described its two components, namely the non-aerodynamic energy cost ($C_{n\text{-}a}$) and the aerodynamic energy cost (C_a) (or, in water, hydrodynamic). We set the linear equation relating C_a to the square of speed, through a proportionality constant k_e.

Constant k_e depends on the area projected on the frontal plane (A_f), a dimensionless coefficient related to the body shape (C_x), the air density (ρ), and the mechanical efficiency of the exercise at stake (η). These principles apply also to all types of locomotion dealt with in the present Chapter, so we invite the reader to refer to Chap. 5 for further details, especially to Eqs. 5.4–5.7. We underline, however, that, at variance with walking and running, all other types of locomotion on land are characterized by a higher, even by far higher, C_a than $C_{n\text{-}a}$.

6.2 A Historical Note

Swimming is not a natural form of locomotion for humans, who are terrestrial mammals, designed to sustain gravity and to take the necessary oxygen to support muscle contraction from the atmosphere through the lungs. Notwithstanding, the wish or the need of crossing rivers and ponds or of entering the sea pushed them to develop, as other terrestrial mammals, an ability to swim. We do not know how far ago humans started to swim, but we know that they did swim, as we have witness of prehistoric rock paintings showing swimming men. For instance, the celebrated paintings of the so-called Cave of Swimmers, in the Egypt desert close to the Libyan border, portray, among others, men with bent limbs in a swimming posture, reminding a kind of breaststroke. These pictures, dating back to some 10,000 years ago, provided the first hint to the revolutionary hypothesis that the Sahara was green and wet in those days (Almasy and Lozach 1936).

The practise of swimming spread in all historical societies. It was soon associated with the practise of diving. Pools were found in ancient palaces in India and in Crete. Several pictures of swimming men can be found in vestiges from all over the ancient world. Literature reports on swimming date back to the Gilgamesh epopee (~2000 B. C.) and the Bible. Various swimming styles evolved in different world areas. The breaststroke was predominant in Europe; a kind of crawl among American, Polynesian, and Hawaiian natives. Everard Digby (~1550–1592), a scholar of St. John's College in Cambridge, described not only the breaststroke, which he regarded as the most economic swimming style, but also the backstroke and the crawl (Digby 1587). He had been the first to put forward the role of buoyancy in swimming.

Swimming became a competitive sport in England in the early nineteenth century. The breaststroke was mostly practised. The crawl was introduced afterward, but was the only style in which athletes competed at the first modern Olympic Games in 1896. The backstroke and the butterfly were introduced later. The latter originated as an evolution of the breaststroke and was recognized as an independent style in 1952. For a more comprehensive history of competitive swimming, please refer to Kehm (2007).

The physiology of swimming was firstly investigated by Liljestrand and Stenström (1919a) and by Liljestrand and Lindhard (1919a), who determined gas exchange, heart rate, blood pressure and cardiac output on subjects swimming the breaststroke at speeds between 0.3 and 0.8 m s^{-1}. They demonstrated that the energy expenditure during swimming increased more than in proportion to the speed, indicating that a swimmer consumes more energy per unit distance at high speed than at low speed. These authors were also the first to investigate the relationship between passive towing force and speed and to observe a lower energy expenditure during swimming, expressed per unit body mass, in women than in men. Karpovich and Millman (1944) investigated the energy cost of front crawl, backstroke, breast-stroke, butterfly, and the now less common sidestroke, at speeds between 0.7 and 1.7 m s^{-1}. Their values, however, are grossly too large, as compared to the presently accepted ones, presumably because of their overestimate of the oxygen debt incurred during the actual swimming period. In a previous study, Karpovich and Pestrecov (1939) had also measured the relationship between passive towing force and speed.

Besides the physiology of swimming, Liljestrand and Stenström 1919b) and Liljestrand and Lindhard (1919b) investigated some aspects of cross-country skiing and of rowing; this last being discussed in some detail also by Henderson and Haggard (1925).

The history of cycling is much shorter and strictly related to the evolution of the bicycle. The first ideas of transforming the cyclic leg movement into a rotational movement of a wheel date back to the fifteenth century, but did not lead to any valuable tool. It is only at the beginning of the nineteenth century that the first two-wheel muscle-propelled machine appeared, in 1817. German Baron Karl von Drais (1785–1851) designed it, whence the name *Draisine*, but he called it *Laufmaschine* (running machine). It was wood-made, weighed about 22 kg, was equipped with iron wheels and was propelled by a walking-like movement of the legs. A similar yet more stylish device was developed in England a couple of years later. The commercial success of these devices prompted a striking mechanical development, which in a hundred years led to the modern road and race bicycles. A detailed account of the history of bicycle development can be found elsewhere (Herlihy 2004). Chief steps were the invention of the rotational crank and pedal and of the rubber tires, in the 1860s. Most fundamental, however, was the design of the so-called safety bicycle, the real predecessor of modern bicycles, by the English engineer Harry John Lawson (1852–1925) in 1876. This bicycle had equal-sized wheels and a chain drive. Indeed the chain drive is a much older invention, dating back to the Hellenistic era, but nobody thought of using it in bicycles before Lawson. The biomechanical and energetic characteristics of historical bicycling was analyzed by Minetti et al. (2001). We shall discuss it later.

From the physiological viewpoint, the energy requirement for recreational cycling was investigated in depth by Zuntz (1899), who also compared the speed attained with a given oxygen consumption during cycling, or walking (Fig. 6.1).

The mechanical efficiency of cycling was investigated by Benedict and Cathcart (1913), who were the first to show that the most economical pedal frequency is of the order of 1 Hz. Yet it was the refinement of the concept of energy cost by Margaria (1938) that, once more, opened the way to the modern era of physiological studies on cycling. The School of Milano played a large part in it since the 1970s. Before that time, only running and walking attracted the curiosity of Margaria and his pupils. Pietro Enrico di Prampero enlarged the perspective of the School by looking into

Fig. 6.1 Measuring oxygen consumption during bicycling at the end of nineteenth century. From Zuntz 1899

other types of locomotion, such as rowing, swimming, skating, and cycling, in collaboration with several colleagues of him in Milano (di Prampero et al. 1971, 1974, 1976, 1979).

Concerning swimming, crucial was the close interaction between Milano and Buffalo, where Hermann Rahn (1912–1990) had promoted the construction of the celebrated circular swimming pool (Fig. 6.2), which was particularly apt to the study of the energetics of swimming. More or less at the same time, Per-Olof Åstrand (1922–2015) in Stockholm promoted the construction of a different device, called the swimming flume (Åstrand and Englesson 1972), on which Ingvar Holmér performed his studies on the oxygen consumption during swimming (Holmér 1974).

6.3 Cycling

On September 6th 1996, in Manchester (UK), Chris Boardman covered 56.375 km in one hour, thus establishing what was then the "One Hour Record" for unaccompanied cycling. This is one of the most remarkable achievements concerning

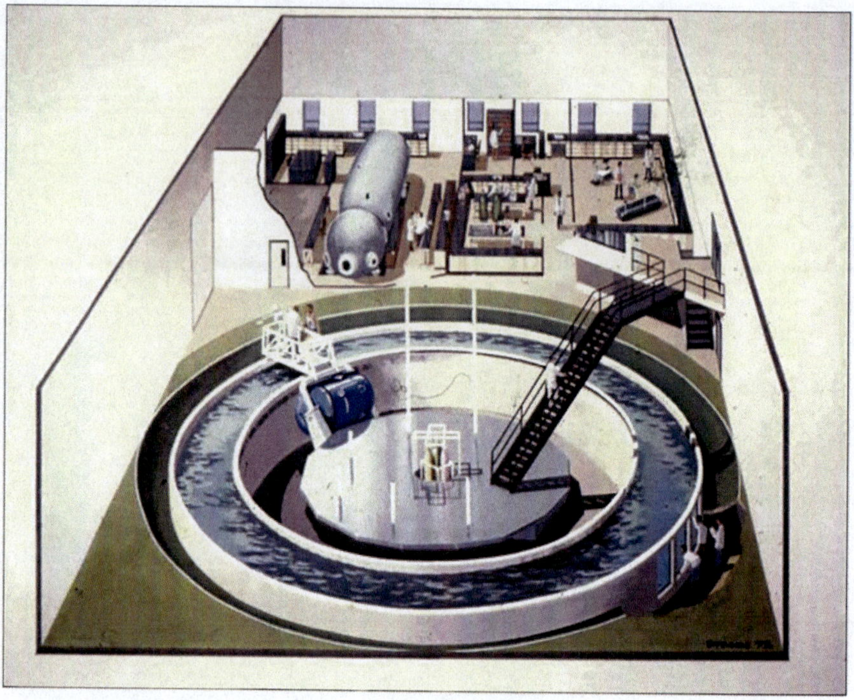

Fig. 6.2 A sketch of the large research installations at the Department of Physiology, State University of New York at Buffalo in the 1970s. The circular swimming pool is on the front plane. The rail-moving chariot following the swimmer is shown. The human centrifuge is on the inner circle. The stairs lead to the centrifuge. On the rear plane, the hyperbaric chambers appear. With the kind permission of Dave Pendergast, Buffalo

locomotion in the history of sport records. That record soon became the "Best One Hour Human Performance," after the obscurantist decision by the *Union Cycliste Internationale* (UCI) to forbid the use of specially designed aerodynamic bicycles for record attempts, and the subsequent cancellation of all records established with that type of bicycles. Since then, the "One Hour Record" for unaccompanied cycling defined only best performances on traditional road-race bicycles. A partial step back was however undertaken in 2014, when the UCI approved the use of aerodynamic bicycles like those for road time trial competitions. With such a bike, in 2019, at Aguascalientes, Mexico (altitude 1900 m above sea level), Victor Campenaerts covered 55.089 km, and Dan Bigham at sea level attained 55.548 km. The preposterous dichotomy between the "One Hour Record" and the "Best One Hour Human Performance" was suddenly made obsolete by the marvellous performance by Filippo Ganna, who, on October 8th 2022 in Grenchen (Switzerland, 450 m above sea level), covered 56.792 km, thus not only beating the official record, but also improving Boardman's performance.

Boardman's and Ganna's extraordinary feats, besides showing their exceptional athletic level, highlight also the formidable characteristics of their bicycles as a means of locomotion on flat terrain. Indeed, thanks to the saddle that supports the body in the vertical plane, and to the pedals that transform the intermittent thrust of the lower limbs in an essentially continuous push, the amount of energy dissipated against gravity and inertial forces at each pedal cycle is reduced to a minimum. This state of affairs allows the cyclist to utilize, during progression on flat terrain at constant speed, his/her metabolic energy entirely (or very nearly so, see Fig. 5.2) against the air drag (di Prampero 2000). Moreover, the C is lower, and therefore, the speed attained with a given metabolic power is higher, than in any other form of human-powered locomotion.

This section is devoted to the main characteristics of cycling: we will deal with mechanical work, energy expenditure and mechanical efficiency of cycling on the level, uphill or downhill and with the effects of the cyclist's body size thereupon.

6.3.1 Mechanical Work and Energy Cost

During progression on flat terrain at constant speed, two forces oppose the cyclist's motion: (i) the rolling resistance, due to the sum of the tires' friction with the terrain and of the (minimal) friction of the gear systems and of the wheel axles and (ii) the air drag (di Prampero et al. 1979). Hence:

$$F_c = F_{n-a} + F_a = F_{n-a} + k_w\, v^2 \tag{6.1}$$

where F_c[1] is the overall force opposing motion (work done per unit of distance), v is the velocity relative to the air, $F_{n\text{-}a}$ and F_a ($= k_w\, v^2$) are, respectively, the non-aerodynamic and the aerodynamic components of the opposing force, and k_w is a proportionality constant (for details on the meaning of k_w, see Eq. 5.5, Chap. 5). The corresponding energy cost of cycling (C_c) depends on the overall efficiency of locomotion (η):

$$C_c = \frac{F_c}{\eta} = \frac{F_{n-a} + k_w\, v^2}{\eta} \tag{6.2}$$

Setting $C_{n-a} = F_{n\text{-}a}/\eta$ and $k_e = k_w/\eta$:

$$C_c = k_e\, v^2 + C_{n-a} \tag{6.3}$$

[1] F_c and C_c have the dimension of a force, thus should be expressed in Newtons. However, in physiological studies, C_c is often expressed in J m^{-1}, since it is obtained as the ratio of metabolic power (usually expressed in L of O_2 per unit of time) to speed. By analogy, F_c is also expressed in the same way.

Equation 6.3 is equivalent to Eq. 5.7 for the specific case of cycling. In other terms, in cycling, the two constants $F_{n\text{-}a}$ and C_{n-a} represent, respectively, the mechanical work dissipated, and the metabolic energy spent, per unit of distance, against the rolling resistance. As such, they depend on the friction of the transmission system and of the wheel axles, but above all, on the characteristics of the terrain, of the type of tires and on the inflation pressure thereof (see Sect. 6.3.3). The terms $k_w\, v^2$ of Eq. 6.1 and $k_e\, v^2$ of Eq. 6.3 are the mechanical work dissipated, or the energy spent, per unit of distance against the air drag. Obviously enough, their ratio is equal to η.

The mechanical ($\dot{w}$) or metabolic ($\dot{E}$) power necessary to proceed at constant v on flat terrain in the absence of wind is given by the product of, respectively, F_c or C_c, and v. Therefore, under these conditions, one obtains:

$$\dot{w} = F_c\, v = k_w\, v^3 + F_{n-a}\, v \tag{6.4}$$

$$\dot{E} = C_c\, v = k_e\, v^3 + C_{n-a}\, v \tag{6.5}$$

where $\dot{w}$ and $\dot{E}$ turn out in W, if F_c and C_c are expressed in J m^{-1}, and v in m s^{-1}. Thus, the power, whether mechanical or metabolic, that is necessary to overcome the air resistance varies with the cube of the speed.[2]

The constants F_{n-a} and k_w of Eq. 6.1 are generally obtained determining the overall traction resistance (R_{tr}, corresponding to F_c) as a function of the speed (Capelli et al. 1993; di Prampero 2000; di Prampero et al. 1979). As shown in Fig. 6.3a, R_{tr} increases linearly with the square of the speed, with positive y-axis intercept; coherently with Eq. 6.1, the slope of the line yields k_w, whereas the y-axis intercept is equal to F_{n-a}. Obviously enough, the relationships reported in Fig. 6.3a are valid only if R_{tr} is measured at constant speed, on flat terrain, in the absence of wind and in standardized conditions in terms of type of bicycle, position of the cyclist, etc.

A similar approach can be utilized to assess the constants C_{n-a} and k_e (Capelli et al. 1993, 1998b). In this case, the dependent variable would be C_c (Fig. 6.3b). However, since C_c is generally obtained from the ratio between the rate of oxygen consumption ($\dot{V}O_2$) at steady state and v, the v range that can be reasonably investigated in this case is more limited than that applying for the two corresponding mechanical constants F_{n-a} and k_w, its upper limit being set by the cyclist's maximal oxygen consumption ($\dot{V}O_2\,^{max}$). It should also be underlined that the equipment utilized for determining $\dot{V}O_2$ should not interfere with the cyclist's movements, nor should it increase in any significant way the geometric determinants of k_w (see Chap. 5, Eq. 5.5).

[2] However, "tradition dies hard": hence it is often more "practical," also from a historical viewpoint, to express C_c in other units such as mlO_2 m^{-1}, since oxygen consumption, which in submaximal aerobic exercise is equal to $\dot{E}$, is usually given in mlO_2 min^{-1}. In this case, v should be expressed in m min^{-1}. Units conversion is easily made, if we consider that of 1 L of O_2 (STPD) yields about 5 kcal or 20.9 kJ of metabolic energy.

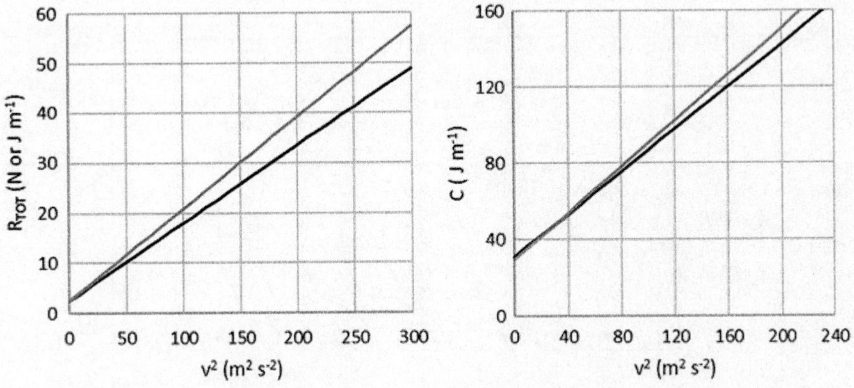

Fig. 6.3 Overall tractional resistance (R_{tot}) (panel **a**) and energy cost per unit of distance (C) (panel **b**) as a function of the square of the speed (v^2) in a cyclist of 70 kg body mass and 1.72 m stature. Traditional (grey line) or aerodynamic (black line) racing bicycle, in fully dropped posture at constant speed, in the absence of wind at sea level. Data from Capelli et al. 1993

The four constants of Eqs. 6.4 and 6.5 were also determined, with similar results, yet with different approaches, by several other authors (Candau et al. 1999; Davies 1980; De Groot et al. 1995; Pugh 1974; Sargeant and Davies 1977).

6.3.2 *Mechanical Efficiency*

The mechanical efficiency η of cycling exercise has been extensively studied in exercise physiology. Indeed, since the original observations of Sylvia Dickinson (1929), a very large series of data has consistently shown that, at least in the 50-to-300 W power range, the optimal pedalling frequency (f_p) increases from $\approx$ 0.7 to $\approx$ 1 Hz. At the optimal frequency, η is about 0.25 (for references, see di Prampero 2000). These data also show that even substantial changes in f_p above or below the optimal one lead only to rather minor changes of η (di Prampero 2000 and Fig. 6.4). Hence, for high values of $\dot{w}$, the relationship between η and f_p is rather flat. This may explain the reason why, during actual competitions, f_p is greater than that yielding the maximal values of η. Indeed, the average f_p of Francesco Moser during his One Hour Record for unaccompanied cycling at Mexico City in 1984 (51.151 km)[3] was 1.75 Hz, corresponding to an efficiency value, estimated from the data of Fig. 6.4 for a $\dot{w}$ of $\approx$ 450 W, smaller than 0.25 but still slightly greater than 0.24. The reduction of the forces exerted on the pedals, due to the f_p increase above the optimal one, may

[3] Moser world record is one of those that were cancelled by the UCI in 2000, because established on specially prepared bicyles, that were instead defined "Best One Hour Human Performances." This distinction, however, has been made obsolete by the recent world record established by Filippo Ganna.

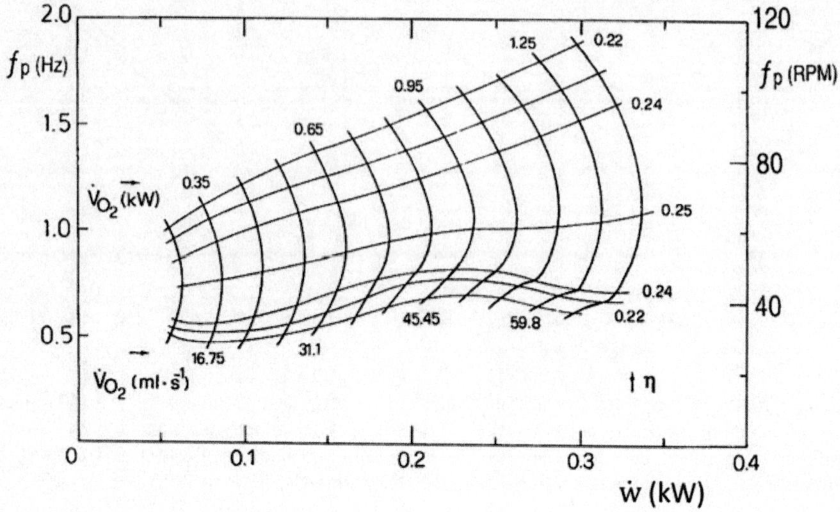

Fig. 6.4 Pedal frequency (f_p, in Hz, left, or cycles per minute, RPM, right) as a function of the external mechanical power ($\dot{w}$) during exercise on a cycle ergometer. Iso-metabolic power lines, convex toward the right, ($\dot{V}O_2$) are also indicated. This allows the attribution of one, and only one, value of mechanical efficiency (η) to any given point on the diagram. The points characterized by $\eta = 0.22$, 0.24 and 0.25 are connected by iso-efficiency functions. Modified after di Prampero, 1986

lead to a small reduction of the static component of the pedal cycle and hence of the corresponding lactic contribution to the energy expenditure, thus compensating for the slight fall of efficiency.

Although this energetic interpretation of the self-selected f_p seems plausible, especially to people issued from the School of Milano, yet it is not the only one. Indeed, other factors, such as the strategy of motor units recruitment, and/or the perceived intensity of the effort (Marsh and Martin 1993, 1995, 1998) and/or the muscle composition in terms of fast and slow fibers (Kohler and Boutellier 2005), may also play a non-negligible role in setting the actual f_p.

6.3.3 *The Rolling Resistance*

The friction of the wheels' axles and of the transmission system of a good-quality bicycle are quite small; therefore, the rolling resistance depends entirely, or very nearly so, on the dimensions, type, and inflation pressure of the tires, and on the characteristics of the terrain.

The rolling resistance is independent of the speed and directly proportional to the overall weight of the moving body (bicycle + cyclist). Hence, it may be convenient to express it as rolling resistance coefficient, i.e., as the ratio between the rolling resistance itself and the overall weight; it is reported, for several sets of experimental conditions, in Table 6.1, wherefrom F_{n-a} can be easily obtained, provided that the overall weight is known.

Table 6.1 The rolling resistance for an overall mass (cyclist + bicycle) of 85 kg, and the corresponding rolling coefficient are reported for different tire types on compact terrain. Wheel diameter, tire contact surface and inflation pressure (1 MPa = 9,87 atm) are also shown. The rolling coefficient is obtained by dividing the rolling resistance by the overall mass and the gravity acceleration (9.81 m s^{-2} on Earth). The energy expenditure per unit of distance against the rolling resistance can be obtained by dividing the rolling coefficient by the mechanical efficiency and multiplying by the overall weight. A e B, sculpted tires; C, pneumatic tires; D, tubulars (180 g); E, tubulars (160 g); F, tubulars (80 g). Data from di Prampero 2000

Tire Type	Tire Characteristics			Wheel diameter	Rolling resistance	Rolling coefficient
	Width	Pressure				
	cm	MPa	Atm	cm	J m-1	
A mountain bike	5.7	0.32	3.16	50.8	14.2	0.017
B Mountain bike	5.7	0.32	3.16	68.6	10.8	0.013
C road, standard	4.5	0.46	4.54	68.6	5.8	0.007
D road, tubulars	3.2	0.85	8.39	50.8	3.8	0.0045
E road, tubulars	1.8	0.85	8.39	68.6	2.8	0.0034
F track, tubulars	1.8	0.85	8.39	68.6	1.8	0.0021

With respect to the data shown in Table 6.1, on linoleum or wooden tracks, the rolling coefficient is reduced to about 60%. The absolute minimal values of the rolling resistance coefficient are those applying for wheels of standard diameter (0.686 m), tubular tires with reduced transverse section and high inflation pressure, on smooth compact terrain. These are about one third of those applying for normal wheels and tires on asphalt roads and about one tenth of those applying for wheels of reduced diameter and knobby tires of large transverse section (mountain bike) on asphalt roads. On irregular, muddy, or grass-covered terrains, the rolling resistance increases further in a manner that is difficult to predict.

The pursuit of a reduction of the rolling resistance accompanied the entire history of bicycle technical development since the eighteenth century (Minetti et al. 2001). The data indicate that this was attained any time that the characteristics of the tires evolved and improved. *Ceteris paribus*, remarkable step changes in rolling resistance occurred when passing from metal to rubber solid tires (about 1870) and from rubber to pneumatic tires (about 1890). Around the same time, the two wheels were set at equal diameter and the basic design of modern bicycles was achieved.

6.3.4 *The Aerodynamic Resistance*

As already mentioned in Sect. 6.3.1, the mechanical work against the air resistance per unit of distance (corresponding to F_a in Eq. 6.1, or the corresponding energy cost C_a, varies with the square of v. The proportionality constants defining these relationships (respectively, k_w of Eqs. 6.1 and k_e of Eq. 6.3) depend on: (i) a dimensionless aerodynamic coefficient (C_x) related to body shape (see Chap. 5 for details on this parameter), which is smaller the more "aerodynamically-shaped" is the moving object, (ii) the area projected on the frontal plane (A_f), (iii) the air density (ρ) and, for k_e only, (iv) the overall mechanical efficiency of movement (η):

$$k_w = 0.5\ C_x\ A_f\ \rho \tag{6.6}$$

$$k_e = \frac{k_w,}{\eta} = \frac{0.5\ C_x\ A_f\ \rho}{\eta} \tag{6.7}$$

In turn, ρ is a function of the prevailing barometric pressure (P_B) and absolute temperature (T):

$$\rho = \rho_0 \frac{273\ P_B}{760\ T} = 0.359\ \rho_0 \frac{P_B}{T} \tag{6.8}$$

where ρ_0 (= 1293 kg m^{-3}) is the air density in standard condition (STPD, P_B = 760 mmHg and T = 273 °K), which depends also on the air humidity, the effects of which, however, are so small that, for all practical purposes, they can be neglected.

Both P_B and T decrease with the altitude above sea level, although P_B values recorded on the field are generally higher than predicted from the Standard Atmosphere, as defined by the International Civil Aviation Organization (Pugh 1957; West et al. 1983). As a first approximation, for T = 273 °K and expressing the altitude in km, di Prampero (2000) used the following empirical relation, established by West et al. (1983) during the American Medical Expedition to Mount Everest:

$$PB = PB_0\ e^{-0{,}124\ km} \tag{6.9}$$

where P_{B0} is the barometric pressure at sea level.

6.3.5 *Of Shape and Size*

An intuitive strategy for reducing A_f is to lean the trunk forward. In race bicycles, particularly in the most recent aerodynamic models, this strategy is pushed to the extreme, so that the cyclist is forced to assume the position yielding the minimal possible value of A_f (Capelli et al. 1993). This is done even if it may lead to a small but statistically significant reduction of the $\dot{V}O_2{}^{max}$ and/or of the cycling efficiency (Ashe et al. 2003; Gnehm et al. 1997; Welbergen and Clijsen 1990). An increased cost of breathing may at least partially explain the latter (Charlton et al. 2017).

This strategy leads also to a reduction of C_x, because of a more streamlined position taken by the body, thus leading to an additional reduction of F_a (Capelli

Table 6.2 The aerodynamic coefficient (C_x) and the area projected on the frontal plane (A_f) of a 70 kg body mass, 1.75 m tall cyclist are indicated for different conditions. The third column (k_w) reports the proportionality constant between the mechanical work per unit of distance and the square of air velocity, at sea level and 20 °C temperature. The fourth column reports the velocity (v) attained on flat compact terrain with an overall mechanical power of 0,735 kW, assuming a rolling resistance applying for road wheels with 180 grams tubular wheels on compact terrain (see Table 6.1). The data of the first three columns are from di Prampero 2000

Bicycle and Position	C_x	A_f (m^2)	k_w ($N\ s^2\ m^{-2}$)	v ($m\ s^{-1}$) for $= \dot{w}$ *0.735 kW*
Traditional, sitting	1.10	0.51	0.337	12.6
Sporting, leaning forward	1.00	0.45	0.271	14.2
Sporting, fully dropped	0.87	0.44	0.230	14.5
Racing (standard) fully dropped	0.80	0.40	0.193	15.4
Racing (aerodynamic) fully dropped	0.65	0.40	0.155	16.8

et al. 1993). However, the best strategy for reducing C_x is to include the bicycle itself with streamlined shapes, allowing for very low C_x values (about 0,09–0,11, Gross et al. 1983; Dal Monte, 1997, personal communication). These considerations are summarized in Table 6.2, where A_f, C_x, and k_w are reported for different experimental conditions.

Another intuitive strategy for reduction of the air resistance is to move in the wake of another cyclist or of a vehicle. At high speeds, the power reduction for cyclists moving in the wake of one another can attain about 30%, and, in the case of more than two cyclists, it depends also on the position in the line. Indeed, as shown by Broker et al. (1999), during a team pursuit competition, at a speed of 60 km h^{-1} (16.67 m s^{-1}), the mean mechanical power was 607 W for the first cyclist in the line, it fell to 70.8% (430 W) in the second position and to 64% (389 W) for the third and fourth cyclist. The extreme example of this state of affairs is the incredible speed of 296.009 km h^{-1} (sic!) attained in 2018 by Denise Mueller-Koronek riding a special bicycle in the wake of a specially prepared American dragster on the Great Salt Lake, in Utah.

As a first approximation, men and women are geometrically similar solids. Hence, their total body surface (A_{tot}) increases with the square of a given linear dimension of the body, whereas the overall body mass increases with the cube of the same dimension. Assuming further that, for a given position on the bicycle, A_f is a constant fraction of A_{tot}, (a somewhat debated question, see e.g., Capelli et al. 1998b; Heil 2001), it necessarily follows that cyclists of great body size are characterized by a smaller A_f per unit body mass, than their smaller colleagues. Hence, at a given speed, the energy expenditure per unit body mass will be less, and conversely, at a given metabolic power per unit body mass, the speed will be greater in cyclists of large body size. For cyclists of extremely different body size (50 kg and 1.50 m vs. 100 kg and 2.00 m), the metabolic power difference, per unit body mass, is about 18% (Fig. 6.5). This type of analysis accounts exclusively for the energy expenditure against the air resistance, neglecting entirely the rolling resistance and the energy expenditure against gravity, both linearly related to the body mass.

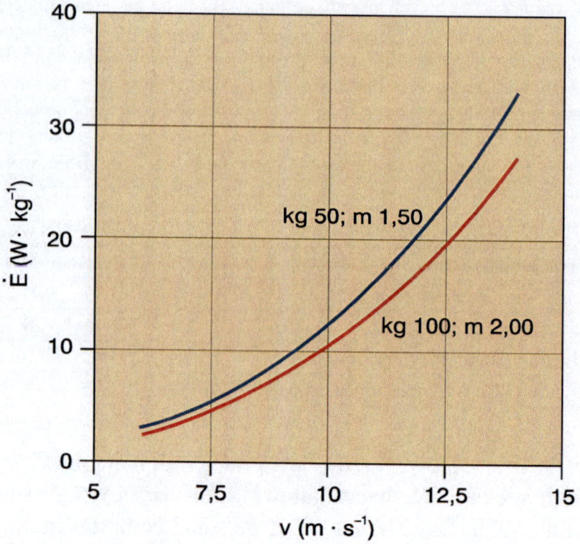

Fig. 6.5 Metabolic power per unit body mass ($\dot{E}$), as a function of the speed (v) in two cyclists of extremely different body size, as indicated. Traditional racing bicycle, fully dropped posture, constant speed, in the absence of wind at sea level. From di Prampero 2015, with the kind permission of Raffaello Grandi, Edi Ermes

6.3.6 Cycling at Altitude

As mentioned above, the F_a at any given velocity depends on the intrinsic characteristics of the moving object, and on ρ, itself a function of P_B and T (Eqs. 6.6–6.8). In turn, P_B and the corresponding ρ decrease with the altitude above sea level (Eq. 6.9), thus leading, *ceteris paribus*, to a corresponding decline of ρ and hence of the energy expenditure (be it mechanical or metabolic) necessary to overcome it. Hence, the speed attained with a given power and at any given temperature will be greater at altitude than at sea level.

However, the fall of P_B with altitude leads to a proportionally equal decrease of the oxygen partial pressure in inspired air (P_{IO_2}), thus leading to a fall of the subject's $\dot{V}O_2{}^{max}$ (Fig. 6.6, see also Chap. 7) and of the power that he/she can sustain in long-distance events (Cerretelli 1980; Ferretti 2014).

Therefore, the final effect of altitude on maximal aerobic performances is the net balance between the fall of ρ, which reduces air resistance, on the one side, and the fall of $\dot{V}O_2{}^{max}$, on the other. For altitudes up to 2500–3000 m, the decrease of the air resistance is greater than that of $\dot{V}O_2{}^{max}$: hence, the maximal aerobic velocity is greater than at sea level. At altitudes above 4000 m, the opposite is true, so that the maximal aerobic velocity starts decreasing with altitude. It follows that there must exist an optimal altitude at which the maximal aerobic speed attains a peak. This optimal altitude: (i) is more elevated, the larger the fraction of energy dissipated against the air resistance; (ii) is lower the more intense is the phenomenon of exercise-induced arterial hypoxemia during maximal exercise at sea level (Dempsey et al. 1984; Dempsey and Wagner 1999); (iii) it can be estimated for any form of locomotion in which these two quantities are known.

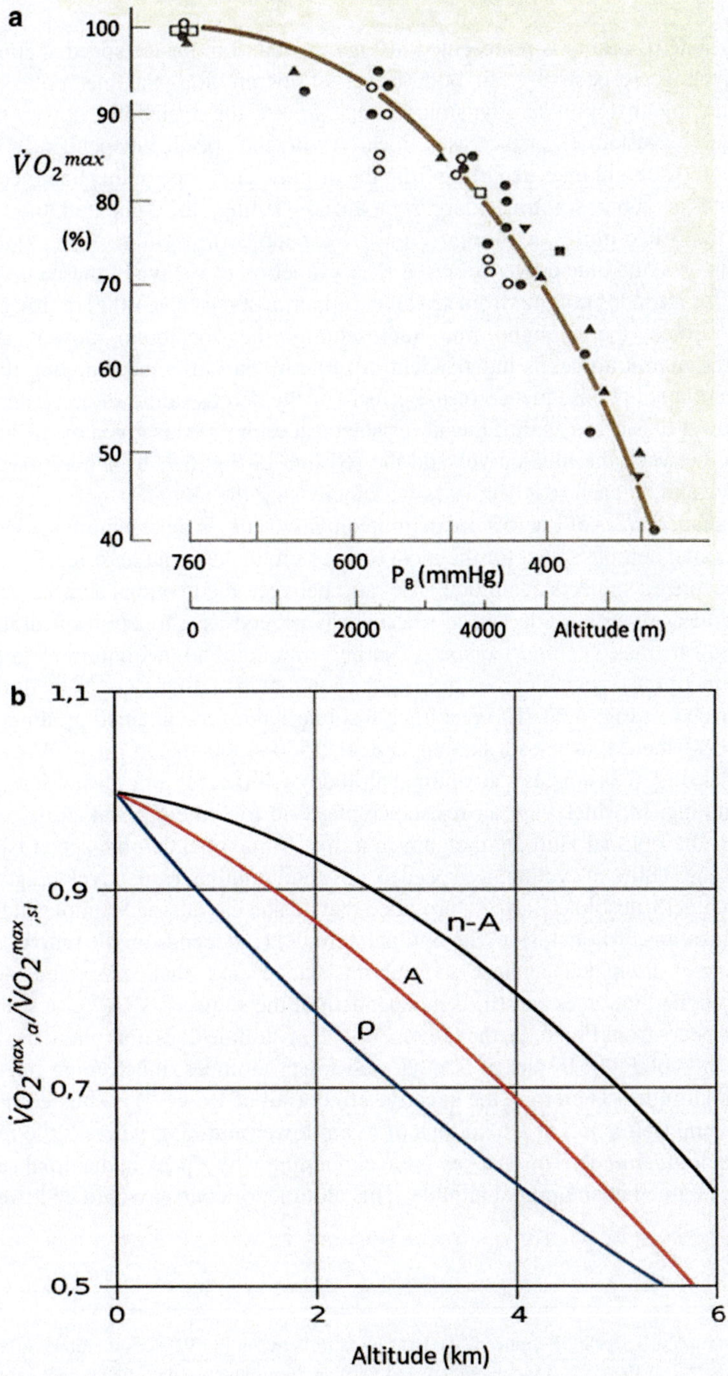

Fig. 6.6 (**a**) Maximal oxygen consumption ($\dot{V}O_2^{max}$, percentage of the sea level value) as a function of the barometric pressure (P_B), or of the corresponding altitude above sea level, in non-athletic well-trained subjects, acclimated (full dots) or non-acclimated (empty circles) to altitude. (From Cerretelli

This state of affairs is represented in Figs. 6.7a and b, for ice speed skating and track cycling, respectively. In both forms of locomotion, the metabolic power dissipated against non-aerodynamic forces, *id est* the metabolic power that is necessary to sustain $C_{n\text{-}a}$, increases linearly with the speed, independently of the altitude (red straight lines departing from the origin). The slope of this line is equal to $C_{n\text{-}a}$, and is about six times larger in speed skating, as compared to cycling (2.2 vs. 0.35 LO_2 min^{-1} at 10 m s^{-1} for a 75 kg subject).

In Fig. 6.7, the blue curves describe $\dot{E}$ as a function of v. Five of these curves are shown, for altitudes ranging from sea level (uppermost curve) to 4000 m (lowermost curve). Since the straight line representing the metabolic power against non-aerodynamic forces is independent of altitude, the differences among the five curves are a sole consequence of the effects of the aerodynamic forces. The metabolic power dissipated against the air resistance at each speed is given by the vertical distance between the blue curves and the red line in Fig. 6.7. It increases with the cube of v and, at each v, it decreases with increasing the altitude.

The shaded areas of Fig. 6.7 are delimited by two curves representing the $\dot{V}O_2{}^{max}$ and maximal aerobic speed for the most (curves on the left) and least (curves on the right) penalized subjects at altitude. The ratio between the maximal aerobic velocity attained at a given altitude and at sea level is plotted as a function of altitude in Fig. 6.8a for track cycling, ice speed skating, running, and swimming for a hypothetical medium-level athlete, on the assumption that his fall of $\dot{V}O_2{}^{max}$ at altitude is as described in Fig. 6.6a. The maximal absolute values are attained at altitudes of 2500–3000 meters in speed skating and of 3500–4000 meters in track cycling. Moreover, Fig. 6.8a suggests an optimal altitude of 400 m for long-distance running. In swimming, in which case air resistance plays no role, the optimal altitude is sea level. At the optimal altitude, the gain in terms of maximal aerobic speed is about 8.5; 4.0 and 0.80% in cycling, ice speed skating and running, respectively, regardless of the subject's absolute $\dot{V}O_2{}^{max}$, provided that he/she has attained complete altitude acclimatization. In conclusion, the optimal altitude: (i) depends on the ratio between the energy dissipated against the air resistance and that dissipated against non-aerodynamic forces and (ii) is independent of the subject's $\dot{V}O_2{}^{max}$ at sea level.

As appears from Fig. 6.6a, the fall of $\dot{V}O_2{}^{max}$ at altitude does not follow the fall of P_B and thus of P_IO_2. In fact, it is modest at lower altitudes, much more intense at elevated altitudes. Therefore, the negative effect (fall of $\dot{V}O_2{}^{max}$) is smaller than the positive one (reduction of ρ, and thus of k_w) at lower altitudes, whereas the reverse occurs at high altitude. This is a key issue explaining why a peak in maximal aerobic speed is attained at an optimal altitude. This altitude, of course, would be lower, the

---

Fig. 6.6 (continued) 1980; Ferretti 2014). (**b**) The ratio between the $\dot{V}O_2{}^{max}$ at altitude and at sea level ($\dot{V}O_2{}^{max}{}_{,a}$ / $\dot{V}O_2{}^{max}{}_{,sl}$) in athletic (A, red line) and non-athletic (n-A, black line) subjects is indicated as a function of the altitude above sea level together with the corresponding air density at 20 °C (ρ, blue line). Data from Capelli and di Prampero 1995 and Ferretti et al. 1997

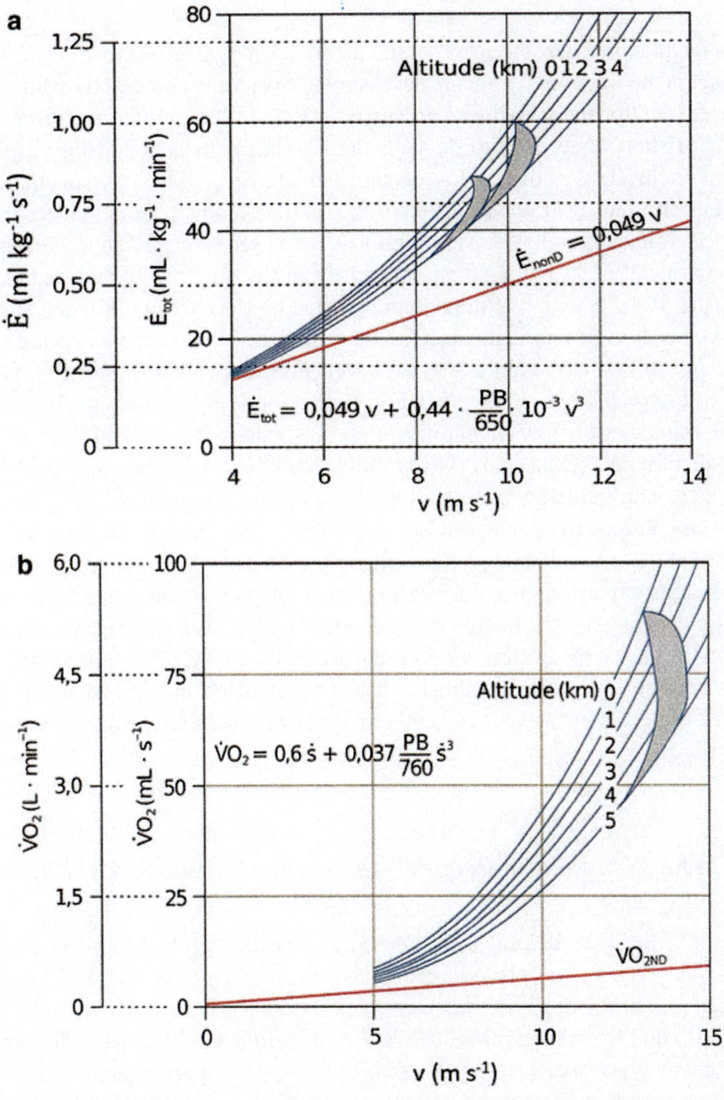

Fig. 6.7 (**a**) Overall metabolic power ($\dot{E}$) as a function of velocity (v) during ice speed skating at the indicated altitudes (blue curves), in the absence of wind at 0 °C. The straight red line refers to the metabolic power dissipated against non-aerodynamic forces. The vertical distance between the red line and the blue curves is the metabolic power dissipated against the air resistance. The uppermost and lowermost shaded areas refer to the decline of maximal oxygen consumption ($\dot{V}O_2{}^{max}$) at altitude in two ideal subjects characterized by a $\dot{V}O_2{}^{max}$ at sea level of 60 or 50 ml kg^{-1} min^{-1} above resting, respectively. Within each shaded area, the "best" acclimatizers lie on the upper curve and the "worst" ones on the lower curve. The equation on the figure allows an estimate of $\dot{E}$ (in ml kg^{-1} s^{-1}) as a function of v at an air temperature (T) of 0 °C. The barometric pressure (P_B) at which the experiments were performed was 650 mmHg. From di Prampero et al. 1976. (**b**) Metabolic power, expressed as oxygen consumption ($\dot{V}O_2$), as a function of v during track cycling at constant speed at the indicated altitudes (blue curves), in the absence of wind at 20 °C in a 70 kg, 1.75 m

lower is the maximal aerobic speed of the locomotion mode at stake at sea level, and the greater is the fraction of $\dot{E}$ that is necessary to overcome non-aerodynamic forces.

The reasons for the non-linear decrease of $\dot{V}O_2{}^{max}$ at altitude are discussed in Chap. 7. In short, they have to do with the non-linear behavior of the respiratory system at sea level, as a consequence of the characteristics of the oxygen equilibrium curve (Ferretti and di Prampero 1995). This carries along a remarkable corollary: endurance athletes, who have a very high $\dot{V}O_2{}^{max}$ at sea level, and thus are subject to exercise-induced arterial hypoxemia (Dempsey and Wagner 1999), are characterized by a greater $\dot{V}O_2{}^{max}$ fall at altitude than reported by Cerretelli (1980) and shown in Fig. 6.6a, because at maximal exercise they operate close to the steep part of the oxygen equilibrium curve already at sea level (Ferretti et al. 1997, see Fig. 6.6b). If this is so, the predictions made after Fig. 6.8a apply only to non-athletic subjects, however trained, with low or midway $\dot{V}O_2{}^{max}$ values at sea level, like top-level Himalayan climbers (Oelz et al. 1986) or high-altitude natives (Marconi et al. 2004). At variance, elite athletes, characterized by extremely high $\dot{V}O_2{}^{max}$ values, are penalized by altitude (percentage-wise) to a greater extent than medium-level sportsmen. Therefore, the advantage that elite athletes can obtain at high altitude is correspondingly reduced, as shown in Fig. 6.8b for track cycling.

Finally, it should be considered that, *ceteris paribus*, ρ is inversely proportional to T (Eq. 6.10); hence, for a reference ambient air T of $\approx$ 20 °C (293 °K), for any given value of P_B, a change of 3 °C leads to a change of about 1% in terms of air resistance. These effects, albeit low, can be crucial when dealing with world-record performances.

6.3.7 *The "One Hour Record" for Unaccompanied Cycling*

The extant "One Hour Record" (56,792 km) was established by Filippo Ganna close to sea level, as were the preceding ones by Chris Boardman (56,375 km) Tony Rominger (55,291 and 53,832 km), Miguel Indurain (53,040 km), Graem Obree (52,713 km) and Boardman himself (52,270 km). However, the metabolic power to proceed at any given speed is smaller (and conversely the speed attained with a given metabolic power is greater) at altitude. As mentioned above, the effects of altitude on cycling performances can be estimated provided the cyclist's $\dot{V}O_2{}^{max}$, together with its decrease with altitude, are known.

Fig. 6.7 (continued) subject. The straight red line and the blue curves have the same meaning as in the left panel. The vertical distance between the red line and the blue curves is the power dissipated against the air resistance. The decline of $\dot{V}O_2{}^{max}$ with altitude in an ideal athlete characterized by a $\dot{V}O_2{}^{max}$ at sea level of 5.17 L min^{-1} above resting is included in the shaded area, for the "best" (upper curve) and the "worst" (lower curve) acclimatizer, respectively. The equation allows an estimate of $\dot{E}$ (in ml s^{-1}) as a function of v at a T of 20 °C. The P_B at which the experiments were performed was 760 mmHg. From di Prampero et al. 1979

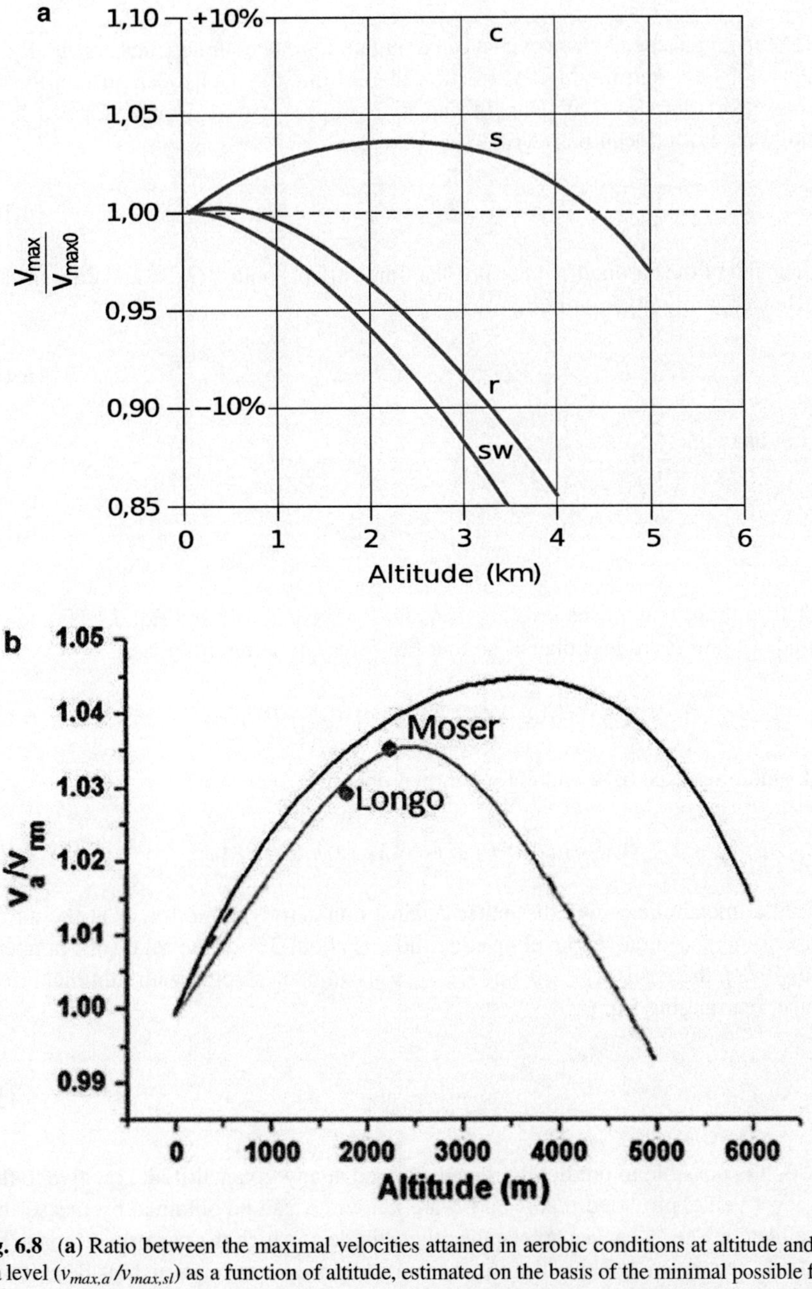

Fig. 6.8 (**a**) Ratio between the maximal velocities attained in aerobic conditions at altitude and at sea level ($v_{max,a}/v_{max,sl}$) as a function of altitude, estimated on the basis of the minimal possible fall of $\dot{V}O_2{}^{max}$ with altitude, in swimming (sw), running (r), speed skating (s) and cycling (c). Modified after di Prampero 2015. (**b**) Ratio between the maximal aerobic speed at altitude and at sea level ($v_{max,a}/v_{max,sl}$), as a function of altitude, in cycling. The black curve is from di Prampero (2000) and was obtained from the fall of $\dot{V}O_2{}^{max}$ at altitude reported by Cerretelli (1980). The blue curve, from Ferretti et al. (2011), is calculated from the $\dot{V}O_2{}^{max}$ fall in hypoxia reported by Ferretti et al. (1997) for athletes. The two experimental points were obtained from the one-hour unaccompanied record

The maximal speed that a cyclist can maintain for one hour depends essentially on his/her $\dot{V}O_2{}^{max}$ and the highest fraction thereof that the cyclist can sustain for the whole effort duration (φ) (Capelli and di Prampero 1995). Hence, for sea level conditions, Eq. 6.5 can be rewritten as follows:

$$\dot{E}_c = \varphi\, \dot{V}O_2^{max} = C_{n-a}\, v + k_e\, v^3 \tag{6.10}$$

The fall of the barometric pressure at altitude affects both $\dot{V}O_2^{max}$ and the constant k_e. Hence, at any given altitude (alt):

$$\dot{E}_{c,alt} = \varphi\, \dot{V}O_2^{max}{}_{,alt} = C_{n-a}\, v_{alt} + k_{e alt}\, v_{alt}{}^3 \tag{6.11}$$

Let us define:

$$A = \frac{\varphi\, \dot{V}O_2^{max}{}_{,alt}}{\varphi\, \dot{V}O_2^{max}} \quad and\; B = \frac{k_{e,alt}}{k_e} \tag{6.12}$$

This implies that, at sea level, both A and B are equal to 1 and Eq. 6.10 holds. At altitude, A and B are less than 1, so that Eq. 6.11 can be rewritten as:

$$\dot{E}_{c,alt} = A\, \varphi\, \dot{V}O_2^{max} = C_{n-a}\, v_{alt} + B\, k_e\, v_{alt}{}^3 \tag{6.13}$$

Combining Eqs. 6.10 and 6.13, we then obtain:

$$A\left(C_{n-a}\, v + k_e\, v^3\right) = C_{n-a}\, v_{alt} + B\, k_e\, v_{alt}{}^3 \tag{6.14}$$

Since the metabolic power dissipated against non-aerodynamic forces at the maximal aerobic speed, at sea level or at altitude, is about 3% of the total (di Prampero et al. 1979), the terms $C_{n-a}\, v$ and $C_{n-a}\, v_{alt}$ can be neglected with minimal error. Hence, rearranging Eq. 6.14:

$$\frac{v_{alt}}{v} = \sqrt[3]{\frac{A}{B}} \tag{6.15}$$

So, it is possible to predict the speed attained at any given altitude, relative to the sea level value, provided that A and B are known. A can be obtained by measuring the athlete's $\dot{V}O_2{}^{max}$ at sea level and at the altitude at which the performance is to be made. In fact it is advisable to refrain from using standard relationships describing the fall of $\dot{V}O_2{}^{max}$ at altitude (Cerretelli 1980), for the reasons discussed above.

Fig. 6.8 (continued) trials by Francesco Moser and Jeannie Longo, carried out at short intervals with the same bicycle at sea level (Bassano del Grappa, Italy; Bordeaux, France) and at altitude (Mexico City, 2.23 km; Colorado Springs, 1.86 km), respectively. From Ferretti and Capelli 2008

B can be computed from the relationship between ρ and altitude (see Fig. 6.6b). In addition, since it was also assumed that the air temperature, the bicycle, the cyclist's position (affecting C_x), and the A_f are the same in both conditions, B is given by the ratio between the P_B at altitude and at sea level (Eqs. 6.7 and 6.8, and Fig. 6.6b).

Thanks to Eq. 6.15, the ratio v_{alt}/v can therefore be estimated for any given altitude. Moreover, Fig. 6.8b shows also that, other things being equal, for people with intermediate $\dot{V}O_2{}^{max}$ values, the ideal altitude is close to 3800 m, not far from that of the Alto Irpavi velodrome (La Paz, Bolivia), as originally suggested by di Prampero et al. (1979). In contrast, for athletes with elevated $\dot{V}O_2{}^{max}$ values at sea level, the optimal altitude would correspond to about 2300 m, which is close to that of the Olympic velodrome in Mexico City. At this altitude, for athletes with very high $\dot{V}O_2{}^{max}$, $v_{alt}/v = 1.035$, so that Ganna, all other variables being equal to those in Grenchen (450 m), could have covered 58.780 km in 1 hour.

6.3.8 Cycling Uphill or Downhill

During uphill cycling at constant speed, part of the work is done to rise the body center of mass against gravity (vertical work, $w_v = Mgh$, where M is the body mass, g is the gravity acceleration and h is the height, see Chap. 5). Expressing w_v per unit of distance along the direction of motion (L), and since $h = L \sin \beta$ (where β is the angle between the terrain and the horizontal plane), the w_v per unit of L is equal to:

$$\frac{w_v}{L} = \frac{Mgh}{L} = \frac{MgL \sin \beta}{L} = Mg \sin \beta = F_v \tag{6.16}$$

In fact $\frac{w_v}{L}$, id est F_v, has the dimension of a force, as all other terms of Eq. 6.16, as well as the energy cost C and the drag. It corresponds to the force that is necessary to overcome the component of the acceleration of gravity.

The sum of F_v and of F_c, as from Eq. 6.1, yields a complete description of the mechanical work per unit of distance, when cycling at constant speed in the absence of wind. Therefore, the sum of Eqs. 6.1 and 6.16 provides the complete equation of motion of a cyclist:

$$F_c = F_{n-a} + F_a + F_v = F_{n-a} + k_w\, v^2 + Mg \sin \beta \tag{6.17}$$

On flat terrain, $Mg \sin \beta = 0 \text{ J m}^{-1}$, so that the equation of motion reduces to Eq. 6.1. On positive slopes, $\sin \beta$ is positive, so that F_c becomes higher than on flat terrain, and the more so the steeper the slope. On negative slopes, $\sin \beta$ is negative, so that F_c becomes lower than on flat terrain, and the more so the steeper (more negative) is the slope. In fact, during downhill cycling, F_v ($= Mg \sin \beta$) is a measure of the mechanical energy made available by gravity and that can be utilized against the other forces opposing motion. Eq. 6.17 can therefore be used for estimating the downhill freewheeling speed for any given incline of the terrain.[4]

For completeness, we also remind that the incline i of the terrain, corresponding to tanβ, affects also $F_{n\text{-}a}$, and hence the work against the rolling resistance, as follows:

$$F_{n-a} = F_{n-a0}\cos\beta = F_{n-a0}\left(\frac{\sin\beta}{\tan\beta}\right) = F_{n-a0}\left(\frac{\sin\beta}{i}\right) \quad (6.18)$$

where F_{n-a0} is the force that is necessary to overcome the rolling resistance on flat terrain. It follows that, on flat terrain, in which case cos $\beta = 1$, $F_{n-a} = F_{n-a0}$, whereas during vertical lift or fall, in which case $\beta = 90$ deg., and cos $\beta = 0$, F_{n-a} is nil. Combining Eqs. 6.17 and 6.18, we thus obtain:

$$F_c = F_{n-a0}\cos\beta + F_a + F_v = F_{n-a} + k_w\, v^2 + Mg\sin\beta \quad (6.19)$$

Within a realistic range of inclines, the term cos β is rather close to 1.0. For instance, for $i = 0.25$, we have $\beta = 14$ deg. and cos $\beta = 0.97$. Within this range of slopes, neglecting Eq. 6.18, as often done in physiological studies, carries along negligible errors.

Since $C_c = F_c/\eta$ (see Eq. 6.2), dividing Eq. 6.19 by η yields C_c:

$$C_c = C_{n-a}\cos\beta + k_e\, v^2 + \frac{Mg\sin\beta}{\eta} \quad (6.20)$$

On flat terrain, this Equation reduces to Eq. 6.3. The corresponding metabolic power $\dot{E}$ becomes then:

$$\dot{E} = C_c\, v = v\left(C_{n-a}\cos\beta + \frac{Mg\sin\beta}{\eta}\right) + k_e v^3 \quad (6.21)$$

On flat terrain, this Equation reduces to Eq. 6.5.

Equation 6.21 can be utilized for estimating the maximal up-slope that a cyclist can climb. To this aim, it is necessary to assign a minimal value to the speed, below which the equilibrium cannot be maintained (v_{min}). In this case, if the maximal $\dot{E}$ of the subject (expressed per kg overall weight, bicycle included) is known, the

[4]Consider for instance a 70 kg cyclist on a traditional racing bike in completely dropped posture, free-wheeling down an asphalt road with an 8.75% incline (the incline being equal to tan β, thus in this specific case, $\beta = 5$ deg). Inserting in Eq. 6.19 the appropriate numerical values of F_{n-a} and k_w, setting $F_c = 0$, sin $\beta = 0.0872$ and cos $\beta = 0.996$, one obtains the corresponding freewheeling speed: 18.44 m s^{-1} (66.4 km h^{-1}).

corresponding incline of the terrain can be easily obtained, by setting $v = v_{min}$ in Eq. 6.21.[5]

6.3.9 Metabolic Power and Body Mass

In this paragraph, we analyze the energetics of road cycling in relation to the cyclist's body size, as a function of the incline of the terrain, after assuming all other pertinent variables to be invariant. Let us first analyze the rolling resistance. Eq. 6.21 indicates that $\dot{E}_{n-a}$, besides being a linear increasing function of the speed, depends on body mass and on the incline of the terrain. On the contrary, the fraction of the metabolic power per unit body of mass dissipated against the air resistance (vertical distance between the straight lines labelled *na* and the colored curves in Fig. 6.9) is greater the smaller the cyclist's size and decreases with increasing the incline of the terrain.

Therefore, for any given metabolic power per unit body mass, the speed is greater the larger the overall body mass. This difference in speed is maximal when cycling on the level and decreases with increasing the incline of the terrain, to become negligible when $i \geq 10\%$, (Fig. 6.9). Thus, the subjects of large body size benefit of an intrinsic structural advantage, as compared to their smaller counterparts. However, the average specific $\dot{V}O_2{}^{max}$ (per kg body mass), is smaller, the larger the subjects' size (Åstrand et al. 2003). This counteracts in part the structural advantage of the cyclists of larger body size. Nevertheless, this does not exclude a priori the existence of potential cyclists characterized, at one and the same time, by a large size and an elevated specific $\dot{V}O_2{}^{max}$, thus allowing them to be highly performant both on flat and up-sloping terrains.

The above considerations are based on a set of assumptions that do not necessarily apply in practice: e.g., with increasing the slope of the terrain, the cyclist's position on the bike and the pedal frequency change, thus leading to inevitable changes of A_f and also, albeit small, of η (see Fig. 6.4). Nevertheless, this type of analysis may provide the trainer and/or the cyclist himself, with a good starting point for the evaluation of the actual performances.

The fall of C_a with increasing the subject's size occurs in all forms of locomotion on land: it is proportional, however, to the fraction of the overall energy expenditure

[5] Assuming $v_{min} = 1$ m s^{-1}, the maximal incline that a cyclist can overcome in aerobic conditions, for $\dot{V}O_2{}^{max} = 43$ ml kg^{-1} min^{-1} (15 W kg^{-1}) above resting, is about 42% ($\beta \approx 23$ deg) and it increases to about 70% ($\beta \approx 35$ deg) for $\dot{V}O_2{}^{max} = 63$ ml kg^{-1} min^{-1} (22 W kg^{-1}), where $\dot{V}O_2{}^{max}$ is expressed per kg of the overall mass (bicycle included). Obviously enough, these extreme performances require an appropriate combination of tires and gear system to match the type and incline of the terrain.

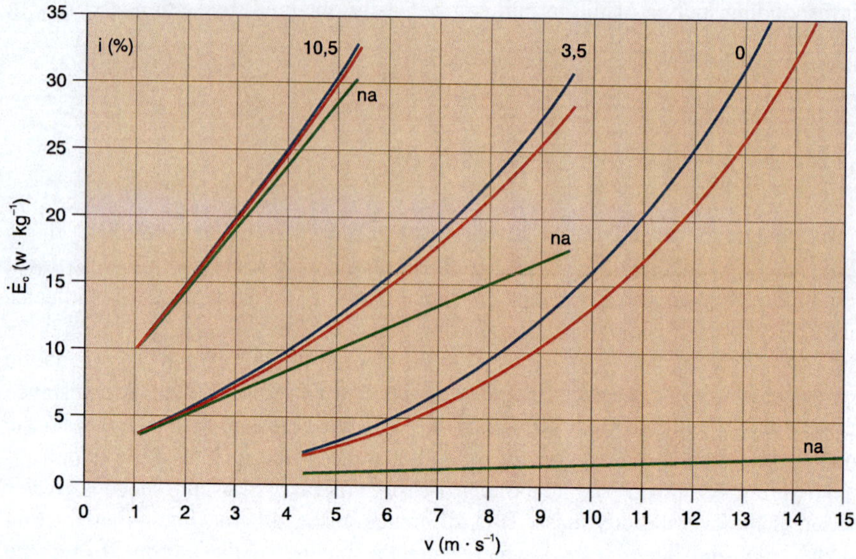

Fig. 6.9 Metabolic power per kg of body mass during cycling ($\dot{E}_c$) as a function of the speed (v) at sea level, on flat terrain or at the indicated inclines (i) for two cyclists of extremely different body size (50 kg, 1.50 m, 100 kg, 2.00 m) in the absence of wind, on a traditional racing bicycle, in fully dropped posture at constant speed. The power dissipated against non-aerodynamic forces (green straight lines, na) increases with i and is independent of the overall mass. On the contrary, the power dissipated against the air resistance, as given by the difference between the green lines and the red (larger subject) and blue (smaller subject) curves, is smaller in the former than in the latter subject. From di Prampero 2015, with the kind permission of Raffaello Grandi, Edi Ermes

dissipated against the air resistance. As such, it attains its maximal value in cycling; it is slightly less in speed skating and very nearly negligible in running.[6]

6.4 Cross-Country Skiing

Cross-country skiing is different from other forms of locomotion, in so far as the conditions in which the performance takes place are more difficult to standardize. Up-slopes and down-slopes, as well as the snow and air temperature, vary substantially among the different circuits and, as concerns temperature, within the same circuit. In addition, the upper limbs contribute substantially to the forward progression, and three techniques are usually employed: Diagonal Pole, Double Pole, and Skating. In the first of these three techniques, the upper limbs motion mimics what

[6]As an example, at speeds close to the maximal aerobic ones in these three forms of locomotion (45, 36 and 21 km/h, respectively), for subjects of extremely different size (50 kg; 1.50 m vs. 100 kg; 2.00 m), the advantage in terms of metabolic power (per kg body mass) for the larger subject is ≈ 14% in cycling, ≈ 9% in speed skating and ≈ 2% in level running.

happens during walking or running: the propelling action of the upper or lower limbs occur in an alternate way (right upper and left lower limb acting simultaneously and vice-versa). In the Double Pole, the push of both upper limbs occurs simultaneously, and in the Skating technique, the lower limbs motion mimics what happens in speed skating. The paragraphs that follow are devoted to a brief analysis of the energetics of cross-country skiing on flat terrain, in the absence of wind, and for standardized and constant snow conditions.

Saibene et al. (1989) showed that, under these conditions, the C of the three techniques is described by the following three empirical equations, which apply to Diagonal Pole, Double Pole and Skating, respectively:

$$C_{sk} = \frac{-8.05}{v} + 3.97 + \frac{0.216\,\mu'\,M\,g}{v} \tag{6.22a}$$

$$C_{sk} = \frac{-10.8}{v} + 4.01 + \frac{0.178\,\mu'\,M\,g}{v} \tag{6.22b}$$

$$C_{sk} = \frac{-11.36}{v} + 3.57 + \frac{0.237\,\mu'\,M\,g}{v} \tag{6.22c}$$

where C_{sk} is the energy cost of skiing, expressed in J kg^{-1} m^{-1}, and v is in m s^{-1}. The third term on the right of these three equations is the energy spent per unit of mass and distance to overcome the friction between snow and ski. This is the product of the overall weight in N ($M\,g$) and a dimensionless[7] friction coefficient (μ'), which is a function of the snow temperature. In the temperature range between 0 and –12 ° C, μ' increases from 0,015 to 0,05.

Under the conditions specified above, the C_{sk}, as predicted by the three Eqs. 6.22, is less than observed for running on the level; at variance with running, however, C_{sk} increases with the speed. In addition, for any given speed, Skating is the most economic technique; its energy cost being ~15% less than that of the Double Pole and of about 30–60% less than that of the Diagonal Pole.[8]

The lower C_{sk}, as compared to the C of running, is due to long sliding phase of the ski at each stride that brings about a reduction of the stride frequency, and thus of the mechanical work against gravity and inertia, per unit of distance. Indeed, the C_{sk} of the Diagonal Pole, the mechanics of which is rather close to that of running, at low snow temperatures ($T \approx -12$ °C, $\mu' \approx 0.05$) approaches the energy cost of running ($C_r \approx 4$ J · kg^{-1} m^{-1}). Conversely, for snow temperatures around 0 °C, in which $\mu' \approx 0.015$, thus allowing for a lesser friction between ski and snow and hence

[7]The coefficient μ' is the ratio beetween the horizontal force necessary to pull the subject forward (FT) and the weight of the subject himself: $\mu' = FT/(M\,g)$.

[8]Equations 6.22 describe translated hyperbolas with negative curvature. This implies that there must be a minimum speed below which C_{sk} becomes negative: hence, it is not possible to ski below such minimum speed. It implies also an asymptotic C_{sk} at high speed, which, however, for any technique, at competition speeds on flat terrain, remains always lower than the C of running (~4 J kg^{-1} m^{-1}), also on very cold snow (see Table 6.3).

Table 6.3 Energy cost (J kg^{-1} m^{-1}) of cross-country skiing on flat terrain at the indicated speeds (v) and at two different snow temperatures (0 °C or − 12 °C) for a 75 kg skier. μ', snow friction coefficient. Data from Saibene et al. 1989

	$T = 0$ °C $\mu' = 0.015$			$T = -12$ °C $\mu' = 0.050$		
	Diagonal	Double Pole	Skating	Diagonal	Double Pole	Skating
v (m s^{-1})						
3.0	2.10	1.08	0.82	3.94	2.63	2.75
5.0	2.84	2.24	1.82	3.95	3.16	3.04
7.0	3.17	2.75	2.32	3.98	3.42	3.22

leading to longer sliding phases, C_{sk} becomes substantially lower than C_r (Table 6.3).

As concerns the two other skiing techniques, at speeds of about 6–7 m s^{-1}, the Recovery, calculated as described in Chap. 5 for walking on the level (see Eq. 5.13), attains 40% in the Skating technique, being substantially lower (15–20%) in the Double Pole technique (Minetti, personal communication). This observation suggests a mechanical analogy between cross-country skiing and walking: in both cases, there exists a pendulum-like mechanism, similar to that observed in natural walking, with cyclic transformation of kinetic into potential energy, and vice-versa, at each stride. Nevertheless, the differences in Recovery values reported above explain the better economy of the Skating technique as compared to the Double Pole.

Finally, it should be noted that the effect of snow temperature on C_{sk} is substantial. Indeed, Eqs. 6.22 show that, in a speed range between 3 and 7 m s^{-1}, an increase of the snow temperature from −12 to 0 °C brings about a reduction of C_{sk} of about 50 to 70% at 3 m s^{-1} and of 20–30% at 7 m s^{-1}, depending on the technique (Table. 6.3).

For further details on human locomotion on snow, the reader is referred to Formenti et al. (2005).

6.5 The Energetics of Locomotion in Water: Introductory Remarks

It is a platitude to state that the velocities attained in swimming or rowing are substantially lower than their running or cycling counterparts. This is due, in part, to the density ρ of the medium (≈ 1000 kg m^{-3} for water at 4 °C, as compared to 1,27 kg m^{-3} for dry air at 760 mmHg and 20 °C), leading to a ≈ 800-fold greater resistance opposing motion in water. In addition, at least in the case of swimming, the anatomical differences between the upper and lower limb structure of humans is such that the overall η of progression in water is substantially less (attaining at most 8–10%) than observed in terrestrial locomotion (e.g., about 25% for cycling). Finally, in all forms of locomotion in water, the recovery of mechanical elastic energy, which plays a substantial role in the economy of running, is precluded, as is the energy exchange between potential and kinetic energy that characterizes walking (see Chap. 5).

The high density of water with respect to air entails another fundamental difference between air and water locomotion: the effect of buoyancy, which is almost nil in the former, substantial in the latter. The buoyant force (F_w) resists the body weight (F_B), which acts in the opposite direction. The net resulting force active on the body (F_{tot}) is therefore equal to:

$$F_{tot} = F_B - F_w = V\,\rho_B\,g - V\,\rho_w\,g \tag{6.23}$$

where V is volume, ρ is density (of course, $V\,\rho = M$) and g is gravity acceleration, whereas the subscripts B and w indicate body and water, respectively. It follows that:

$$\frac{F_B}{F_w} = \frac{\rho_B}{\rho_w} = \Phi \tag{6.24}$$

When $\Phi = 1$, $F_{tot} = 0$: the effects of the two forces F_B and F_w cancel out. This occurs when the body and the fluid have the same density: if no external forces intervene (for instance, muscle contraction force in humans), the body stays still where it is. Consider now a 1.7 m tall man, who is standing in air: his body density ρ_B is, on average, 1068 kg m^{-3} (Pollock et al. 1976). In this case, F_B is 99.88% of F_{tot} and $\Phi = 841$: there is practically no buoyancy and this man is exposed to the same external pressure at the head as at the feet, equal to P_B, and can stand only thanks to the contraction of his lower limb muscles. Consider now the same man, who takes the vertical position when fully immersed in water at 20 °C ($\rho_w = 998.2$ kg m^{-3}). In this case, $\Phi = 1.07$: the buoyant force of water F_w is just smaller than body weight: the body slowly sinks, and is exposed to a 129 mmHg higher pressure at the feet than at the head.

The average ρ_B value reported above varies with body composition, since the body consists of biological materials of remarkably different densities. For instance, muscle density $\approx$1000 kg m^{-3}, fat density $\approx$800 kg m^{-3} and bone density $\approx$3000 kg m^{-3}. Therefore, overweight people are more buoyant than lean individuals are, as are females with respect to males. The latter difference has an impact on the energy cost and the mechanical efficiency of swimming, as reported in Table 6.4 (see below).

Table 6.4 Energy cost of swimming the crawl at constant speed between 0.6 and 1.0 m · s^{-1}, expressed in absolute values (kJ m^{-1}) or per m^2 of body surface area (kJ m^{-1} m^{-2}), in males or females of medium-high technical level. From Pendergast et al. 1977

	Males	Females
kJ · m^{-1}	1,10	0,78
kJ · m^{-1} · m^{-2}	0,57	0,46

The above analysis neglects the effects of lung air filling. Indeed, in a static human body immersed into water, the center of application of F_B is more caudal than that of F_w, because of air in the lungs. Therefore, if they are aligned on the vertical axis, along which g acts, the body keeps the vertical position, because the lever arms of F_B and F_w are nil. If they are not aligned, the body starts rotating toward the vertical position yielding the static equilibrium. If our immersed human moves the arms, or one leg, or bends forward, his body center of mass is displaced, which also causes rotation of the body. If the arms are brought out of the water, F_w is reduced and its center of application changes. Breathing induces cyclic movements of the center of application of F_w, due to the cyclic changes in lung volumes. Therefore, in order to maintain the chosen position (e.g., horizontal in swimming), the body has to do work, and thus spend metabolic energy (Pendergast et al. 2015). During swimming, this is not work done to move along the swimming direction, so it can be assimilated to w_i, which contributes to the low values of the overall η of progression in water.

Strategies have been developed to modify F_w or F_B. For instance, breath-hold divers carry added weights to facilitate descent: this increases F_B, and thus F_{tot} and Φ. Suicides by drowning also do the same. During ascent, extreme breath-hold divers, who employ the free-diving technique (see Chap. 12), use inflatable diving suits, with a large sealed pocket, which is filled with gas from a small tank on the bottom: this reduces F_B remarkably, so that $F_B < F_w$, $F_{tot} < 0$ and $\Phi < 1$, the buoyant force prevails and pushes the diver up to the surface. F_w is larger, the higher the salt concentration in water: this is why we are more buoyant in sea water than in lake or river water.

When a human walks in the air, the drag of air is so small that it can reasonably be neglected. When a human walks in water, the drag (fluid resistance) increases substantially in proportion to the level of immersion. So, *ceteris paribus*, the energy cost of walking in water is higher and the maximal walking speed is lower, the higher the level of immersion. This is a key factor in assisted water locomotion, wherein the speed attained by a human-powered or a wind-powered boat is higher, the smaller is the level of immersion, and thus the frontal area of the boat's keel exposed to water. This factor has been downplayed to its extreme consequences by the foil systems used on the sailing yachts competing for the 2021 America's Cup in Auckland, New Zealand. This allowed attaining the highest speeds ever reported in a sailing competition. ~ In these mono-hull yachts, if the wind is above 6 knots (~3 m s^{-1}), the lateral foils can keep the keel mostly out of the water, so that the drag is minimized. The sailing speeds reported in Auckland (up to 25 m s^{-1}) have been the highest ever attained in a sailing competition.

In the paragraphs that follow, we briefly analyze the contributions of the School of Milano to our understanding of the bioenergetics and biomechanics of swimming and of some forms of assisted locomotion in water. Concerning swimming, special attention will be given to the differences among: (i) swimmers of medium and high technical level, (ii) the four classical swimming styles, and (iii) men and women. For a more detailed analysis of the biomechanics of swimming, the reader is referred to Seifert et al. (2011).

6.6 Energetics and Biomechanics of Swimming

The energy cost of swimming (C_{sw}) is greater than in any other form of human locomotion. To give an idea of the C_{sw} range, average values of C_{sw} for medium-level male and female front crawl swimmers are reported in Table 6.4. Obviously enough, because of the high values of C_{sw}, the maximal speeds attained in swimming are substantially lower than in all other forms of human locomotion; and this is even more true, if we consider that C_{sw} increases with the speed. The lower C_{sw} of females than of males, due to differences in body composition and structure, is also noteworthy.

The data reported in Table 6.4 are not representative of the entire range of C_{sw} values that may be observed in swimming. In no other form of natural locomotion, technical skill has a greater impact on the energetics of performance than in swimming. Top-level swimmers have a definitely lower C_{sw} than the medium-level subjects of Table 6.4 (see, e.g., Zamparo et al. 2005a). In addition, the four classical competition styles have different C_{sw} at any given swimming speed (Capelli et al. 1998a). These aspects will be analyzed separately in the next paragraphs.

6.6.1 Technical Skill

The C_{sw} of a group of medium-level college swimmers (Pendergast et al. 1977) and of elite swimmers (Holmér 1974) are reported in Fig. 6.10 as a function of speed. In both groups, at speeds $> \approx 1.0 \text{ m} \cdot \text{s}^{-1}$, C_{sw} increases substantially; in addition, regardless of the speed, C_{sw} is about 20–40% smaller in elite swimmers. This highlights the fact that these are characterized by a more economical technique as compared to their medium to low-level colleagues. Thus, they spend less metabolic energy to perform a given amount of work, which reduces C_{sw} and increases η. Therefore, they are able to reach greater speeds with the same metabolic power output.

6.6.2 Swimming Style

The C_{sw} of the different swimming styles as a function of speed is reported in Fig. 6.11 for a group of male swimmers of high technical level (Capelli et al. 1998a). These authors have condensed a large set of data in a series of empirical equations, reported in the Figure's legend. These equations allow an estimate of C_{sw}, provided that the speed is known. It should be noted that these equations refer to swimming in a straight line at constant speed; as such, they do not take into account the diving start and the push on the wall of the pool, occurring during swimming competitions, both factors leading to a substantial reduction of the average C_{sw}.

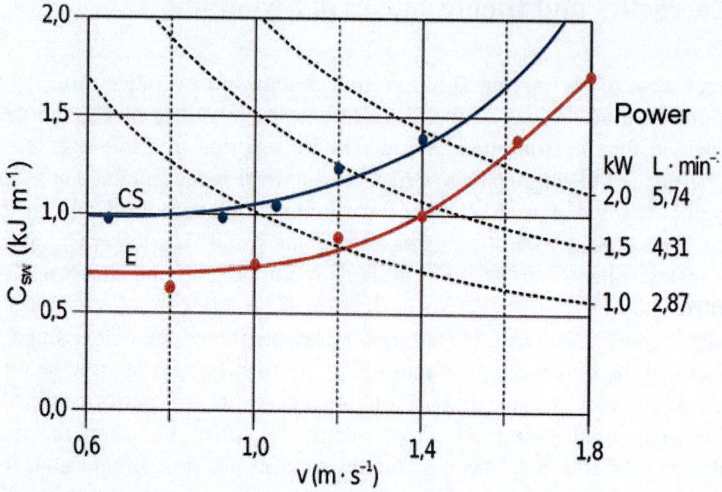

Fig. 6.10 Energy cost of swimming the crawl (C_{sw}) as a function of speed (v) in medium-level (College swimmers, CS, $n = 5$) or elite swimmers (E, $n = 9$; for the speeds of 1.6 and 1.8 m s^{-1}, $n = 3$). The three hyperbolic functions (corresponding to the indicated metabolic power values, kW or L min^{-1}) show that, for any given power, the speed attained by an élite swimmer is about 0.4 m s^{-1} greater than that attained by a College Swimmer. Modified after di Prampero 2015; data from Holmér 1974 and Pendergast et al. 1977

These data show that the most economical style is the overarm crawl, followed by the backstroke, whereas the breaststroke, at all speeds, is the more energy-demanding style. In this context, it is interesting to note that, in swimmers of low technical level at low speeds (0.4–0.6 m s^{-1}), the breaststroke is characterized by C_{sw} values similar to those observed, in these same swimmers, during the front crawl (Holmér 1974). This is likely because the breaststroke is a more "spontaneous" swimming technique; therefore, a low-level swimmer is somehow less so when swimming the breaststroke than the front crawl at low speeds.

6.6.3 *Of Women and Men*

The data reported in Figs. 6.10 and 6.11 were obtained in male swimmers (Capelli et al. 1998a; Holmér 1974; Pendergast et al. 1977); however, in medium-level swimmers, the C_{sw} of front crawl in women is substantially less than in men (Table 6.4). This difference between women and men is due to the different anthropometric characteristics of the female and male bodies. Indeed, the density of a human body is unbalanced, being higher below (toward the feet) than above the body center of mass, due to the air inside the lungs. Thus, a body immersed horizontally in water is subjected to a push in the upper direction at the level of

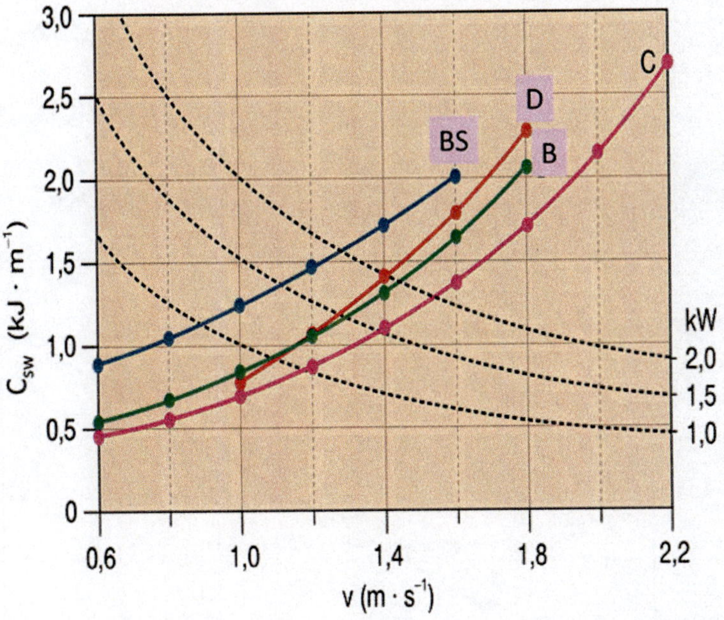

Fig. 6.11 Energy cost of different swimming styles (C_{sw}) as a function of the speed (v) in elite swimmers: crawl (C: y = 0.228 * $10^{0.488x}$, $n = 8$); back-stroke (B: y = 0.270 * $10^{0.491x}$, $n = 3$); breaststroke (BS: y = 0.548 * $10^{0.355x}$, $n = 6$); dolphin (D: y = 0.234 * $10^{0.547x}$, $n = 3$). The three hyperbolic functions (corresponding to the indicated metabolic power values, kW) show that, for any given power, the speed attained swimming the crawl is about 0.4 m s^{-1} greater than in the breaststroke. From di Prampero 2015, with the kind permission of Raffaele Grandi, Edi Ermes; data from Capelli et al. 1998a

the lungs, which, being characterized by $F_B < F_w$, tend to float, whereas the lower limbs, in which $F_B > F_w$, tend to sink. Pendergast et al. (1977) defined center of air the point of application of the upward pushing force, generally located in the lungs at the level of the mammillary line. The floating and the sinking forces generate a torque that, in the absence of other forces, leads to a rotation of the body around its center of mass until a vertical position is reached, wherein the center of air, where F_w is applied, is positioned exactly along the same vertical axis as the center of mass, where F_B is applied. Incidentally, this state of affairs is empirically very familiar to everybody who has ever attempted to stay still in water in a horizontal position. Nevertheless, in swimming, the most favorable position is the horizontal one, because it minimizes the frontal surface area in the direction of movement. It follows that a fraction of the overall energy expenditure must be dissipated for maintaining the body in a horizontal position.

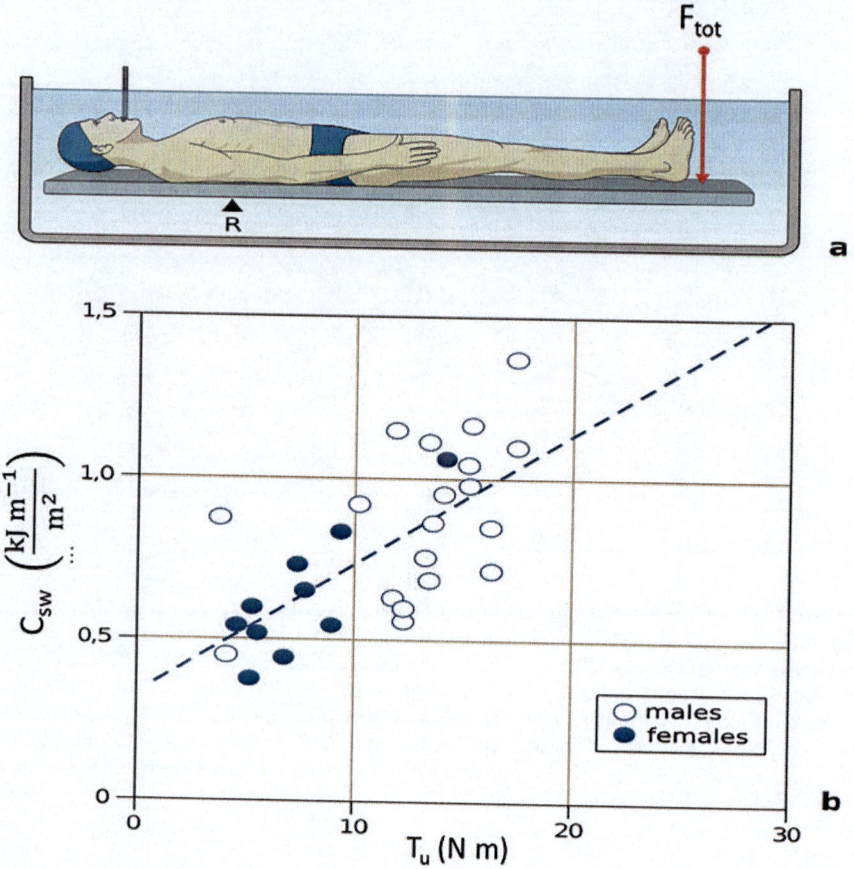

Fig. 6.12 (**a**) The underwater torque (T_u, N m) is the product of the sinking force (F_{tot}, N) as measured at the feet of a subject immersed in water ad supported by a horizontal frame free to rotate around an axis (R in the figure) positioned at the level of the center of air, times the distance between R and the point of application of the sinking force itself (the feet). (**b**) Energy cost of swimming the crawl per unit of distance and per square meter of body surface area (C_{sw}) in male (open dots) and female (full dots) swimmers of medium technical level at the speed of 1.0 m s^{-1}, as a function of T_u. Modified after Pendergast et al. (1977)

The morphologies of the female and male bodies are different from each other, among other things because of a larger fraction of fat in females and of a different distribution of fat and of muscles at the level of buttocks, hips, and thighs. Indeed, in general, women are characterized by larger fat depots in these areas and by shorter lower limbs with less developed muscles. As a consequence, ceteris paribus, women need a lesser amount of energy to maintain a horizontal position in water, a fact that per se leads to a reduction of C_{sw}.

The reasons for this are quantitatively represented in Fig. 6.12, where C_{sw}, expressed per unit of body surface area, is reported as a function of the tendency

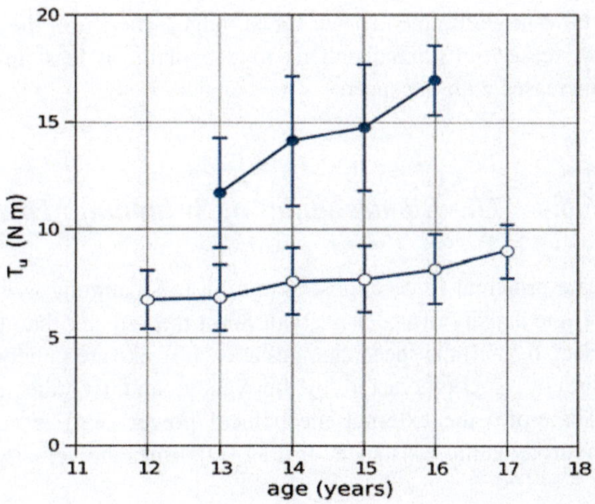

Fig. 6.13 Underwater torque (T_u, Nm) +/− 1 standard deviation as a function of the age (years) in 67 girls (open dots) and 43 boys (full dots). Modified after di Prampero 2015; data from Zamparo et al. 1996

of the feet to sink, for men and women. This, appropriately defined "torque" in the original paper by Pendergast et al. (1977), has been assessed by means of the so-called "swimming scale" (Fig. 6.12a), allowing a measure of the force with which the feet of a subject immersed horizontally in water tend to sink. The product of this force times the distance between the center of air and the application point of the sinking force itself (the feet) yields the so-called underwater torque. Fig. 6.12b shows that women, as compared to men of a similar technical level, are characterized by a lesser torque and, at one and the same time, by a lesser value of C_{sw}.

These differences appear after puberty, as long as pre-pubertal boys and girls are characterized by essentially equal torque and C_{sw} values. At puberty, the increase of the lower limbs muscle mass and length is smaller in girls than in boys, so that the underwater torque and C_{sw} become smaller in the former (Zamparo et al. 1996). The relationship between underwater torque and age in adolescence is shown in Fig. 6.13.

Thus, for anthropometric reasons, the female's body, as compared to the male's one, is structurally more suited to swimming. Within the same sex, there exists a large variability in terms of torque; hence, a simple measure of torque informs on the structural swimming attitude of the subject in question, regardless of any other technical or physiological characteristics.

The role of the underwater torque on the energy cost of swimming has been unambiguously determined only for the front crawl, even if it seems likely that it may be relevant also in other swimming strokes. Moreover, it is noteworthy that Zamparo et al. (2000) showed that the effect of the torque on C_{sw} is particularly relevant within the aerobic speed range, whereas at the maximal absolute speeds, it decreases substantially. This may explain, at least in part, why in swimming η increases with the speed.

6.6.4 *The Biomechanics of Swimming: Drag and Efficiency*

The principal force opposing motion in swimming is the water resistance. Indeed: (i) the density of water is about 800 times greater than that of air, as pointed out in Sect. 6.5; (ii) the speed changes at each stroke are rather minor; (iii) the work against gravity is counteracted by buoyancy; and (iv) the air resistance is negligible. Therefore, the external mechanical power ($\dot{w}_e$) is equal to the product of the hydrodynamic resistance, or drag (D), times the velocity in respect to the water (v):

$$\dot{w}_e = D\,v \tag{6.25}$$

In swimming, as in the other forms of locomotion considered so far, a fraction of the overall work is utilized to move the upper and lower limbs in respect to the center of mass of the subject. This fraction, generally defined internal work (see Chap. 5), although necessary for structural reasons, does not contribute directly to progression, even if it requires an additional energy expenditure; it will not be explicitly considered in the paragraphs that follow, devoted, as they are, to an analysis of $\dot{w}_e$ in swimming, as described by Eq. 6.25. The interested reader is referred to Seifert et al. (2011).

Karpovich and Pestrecov (1939) were the first to determine D. They measured in fact the force necessary to pull a swimmer motionless in water at constant speed, by means of a dynamometer. In this type of measurements, however, the subject maintains a predetermined position, whereas in real swimming conditions, the movements of the head, trunk, upper and lower limbs lead to an increase of the hydrodynamic resistance. Therefore, the values of D obtained by this approach are smaller than those observed under real swimming conditions, and are generally defined as passive drag (see below).

A more realistic method to assess D under real swimming conditions, apt to be used in the circular swimming pool at Buffalo (see Fig. 6.2), was proposed by di Prampero et al. (1974). In this case, an appropriate system of cables and pulleys allows application of known forces, acting along the axis of motion, to the swimmer's body. These forces may either aid (negative added drag) or hinder (positive added drag) the swimmer's motion, depending on the direction in which they are applied. At any given speed, the relationship between net metabolic power and added drag is linear (Fig. 6.14).

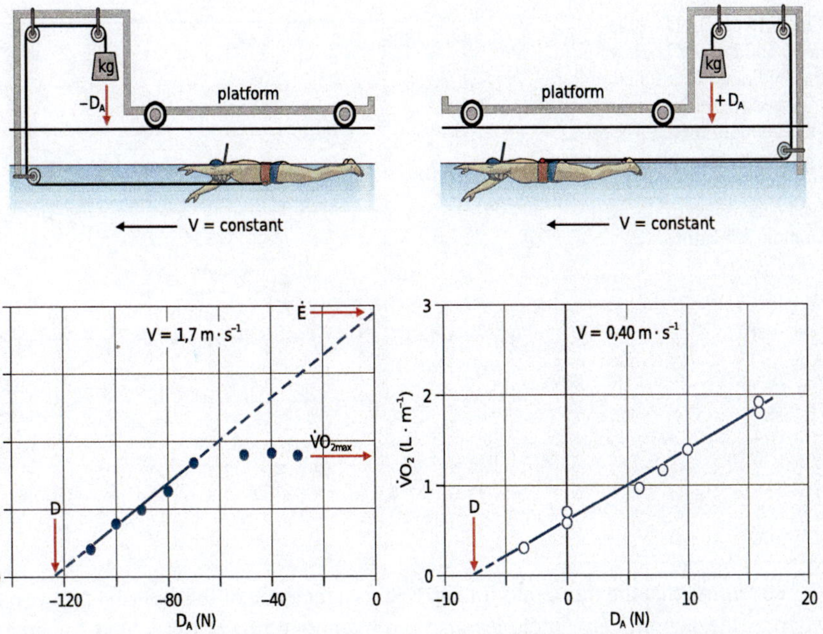

Fig. 6.14 Oxygen consumption above resting ($\dot{V}O_2$, L min^{-1}) during crawl swimming at constant speed (1.7 or 0.4 m s^{-1}) as a function of the force applied to the swimmer's body in the direction of motion (added drag, D_A, N). The two upper figures illustrate schematically the set-up for applying negative (left panel) or positive (right panel) D_A, and thus assessing the swimmer's natural drag (D). Note that the platform moves at the same speed as the swimmer. D is obtained by back-extrapolation of the relationship $\dot{V}O_2/D_A$ to the resting $\dot{V}O_2$, as measured motionless in water. When the exercise intensity is greater than the swimmer's $\dot{V}O_2{}^{max}$ (lower horizontal red arrow) the extrapolation of the $\dot{V}O_2/D_A$ function to $D_A = 0$ yields the metabolic power (in oxygen equivalents) for swimming at the speed in question without any external forces ($\dot{E}$,upper horizontal red arrow). After di Prampero 2015, with the kind permission of Raffaello Grandi, Edi Ermes

Back extrapolation of these relationships to the subject's resting energy expenditure, as determined motionless in water, yields the value of added drag that is necessary to counteract the natural drag at that speed. Knowledge of the swimmer's D and of v allows an estimate of the external mechanical power developed by the swimmer (Eq. 6.25). Hence, if the corresponding metabolic power $\dot{E}$ is also known, η can easily be obtained as the ratio between $\dot{w}_e$ and $\dot{E}$:

$$\eta = \frac{\dot{w}_e}{\dot{E}} = \frac{D\,v}{\dot{E}} \tag{6.26}$$

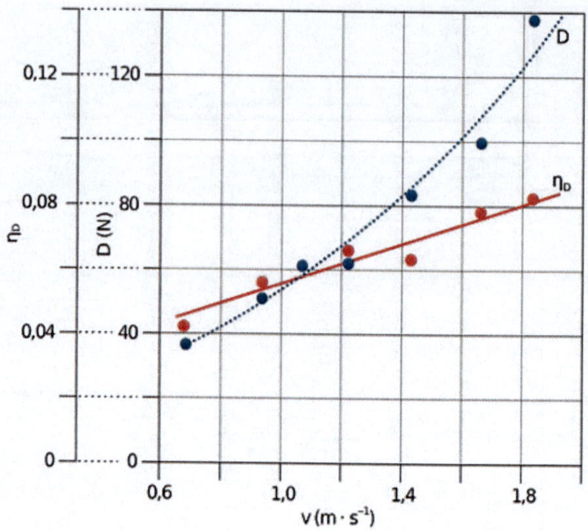

Fig. 6.15 Drag (*D, blue dots*) and drag efficiency (η_D, red dots) as a function of speed (v) during crawl swimming in medium-level swimmers. From di Prampero 2015, with the kind permission of Raffaello Grandi, Edi Ermes

We remind that the numerator of Eq. 6.26 is a measure of the external power only, whereas the denominator includes also the metabolic power dissipated for internal work performance; as such the resulting efficiency values are smaller than the "true muscular" efficiency, also because the effect of the so-called "propelling efficiency" (see Sect. 6.6.5) is neglected. They nevertheless yield a measure of the swimming efficiency under real conditions. For a more detailed analysis, the reader is referred to di Prampero et al. (2011).

The average η of the overarm crawl, calculated according to Eq. 6.26 in a group of swimmers of medium technical level, increased with the speed from $\approx$ 4% at 0.6 m s^{-1} to $\approx$ 8% at 1.9 m s^{-1} (Fig. 6.15). These values are remarkably lower than observed in fishes, in which η during swimming at the optimal speed is about 20–25% (Schmidt-Nielsen 1997).

One of the main reasons of the low η of swimming in humans is the limited propelling action of the lower limbs. Indeed, at least as concerns the overarm crawl, when the propulsion is due to the lower limbs only, because of their relatively large muscle mass and low propelling action, the overall metabolic power requirement is two to four times larger than it would be, at the same speed, utilizing the upper limbs only. When the efficacy of the lower limbs action is increased by means of appropriate fins, the overall mechanical efficiency of swimming can attain 13% (Zamparo et al. 2002).

In addition to the efficiency values mentioned above, Fig. 6.15 reports also the D values determined during front crawl swimming. In the same group of swimmers, D increases from $\approx$ 30 N at 0.6 m s^{-1} to $\approx$ 120 N at 1.8 m s^{-1}, as empirically described by:

$$D = 58v^{1.2} \tag{6.27}$$

where D is in N and v in m s^{-1}.

The overall D encountered by an object moving on the water surface depends on three main factors: (i) the physical characteristics of the water and of the object's wetted surface; (ii) the shape of object's wetted surface; (iii) the generation of waves in the wake of the moving objects, also due to the shape of the wetted surface. Functions such as that described by Eq. 6.27 do not allow disentangling the role of the three factors summarized above; therefore, they must be considered as purely phenomenological descriptions.

The drag due to wave formation becomes vanishingly small when swimming underwater, at least 1 m below the surface and 1 m above the bottom of the pool. Hence, other things being equal, swimming underwater is more economical than swimming on the surface, by about 10–15% on the average. Indeed this is the rationale for forbidding any form of underwater swimming during competitions (Pendergast et al. 1996).

It seems also interesting to note that Pendergast et al. (1989) applied this same approach to estimate the drag during kayaking. The so obtained values turned out to be substantially lower than observed when swimming the crawl: ranging from ≈ 15 N at 1.0 m s^{-1}, to ≈ 60 N at 2.2 m s^{-1} in elite athletes and about 10–20 N larger in lower level athletes (for crawl swimming see Eq. 6.27).

6.6.5 Propelling Efficiency

In swimming, as in other forms of water locomotion, $\dot{w}_e$ consists of two components: one does contribute to the actual propulsion of the swimmer along the direction of motion by overcoming D, the other does not, although it also transmits kinetic energy to water. Zamparo et al. (2002) defined the former as $\dot{w}_D$ and the latter as $\dot{w}_k$. Of course, as in any kind of locomotion (see Chap. 5), the total mechanical power $\dot{w}_{tot}$ is the sum of internal power ($\dot{w}_i$) and $\dot{w}_e$. So, we can write:

$$\dot{w}_{tot} = \dot{w}_i + \dot{w}_e = \dot{w}_i + \dot{w}_D + \dot{w}_k \tag{6.28}$$

The fraction of $\dot{w}_{tot}$ represented by $\dot{w}_D$ is defined as propelling efficiency (η_p), which is equal to:

$$\eta_p = \frac{\dot{w}_D}{\dot{w}_{tot}} \tag{6.29}$$

To give a plain visual idea of the concept, consider that the propulsive effect of the limbs in swimming is due to the amount of water that they are able to push backwards. A trivial example may help illustrate the matter at stake. If in swimming, the swimmer's limbs could push against pillars fixed to the bottom of the pond, rather than on the water, the overall amount of work performed could be utilized for moving the swimmer's body forwards. In other words, the swimmer's limb would become an ideal lever, the fulcrum of which is the motionless pillar: if this is so, η_p would be 100%. This is not the case in actual swimming, because the swimmer's limbs push against the water, which is not fixed, so that the fulcrum of the lever moves also backwards and thus the propelling efficiency turns out lower than 100%.

Equation 6.29 depicts a purely mechanical relationship and does not account for the metabolic power $\dot{E}$. In fact, the efficiency with which $\dot{E}$ is transformed into $\dot{w}_D$ is termed drag efficiency (η_D). Thus:

$$\eta_D = \frac{\dot{w}_D}{\dot{E}} \tag{6.30}$$

It is noteworthy that Eq. 6.30 is conceptually different from Eq. 6.26, as long as $\dot{w}_D$ is only a fraction of $\dot{w}_e$: therefore, η_D is lower than the "spurious" η provided by Eq. 6.26.

Dividing Eq. 6.30 by Eq. 6.29, and combining with Eq. 6.28, we obtain:

$$\frac{\eta_D}{\eta_p} = \frac{\dot{w}_{tot}}{\dot{E}} = \frac{\dot{w}_i + \dot{w}_e}{\dot{E}} = \eta_O \tag{6.31}$$

where η_O is the overall (or gross) mechanical efficiency. Of course, η_O is higher, not only than η_D, but also than the η provided by Eq. 6.26. For more details on efficiency in swimming, see Zamparo et al. 2020 and Zamparo et al. 2002, 2011.

The earliest attempts to compute efficiency in swimming made use of an analogous of Eq. 6.30, after assuming $\eta_O = \eta_D$, and thus $\eta_p = 1$. Very low values of η_D were obtained (0.01–0.02, Karpovich and Pestrecov 1939). The assumption of the equivalence of Eqs. 6.26 and 6.30 provided η_D values ranging from 0.03 to 0.09, depending on speed and on the method used to measure D (active or passive drag) (di Prampero et al. 1974; Holmér 1974; Pendergast et al. 1977).

Toussaint et al. (1988) first realized that only a fraction of the kinetic energy transmitted to water was indeed used to propel the body in the direction of movement, so that $\dot{w}_e > \dot{w}_D$, and thus that $\eta_O > \eta_D$. In fact, they set out experiments to determine η_p independently. They obtained values ranging between 0.45 and 0.75, which confuted the hypothesis that $\eta_p = 1$ and $\dot{w}_e = \dot{w}_D$ and demonstrated that $\dot{w}_k$ is large indeed. Differences in η_p depend on sex and technical skill and carry along differences in η_O, and thus in C_{sw}. However, the method used by Toussaint et al. (1988) did not account for $\dot{w}_i$, so that their η_p corresponded to $\dot{w}_D/\dot{w}_e$ rather than to $\dot{w}_D/\dot{w}_{tot}$. Zamparo et al. (2005b) and Zamparo (2006), who treated the arm stroke as if it were a rotating paddle wheel (Martin et al. 1981), obtained η_p values of 0.45 instead of 0.75 for competitive swimmers. This difference is fully accounted for by considering $\dot{w}_i > 0$ and thus $\dot{w}_{tot} > \dot{w}_e$, contrary to Toussaint et al. (1988).

According to Zamparo et al. (2008), after accounting for $\dot{w}_i$, η_p is similar in males as in females of equivalent age and skill, being about 0.30 before puberty,

0.38–0.40 at 20 years and around 0.25 at age above 40 years. Obviously enough, the higher η_p, and thus η_O, the lower C_{sw}. η_p is a critical determinant of C_{sw} indeed.

6.7 Energetics and Biomechanics of Assisted Locomotion in Water

In swimming, the propelling force of the upper and lower limbs is directly applied to the water; on the contrary, in assisted locomotion, the force generated by the limbs is transmitted to the water via a system of levers (oars, paddles, etc.), leading to the progression of the boat supporting the oarsmen. The major advantage of the several forms of assisted locomotion in water, in addition to a more favorable hydrodynamics, is an overall efficiency of progression greater than in swimming. *Ceteris paribus*, this leads to a substantial increase of the speed.

The aim of the paragraphs that follow is to discuss the energy cost of some forms of assisted locomotion (rowing, kayaking, canoeing, sculling a Venetian gondola or a "bissa"[9]) and to describe briefly the biomechanical characteristics of some of them.

6.7.1 The Energy Cost

The energy cost of rowing, kayaking, canoeing, sculling a venetian gondola or a "bissa" is summarized in Fig. 6.16 as a function of the speed in still water (Buglione et al. 2011; Capelli et al. 1990; Capelli et al. 2009; di Prampero et al. 1971). All data were obtained on subjects of high technical skill. As concerns rowing, the data refer to the energy cost of one rower on a "two oars with coxswain": as a consequence, the energy cost for moving the boat itself is twice the value reported in Fig. 6.16; on the contrary, for the other forms of locomotion implying one rower only, the energy cost reported in the figure is the total necessary for the progression of the boat.

This figure shows that canoeing is the most demanding among the sporting forms of locomotion considered (rowing, kayaking, canoeing), whereas rowing is the most economical. The energy cost determined by Zamparo et al. (1999) on medium-level athletes (not shown in Fig. 6.16) is slightly larger than determined by Buglione et al. (2011) on 69 elite kayakers and reported in the figure itself. The average energy cost determined by these authors on male and female athletes were the same, the data reported in Fig. 6.16 being the grand average over males and females.

The energy cost of the other two forms of assisted locomotion under consideration (gondola and "bissa": flat skull boats of a more practical "flavor," traditionally used in still waters: lagoons, lakes, and canals), at low speeds, is rather low,

[9] "Bissa" is the Italian term defining the flat skull boats traditionally used in the lakes of Northern Italy.

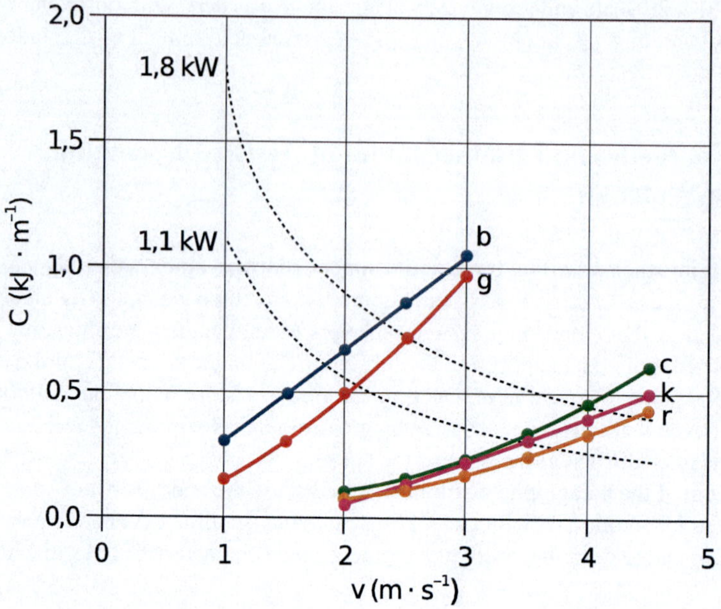

Fig. 6.16 Energy cost (C, kJ m^{-1}) as a function of the speed (v, m s^{-1}) in some forms of assisted locomotion in water (b, “bissa”; g gondola; c, canoe; r, rowing two oars with coxswain; k kayak). The two iso-metabolic power hyperbolae (1.1 and 1.8 kW,) correspond to the $\dot{V}O_2^{max}$ above resting of non-athletic (3.16 L min^{-1}) and athletic (5.17 L min^{-1}) subjects, respectively. The points of intersection between a given function and an iso-power hyperbola yield the maximal aerobic speed for the metabolic power and the type of locomotion considered. From di Prampero 2015, with the kind permission of Raffaello Grandi, Edi Ermes; data from Buglione et al. 2011; Capelli et al. 1990, 2009; di Prampero et al. 1971

particularly when considering their mass and dimension. As an example, at a speed of about 1.2 m s^{-1}, the energy cost of sculling a gondola is about 0,20 kJ m^{-1}, similar to that of an average size man walking at the same speed on flat terrain. In other words, at this speed, the oarsman transports himself, the gondola (about 400 kg mass) and an about equal load of goods and/or persons, with the same overall energy expenditure per unit of distance that, on land, would let him transport only his own body mass.

However, the energy cost of sculling a gondola or a bissa increases markedly with the speed (Fig. 6.16). This explains why this type of boats are traditionally utilized for transporting persons and/or goods only in still waters. Indeed, as soon as the boat moves against even moderate currents, the increase of the speed in respect to water, the resulting higher energy cost, and the difficulty to control, under these conditions, a flat-skull boat, would make the progression slow, uneconomical, and unsafe.

Every point of the plane delimited by the coordinates of Fig. 6.16 is characterized by one value of their product. Since the product of C times v is $\dot{E}$, every point of this Figure is characterized by one value of $\dot{E}$ and all points characterized by a given values of $\dot{E}$ lie on the same hyperbolic function. The two hyperbolae reported in Fig. 6.16 refer to $\dot{E}$ values of 1.1 and 1.8 kW (= 3.16 and 5.17 LO_2 min^{-1} above resting). These values correspond to the average maximal aerobic power of young non-athletic subjects or of high-level athletes, respectively. The point of intersection between the function C versus v of a given form of locomotion and any iso-power hyperbola define the maximal aerobic speed that a hypothetical subject characterized by the $\dot{E}$ in question could attain in the considered locomotion mode. It is immediately apparent that the maximal v attained with a given maximal aerobic power is approximately 1.6–1.9 m s^{-1} higher in the "sporting" than in the "practical" forms of water locomotion. Moreover, for any given $\dot{E}$, the maximal absolute v values are reached in rowing.[10]

Finally, on the practical side, once the relationship between C and v is known, together with the maximal metabolic power of the subjects considered, it becomes possible to estimate the resulting maximal v, a fact to be discussed in more details in the next section.

6.7.2 *The Hydrodynamic Resistance*

The D of three of the boats mentioned above (gondola, bissa and two oars with coxswain rowing boat) is summarized in Fig. 6.17 as a function of v, together with the corresponding values for swimming the crawl, and is reported for some specific speeds in Table 6.5, together with the appropriate C. At each speed, the ratio between D and C is a measure of the drag efficiency η_D that, as indicated in the same Table, varies from a minimum of $\approx$ 0.04 (bissa at 1.0 m s^{-1}), to a maximum of $\approx$ 0.19 (rowing at 2.0 m s^{-1}).

6.7.3 *Efficiency*

The analysis of efficiency carried out in Table 6.5 deals with η_D only. However, as already discussed for swimming (see Sect. 6.6.5), (i) the total work done is the sum of w_i and w_e, (ii) only the latter is useful to progression, and (iii) only a fraction of w_e leads to progression along the direction of movement, the remainder being useless under this respect, although it transmits kinetic energy to the surrounding water too.

[10]The energy cost of rowing reported in Figs. 6.16 and 6.17 refers to only one rower on a "two oars with coxswain"; as a consequence the overall energy cost for moving the boat is twice the reported value. However, also the iso-power functions, when considering both oarsmen are greater than the value reported in the figure by a factor of two. It follows that the point of intersection between the two functions (energy cost and metabolic power) occurs at the same speed, regardless of the number of oarsmen considered (one or two). Hence, for any given value of power (per oarsman) the speed attained will be larger in rowing than in kayaking or canoeing.

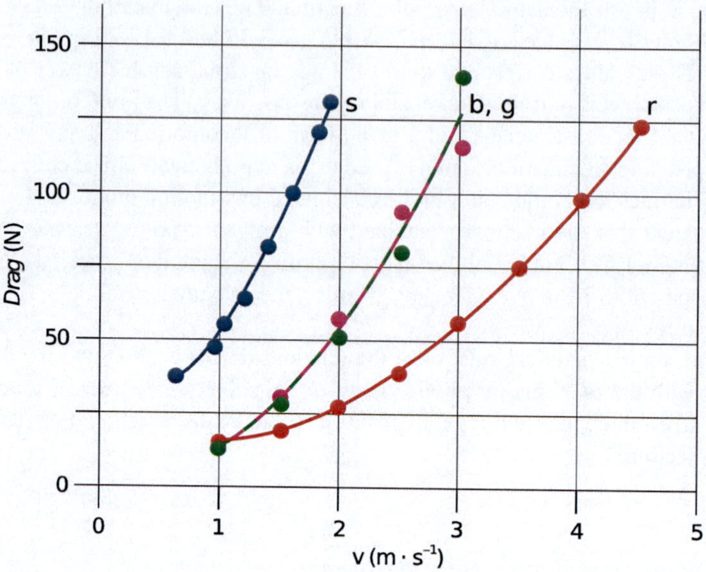

Fig. 6.17 Drag (D, N) as a function of the speed (v, m s^{-1}) for the indicated forms of locomotion: s, swimming the crawl; b "bissa"; g, gondola, r, rowing "two oars with coxswain." The data for the "bissa" and gondola have been interpolated by the same function. From di Prampero 2015, with the kind permission of Raffaello Grandi, Edi Ermes; data from Capelli et al. 1990, 2009; di Prampero et al. 1971, 1974

Table 6.5 Drag (D, N), energy cost (C, J m^{-1}) and drag efficiency (η_D) of rowing "two oars with coxswain," gondola and "bissa" at the indicated speeds. D and C data refer to one rower only: thus, the overall D or C for moving the boat is the sum of the values of all intervening rowers. Data from Capelli et al. 1990, 2009; di Prampero et al. 1971

	ROWING			GONDOLA			BISSA		
v	D	C	η_D	D	C	η_D	D	C	η_D
M s^{-1}	N	J m^{-1}		N	J m^{-1}		N	J m^{-1}	
1.0				12.3	155	0.079	12.8	306	0.042
1.5				30.1	306	0.098	28.8	490	0.059
2.0	27.4	146	0.188	56.9	494	0.115	51.2	675	0.076
2.5	39.2	238	0.165	93.2	717	0.130	80.0	860	0.093
3.0	56.0	356	0.157	139.4	972	0.143	115.2	1044	0.110
3.5	75.8	500	0.151						
4.0	98.2	670	0.147						
4.5	123.6	870	0.142						

The resulting efficiencies are discussed herewith for the case of assisted locomotion in water.

In assisted water locomotion, as well as in swimming, an important component of the overall efficiency is due to the propelling efficiency (η_p), defined by Eq. 6.29 as the ratio between w_D and w_{tot}. In other words, using the jargon of aeronautical and naval sciences, η_p quantifies the "net efficiency" of the propeller. Indeed, in both cases, the propulsive action is due to the amount of fluid (air or water) that the propeller pushes in the direction opposite to that of motion.

To better focus the concept of efficiency as applied to assisted locomotion in water, it is useful to start from an analysis of the forces acting on the oar or on the pulling pole. Let us take rowing as a paradigmatic example. A schematic representation of the forces acting on the oar is reported in Fig. 6.18. According to Celentano et al. (1974), four forces act upon each oar: (i) the reaction p of the water against the blade. Of this force, one component (p_a) has the same direction of the shell movement and is therefore useful for progression; a second component (p_t) is perpendicular to this direction: its effect is a strain on the shell, which is of no use for progression; (ii) the traction force exerted by the rower (T_r), which may be considered as having the same direction as the boat movement; (iii) the resistance to progression (D) for each oar, considered as being applied to the oarlock pin, which is rigidly bound to the shell; (iv) the oarlock reaction (O_t) to the force p_t.

Celentano et al. (1974) measured the overall propelling force acting on the oarlock pin (F), which is equal to the sum of T_r and p_a. Thus the work performed by the oarsman per unit of distance (F_r), when the boat covers a given distance L, is given by:

$$F_r = \frac{1}{L}\int_o^L F\, d\mathrm{L} = \frac{v}{L}\int_0^t F\, dt \tag{6.32}$$

where t is the time necessary to cover the distance L and v is the average speed. It is noteworthy that, as defined, w_r turns out equivalent to w_e. Since during the stroke the blade of the oar moves back, only ≈ 70% of F_r (up to a speed of ≈ 5 m s^{-1}) is utilized for the boat progression, the remaining fraction being wasted. In other words, this means that, under real rowing conditions, $\eta_p \approx 0.7$. This fraction of F_r represents F_D:

$$F_D = \eta_p F_r = \eta_p \frac{v}{L}\int_0^t F\, dt \tag{6.33}$$

Celentano et al. (1974) showed that F_r increases approximately with the square of the speed:

$$F_r = 4.7v^{1.95} \tag{6.34}$$

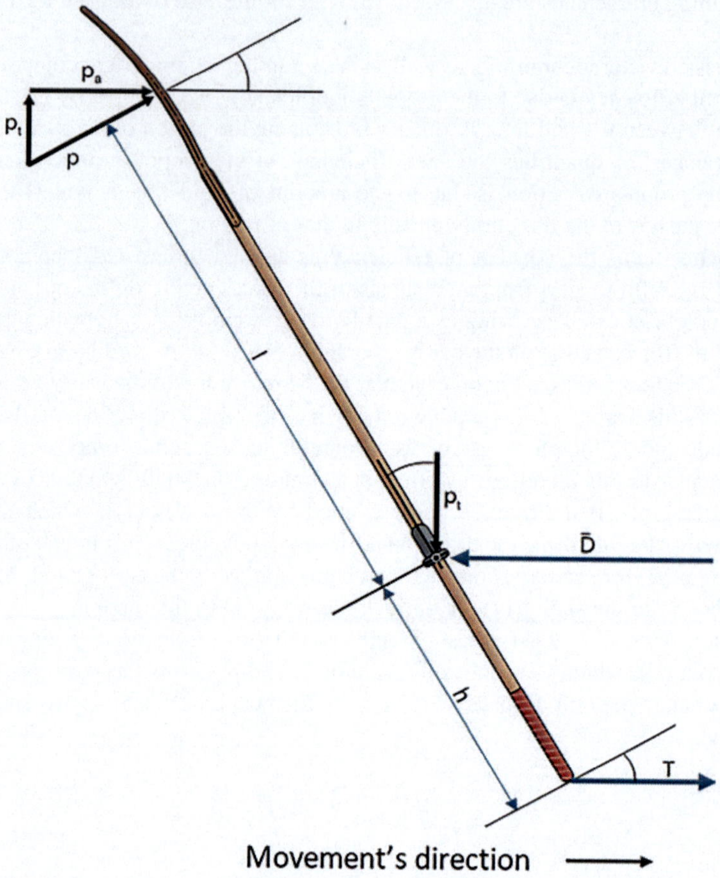

Fig. 6.18 Schematic representation of the forces acting on the oar. T, oarsman pull; D, resistance to progression or drag; p, water reaction upon the blade, which has been decomposed into the components, pa and pt. l and h represent the two arms of the oar. See text for details. From di Prampero, 2015, with the kind permission of Raffaele Grandi, Edi Ermes

where F_r is expressed in J m^{-1} and v in m s^{-1}. This implies that, since F_D varies with the square of speed, η_p is invariant and independent of the speed. The corresponding mechanical power ($\dot{w}_r$, in W) developed by the oarsman to keep the boat in motion is therefore given by:

$$\dot{w}_r = F_r\ v = 4.7\ v^{2.95} \tag{6.35}$$

Equations 6.35 and 6.36 substantially agree with the prediction that the work against hydrodynamic forces varies with the square of speed, whereas the corresponding power varies with the cube of speed. A simultaneous determination of the metabolic

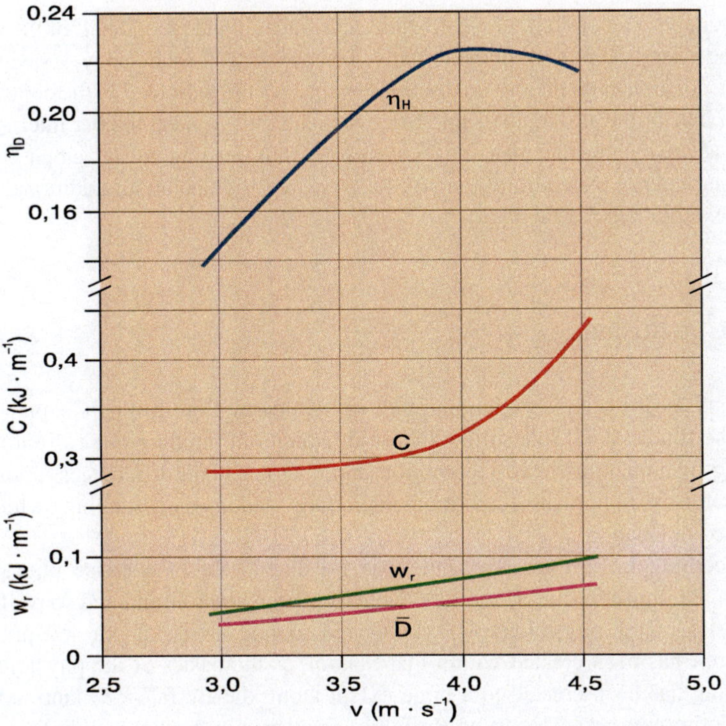

Fig. 6.19 Average values of the mechanical work performed on the oar (w_r) and of the energy cost per unit of distance (C) (both referring to one oarsman on a two oars with coxswain boat), as a function of the speed (v). The propelling efficiency of the oar is about 0.70, hence the average drag (pink line, D) amounts to about 70% of w_r. Since C includes to overall energy expenditure, the hydraulic efficiency (η_H) can also be obtained (blue line). From di Prampero 2015, with the kind permission of Raffaello Grandi, Edi Ermes; data from Celentano et al. 1974; di Prampero et al. 1971

power $\dot{E}$ and of the above-mentioned mechanical parameters allowed the same authors to assess the mechanical efficiency of rowing. Since $\dot{w}_r = \dot{w}_e$, the efficiency, which they obtained, corresponds to the so-called hydraulic efficiency (η_H), defined as (Zamparo et al. 2002):

$$\eta_H = \frac{\dot{w}_e}{\dot{E}} = \frac{\dot{w}_r}{\dot{E}} \tag{6.36}$$

which turned out to be about 0.15 at a speed of 3 m s^{-1}, and 0.20–0.22 at 4–4.5 m s^{-1} (Fig. 6.19). The η_H attained at the highest speeds corresponds well to the overall efficiency observed in cycling. Moreover, it corresponds to that reported for walking or running uphill, suggesting that almost all the work performed by the rower is positive. Hence, the mechanics of rowing, as described in Fig. 6.18, leads to an optimization of the upper limbs action.

The analysis reported in Fig. 6.18 shows also that, in the second half of the stroke, the force p_t tends to pull the oar externally. To counteract this tendency, a mechanical collar is positioned on the oar internally in respect to the oarlock. On the contrary, in the first half of the stroke, the same force, which tends to push the oar internally, is counteracted by the oarsman muscle activity, thus leading to an, albeit limited, additional energy expenditure that could be spared by placing an additional collar on the external side of the oarlock.

6.8 Conclusions

The analysis of cycling, swimming, and rowing carried out in this Chapter, highlights the fundamental role of the School of Milano in building our current understanding of human-powered locomotion and analogies and differences among the various forms of human locomotion, including walking and running, which are discussed in Chap. 5.

Concerning the former issue, there is no other field in exercise physiology, wherein the impact of the School of Milano has been so profound and so exclusive. The basics of the physiology of cycling and rowing, as is known and practiced nowadays, has been created within that School, as the basics of the physiology of swimming has been created to a large extent along the Buffalo—Milano axis. An important tool has revealed to be Buffalo's circular pool, shown in Fig. 6.2, where most of the data on swimming presented and discussed in this Chapter had been obtained. In Milano, under the original impulse of Margaria, Pietro Enrico di Prampero was the motor of these studies, which he pursued along his entire scientific life. After him, his pupils took the relay, especially Carlo Capelli and Paola Zamparo. In Buffalo, a central role, especially in studying swimming, has been played by Dave Pendergast and Don Rennie, who collaborated with di Prampero, when he was in Buffalo for the earliest studies. Pendergast collaborated also with Capelli and Zamparo in several of the most recent studies. It is not by hazard that Pendergast and Zamparo are co-authors, with others, of a major review on human physiology in aquatic environments (Pendergast et al. 2015). Alberto Minetti is reputed for his work on walking and running, along the tradition of Giovanni Cavagna and Franco Saibene, but his spotty excursus on the study of swimming and cycling, especially in collaboration with Zamparo, deserve high consideration.

Notwithstanding the roles and the merit of each of the aforementioned scientists, it is fair to point out that the School of Milano has a debt toward Huub Toussaint from Amsterdam. Toussaint is issued from a School that is perhaps less celebrated than others in exercise physiology, but that had remarkable impact in human muscle physiology and locomotion (especially as far as speed skating is concerned). Toussaint and his colleagues were in fact the first to realize that in swimming, $\dot{w}_D < \dot{w}_e$ and thus $\eta_P < 1$. Although the method that they employed is imperfect, the concept stands luminous, and has been absorbed and integrated in the way of thinking of the School of Milano: Paola Zamparo in particular generated original contributions about it.

Concerning analogies and differences, we highlight that, contrary to walking and running, cycling on flat terrain is almost exclusively a matter of air resistance, and thus of drag, whereas gravity plays no role. All the analysis of cycling at altitude in this Chapter started from this simple concept. It is noteworthy that among the first who established the "One Hour Record" for unaccompanied cycling at altitude, well before the appearance of aerodynamic bicycles, was the multiple winner of the Tour de France and Giro d'Italia Eddy Merckx from Belgium: during training for that record trial, all his physiological evaluation took place at the Institute of Physiology in Milano by Margaria's pupils, Margaria being still active. And this was before the analysis of cycling at altitude started. That analysis, initiated by the classical paper by di Prampero et al. (1979), was progressively refined, especially accounting for the effect of exercise-induced arterial hypoxemia, a phenomenon that in 1979 was not yet considered.

For swimming, as for all types of water locomotion, another actor enters the game: buoyancy. In water locomotion, the mechanics and the energetics depend thus on two crucial factors: drag and buoyancy. This is at variance with air locomotion, where buoyancy is negligible, because of the extremely low air density as compared to water.

So, for cycling, neglecting $\dot{w}_i$, we can state that $\dot{w}_e = \dot{w}_D + \dot{w}_{n-D}$, where subscripts D and n-D indicate that the power is used to overcome drag or non-drag forces, respectively. This implies $\eta_P = 1$ and $\eta_O = \eta_H$, although, since cycling takes place in air, η_H is more an "aeraulic" rather than a hydraulic efficiency. On the other extreme, for swimming, $\dot{w}_{tot} > \dot{w}_e > \dot{w}_D$, which implies $\eta_P < 1$ and $\eta_O < \eta_H$. All other forms of human locomotion lay between these two extremes, differences depending on (i) the quantity of $\dot{w}_i$ and $\dot{w}_e$, and (ii) the existence of mechanisms allowing recovery of mechanical energy from other sources—e.g., the recovery of elastic energy in running –, thereby affecting η_O.

All these, however, can be considered particular cases of a more general equation that, using the symbolism adopted in this Chapter, can be expressed as follows:

$$\eta_O\,\dot{E} = \dot{w}_{tot} = \dot{w}_i + \dot{w}_e = \dot{w}_i + \frac{\dot{w}_D}{\eta_P} + \dot{w}_{n-D} = \dot{w}_i + \frac{k_w\,v^2}{\eta_P} + \dot{w}_{n-D} \qquad (6.37)$$

One may call Eq. 6.37 the general equation of human motion on Earth. Incorporating Eq. 6.37 in the general equation of the energetics of muscular exercise (Chap. 3, Eq. 3.14), we obtain:

$$\eta_O\,(c\,\dot{V}O_2 + b\,\dot{La} + a\,\dot{PCr}) = \eta_O\,\Delta G^*{}_{ATP}\,\overrightarrow{\dot{ATP}} = \dot{w}_{tot} = \dot{w}_i + \dot{w}_e =$$

$$= \dot{w}_i + \dot{w}_e = \dot{w}_i + \frac{\dot{w}_D}{\eta_P} + \dot{w}_{n-D} = \dot{w}_i + \frac{k_w\, v^2}{\eta_P} + \dot{w}_{n-D} \quad (6.38)$$

where $\dot{V}o_2$, $\dot{La}$, and $\dot{PCr}$ represent, respectively, the oxygen consumption, the rate of blood lactate accumulation and the rate of phosphocreatine (PCr) splitting, the constants *a, b,* and *c* indicate the moles of ATP provided by one mole of consumed oxygen, accumulated lactate and split PCr; ΔG^{*}_{ATP} is the free energy liberated by the splitting of one mole of ATP and $\overrightarrow{\dot{ATP}}$ is the rate of ATP hydrolysis.

Equation 6.38 describes the overall energy exchanges (metabolic and mechanical) at the whole body level, that a moving human undergoes, whatever the type of locomotion he/she adopts, the fluid in which he/she moves and the energetic metabolism he/she relies on, in a holistic vision of exercise physiology.

References

Almasy L, Lozach J (1936) Récentes explorations dans le Désert Libyque. Société Royale de Géographie d'Egypte, Cairo, Egypt

Ashe MC, Scroop GC, Frisken PI, Amery CA, Wilkins MA, Khan KM (2003) Body position affects performance in untrained cyclists. Br J Sports Med 37:441–444

Åstrand PO, Englesson S (1972) A swimming flume. J Appl Physiol 33:514

Åstrand PO, Rodahl K, Dahl HA, Strømme SB (2003) Textbook of work physiology. Human Kinetics, Champaign (IL, USA)

Benedict FG, Cathcart EP (1913) A metabolic study with special reference to the efficiency of the human body as a machine, Publication no. 187. Washington, D.C., Carnegie Institution

Broker JR, Kyle CR, Burke ER (1999) Racing cyclist power requirements in the 4000-m individual and team pursuits. Med Sci Sports Exerc 31:1677–1685

Buglione A, Lazzer S, Colli R, Introini E, Prampero (di) PE (2011) Energetics of best performances in elite kayakers and canoeists. Med Sci Sports Exerc 43:877–884

Candau RB, Grappe F, Ménard M, Barbier B, Millet GY, Hoffman MD, Belli AR, Rouillon JD (1999) Simplified deceleration method for assessment of resistive forces in cycling. Med Sci Sports Exerc 31:1441–1447

Capelli C, di Prampero PE (1995) Effects of altitude on top speeds during 1 h unaccompanied cycling. Eur J Appl Physiol 80:298–307

Capelli C, Donatelli C, Moia C, Valier C, di Prampero PE (1990) Energy cost and efficiency of sculling a venetian gondola. Eur J Appl Physiol 60:175–178

Capelli C, Pendergast DR, Termin B (1998a) Energetics of swimming at maximal speeds in humans. Eur J Appl Physiol 78:385–395

Capelli C, Rosa G, Butti F, Ferretti G, Veicsteinas A, di Prampero PE (1993) Energy cost and efficiency of riding aerodynamic bicycles. Eur J Appl Physiol 67:144–149

Capelli C, Schena F, Zamparo P, Dal Monte A, Faina M, di Prampero PE (1998b) Energetics of best performances in track cycling. Med Sci Sports Exerc 30:614–624

Capelli C, Tarperi C, Schena F, Cevese A (2009) Energy cost and efficincy of venetian rowing on a traditional, flat hull boat (Bissa). Eur J Appl Physiol 105:653–661

Celentano F, Cortili G, Prampero (di) PE, Cerretelli P (1974) Mechanical aspects of rowing. J Appl Physiol 36:642–647

Cerretelli P (1980) Gas exchange at high altitude. In: "Pulmonary gas exchange", a cura di JB West, vol II. Academic Press, New York, pp 97–147

Charlton JM, Ramsook AH, Mitchell RA, Hunt MA, Puyat JH, Guenette JA (2017) Respiratory mechanical and cardiorespiratory consequences of cycling with aerobars. Med Sci Sports Exerc 49:2578–2584

Davies CTM (1980) Effect of air resistance on the metabolic cost and performance of cycling. Eur J Appl Physiol 45:245–254

De Groot G, Sargeant A, Geysel J (1995) Air friction and rolling resistance during cycling. Med Sci Sports Exerc 27:1090–1095

Dempsey JA, Hanson PG, Henderson KS (1984) Exercise-induced arterial hypoxemia in healthy human subjects at sea level. J Physiol Lond 355:161–175

Dempsey JA, Wagner PD (1999) Exercise – induced arterial hypoxemia. J Appl Physiol 87:1997–2006

di Prampero PE (2000) Cycling on earth, in space, on the Moon. Eur J Appl Physiol 82:345–360

di Prampero PE (2015) La locomozione umana su terra, in acqua, in aria: fatti e teorie. Edi Ermes, Milano

di Prampero PE, Cortili G, Celentano F, Cerretelli P (1971) Physiological aspects of rowing. J Appl Physiol 31:853–857

di Prampero PE, Cortili G, Celentano F, Cerretelli P (1976) The energy cost of speed skating and the efficiency of work against the air resistance. J Appl Physiol 40:584–591

di Prampero PE, Cortili G, Mognoni P, Saibene F (1979) The equation of motion of a cyclist. J Appl Physiol 47:201–206

di Prampero PE, Pendergast DR, Wilson D, Rennie DW (1974) Energetics of swimming in man. J Appl Physiol 37:1–5

di Prampero PE (1986) The energy cost of human locomotion on land and in water. Int J Sports Med 7:55–72

di Prampero PE, Pendergast DR, Zamparo P (2011). Swimming economy (energy cost) and efficiency. In "World book of swimming: from science to performance", di Cura A L Seifert, D Chollet, Mujika I. Nova Science Publishers, New York, pp: 297–312

Dickinson S (1929) The efficiency of bicycle-pedalling, as affected by speed and load. J Physiol Lond 67:242–255

Digby E (1587) De arte natandi libri duo. Thomas Dawson, London

Ferretti G (2014) Maximal oxygen consumption in healthy humans: theories and facts. Eur J Appl Physiol 114:2007–2036

Ferretti G, Bringard A, Perini R (2011) An analysis of performance in human locomotion. Eur J Appl Physiol 111:391–401

Ferretti G, Capelli C (2008) Dagli abissi allo spazio: ambienti e limiti umani. Edi Ermes, Milano

Ferretti G, di Prampero PE (1995) Factors limiting maximal O_2 consumption: effects of acute changes in ventilation. Respir Physiol 99:259–271

Ferretti G, Moia C, Thomet J, Kayser B (1997) The decrease of maximal oxygen consumption during hypoxia in man: a mirror image of the oxygen equilibrium curve. J Physiol Lond 498: 231–237

Formenti F, Ardigò LP, Minetti AE (2005) Human locomotion on snow: determinants of economy and speed of skiing across the ages. Proc Roy Soc Lond B 272:1561–1569

Gnehm P, Reichenbach S, Altpeter E, Widmer H, Hoppeler H (1997) Influence of different racing positions on metabolic cost in elite cyclists. Med Sci Sports Exerc 29:818–823

Gross AC, Kyle CR, Malewicki DJ (1983) The aerodynamics of human-powered land vehicles. Sci Am 249:124–134

Heil DP (2001) Body mass scaling of projected frontal area in competitive cyclists. Eur J Appl Physiol 85:358–366

Henderson Y, Haggard HW (1925) The maximum of human power and its fuel. Am J Phys 72:264–282

Herlihy D (2004) Bicycle: the history. Yale University Press, New Haven CT, USA

Holmér I (1974) Physiology of swimming man. Acta Physiol Scand Suppl 407:1–55

Karpovich PV, Millman M (1944) Energy expenditure in swimming. Am J Phys 142:140–144

Karpovich PV, Pestrecov K (1939) Mechanical work and efficiency in swimming crawl and back stroke. Arbeitsphysiol 10:504–514

Kehm G (2007) Olympic swimming and diving. Rosen Publishing Group, New York

Kohler G, Boutellier U (2005) The generalized force-velocity relationship explains why the preferred pedaling rate of cyclists exceeds the most efficient one. Eur J Appl Physiol 94:188–195

Liljestrand G, Lindhard J (1919a) Über das Minutenvolumen des Herzens bei Schwimmen. Skand Arch Physiol 39:64–77

Liljestrand G, Lindhard J (1919b) Zur Physiologie des Ruderns. Skand Arch Physiol. 39:215–235

Liljestrand G, Stenström M (1919a) Studien über die Physiologie des Schwimmens. Skand Arch Physiol. 39:1–63

Liljestrand G, Stenström M (1919b) Respirationsversuche beim Gehen, Laufen, Ski- und Schlittschuhlaufen. Skand Arch Physiol 39:167–206

Marconi C, Marzorati M, Grassi B, Basnyat B, Colombini A, Kayser B, Cerretelli P (2004) Second generation Tibetan lowlanders acclimatize to high altitude more quickly than Caucasians. J Physiol Lond 556:661–671

Margaria R (1938) Sulla fisiologia e specialmente sul consumo energetico della marcia e della corsa a varia velocità ed inclinazione del terreno. Atti Acc Naz Lincei 7:299–368

Marsh AP, Martin PE (1993) The association between cycling experience and preferred and most economical cadences. Med Sci Sports Exerc 25:1269–1274

Marsh AP, Martin PE (1995) The relationship between cadence and lower extremity EMG in cyclists and noncyclists. Med Sci Sports Exerc 27:217–225

Marsh AP, Martin PE (1998) Perceived exertion and the preferred cycling cadence. Med Sci Sports Exerc 30:942–948

Martin RB, Yeater RA, White MK (1981) A simple analytical model for the crawl stroke. J Biomech 14:539–548

Minetti AE, Pinkerton J, Zamparo P (2001) From bipedalism to bicyclism: evolution in energetics and biomechanics of historic bicycles. Proc Roy Soc Lond B 268:1351–1360

Oelz O, Howald H, di Prampero PE, Hoppeler H, Claassen H, Jenni R, Bühlmann A, Ferretti G, Brückner J-C, Veicsteinas A, Gussoni M, Cerretelli P (1986) Physiological profile of world class high altitude climbers. J Appl Physiol 60:1734–1742

Pendergast DR, Bushnell D, Wilson DW, Cerretelli P (1989) Energetics of kayaking. Eur J Appl Physiol 59:342–350

Pendergast DP, di Prampero PE, Craig AB Jr, Wilson DW, Rennie DW (1977) Quantitative analysis of the front crawl in men and women. J Appl Physiol 43:474–479

Pendergast DR, Moon RE, Krasney JJ, Held HE, Zamparo P (2015) Human physiology in an aquatic environment. Compr Physiol 20:1705–1750

Pendergast DR, Tedesco M, Nawrocki DM, Fisher NM (1996) Energetics of underwater swimming with SCUBA. Med Sci Sports Exerc 28:573–580

Pollock ML, Hickman T, Kendrick Z, Jackson A, Linnerud AC, Dawson G (1976) Prediction of body density in young and middle-aged men. J Appl Physiol 40:300–304

Pugh LGCE (1957) Resting ventilation and alveolar air on Mount Everest: with remarks on the relation of barometric pressure to altitude in mountains. J Physiol Lond 135:590–610

Pugh LCGE (1974) The relation of oxygen intake and speed in competition cycling and comparative observations on the bicycle ergometer. J Physiol Lond 241:795–808

Saibene F, Cortili G, Roi G, Colombini A (1989) The energy cost of level cross-country skiing and the effect of the friction of the ski. Eur J Appl Physiol 58:791–795

Sargeant AJ, Davies CTM (1977) Forces applied to cranks of a bicycle ergometer during one- and two-leg cycling. J Appl Physiol 42:514–518

Schmidt-Nielsen K (1997) Animal physiology: adaptation and environment. Cambridge University Press, Cambridge UK

Seifert L, Chollet D, Mujika I (eds) (2011) World book of swimming: from science to performance. Nova Science Publishers, New York, p 550. + XXVI

Toussaint HM, Beleen A, Rodenburg A, Sargeant AJ, de Groot G, Hollander AP, van Ingen Schenau GJ (1988) Propelling efficiency of front crawl swimming. J Appl Physiol 65:2506–2512

Welbergen E, Clijsen LP (1990) The influence of body position on maximal performance in cycling. Eur J Appl Physiol 61:138–142

West JB, Lahiri S, Maret KH, Peters RM Jr, Pizzo CJ (1983) Barometric pressure at extreme altitudes on Mount Everest: physiological significance. J Appl Physiol 54:1188–1194

Zamparo P (2006) Effects of age and gender on the propelling efficiency of the arm stroke. Eur J Appl Physiol 97:52–58

Zamparo P, Antonutto G, Capelli C, Francescato MP, Girardis M, Sangoi R, Soule RG, Pendergast DR (1996) Effects of body size, body density, gender and growth on underwater torque. Scand J Med Sci Sports 6:273–280

Zamparo P, Bonifazi M, Faina M, Milan A, Sardella F, Schena F, Capelli C (2005a) Energy cost of swimming of elite long-distance swimmers. Eur J Appl Physiol 94:697–704

Zamparo P, Capelli C, Cautero M, Di Nino A (2000) Energy cost of front-crawl swimming at supramaximal speeds and underwater torque in young swimmers. Eur J Appl Physiol 83:487–491

Zamparo P, Capelli C, Guerrini G (1999) Energetics of kayaking at submaximal and maximal speeds. Eur J Appl Physiol 80:542–548

Zamparo P, Capelli C, Pendergast DR (2011) Energetics of swimming: a historical perspective. Eur J Appl Physiol 111:367–378

Zamparo P, Lazzer S, Antoniazzi C, Cedolin S, Avon R, Lesa C (2008) The interplay between propelling efficiency, hydrodynamic position and energy cost of front crawl in 8 to 19 year-old swimmers. Eur J Appl Physiol 104:689–699

Zamparo P, Pendergast DR, Mollendorf J, Termin A, Minetti AE (2005b) An energy balance of front crawl. Eur J Appl Physiol 94:134–144

Zamparo P, Pendergast DR, Termin B, Minetti AE (2002) How fins affect the economy and efficiency of human swimming. J Exp Biol 205:2665–2676

Zamparo P, Cortesi M, Gatta G (2020) The energy cost of swimming and its determinants. Eur J Appl Physiol 120:41–66

Zuntz L (1899) Untersuchungen über den Gaswechsel und Energie-umsatz des Radfahrers. August Hirchwald Verlag, Berlin (Germany)

Chapter 7
Maximal Oxygen Consumption

Guido Ferretti and Pietro Enrico di Prampero

Abstract This chapter is on maximal oxygen consumption and particularly on its limitation. The history of maximal oxygen consumption and the unifactorial vision of its limitation are firstly analyzed. The oxygen cascade theory of oxygen flow set the basis for the subsequent creation of multifactorial models. Two variations around the oxygen cascade theory led to two competing models, the one by Pietro Enrico di Prampero and Guido Ferretti within the School of Milano, the other by Peter Wagner within the School of San Diego. Both models are analyzed in detail and critically discussed. A synthesis of both models leading to a common conclusion is then proposed. Two specific cases, with remarkable theoretical implications, are finally discussed in detail, namely the effects of hypoxia and the effects of resuming a standing posture at the end of prolonged bed rest.

G. Ferretti (✉)
Department of Molecular and Translational Medicine, University of Brescia, Brescia, Italy
e-mail: Guido.Ferretti@unibs.it

P. E. di Prampero
Emeritus Professor of Physiology, University of Udine, Udine, Italy
e-mail: Pietro.Prampero@uniud.it

G. Ferretti (ed.), *Exercise, Respiratory and Environmental Physiology*, Perspectives in Physiology, https://doi.org/10.1007/978-3-031-19197-8_7

Measuring maximal oxygen consumption at Kipchoge Keyno stadium, Eldoret, Kenya, on elite Kalenjin marathon runners. Enrico Tam is on the bike, dictating the running pace. The athlete, running on Tam's right, slightly behind him, carries a portable system for measuring oxygen consumption. Eldoret, Kenya, January 30th, 2007. With the kind permission of Guido Ferretti

7.1 The Early History of Maximal Oxygen Consumption

When oxygen was identified as a specific single gas sustaining combustion, in the eighteenth century, it soon became clear that living animals breathe and consume oxygen and that its consumption increases when animals exercise. As already pointed out in Chap. 2, Antoine Laurent Lavoisier (1743–1794) clearly stated that respiration is a combustion process using oxygen as oxidizer. No life would be possible in the absence of oxygen. The question soon became: how much can oxygen consumption increase during exercise? Is there a limit to its increase or can it increase indefinitely? Verzar (1912) was the first to realize that perhaps oxygen consumption could not increase above a certain level in exercising animals, thus being limited. Soon afterward, it was demonstrated that, in humans exercising on a cycle ergometer, the linear relationship between oxygen uptake ($\dot{V}O_2$) and mechanical power ($\dot{w}$) attains a plateau which cannot be overcome, despite further increases

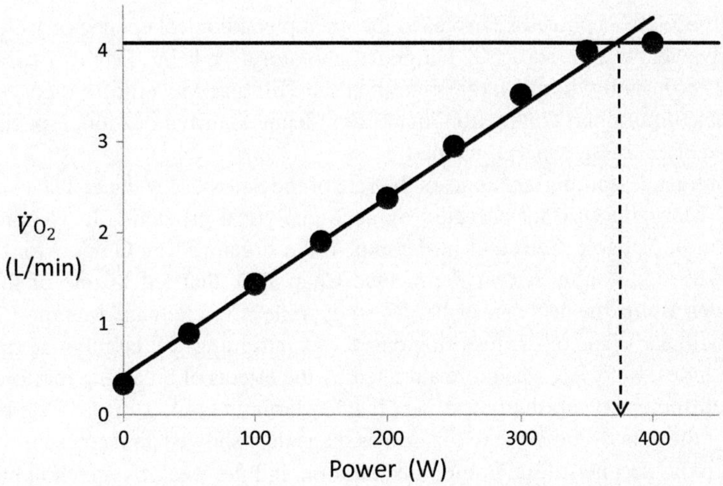

Fig. 7.1 Linear relationship between oxygen consumption ($\dot{V}O_2$) and power during a classical discontinuous protocol for $\dot{V}O_2^{max}$ measurements. Data refer to a trained top-level cyclist tested in Geneva. The regression line was calculated from the submaximal $\dot{V}O_2$ values. The horizontal line indicates the $\dot{V}O_2^{max}$ plateau. The vertical dashed arrow indicates the maximal aerobic power. From Ferretti 2014, 2015

of $\dot{w}$ (Herbst 1928; Hill and Lupton 1923). This led to the creation of the concept of maximal oxygen consumption ($\dot{V}O_2^{max}$) (Fig. 7.1).

The descriptive physiology of $\dot{V}O_2^{max}$ was widely investigated in the following 50 years, especially by Scandinavian and American scientists. The effects of gender, age, altitude exposure, training, hyperoxia, exercise mode, size of active muscle mass, anemia and polycythemia, disuse, prolonged bed rest were investigated. Endurance athletes were studied, as were differences among ethnic groups. The relationship between $\dot{V}O_2^{max}$ and maximal cardiac output ($\dot{Q}^{max}$) was established. Indirect methods for $\dot{V}O_2^{max}$ estimation, based on heart rate (f_H) measurements, were conceived. Several animals, of extremely variable body size, from mice to elephants, were studied, and allometric relationships for $\dot{V}O_2^{max}$ were set based on those studies. The results were summarized in numerous books and review articles (Asmussen 1965; Åstrand 1956; Åstrand et al. 2003; Blomqvist and Saltin 1983; Cerretelli and di Prampero 1987; Cerretelli and Hoppeler 1996; Clausen 1977; Ferretti 2014, 2015; Lacour and Flandrois 1977; Saltin 1977; Skattebo et al. 2020). In the last 50 years, also the reduction of $\dot{V}O_2^{max}$ in a variety of cardiopulmonary diseases and the associated reduction of exercise capacity were widely studied (see, e.g., Esposito et al. 2010a, b; Hanson 1994 for heart failure; for chronic obstructive pulmonary disease, see Sect. 7.8).

The history of exercise physiology is strongly indebted to the Scandinavian School, for having generated a large fraction of the huge amount of data defining the descriptive physiology of $\dot{V}O_2^{max}$. Erik-Höhwu Christensen (1904–1996), Erling Asmussen (1907–1991), Per-Olof Åstrand (1922–2015), Björn Ekblom and Bengt

Saltin (1935–2014) played key roles in the great physiological epopee of $\dot{V}O_2^{max}$. In the early years, the Harvard Fatigue Laboratory, led by David Bruce Dill (1891–1986), with whom Margaria refuted the Hill and Meyerhof's theory of the energetics of muscular contraction, made significant contributions too, especially as far as the effect of age was concerned.

In contrast, less important appears the role of the School of Milano. The synthetic mind of Margaria was not attracted by such analytical problems. It was only the participation of Paolo Cerretelli in the expedition organized by Guido Monzino to conquer Mount Kanjut-Sar, in Pamir (see Chap. 10), that led to one of the first studies describing the decrease of $\dot{V}O_2^{max}$ in hypoxia (Cerretelli and Margaria 1961). The curvilinear shape of the resulting curve was intriguing, yet largely unexplained in those times. Margaria was more attracted by the effects of breathing mixtures that contained more oxygen than air at sea level (Margaria et al. 1961, 1972), but the results of those two studies were bizarre, as far as they showed an increase in $\dot{V}O_2^{max}$ in hyperoxia, despite full hemoglobin saturation and the weak oxygen solubility in blood.

More impact had the methodological studies, promoted by Margaria's strive toward practical applications of physiological knowledge in sport medicine. His indirect method for $\dot{V}O_2^{max}$ estimation (Margaria et al. 1965a) is still used, as well as that of Åstrand (Åstrand and Ryhming 1954), when large cohorts are to be investigated. Margaria and colleagues assumed that, if we use an exercise type, which we know the mechanical efficiency of, a simple determination of steady state f_H and of the mechanical power developed during given submaximal exercises is enough to estimate $\dot{V}O_2^{max}$. Mechanical power can be transformed into metabolic power, this is plotted as a function of f_H, and the resulting linear relationship is extrapolated up to the maximal heart rate (f_H^{max}). To this aim, they used the empirical relationship of Åstrand, describing the variations of f_H^{max} with age (Åstrand 1956). Margaria et al. (1965a) constructed specific nomograms for $\dot{V}O_2^{max}$ determination with such a method (Fig. 7.2). Yet Margaria's approach suffered of two weaknesses: (i) the Åstrand's relationship describing the f_H changes with age is subject to large variability, and thus is a cause of error, and (ii) Margaria et al. (1965a) used only two points to construct the f_H versus metabolic power line. In spite of this, the method, which we do not endorse for laboratory studies, had a remarkable success in terrain studies on large population screenings, for its simplicity.

7.2 The Unifactorial Vision of Maximal Oxygen Consumption Limitation

The $\dot{V}O_2$ plateau of Fig. 7.1 implies $\dot{V}O_2^{max}$ limitation, whence the inevitable question: what limits $\dot{V}O_2^{max}$? Initially, and for long, physiologists have been in quest of the single factor limiting $\dot{V}O_2^{max}$. Two opposed fields emerged, that of central (cardiovascular) limitation and that of peripheral (muscular) limitation. The

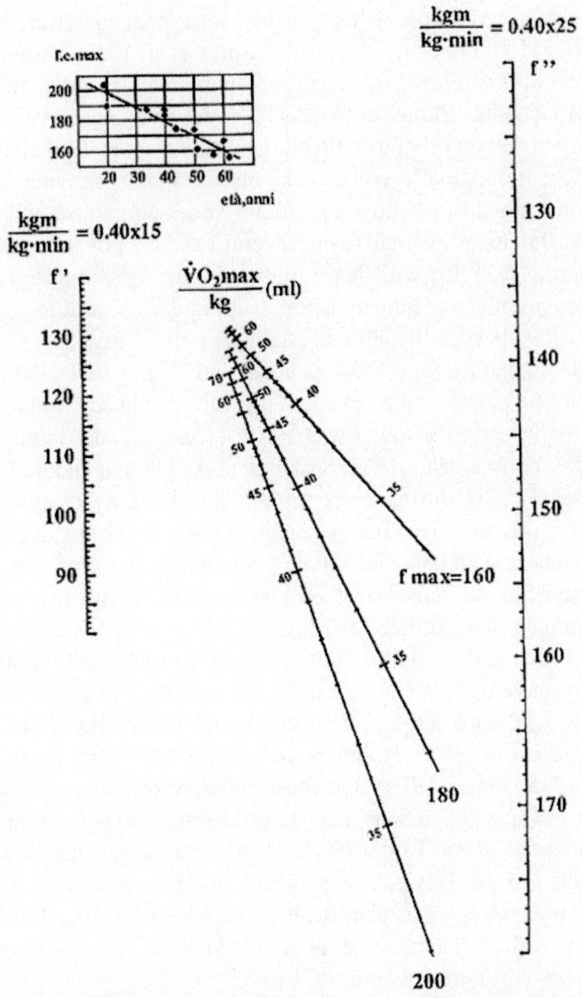

Fig. 7.2 Nomogram for indirect $\dot{V}O_2^{max}$ determination with Margaria's method. The two vertical lines indicate the heart rate attained at steady state when climbing up and down a 40 cm tall step at the rate of 15 (left side) and 25 (right side) cycles per minute. The three scales in the middle report the expected $\dot{V}O_2^{max}$ values for a predicted maximal heart rate of 160, 180, and 200 beats per minute. The upper left inset reports Åstrand's diagram relating heart rate to age. From Margaria et al. 1965a

stirred debate among these fields that went on for more than 50 years was such that no consensual conclusion was possible for such a long time.

Several observations suggested a cardiovascular limitation of $\dot{V}O_2^{max}$. The largely different values of $\dot{V}O_2^{max}$ that can be found in the literature are associated with consensual, linear and quantitatively similar changes in $\dot{Q}_{max}$ (Blomqvist and Saltin 1983; Cerretelli and di Prampero 1987; Ekblom et al. 1968). Moreover, $\dot{V}O_2^{max}$ is (i) lower in acute anemia than in normaemia (Celsing et al. 1987; Woodson et al.

1978); (ii) higher in acute polycythaemia than in normaemia (Buick et al. 1980; Celsing et al. 1987; Ekblom et al. 1975, 1976; Spriet et al. 1986; Turner et al. 1993); (iii) lower when a small fraction of carbon monoxide is added to inspired air (Ekblom and Huot 1972; Pirnay et al. 1971; Vogel and Gleser 1972). Finally, if we increase oxygen delivery to the contracting muscle mass by artificially increasing muscle blood flow at maximal exercise, local muscle $\dot{V}O_2^{max}$ can increase above the levels attained in normal blood flow condition (Andersen and Saltin 1985; Rowell et al. 1986). On this experimental basis, several exercise physiologists concluded that, at least during exercise with large muscle groups, the single factor limiting $\dot{V}O_2^{max}$ is cardiovascular oxygen transport (Blomqvist and Saltin 1983; Clausen 1977; Ekblom 1969, 1986; Mitchell and Blomqvist 1971; Rowell 1974; Saltin and Rowell 1980; Saltin and Strange 1992; Scheuer and Tipton 1977; Sutton 1992).

Some evidence however seemed to contradict this conclusion. In fact, the smaller is the active muscle mass, the lower was $\dot{V}O_2^{max}$ (Åstrand and Saltin 1961; Bergh et al. 1976; Davies and Sargeant 1974; Secher et al. 1974); and endurance training of one leg increased $\dot{V}O_2^{max}$ during exercise with that leg only (Saltin et al. 1976). Moreover, looking into muscle structure, a greater fraction of oxidative type I muscle fibers, a higher muscle capillary density, and a higher activity of muscle oxidative enzymes characterized the muscles of endurance athletes with respect to those of sedentary individuals (Brodal et al. 1977; Costill et al. 1976; Gollnick et al. 1972; Howald 1982; Tesch and Karlsson 1985; Zumstein et al. 1983). Finally, aerobic physical training increased capillary density, mitochondrial volume and oxidative enzyme activities of the trained muscles (Andersen and Henriksson 1977; Gollnick et al. 1972; Henriksson 1977; Holloszy and Coyle 1984; Hoppeler et al. 1985; Howald et al. 1985; Ingjer 1979). On these bases, some authors concluded that muscle oxidative capacity, rather than cardiovascular oxygen transport, limits $\dot{V}O_2^{max}$ (Lindstedt et al. 1988; Taylor 1987; Weibel 1987), especially during exercise with small muscle groups (Davies and Sargeant 1974; Kaijser 1970; Saltin 1977).

In a minority of cases, such as extreme hypoxia (West 1983) and in athletes with very high $\dot{V}O_2^{max}$ values (Dempsey et al. 1984; Dempsey and Wagner 1999), the lungs as well were considered as limiting $\dot{V}O_2^{max}$

The School of Milano remained quite aside this debate. Yet, Margaria once expressed his viewpoint in a quite bizarre and collateral way. In 1965, he and some of his collaborators published a paper on the kinetics of $\dot{V}O_2$ at the onset of square-wave exercises of various intensities (Margaria et al. 1965b). They described this kinetics as being mono-exponential, with a half response time of some 30 s, independent of the exercise intensity. The technique by which they obtained this result was quite primitive, typical of the time that preceded gas exchange analysis on single-breath basis. They concluded that, we quote, *intracellular oxidation processes are coupled with the splitting and resynthesis of the high-energy phosphate processes in muscle*. This sentence reiterates Margaria's concept of alactic oxygen deficit. Then they added: *The speed of oxidative processes in muscle does not seem to be a limiting factor to the oxygen uptake in muscular exercise*. By this, they mean that oxidative processes may be accelerated up to a speed higher than the one that can be sustained by oxygen delivery. Finally, with a typical logical gap of a visionary mind, the fireworks exploded: *The maximum oxygen consumption level is presumably set by the capacity of the oxygen transport from the lungs to the active tissues*. This statement is reckless, risky, even if mitigated by the use of the term *presumably*. The experiment was made in a different context, had a different aim and the results did not inform directly on $\dot{V}O_2^{max}$ limitation. So, Margaria took a hazard, of a type that very few scientists would dare to take, especially nowadays. Yet this statement tells very clearly on which side Margaria was, if he had to express his view in a debate, which did not excite him. That paper was often cited as a classical publication on exercise transients, yet largely forgotten for its spectacular conclusion.

More impact, from the School of Milano, had a paper by Paolo Cerretelli reporting data collected during the 1973 Italian expedition to Mount Everest. That paper reports that the $\dot{V}O_2^{max}$ of altitude-acclimatized subjects, who are living in chronic hypoxia, does not return to the pre-acclimatization levels after sudden exposure to normoxic gas mixtures (Cerretelli 1976, see Chap. 10 for a more detailed description of the study). Cerretelli's conclusion was that $\dot{V}O_2^{max}$ could not increase as predicted because, we quote, *the limit to $\dot{V}O_2^{max}$ at altitude is at least in part peripheral. Whether the reduced oxygen flow to the muscles is attributable to a limitation of tissue perfusion or to changes in the oxygen diffusion rate within the capillary or at the capillary-tissue barrier is still a matter of investigation*. Thus, he directed his focus to the periphery. Even if he tended to exclude muscle structural changes as possible determinants of his results, Cerretelli's conclusion was against wrong yet deeply established concepts in those days, namely that an increase in muscle oxidative capacity was a necessity in order to face the lack of oxygen at altitude (Reynafarjee 1962).

The impact of that paper was huge. On one side, it reinforced the field of muscle limitation of $\dot{V}O_2^{max}$. On the other side, and despite the stated conclusion, the mere fact of pointing to the periphery originated a series of fundamental studies on muscle structure at altitude. These studies were issued from the collaborations promoted by Paolo Cerretelli, who meanwhile had moved from Milano to Geneva, and Hans Hoppeler from the Institute of Anatomy of Bern, that contributed largely to the success of Weibel's concept of symmorphosis (Cerretelli and Hoppeler 1996). One of us criticized Cerretelli's 1976 paper in the perspective of a more recent, deeper knowledge of oxygen flow in the respiratory system (Ferretti 2003), as outlined below. We tell this story more in detail in Chap. 10.

7.3 The Oxygen Cascade and the Origin of Multifactorial Models

A revolution in the approach to the subject of $\dot{V}O_2^{max}$ limitation started in the years 1980s. Dick Taylor (1939–1995) from Harvard and Ewald Weibel from Bern (1929–2019) inadvertently promoted it, when, in 1981, they resumed the oxygen cascade theory as a tool for describing oxygen transfer from ambient air to the mitochondria in mammals. The oxygen cascade theory states that, in analogy with water flow in pipes or current in high-resistance electric lines, oxygen pressure gradients drive oxygen flow along the respiratory system against several in-series resistances. In such a conceptual design, the total resistance to oxygen flow is equal to the sum of the numerous resistances displayed along the system itself. In principle, each of these resistances may provide a given measurable fraction of the overall limitation to oxygen flow. The concept was not new. For instance, it was present *in nuce* in the years 1930s (see, e.g., Haldane and Priestley 1935) and it was refined elegantly by Roy Shephard (1969), who expressed it in the form of an oxygen conductance equation, which he provided a non-linear solution of. Notwithstanding Shephard's work, the concept remained in the background of physiological thinking for long and was essentially forgotten by the mainstream of physiological research.

Taylor and Weibel's aim was to analyze the structural constraints of respiratory systems under maximal stress in animals encompassing a wide range of body size (Taylor and Weibel 1981). They were not, or rather not yet, interested to the physiological limitation of $\dot{V}O_2^{max}$ in humans. In spite of this, their idea turned a light on. A few exercise physiologists, stimulated by that remarkable series of papers, started to think in terms of multiple factors that together contribute to $\dot{V}O_2^{max}$ limitation under a holistic perspective. All of a sudden, the way to the creation of the multifactorial models of $\dot{V}O_2^{max}$ limitation was open.

That was when the School of Milano really entered the game. Pietro Enrico di Prampero was the first to propose a multifactorial model of $\dot{V}O_2^{max}$ limitation (di Prampero 1985). He identified a set of five resistances in-series, which he could calculate or estimate, and devised a simple solution of the oxygen conductance equation for the case in which only one resistance was induced to change by an acute maneuver (he took blood withdrawal and reinjection as a paradigmatic example). He understood that the system has a non-linear response and attributed that to the characteristics of the oxygen equilibrium curve. He calculated that most of the limitation at sea level was due to cardiovascular oxygen transport. He perceived that the lungs were not limiting $\dot{V}O_2^{max}$ at sea level. He tried to separate peripheral oxygen diffusion and muscle oxygen consumption as $\dot{V}O_2^{max}$ determinants, by introducing an unnecessary proportionality constant defining the ratio of the associated resistances.

The model was later refined and simplified by di Prampero and Ferretti (1990), who lumped the two muscular resistances and set a system in which, at least at sea level, the fraction of $\dot{V}O_2^{max}$ limitation that is not due to cardiovascular oxygen transport is attributed to that lumped muscular resistance. Then Ferretti and di Prampero (1995) expanded the model to include hypoxia and estimated the progressive role taken up by the lungs as limiting factor. di Prampero (2003) published a further summary of the model.

Meanwhile, Peter Wagner from San Diego (Wagner 1992, 1993, 1996a, b) designed an apparently different system, in which, instead of having a purely hydraulic system with four resistances in-series, he devised two convective steps separated by two conductive steps.

These models have characterized the last 30 years of debate on $\dot{V}O_2^{max}$ limitation. The quest for the single limiting step was over, and the revolution generated by the introduction of multifactorial models of $\dot{V}O_2^{max}$ limitation transformed the inconclusive debate on the cardiovascular vs. muscular limitation of $\dot{V}O_2^{max}$ into a historical item. In spite of this, it is flattening to realize how deeply the unifactorial view of limitation was rooted, as a dogma, in physiological thinking: for example, consider that still more than 20 years after the appearance of the first multifactorial model, a supposedly important review on $\dot{V}O_2^{max}$ reiterated a preposterous unifactorial view of the subject (Levine 2008).

7.4 Introducing the Multifactorial Models

The oxygen cascade theory is the axiom behind multifactorial models of $\dot{V}O_2^{max}$ limitation. The overall driving pressure is the pressure gradient between inspired air and the mitochondria. The oxygen partial pressure that can be measured at various sites decreases as long as we proceed from ambient air to the mitochondria. The resulting pressure gradients allow definition of clear and physiologically relevant resistances to flow. A representation of the oxygen flow from ambient air to the mitochondria is depicted in Fig. 7.3. A set of simple equations describing the flow of oxygen across each resistance is also reported.

At steady state, all these equations have equal solutions for oxygen flow. The highest value that this flow can attain corresponds to $\dot{V}O_2^{max}$. Thus, in this context, $\dot{V}O_2^{max}$ can be defined as the maximal flow of oxygen that the incurring overall pressure gradient can sustain against the total resistance (R_T). R_T decreases as long as exercise intensity is increased, because of the concomitant elevation of ventilation, lung diffusion capacity, cardiovascular oxygen flow, muscle blood flow, peripheral diffusion capacity and rate of oxidative metabolism. R_T at maximal exercise, i.e., at $\dot{V}O_2^{max}$, is the minimal resistance that the respiratory system can oppose to oxygen flow.

When using the oxygen cascade theory at maximal exercise as the axiom upon which to construct a multifactorial model of $\dot{V}O_2^{max}$ limitation, two options lie before us. The choice depends on whether one considers cardiovascular oxygen transport a merely convective step or not. In the former case, the driving force of oxygen flow across this step is the pressure difference between the mean capillary oxygen partial pressures ($P_{\bar{c}}O_2$) in the lungs and in muscles. $P_{\bar{c}}O_2$ can be calculated by Bohr's integration (Bohr 1909). Since this pressure difference is negligible, this option considers the cardiovascular system as a convective element connecting two distant resistances, not as a resistive step *in se*. In the latter case, the driving force across the cardiovascular oxygen transport would be the difference between arterial and mixed venous oxygen partial pressures (P_aO_2 and $P_{\bar{v}}O_2$ respectively). Since the difference between P_aO_2 and $P_{\bar{v}}O_2$ is remarkable (some 80 mmHg at maximal exercise at sea level) this option looks at cardiovascular oxygen transport as a resistive element, providing one of the many in-series resistances along a hydraulic system.

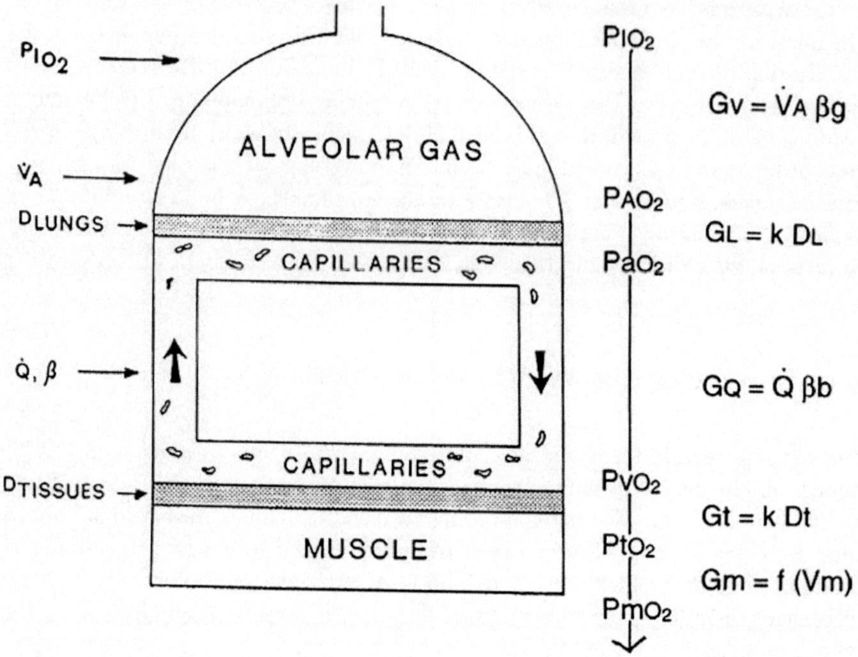

Fig. 7.3 Schematic representation of the oxygen cascade from ambient air to the mitochondria. Five steps are identified, each being characterized by a corresponding conductance: (i) from ambient to alveolar air (G_V), (ii) from alveolar air to arterial blood (G_L), (iii) from arterial to mixed venous blood (G_Q); (iv) from mixed venous blood to the cells (G_t), (v) from cells to mitochondria (G_m). As long as oxygen flows in the respiratory system, its partial pressure drops, for energy is lost to overcome the in-series resistances opposing oxygen flow. P_IO_2, P_AO_2, P_aO_2, $P_{\bar{v}}O_2$, P_tO_2, and P_mO_2 are the partial pressures of oxygen, respectively, in inspired air, alveolar air, arterial blood, mixed venous blood, peripheral tissues and mitochondria. $\dot{V}_A$, alveolar ventilation; βg, oxygen transfer coefficient in a gas phase; D_L, lung diffusing capacity; $\dot{Q}$, cardiac output; βb, oxygen transfer coefficient in blood; D_t, tissue diffusing capacity; V_m, mitochondrial volume. k designates proportionality constants. From Ferretti (2014), modified after Taylor and Weibel (1981)

Peter Wagner took the convective option, Pietro Enrico di Prampero the conductive option. This generated some apparent discrepancies in the respective formulations. Therefore, the two models were often seen as conflicting. In fact, this is not the case, at least as long as the analysis is restricted to the trait of the respiratory system downstream of the lungs (Ferretti 2014). This restriction is acceptable in normoxia, when we operate on the flat part of the oxygen equilibrium curve and the respiratory system has a non-linear behavior, whereby the lungs do not limit $\dot{V}O_2^{max}$ (Ferretti and di Prampero 1995).

At present, there is no solid reason to consider one option better than the other. Both can be considered acceptable. In fact, there is a still unresolved step in the oxygen cascade theory: the effects of the heterogeneity of distribution of the ventilation/perfusion ratio ($\dot{V}_A/\dot{Q}$) in the lungs (Rahn and Fenn 1955), which generates the difference between mean alveolar oxygen partial pressure (P_AO_2) and P_aO_2. The best tool for the analysis of oxygen transfer between alveoli and arterial blood is still the diffusion-perfusion interaction equation for the lung (Piiper and Scheid 1981). Although it does not include an analysis of the effect of $\dot{V}_A/\dot{Q}$ heterogeneity on P_aO_2, it introduces two concepts, that of perfusion limitation and that of diffusion limitation, which set the basis for understanding what happens in alveoli with low or high $\dot{V}_A/\dot{Q}$. Incidentally, let us underline that $\dot{V}_A/\dot{Q}$ heterogeneity is a brilliant creation of the Rochester—Buffalo School, whereas the diffusion-perfusion interaction equations are creations of the Göttingen School. Both schools had strong relationships to Milano, originated by the acquaintance that Margaria had with Wallace Fenn, whom he met when he was in England. Hermann Rahn was a pupil of Fenn, and Johannes Piiper started his acquaintance with Milano when he was a visiting professor in Buffalo at the end of the 1950s. An analysis of $\dot{V}_A/\dot{Q}$ heterogeneity and of the diffusion-perfusion interaction equation for the lung, as viewed from the School of Milano, has been recently published (Ferretti et al. 2022).

7.5 An Analysis of di Prampero's Model

di Prampero's model is a hydraulic model of in-series resistances and assumes steady state condition at maximal exercise. If this is so, and we have n resistances in series, we can write:

$$\dot{V} = \frac{\Delta P_1}{R_1} = \frac{\Delta P_2}{R_2} = \ldots = \frac{\Delta P_n}{R_n} = \frac{\Delta P_T}{R_T} \tag{7.1}$$

where $\dot{V}$ is the gas flow, ΔP is the pressure gradient sustaining $\dot{V}$ across the ith resistance R, and ΔP_T is the overall pressure gradient, namely the difference between the inspired and the mitochondrial partial pressure of oxygen, $P_IO_2 - P_mO_2$. Since P_mO_2 tends to 0 mmHg (Gayeski and Honig 1986; Wagner 2012), ΔP_T can be set equal to P_IO_2. At maximal exercise, $\dot{V}$ is $\dot{V}O_2^{max}$. ΔP_T is equal to the sum of the pressure gradients across each resistance:

$$\Delta P_T = \Delta P_1 + \Delta P_2 + \ldots + \Delta P_n \tag{7.2}$$

and, if the systems behaves linearly, the fraction of the overall limitation imposed by the ith resistance to oxygen flow is given by:

$$F_i = \frac{R_i}{R_T} \tag{7.3}$$

so that

$$\frac{R_1}{R_T} + \frac{R_2}{R_T} + \ldots + \frac{R_n}{R_T} = F_1 + F_2 + \ldots + F_n = 1 \tag{7.4}$$

Figure 7.3 describes five resistances of clear physiological meaning. Following the oxygen pathway from ambient air to the mitochondria, these resistances are (i) the ventilatory resistance (R_V), (ii) the lung resistance (R_L), including the oxygen transfer from the alveoli to the arterial blood, (iii) the cardiovascular resistance (R_Q), (iv) the tissue resistance (R_t), including the oxygen transfer from muscle capillaries to muscle fibers, and v) the mitochondrial resistance (R_m), related to mitochondrial oxygen flow and utilization. Although di Prampero (1985) tried to separate R_t and R_m, by introducing what he called "constant α," which describes their relationship, discrimination between R_t and R_m is practically impossible, because they are not really in-series, but they have a strong structural interrelation. This is why di Prampero and Ferretti (1990) merged them to form a lumped peripheral resistance (R_p). Thus for $\dot{V}O_2^{max}$, and on the basis of four in series resistances, Eq. (7.1) takes becomes:

$$\dot{V}O_2^{max} = \frac{(P_IO_2 - P_AO_2)}{R_V} = \frac{(P_AO_2 - P_aO_2)}{R_L} = \frac{(P_aO_2 - P_{\bar{v}}O_2)}{R_Q} = \frac{P_{\bar{v}}O_2}{R_p} = \frac{P_IO_2}{R_T} \tag{7.5}$$

Of these resistances, only R_V and R_Q are defined as the product of clear physiological variables (see Fig. 7.3), so that they are equal to, respectively:

$$R_V = \frac{1}{G_V} = \frac{1}{\dot{V}_A \cdot \beta_g} \tag{7.6a}$$

$$R_Q = \frac{1}{G_Q} = \frac{1}{\dot{Q} \cdot \beta_b} \tag{7.6b}$$

where $\dot{V}_A$ is alveolar ventilation, $\dot{Q}$ is cardiac output, constant β_g is the oxygen transfer coefficient for air (in STPD condition, it is equal to 1.16 ml mmHg^{-1}) and constant β_b is the oxygen transfer coefficient for blood. This last is equal to:

$$\beta_b = \frac{(CaO_2 - C\bar{v}O_2)}{(PaO_2 - P\bar{v}O_2)} \tag{7.7}$$

that corresponds to the average slope of the oxygen equilibrium curve. Therefore, β_b does not take an invariant value, but depends on the actual oxygen pressure range, and thus on the portion of the oxygen equilibrium curve on which blood operates.

Concerning R_L, di Prampero and Ferretti (1990) set it proportional to a factor that includes lung diffusing capacity and the effect of $\dot{V}_A/\dot{Q}$ heterogeneity. In turn, R_p was set proportional to muscle capillary density and to muscle mitochondrial volume.

When a manipulation affects $\dot{V}O_2^{max}$ without varying P_IO_2, and thus ΔP_T, the observed changes in $\dot{V}O_2^{max}$ result from changes in one or more of the resistances in series. The effect of each resistance depends on the fraction of the overall $\dot{V}O_2^{max}$ limitation, which it is responsible for. Aerobic training, for instance (Ekblom et al. 1968; Hoppeler et al. 1985), leads to a given $\dot{V}O_2^{max}$ increase, $\Delta\dot{V}O_2^{max}$, because at least three resistances, namely R_Q, R_t and R_m (or the lumped resistance R_p) have decreased, and so has R_T, as a consequence of the increase in $\dot{Q}$, muscle capillary density and muscle mitochondrial volume density. Thus, after training, Eq. 7.5 can be rewritten as follows:

$$\dot{V}O_2^{max} + \Delta\dot{V}O_2^{max} = \frac{(P_IO_2 - P_AO_2)}{(R_V + \Delta R_V)} = \frac{(P_AO_2 - P_aO_2)}{(R_L + \Delta R_L)}$$
$$= \frac{(P_aO_2 - P_{\bar{v}}O_2)}{(R_Q + \Delta R_Q)} = \frac{P_{\bar{v}}O_2}{(R_p + \Delta R_p)} = \frac{P_IO_2}{(R_T + \Delta R_T)} \tag{7.8}$$

After dividing Eq. 7.5 by Eq. 7.8, and taking into account Eq. 7.4 (see Ferretti 2014 for details), through a number of algebraic transformations we obtain:

$$\frac{\dot{V}O_2^{max}}{\left(\dot{V}O_2^{max} + \Delta\dot{V}O_2^{max}\right)} = 1 + F_V\frac{\Delta R_V}{R_V} + F_L\frac{\Delta R_L}{R_L} + F_Q\frac{\Delta R_Q}{R_Q} + F_p\frac{\Delta R_p}{R_p} \tag{7.9}$$

Training generates a complex effect, leading to an Equation with four unknowns, which we cannot solve. However, if we take a condition wherein an experimental manipulation affects only one resistance, as is the case, according to di Prampero and Ferretti (1990), for R_Q after acute blood reinfusion or withdrawal, three terms of Eq. 7.9 annihilate, thus allowing simplification of Eq. 7.9, as follows:

$$\frac{\dot{V}O_2^{max}}{\left(\dot{V}O_2^{max} + \Delta\dot{V}O_2^{max}\right)} = 1 + F_Q\frac{\Delta R_Q}{R_Q} \tag{7.10}$$

Equation 7.10 can be solved for F_Q. Suffice it to know the $\dot{V}O_2^{max}$ before and after, the R_Q before and the absolute change in R_Q induced by the maneuvre of withdrawing or injecting blood. Fig. 7.4 reports an analytical solution of Eq. 7.10, using data from different sources in the literature (di Prampero and Ferretti 1990). In Fig. 7.4, the ratio between the $\dot{V}O_2^{max}$ values before and after the maneuver (left-hand branch of Eq. 4.13) is plotted as a function of the ratio between ΔR_Q and R_Q. Eq. 7.10 predicts that this relationship must be linear, with y-intercept equal to 1 and slope equal to F_Q.

From linear regression analysis of the data reported in Fig. 7.4, di Prampero and Ferretti (1990) obtained $F_Q = 0.70$. This means that R_Q provides 70% of the fractional limitation of $\dot{V}O_2^{max}$, at least at sea level.

A F_Q value equal to 0.70 implies that the respiratory system does not behave linearly. In fact, in a linear system, the ratio of any R_i to R_T would correspond to the

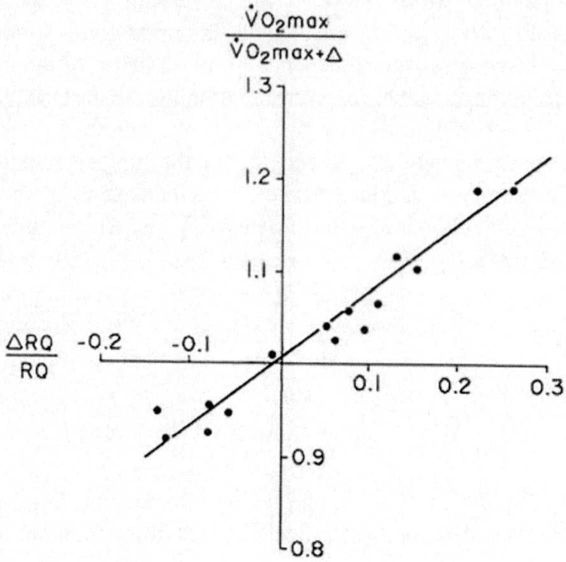

Fig. 7.4 Graphical representation of Eq. 7.10. The changes in $\dot{V}O_2^{max}$ after an induced acute change of the cardiovascular resistance to oxygen flow (R_Q) are expressed as the ratio of the $\dot{V}O_2^{max}$ before to the $\dot{V}O_2^{max}$ after ($\dot{V}O_2{}^{max} + \Delta\dot{V}O_2{}^{max}$). This ratio is plotted as a function of the ratio between the induced change in R_Q (ΔR_Q) and the R_Q before the maneuver. Points are mean values from different sources in the literature. The continuous straight line is the corresponding regression equation ($y = 1.006 + 0.7\,x$, $r = 0.97$, $n = 15$). The slope of the line, equal to 0.7, indicates that 70% of the overall limitation to $\dot{V}O_2{}^{max}$ is imposed by cardiovascular oxygen transport. From Ferretti, 2014, modified after di Prampero and Ferretti (1990)

ratio of the pressure gradient across that R_i to the overall pressure gradient overcoming R_T, so that:

$$F_Q = \frac{(P_aO_2 - P_{\bar{v}}O_2)}{P_IO_2} \tag{7.11}$$

In this case, for P_IO_2= 150 mmHg, P_aO_2= 95 mmHg and $P_{\bar{v}}O_2$ = 20 mmHg at sea level (di Prampero and Ferretti 1990), Eq. 7.11 would provide F_Q= 0.50 instead of 0.70.

The effects of the oxygen equilibrium curve on β_b represent the source of non-linearity of the system, as shown in Fig. 7.5, and thus explains this discrepancy. Consider, for example, that an acute maneuver (for instance, replacing nitrogen with helium in the inspired gas mixture) reduces R_V. This would increase $\dot{V}_A$ and P_AO_2, and thus also P_aO_2. However, since in normoxia blood operates on the flat part of the oxygen equilibrium curve, this would not imply changes in C_aO_2. Therefore, since in this condition $P_{\bar{v}}O_2$ undergoes only negligible changes, β_b would be reduced (see Fig. 7.5), and thus R_Q would be increased. As a consequence, because of the shape of

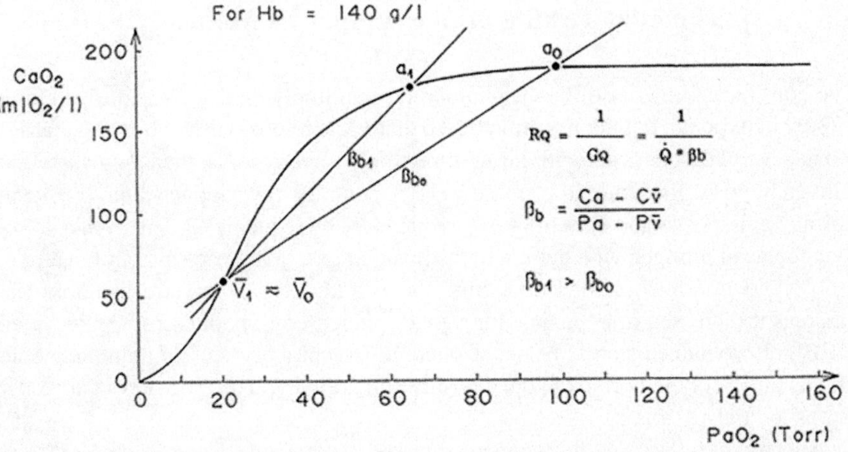

Fig. 7.5 Graphical representation of an oxygen equilibrium curve. Two arterial and mixed venous points are reported, applying to normoxia (a_0, $\overline{v}_0$) and hypoxia (a_1, $\overline{v}_1$). The two straight lines connecting the two couples of points have a slope equal to the respective oxygen transport coefficients for blood (β_b), which is higher in hypoxia than in normoxia. Thus, the ventilatory resistance R_V increases and entails a decrease in arterial oxygen partial pressure, β_b becomes higher and the cardiovascular resistance R_Q lower. An opposite change in R_Q compensates for an induced change in R_V, so that no changes in $\dot{V}O_2^{max}$ can be induced by an acute change in R_V. The lungs do not limit $\dot{V}O_2^{max}$ in normoxia. From Ferretti (2014), modified after di Prampero (1985)

the oxygen equilibrium curve, as long as we are in normoxia, any maneuver acting on R_V only has no effects on $\dot{V}O_2^{max}$, because any change in R_V would be counteracted by an opposite change in R_Q. This line of reasoning would apply also to acute interventions, altering R_L. In conclusion, R_V and R_L do not limit $\dot{V}O_2^{max}$ in normoxia.

If this is so, a two-site model is sufficient to describe the effects of the respiratory system on $\dot{V}O_2^{max}$ in normoxia. In fact all the limitation is located distally to P_aO_2. This is why $F_Q = 0.7$ instead of 0.5, so that $F_p = 0.3$. Direct experimental assessment of the parameters of Eq. 7.10 provided F_Q values in normoxia comprised between 0.65 and 0.76 (Bringard et al. 2010; Ferretti et al. 1997a, b; Turner et al. 1993), thus further supporting the concept that cardiovascular oxygen transport is the main determinant of $\dot{V}O_2^{max}$ in normoxia, but not the unique.

It is important to stress that the R_Q increases consequent to a fall in R_V (or in R_L), preventing $\dot{V}O_2^{max}$ limitation by R_V and R_L, occurs only when we operate on the flat part of the oxygen equilibrium curve. Thus, it does not occur in hypoxia, as long as we operate on the steep part of the oxygen equilibrium curve. Nor it occurs in subjects with elevated $\dot{V}O_2^{max}$ in normoxia, who undergo exercise-induced arterial hypoxaemia.

7.6 Experimental Testing of di Prampero's Model

The non-linear behavior of the respiratory system, implying that R_V and R_L do limit $\dot{V}O_2^{max}$ in hypoxia, but not in normoxia, so that F_Q in hypoxia is less than 0.7, carries along several corollaries. Firstly, if R_V limits $\dot{V}O_2^{max}$ in hypoxia, then reducing R_V at a given hypoxic P_IO_2 should increase $\dot{V}O_2^{max}$, contrary to what one would predict in normoxia. Esposito and Ferretti (1997) reduced R_V by reducing air density. To do so, they replaced nitrogen with helium in the inspired gas. With respect to air breathing, $\dot{V}O_2^{max}$ did not change when breathing the helium-oxygen mixture in normoxia, whereas it increased in hypoxia, although $\dot{V}_A$ increased in both cases. Ogawa et al. (2010) also obtained similar results. Coherently, respiratory muscle training had no effects on $\dot{V}O_2^{max}$ in normoxia, but a positive effect in hypoxia (Downey et al. 2007; Esposito et al. 2010a, b).

Another corollary of di Prampero's model is that the decrease of $\dot{V}O_2^{max}$ in hypoxia should be larger in athletic subjects, with high $\dot{V}O_2^{max}$ values in normoxia, than in non-athletic subjects, because the former subjects undergo exercise-induced arterial hypoxemia (Dempsey et al. 1984), and thus operate near the steep part of their oxygen equilibrium curve at maximal exercise. On two groups of subjects, one with high, the other with low $\dot{V}O_2^{max}$ in normoxia, Ferretti et al. (1997b) found that in the former with respect to the latter group, (i) the decrease in $\dot{V}O_2^{max}$ in hypoxia was larger, and (ii) a significant increase in $\dot{V}O_2^{max}$ in hyperoxia occurred. A $\dot{V}O_2^{max}$ increase in hyperoxia only in subjects with high $\dot{V}O_2^{max}$ in normoxia was found also by Grataloup et al. (2005). In fact, the higher is the $\dot{V}O_2^{max}$ in normoxia, the lower is the arterial oxygen saturation (S_aO_2) at maximal exercise (Grataloup et al. 2005; Harms et al. 2000; Powers et al. 1989; Rasmussen et al. 1991; Williams et al. 1986). Moreover, in hypoxia, Ferretti et al. (1997b) found a linear relationship between $\dot{V}O_2^{max}$, expressed relative to the value in hyperoxia set equal to 100%, and S_aO_2. This relationship was identical in both groups of subjects, in agreement with the above predictions. According to Wehrlin and Hallén (2006), the $\dot{V}O_2^{max}$ of endurance athletes is so high, that the decrease of $\dot{V}O_2^{max}$ in hypoxia becomes linear.

7.7 An Analysis of Wagner's Model

Peter Wagner (1993) constructed a three-equation system with three unknowns (P_AO_2, P_aO_2 and $P_{\bar{v}}O_2$). His vision of the oxygen cascade implied two mass balance equations, describing convective oxygen transfer in the airways and in the blood circulation, connected by two conductive components, described by the diffusion-perfusion interaction equations (Piiper and Scheid 1981; Piiper et al. 1984). Thus, he constructed his system by combining the mass conservation equation for blood (Fick principle) and the diffusion-perfusion interaction equations. At steady state, all these equations must have equal solutions for $\dot{V}O_2^{max}$. Further algebraic development of the three-equation system generated three separate equations allowing for a solution for P_AO_2, P_aO_2 and $P_{\bar{v}}O_2$. The obtained values are compatible with one, and only one $\dot{V}O_2^{max}$ value, for any combination of known values of P_IO_2, $\dot{V}_A$, lung oxygen diffusing capacity (D_LO_2), $\dot{Q}$, β_b and tissue oxygen diffusing capacity (D_tO_2) at maximal exercise.

At the lung level, the interaction of a convective component (ventilation) with a diffusive component (described by the diffusion-perfusion interaction equation setting oxygen flow from alveoli to lung capillaries) sets the maximal oxygen flow in arterial blood ($\dot{Q}_aO_2^{max}$), and this is the first step in the system. At muscular level, the interaction of a convective component (Fick principle) with a diffusive component (the diffusion-perfusion interaction equation setting oxygen flow from peripheral capillaries to the muscle fibers), sets $\dot{V}O_2^{max}$, and this is the second step in the system.

The Fick equation at maximal exercise can be formulated as follows:

$$\dot{V}O_2^{max} = \dot{Q}^{max} \cdot (C_aO_2 - C_{\bar{v}}O_2) = \dot{Q}^{max} \cdot \beta_b \cdot (P_aO_2 - P_{\bar{v}}O_2) \tag{7.12}$$

The term β_b in Eq. 7.12 carries along a non-linear negative relationship between $\dot{V}O_2^{max}$ and $P_{\bar{v}}O_2$ (convective curve). Its algebraic formulation depends on the mathematical solution that we give to the oxygen equilibrium curve. The diffusive component is described by the following equation:

$$\dot{V}O_2^{max} = D_tO_2 \cdot (P_{\bar{c}}O_2 - P_mO_2) \tag{7.13}$$

where P_mO_2 cancels out, because it is considered equal to 0 mmHg. The problem is that the right branches of Eqs. 7.12 and 7.13 do not share any term. Therefore, Wagner assumed that $P_{\bar{v}}O_2$ is directly proportional to $P_{\bar{c}}O_2$, arguing that the segment of the oxygen equilibrium curve between these two pressure values is essentially linear. On this basis, he rewrote Eq. 7.13 as follows:

$$\dot{V}O_2^{max} = D_tO_2 \cdot K_p \cdot P_{\bar{v}}O_2 \tag{7.14}$$

where K_p is the dimensionless constant relating $P_{\bar{v}}O_2$ to $P_{\bar{c}}O_2$. Eq. 7.14 tells that there must be a positive linear relationship between $\dot{V}O_2^{max}$ and $P_{\bar{v}}O_2$ (diffusion line). The slope of the line is equal to the product $D_tO_2 \cdot K_p$, which Ferretti (2014) called Wagner's constant, K_{Wa}. Eq. 7.14 shares the term $P_{\bar{v}}O_2$ with Eq. 7.12. So, if we plot $\dot{V}O_2^{max}$ as a function of $P_{\bar{v}}O_2$, Eqs. 7.12 and 7.14 can be directly compared on the same graph (Fig. 7.6). In this Figure, the necessary $\dot{V}O_2^{max}$ values, resulting from any given combination of $\dot{Q}_aO_2^{max}$ and K_{Wa}, is provided by the intersection of the two functions, which occurs at a precise value of $P_{\bar{v}}O_2$.

7.8 Experimental Testing of Wagner's Model

When Wagner published his model (Wagner 1993), Roca et al. (1989) had already demonstrated the linear relationship between $\dot{V}O_2^{max}$ and $P_{\bar{v}}O_2$, indicating that K_{Wa} is a constant that is independent of $P_{\bar{v}}O_2$. Since the line intercepted the y-axis at the

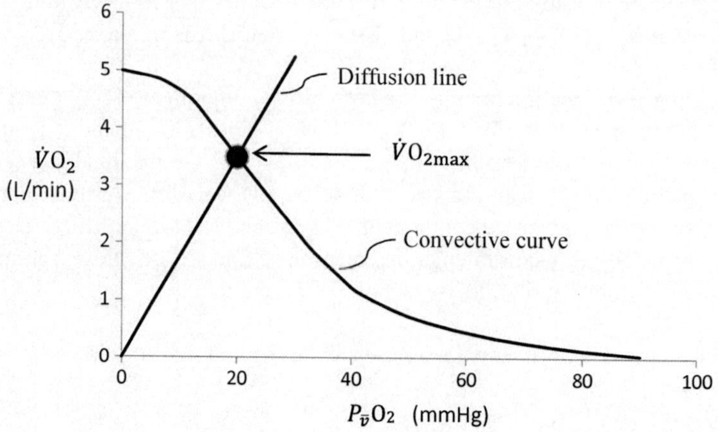

Fig. 7.6 Graphical representation of Wagner's model. Oxygen uptake ($\dot{V}O_2$) is plotted as a function of mixed venous oxygen pressure ($P_{\bar{v}}O_2$). The curve with negative slope is the convective curve. The straight line with positive slope is the diffusion line: its slope is equal to Wagner's constant, K_{Wa}. The convective curve intercepts the y-axis at $\dot{V}O_2$ equal to the arterial oxygen flow ($\dot{Q}_aO_2$), and intercepts the x-axis at $P_{\bar{v}}O_2$ is equal to the arterial oxygen pressure. The $\dot{V}O_2^{max}$ value is found on the crossing of the convective curve with the diffusion line (full dot). From Ferretti (2014), modified after Wagner (1996a)

origin of the axes, we can say that $\dot{V}O_2^{max}$ is directly proportional to $P_{\bar{v}}O_2$. Since K_{Wa} is the product between K_p and D_tO_2 and K_p is an invariant constant, K_{Wa} turns out directly proportional to D_tO_2. Thus, any decrease in K_{Wa} implies a fall of $\dot{V}O_2^{max}$ and an increase in $P_{\bar{v}}O_2$. The opposite occurs when D_tO_2 increases.

Patients affected by chronic obstructive pulmonary disease (COPD) are characterized by low $\dot{V}O_2^{max}$ values (Bauerle and Younes 1995; Henriquez et al. 1986; Mathur et al. 1995; Punzal et al. 1991; Servera and Gimenez 1984). Their skeletal muscles are atrophic and undergo morphological and functional changes that are typical of disuse (Gea et al. 2013; Gosker et al. 2007; Seay et al. 2005). Wagner (1996a) predicted quite accurately the effects of a reduced D_tO_2 on $\dot{V}O_2^{max}$ and $P_{\bar{v}}O_2$ in COPD patients. A refinement of his analysis was recently published (Broxterman et al. 2020). Muscle capillary density is a main determinant of D_tO_2 (Bicudo et al. 1996; Chilibeck et al. 1997). Assuming that K_{Wa} is directly proportional to muscle capillary density, an analysis of literature data of muscle morphometry and $\dot{V}O_2^{max}$ of altitude-acclimatized climbers (Hoppeler et al. 1990; Kayser et al. 1991; Oelz et al. 1986) or endurance-trained subjects (Hoppeler et al. 1985) yielded $P_{\bar{v}}O_2$ values coherent with Wagner's predictions (Ferretti 2014). The same was the case for the data of Operation Everest II (Wagner 2017).

Concerning the convective curve, an increase in G_Q, i.e., of the product between $\dot{Q}$ and β_b, carries along an increase in both $\dot{V}O_2^{max}$ and $P_{\bar{v}}O_2$. The convective curve is displaced upward and its slope becomes steeper (lower). The intercept on the x-axis of the convective curve corresponds to the P_aO_2 point, id est to the point at which

$P_{\bar{v}}O_2 = P_aO_2$. This is how Wagner's model explains the effects of acute polycythemia and anemia on $\dot{V}O_2^{max}$. Hyperoxia displaces the P_aO_2 point rightward, thus increasing (making less negative) the slope of the convective curve. Therefore, $\dot{V}O_2^{max}$ should slightly increase in hyperoxia, because the convective curve intercepts the diffusion line at a higher $\dot{V}O_2^{max}$ value. In spite of this prediction, such an increase was rarely observed in humans (Bannister and Cunningham 1954; Esposito and Ferretti 1997; Fagraeus et al. 1973; Margaria et al. 1961, 1972; Welch and Pedersen 1981). As already pointed out, an increase in $\dot{V}O_2^{max}$ in hyperoxia was clearly observed only in subjects with elevated $\dot{V}O_2^{max}$ (Ferretti et al. 1997b; Grataloup et al. 2005; Harms et al. 2000). Wagner's predictions are in contrast with those of di Prampero's model and the apparent discrepancy between theoretical predictions and experimental data is hard to explain. The thoroughbred horse, an athletic mammal subject to exercise-induced arterial hypoxemia, and with big muscle masses, is the best example supporting Wagner's prediction of a $\dot{V}O_2^{max}$ increase in hyperoxia (Wagner et al. 1989, 1996).

7.9 A Critical Comparison of the Multifactorial Models

Both di Prampero and Wagner models exclude that pulmonary ventilation and oxygen diffusion capacity limit $\dot{V}O_2^{max}$ in healthy non-athletic humans in normoxia. The way di Prampero and Ferretti (1990) arrived at that conclusion is analyzed above, when discussing of the non-linear response of the system and of the role of β_b. Wagner merely excluded the lungs from his analysis and concentrated on the comparison between the convective curve and the diffusive line. Thus, at least in normoxia, both looked at the distal part of the respiratory system, thus making a comparison of the two models possible. The present analysis is a summary of that proposed by Ferretti (2014), which is the most recent contribution to the issue by the School of Milano.

Both models share Eqs. 7.12 and 7.14. Eq. 7.12 defines R_Q, since, according to Eq. 7.6b:

$$\dot{V}O_2^{max} = \dot{Q}\,\beta_b\,(P_aO_2 - P_{\bar{v}}O_2) = \frac{1}{R_Q}\,(P_aO_2 - \mathrm{P}_{\bar{v}}O_2) \tag{7.15}$$

Equation 7.14, for $P_mO_2 = 0$ mmHg, defines R_p, 7.since:

$$\dot{V}O_2^{max} = D_tO_2\,K_p\,P_{\bar{v}}O_2 = K_{Wa}\,P_{\bar{v}}O_2 \tag{7.16}$$

and since:

$$\frac{\dot{V}O_2^{max}}{P_{\bar{v}}O_2} = G_P = K_{Wa} \tag{7.17}$$

Wagner's constant K_{Wa} is equal to di Prampero's G_p, the reciprocal of R_p. Eqs. 7.15 and 7.17 place the two models under a different light, whereby the slope of the diffusive line is G_p, or $1/R_p$ and the slope of the convective curve, is equal to $-\dot{Q} \cdot \beta_b$, i.e. $-G_Q$, or $-1/R_Q$. In this case, the y-axis intercept of the convective curve is at $\dot{V}O_2^{max} = \dot{Q}_aO_2^{max}$. The diffusive line intercepts the convective curve at that point when $K_{Wa} = \infty$, so that $G_p = \infty$, $R_p = 0$ and, according to di Prampero's model, $F_Q = 1$ and thus $F_p = 0$: all oxygen delivered to peripheral capillaries is consumed by mitochondria. On the opposite side, the x-axis intercept of the convective curve is at $P_{\bar{v}}O_2 = P_aO_2$. The diffusive line intercepts the convective curve at that point where $K_{Wa} = 0$. In this case, $G_p = 0$, $R_p = \infty$ and, according to di Prampero's model, $F_Q = 0$ and thus $F_p = 1$, so that $\dot{V}O_2\ max = 0$ L min−1, no oxygen flows from capillaries to mitochondria, all the oxygen delivered by arterial blood comes back to the heart through the veins and the composition of venous blood would be equal to that of arterial blood. Thus, Wagner's model includes F_Q and di Prampero's model includes K_{Wa}.

Moreover, neglecting R_V and R_L at sea level, a simplified, linear version of di Prampero's model, describing only the oxygen flow downstream of the lungs, can be written:

$$F_Q = \frac{(P_aO_2 - P_{\bar{v}}O_2)}{P_aO_2} = \frac{R_Q}{(R_Q + R_p)} \tag{7.17}$$

whence

$$\frac{1}{F_Q} = \frac{(R_Q + R_p)}{R_Q} = 1 + \frac{R_p}{R_Q} = 1 + \frac{G_Q}{G_p} \tag{7.18}$$

Since:

$$\dot{Q}_aO_2^{max} = \dot{Q}\,\beta_b\,P_aO_2 \tag{7.19}$$

dividing Eq. 7.15 by Eq. 7.19, one obtains an algebraic definition of the oxygen extraction coefficient, as follows:

$$\frac{\dot{V}O_2^{max}}{\dot{Q}_aO_2^{max}} = \frac{\dot{Q}\,(C_aO_2 - C_{\bar{v}}O_2)}{\dot{Q}\,C_aO_2} = \frac{\dot{Q}\,\beta_b\,(P_aO_2 - P_{\bar{v}}O_2)}{\dot{Q}\,\beta_b\,P_aO_2}$$
$$= \frac{(P_aO_2 - P_{\bar{v}}O_2)}{P_aO_2} = F_Q \tag{7.20}$$

This implies that F_Q in normoxia is equal to the oxygen extraction coefficient! Since, according to di Prampero and Ferretti (1990), $F_Q = 0.7$ in normoxia, we also have that $\dot{V}O_2^{max} = 0.7\ \dot{Q}_aO_2^{max}$ and $P_{\bar{v}}O_2 = 0.3\ P_aO_2$. Therefore, we have:

$$K_{Wa} = \frac{\dot{V}O_2^{max}}{0.3\ P_aO_2} \tag{7.21}$$

Equation 7.21 implies that a simple measure of $\dot{V}O_2^{max}$ and P_aO_2 is enough to know K_{Wa}. For a young non-athletic subject, whose $\dot{V}O_2^{max}$ is 3 L min^{-1}, and who has a normal P_aO_2 of 100 mHg, the $P_{\bar{v}}O_2$ at maximal exercise would be 30 mmHg and K_{Wa} would turn out equal to 100 ml min^{-1} mmHg^{-1}. The $\dot{V}O_2^{max}$ and the $P_{\bar{v}}O_2$ values set the coordinates of the intersection between the convective curve and the diffusive line on Wagner's plot. The closer is the intersection to the P_aO_2 point, the lower is K_{Wa}, and thus the higher are R_p and F_p and the lower is F_Q. In di Prampero's model, with the same values we would have: R_Q= 23.33 mmHg min L^{-1} and R_p= 10 mmHg min L^{-1}.

This analysis justifies the statement that di Prampero's model and Wagner's model converge on the same conclusion, that both cardiovascular oxygen transport and muscle oxygen diffusion and utilization are necessary determinants of $\dot{V}O_2^{max}$. According to the preceding analysis, the former is responsible for the larger fraction of the overall $\dot{V}O_2^{max}$ limitation, 70%, according to di Prampero and Ferretti (1990). In addition, among the greats of the past, who participated in the useless and annoying (according to Margaria) debate on the single factor that limits $\dot{V}O_2^{max}$, those in favor of the cardiovascular limitation were less wrong than their opponents, who favored muscular limitation. Eventually, the reckless conclusion of Margaria (Margaria et al. 1965b) was eventually not that bad.

7.10 Of Maximal Oxygen Consumption in Hypoxia

The School of Milano played an important role in the history of our knowledge on $\dot{V}O_2^{max}$ in hypoxia indeed. That $\dot{V}O_2^{max}$ decreases in hypoxia, both acute and chronic, is a well-known notion since long (see for review Cerretelli 1980; Cerretelli and Hoppeler 1996; Ferretti 1990; Ferretti 2014, just to stay within the School of Milano). This decrease is smaller, the smaller is the active muscle mass (Cardus et al. 1998). The cause of the $\dot{V}O_2^{max}$ decrease in hypoxia is the drop of P_IO_2. However, the $\dot{V}O_2^{max}$ decrease is small at altitudes below 3000 m, much smaller than one would expect after the curves of barometric pressure decrease at altitude (West 1983). In discussing di Prampero's model, we highlighted the non-linear behavior of the respiratory system because of the shape of the oxygen equilibrium curve, and the linear relationship between $\dot{V}O_2^{max}$ and S_aO_2. As long as we operate on the flat portion of the oxygen equilibrium curve, a decrease in P_aO_2 at altitude, which is not accompanied by a decrease in C_aO_2, entails an increase in β_b, and thus in G_Q. This

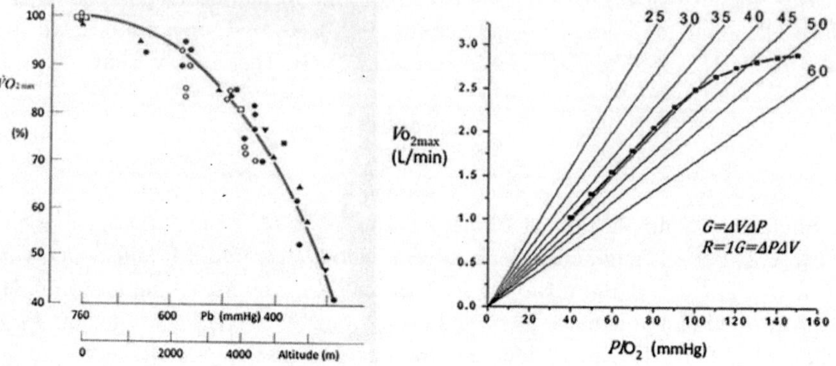

Fig. 7.7 *Left:* Fall of maximal oxygen consumption ($\dot{V}O_2^{max}$) at altitude. $\dot{V}O_2^{max}$ is expressed relative to the value observed at sea level, set equal to 100%. Two x-axis are shown, one for barometric pressure (P_B), the other for altitude. Open and full dots refer to acute and chronic hypoxia, respectively. Data from Cerretelli (1980). *Right:* Same curve as on left, calculated for a sea level $\dot{V}O_2^{max}$ of 2.9 L min^{-1} (Cerretelli and di Prampero 1987), where P_B is replaced by the inspired oxygen pressure (P_IO_2). The straight lines converging on the origin of the axes have a slope ($\Delta\dot{V}/\Delta P$) equal to the overall oxygen conductance of the respiratory system (G_T). The modest $\dot{V}O_2^{max}$ decrease at low altitude is a consequence of the simultaneous increase in G_T, due to the effects of the shape of the oxygen equilibrium curve. From Ferretti (2014)

reduces R_Q and thus R_T, keeping up $\dot{V}O_2^{max}$. This explains the curve describing the $\dot{V}O_2^{max}$ decrease in hypoxia, reported in Fig. 7.7.

In this Figure, the left panel reports the classical representation of the $\dot{V}O_2^{max}$ decrease at altitude (Cerretelli 1980). The right panel has P_IO_2 on the x-axis, whereas $\dot{V}O_2^{max}$ is expressed in absolute terms, rather than relative to its value at sea level (Ferretti 2014). This has the advantage of allowing the construction of overall conductance (G_T) isopleths, showing that, as we move to altitude and P_IO_2 declines, the $\dot{V}O_2^{max}$ curve crosses isopleths of progressively higher G_T values (lower R_T values). At low P_IO_2, when we attain the steep part of the oxygen equilibrium curve, and P_aO_2 and thus C_aO_2 start decreasing, the $\dot{V}O_2^{max}$ curve enters its linear part, coinciding with a specific G_T isopleth. The function describing the $\dot{V}O_2^{max}$ decrease at altitude is a kind of mirror image of the oxygen equilibrium curve (Ferretti et al. 1997b).

Ferretti and di Prampero (1995) analysed the interrelations between R_Q, R_V and R_L in hypoxia. They demonstrated that, as hypoxia gets deeper and we approximate the steep part of the oxygen equilibrium curve, the fraction of $\dot{V}O_2^{max}$ limitation due to R_V and R_L becomes more important. They calculated that in extreme hypoxia F_Q may decrease down to 0.20 with F_V and F_L going up to about 0.40 and 0.35, respectively. This expansion of di Prampero's model led to some predictions that received experimental support. The effects of hypoxia on $\dot{V}O_2^{max}$ are larger the higher is the subject's $\dot{V}O_2^{max}$ in normoxia (Ferretti et al. 1997b; Gavin et al. 1998; Koistinen et al. 1995; Wehrlin and Hallén 2006), because of exercise-induced arterial hypoxemia. They are also smaller, the more intense is the ventilatory response to hypoxia (Benoit et al. 1995; Gavin et al. 1998; Giesbrecht et al. 1991; Marconi et al. 2004; Ogawa et al. 2007). However, the ventilatory response to hypoxia is unrelated to the absolute $\dot{V}O_2^{max}$ in normoxia (Sheel et al. 2006), suggesting that the two phenomena are dissociated.

The analysis of the factors limiting $\dot{V}O_2^{max}$ in hypoxia, carried out by Ferretti and di Prampero (1995), set the basis for the reinterpretation of the fundamental work on the determinants of $\dot{V}O_2^{max}$ at altitude carried out by Cerretelli (1976), which we analyze in detail in Chap. 10. We just remark here that Cerretelli took his findings as a demonstration that the limitation of $\dot{V}O_2^{max}$ at altitude is peripheral, rather than central. In the context of di Prampero's model of $\dot{V}O_2^{max}$ limitation, Ferretti (2003) argued that administering oxygen at 150 mmHg to Cerretelli's subjects at Everest base camp changed three terms of the oxygen conductance equation. The overall oxygen gradient was elevated, because P_IO_2 was increased; maximal $\dot{Q}$ was increased; β_b was lowered. In fact, the increase in P_IO_2 induced a subsequent increase in P_AO_2 and in P_aO_2, so that the arterial blood point on the oxygen equilibrium curve moved toward and along the flat part of it. Since $P_{\bar{v}}O_2$ was only slightly affected, β_b turned out remarkably reduced. This means a reduction of G_Q (increase in R_Q) in subjects whose R_p was higher than in Milano before departure, because muscle oxidative capacity is reduced in chronic hypoxia. Within Wagner's model, the experiment altered the convective curve by elevating both the $\dot{Q}_aO_2^{max}$ point and the P_aO_2 point, while reducing K_{Wa}. Indeed Cerretelli's experiment provided a brilliant confirmation of the predominant role of R_Q in limiting $\dot{V}O_2{}^{max.}$

Whereas the School of Milano integrated the effects of R_V and R_L in the multifactorial model of $\dot{V}O_2^{max}$ limitation, Wagner looked at the respiratory system distally to the lungs also in hypoxia. Within Wagner's model, hypoxia shrinks the convective curve, because both $\dot{Q}_aO_2^{max}$ and P_aO_2 are decreased and the curve lacks its flat part at high $P_{\bar{v}}O_2$, because we operate only on the steep part of the oxygen equilibrium curve. The slope of the diffusive line is reduced because of the decreased muscle oxidative capacity, despite the increased capillary density (Fig. 7.8) (Wagner 2017). Using data from Operation Everest II, Wagner (1996b) demonstrated the linearity of the oxygen equilibrium curve in hypoxia in the $P_aO_2 - P_{\bar{v}}O_2$ pressure range. If this is so, and thus β_b is invariant, the oxygen conductance equation becomes linear. Thus, the effective oxygen pressure gradient is equal to P_IO_2 instead of P_aO_2, so that Eq. 7.20 can be rewritten as follows:

$$F_Q = \frac{\dot{V}O_2^{max}}{\dot{Q}_aO_2^{max}} \frac{P_aO_2}{P_IO_2} \tag{7.22}$$

If we use the data of Operation Everest II to solve Eq. 7.22 (Wagner 1996b), F_Q turns out equal to 0.19, a value that is very close to the theoretical value of 0.20 obtained by Ferretti and di Prampero (1995). On the other hand:

$$F_p = (1 - \frac{\dot{V}O_2^{max}}{\dot{Q}_aO_2^{max}}) \frac{P_aO_2}{P_IO_2} \tag{7.23}$$

whence, using the same data, $F_p = 0.22$ and $K_{Wa} = 73$ ml min^{-1} mmHg^{-1}.

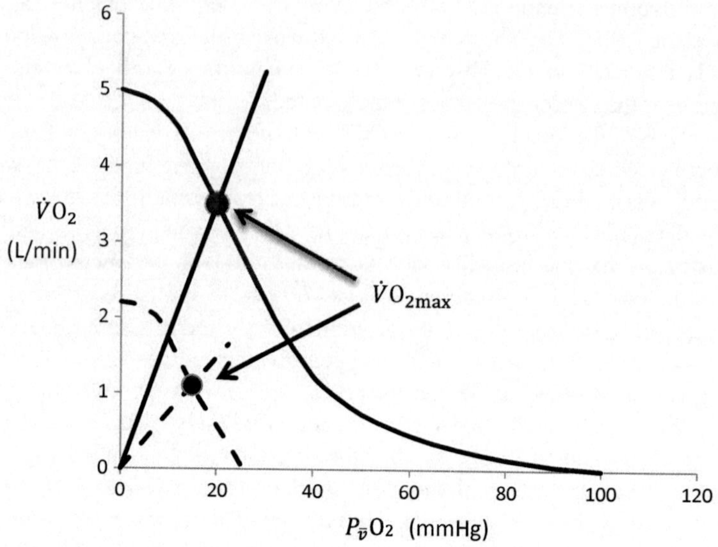

Fig. 7.8 Graphical representation of Wagner's model in hypoxia. Oxygen uptake ($\dot{V}O_2$) is plotted as a function of mixed venous oxygen pressure ($P_{\bar{v}}O_2$). Continuous lines represent the convective curve and the diffusion line, as from Fig. 7.6. Dashed lines refer to the convective curve and the diffusion line in hypoxia. Concerning former, it lacks the flattening part at high $P_{\bar{v}}O_2$ values, because we operate exclusively of the steep part of the oxygen equilibrium curve. The diffusion line has lower slope in hypoxia than in normoxia, indicating the decrease of Wagner's constant in hypoxia. In normoxia, arterial oxygen partial pressure was assumed equal to 100 mmHg, and $P_{\bar{v}}O_2$ was assumed equal to 20 mmHg. The data of Operation Everest II (Wagner 1996b) were used for the construction of the convective curve and of the diffusion line in hypoxia. $\dot{V}O_2^{max}$, maximal oxygen consumption. After Ferretti (2014)

In conclusion, in extreme hypoxia, R_V and R_L provide most of $\dot{V}O_2^{max}$ limitation. Despite its increase, witnessed also by the lower K_{Wa}, R_p plays a smaller role than in normoxia in limiting $\dot{V}O_2^{max}$. The role of R_Q is minimized, because of both its absolute decrease and the greater role of R_V and R_L as limiting factors.

7.11 Of Maximal Oxygen Consumption at the End of Bed Rest

$\dot{V}O_2^{max}$ decreases after bed rest (Bringard et al. 2010; Capelli et al. 2006; Convertino et al. 1982; Ferretti et al. 1997a; Greenleaf et al. 1989; Kashihara et al. 1994; Saltin et al. 1968; Shibata et al. 2010; Smorawinski et al. 2001; Stremel et al. 1976; Trappe et al. 2006, just to cite some of the most significant: more than 80 studies were reckoned by Ried-Larsen et al. 2017). The size of the $\dot{V}O_2^{max}$ decrease is larger, the longer the bed rest duration. The absolute size of the bed rest decrease is also larger the higher the $\dot{V}O_2^{max}$ before bed rest (Ried-Larsen et al. 2017). The rate of $\dot{V}O_2^{max}$

decrease is fast in the first days, becoming slower as bed rest duration is prolonged. This means that the $\dot{V}O_2^{max}$ change at the end of bed rest, as a function of bed rest duration, tends to an asymptote (Capelli et al. 2006).

These statements apply only to the $\dot{V}O_2^{max}$ determined in upright posture, measured shortly after the end of the bed rest period. This is not so during bed rest (or space flight), or in supine posture after bed rest. Very small changes, if any, in $\dot{V}O_2^{max}$ were observed in these conditions (Bringard et al. 2010; Greenleaf et al. 1989; Levine et al. 1996; Trappe et al. 2006).

Ferretti et al. (1997a) investigated the overall behavior of the respiratory system before and after a 42-day bed rest without countermeasure. They used the data to analyze the factor limiting $\dot{V}O_2^{max}$ in the context of di Prampero's model. They estimated the changes in R_Q, R_t and R_m (or the lumped peripheral resistance R_p) from the measured changes in $\dot{Q}_aO_2^{max}$, muscle capillary length and muscle oxidative capacity. On this basis, they obtained $F_Q = 0.73$ and $F_p = 0.27$. So, they provided the first evidence that the prediction (di Prampero and Ferretti 1990) that $F_Q = 0.70$ was essentially correct also in case of an adaptive response of the entire respiratory system.

The time course of the $\dot{V}O_2^{max}$ changes in upright posture at the end of head-down tilt bed rest without countermeasures is shown in Fig. 7.9. Assuming an exponential $\dot{V}O_2^{max}$ decay, Ferretti and Capelli (2009) linearized the relationship by expressing it in logarithmic form. The resulting representation is shown in the bottom panel of Fig. 7.9. Two components of the $\dot{V}O_2^{max}$ decline appeared. This means that the decline of $\dot{V}O_2^{max}$ due to bed rest has at least two causes that simultaneously act on the respiratory system, and that at least two capacitances are involved. We can express this concept in the following double exponential Equation:

$$\dot{V}O_2^{max}{}_t - \dot{V}O_2^{max}{}_a = \left(\dot{V}O_2^{max}{}_0 - \dot{V}O_2^{max}{}_a\right)\left(e^{-k_1 t} + e^{-k_2 t}\right) \tag{7.24}$$

where suffixes 0, a and t indicate the $\dot{V}O_2^{max}$ values before bed rest (time 0 days), at the asymptote of the $\dot{V}O_2^{max}$ decrease and at bed rest time t, respectively. k_1 and k_2 are the rate constants of the two components of the $\dot{V}O_2^{max}$ decrease during bed rest (slope of the straight lines of the bottom panel of Fig. 7.9). Ferretti and Capelli (2009) obtained $k_1 = 0.083$ day^{-1}, and $k_2 = 0.0098$ day^{-1}. The corresponding time constants were equal to 12.1 and 102.0 days, respectively.

The $\dot{V}O_2^{max}$ changes appear as the epiphenomenon reflecting long-term adaptive changes that affect the overall respiratory system. In view of the reported differences in time constants, the two postulated capacitances must have remarkably different size. The effects on the smaller capacitance initially prevail, because of its minor time constant, and dictate a rapid change in $\dot{V}O_2^{max}$ already in the first days. According to the analysis of Ferretti and Capelli (2009), this fast component attains its asymptote within approximately one month. Afterward, the second, larger capacitance takes over, dictating a further, though slower, $\dot{V}O_2^{max}$ decline. Ferretti (2014) attributed the fast component of the $\dot{V}O_2^{max}$ decrease after bed rest to changes in R_Q, and thus to the reduction of $\dot{Q}_aO_2^{max}$, and the slow component to changes in R_p, and thus to the development of lower limb muscle hypotrophy.

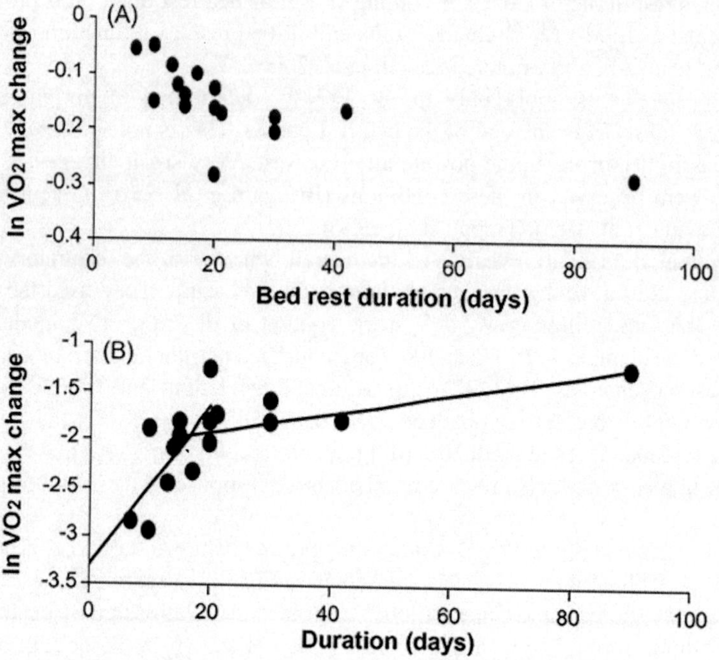

Fig. 7.9 *Top panel.* The change in maximal oxygen consumption ($\dot{V}O_2^{max}$) in upright posture, at the end of bed rest or space flight, is expressed as the absolute change in $\dot{V}O_2^{max}$ with respect to the corresponding pre-bed rest value, and plotted as a function of bed rest duration. *Bottom panel.* Same as on top, except that the change in $\dot{V}O_2^{max}$ is expressed in logarithmic form. The lines are regression lines calculated for bed rests lasting less than 20 days and longer than 20 days, respectively. The slopes of the two lines indicate the velocity constant of the fast (0.083 day^{-1}) and the slow (0.0098 day^{-1}) components of the $\dot{V}O_2^{max}$ decrease. The corresponding time constants are 12.1 and 102.0 days (half-times of 8.4 and 70.7 days), respectively. From Ferretti and Capelli 2009

Bringard et al. (2010) had been the first and only ones, who compared the effects of bed rest on $\dot{V}O_2^{max}$ and $\dot{Q}_aO_2^{max}$ in upright and supine posture. Before that study, it is noteworthy that Levine et al. (1996) reported a $\dot{V}O_2^{max}$ decrease in upright posture upon return from a 17-day space flight that corresponds well to those that we may predict from Fig. 7.9. Conversely, they found no changes in $\dot{V}O_2^{max}$ on the same subjects in space. They attributed the $\dot{V}O_2^{max}$ decline upon return to the effects of blood volume redistribution upon resumption of gravity exposure. According to Fig. 7.9, a 17-day space flight duration was just too short to evidence the effects of muscle hypotrophy. This component was just visible, after the same time, in the study by Trappe et al. (2006) in space and in supine posture after bed rest. Significant $\dot{V}O_2^{max}$ reductions were in fact observed during long-duration space flight in the International Space Station (Ade et al. 2017). Concerning bed rest, Greenleaf et al. (1989) had reported a greater $\dot{V}O_2^{max}$ decrease in upright than in supine posture.

At the end of a 35-day bed rest without countermeasures, on the day of reambulation, not 2–3 days after, as in most studies, Bringard et al. (2010) found a

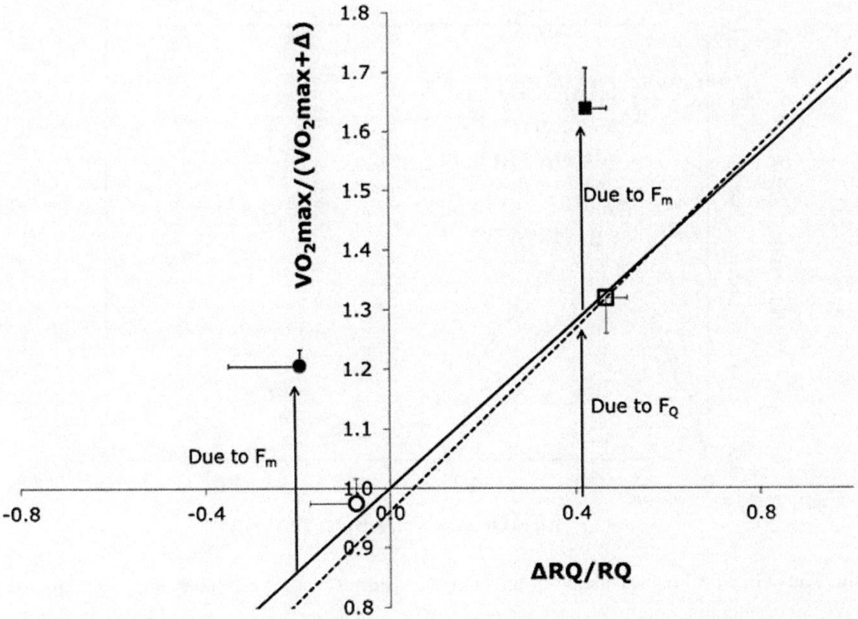

Fig. 7.10 The ratio between maximal oxygen consumption ($\dot{V}O_2^{max}$) before and after a given maneuver [$\dot{V}O_2^{max}/(\dot{V}O_2^{max}+\Delta)$, y-axis] is plotted as a function of the relative change in the cardiovascular resistance to oxygen flow ($\Delta R_Q/R_Q$, x-axis). The continuous line, with a slope of 0.7, is the theoretical line obtained by di Prampero and Ferretti (1990). The open symbols refer to the effects of postural changes from supine to upright before (open dot) and after (open square) bed rest. The dashed line is the regression equation calculated on the data of Bringard et al. (2010) after bed rest ($y = 0.76x + 0.96$). Statistical analysis showed that the slope of this line did not differ from that of the theoretical one, and that its y-intercept of the experimental line did not differ from 1. The symbols located above the lines refer to the effects of bed rest in supine (filled dot) and upright (filled square). Error bars indicate standard error. The arrows highlight the effect on $\dot{V}O_2^{max}$ due to cardiovascular (F_Q) and peripheral (F_p) $\dot{V}O_2^{max}$ limitation. From Ferretti (2014)

44% reduction of stroke volume at maximal exercise in upright posture, with no changes in f_H^{max}: the consequence was a 45% decrease in $\dot{Q}_{max}$. Considering a 13% increase in C_aO_2 due to higher hemoglobin concentration, this resulted in a 38% decrease in $\dot{Q}_aO_2^{max}$. Conversely, in the same study and on the same subjects, no changes in $\dot{Q}_{max}$ were observed in supine posture after bed rest, thus implying a slight increase in $\dot{Q}_aO_2^{max}$. Therefore, after bed rest, $\dot{Q}_aO_2^{max}$ was 56% lower, and R_Q 78% higher, upright than supine: an acute postural change from supine to upright entailed a $\dot{V}O_2^{max}$ decrease only due to changes in R_Q. These results are similar to those obtained by Levine et al. (1996), when they compared the $\dot{V}O_2^{max}$ data in flight with those obtained upright shortly after landing.

Figure 7.10 reports a representation of di Prampero's model for the case of prolonged bed rest, using the data of Bringard et al. (2010). The continuous line reports the theoretical F_Q value of 0.7 (di Prampero and Ferretti 1990). The open

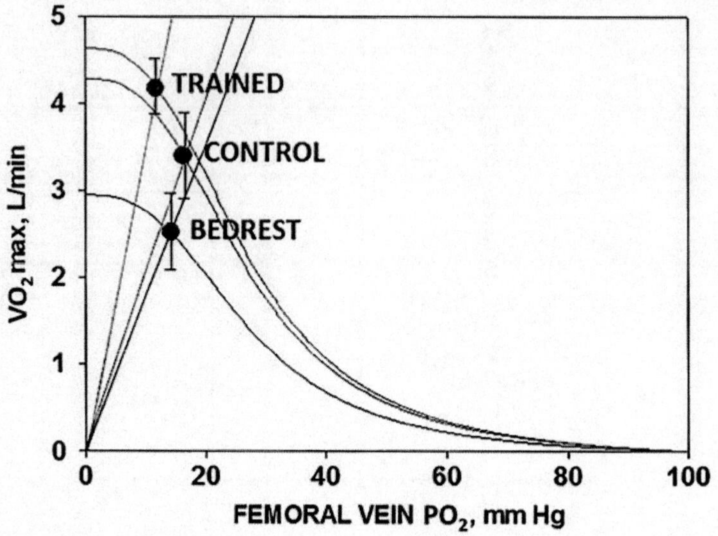

Fig. 7.11 Graphical representation of the effects of bed rest and of aerobic training on maximal oxygen consumption within Wagner's model. Oxygen uptake ($\dot{V}O_2$) is plotted as a function of femoral venous oxygen pressure. Three convective curves (from top to bottom) and three diffusion lines (from steepest to flattest) are reported, respectively for training, control, and bed rest. Concerning the convective curves, the one after bed rest is flatter than the one before bed rest, because of the dramatic decrease in arterial oxygen flow after bed rest. From Wagner (2015), data from Saltin et al. (1968)

symbols lying on this line refer to the acute maneuvre of moving from supine to upright, before and after bed rest. The full dots lying above this line refer to the overall effect of bed rest, in supine—lower left point—and upright—upper right point—posture. The size of the upward shift of filled symbols with respect to reference line is the same for both postures. This means that the factor that caused the $\dot{V}O_2^{max}$ decrease supine (muscle hypotrophy, according to Ferretti and Capelli (2009) after bed rest acted by the same extent also in the upright posture. This factor is independent of the acute postural change. According to Bringard et al. (2010), the vertical distance between the filled symbols and the reference line represents the effects of the change in R_p on $\dot{V}O_2{}^{max.}$

The increase in R_p due to muscle hypotrophy implies a decrease in K_{Wa} in Wagner's model, whereas the increase in R_Q generates a downward shift of the $\dot{Q}_aO_2$ point, with consequent decrease in the slope of the convective curve. This appears clearly in the plot that Wagner (2015) constructed using the bed rest data of Saltin et al. (1968). These last authors investigated also the effects of aerobic training. Wagner (2015) plotted also these data on the same Figure. His Figure, reproduced here as Fig. 7.11, has therefore become a most impressive, plastic representation of the overall adaptive response of the respiratory system after training and bed rest. The reported changes in the convective curve and in the diffusive line generated by training and bed rest are the consequence, respectively, of a consensual decrease or increase of R_p and R_Q. Their ratio to R_T remains unchanged, so that $F_Q = 0.7$ in all cases.

7.12 Conclusions

Although in the times of Margaria the School of Milano contributed less to our understanding of $\dot{V}O_2^{max}$ than other major schools, especially the Scandinavian, nevertheless it generated the fertile cultural environment into which the multifactorial models of $\dot{V}O_2^{max}$ limitation were created. This is a fundamental achievement of the School of Milano in the last forty years of research on the subject of $\dot{V}O_2^{max}$ limitation. The authors of this chapter played a key role in this story, especially in theoretical analysis, but the significant contributions of Guglielmo Antonutto, Carlo Capelli and Fabio Esposito in the experimental part of the story is also to be acknowledged. Paolo Cerretelli and his pupil Claudio Marconi had very important roles in altitude studies on $\dot{V}O_2^{max}$. Of the younger fellows, also Alessandra Adami, who trained with Capelli in Verona and Ferretti in Geneva, and now is in the USA, Aurélien Bringard and Frédéric Lador from Geneva, Silvia Pogliaghi from Verona, Enrico Tam, who trained in Udine with di Prampero and then followed Capelli to Verona, deserve a mention for their roles in this story. These achievements, however, would have been impossible without the strong relationships with the physiological Schools of Göttingen, and with the anatomical school of Bern.

Peter Wagner followed different pathways, yet he acknowledges his debt to Johannes Piiper and the School of Göttingen for the diffusion–perfusion interaction equations. To our mind, he was influenced, at least in part, also by Milano, under several respects. But he does not acknowledge this in written texts and is reticent to cite the relative works. Three major theoretical papers were published in the last eight years on $\dot{V}O_2^{max}$ limitation (Ferretti 2014; Wagner 2015; Wagner 2017). Ferretti dedicated an entire section of his review to analyze Wagner's model critically. Wagner, although he is certainly aware of the contributions of the School of Milano to the subjects of $\dot{V}O_2^{max}$ limitation and of the effects of altitude and bed rest on $\dot{V}O_2^{max}$, simply neglected that work. A portrait of Wagner, di Prampero and Ferretti at a meeting of the Italian Physiological Society in Sorrento, Italy, in 2011 appears in Fig. 7.12.

Fig. 7.12 From left to right, Peter Wagner, Fabio Benfenati (then President of the Italian Physiological Society, current President of the European Federation of Physiological Societies), Arsenio Veicsteinas, Pietro Enrico di Prampero, Guido Ferretti, Fabio Esposito, Carlo Capelli and Paolo Cerretelli. Wagner and di Prampero are the creators of the two multifactorial models of $\dot{V}O_2^{max}$ limitation. Picture taken at the 62° annual meeting of the Italian Physiological Society, which took place in Sorrento, Italy, on September 25–27, 2011. With the kind permission of Guido Ferretti

References

Ade CJ, Broxterman RM, Moore AD, Barstow TJ (2017) Decreases in maximal oxygen uptake following long-duration spaceflight: role of convective and diffusive O_2 transport mechanisms. J Appl Physiol 122:968–975

Andersen P, Henriksson J (1977) Capillary supply of the quadriceps skeletal muscle of man: adaptive response to exercise. J Physiol 270:677–690

Andersen P, Saltin B (1985) Maximal perfusion of skeletal muscle in man. J Physiol 366:233–249

Asmussen E (1965) Muscular exercise. In: Fenn WO, Rahn H (eds) Handbook of Physiology Respiration, vol 2. Am Physiol Soc, Washington DC, pp 939–978

Åstrand PO (1956) Human physical fitness with special reference to sex and age. Physiol Rev 36: 307–335

Åstrand PO, Rodahl K, Dahl HA, Strømme SB (2003) Textbook of work physiology. Physiological bases of exercise, 4th edn. Human Kinetics, Champaign IL, USA

Åstrand PO, Ryhming I (1954) A nomogram for calculation of aerobic capacity (physical fitness) from pulse rate during sub-maximal work. J Appl Physiol 7:218–221

Åstrand PO, Saltin B (1961) Maximal oxygen uptake and heart rate in various types of muscular activity. J Appl Physiol 16:977–981

Bannister RG, Cunningham DJC (1954) The effects on the respiration and performance during exercise of adding oxygen to the inspired air. J Physiol 125:118–137

Bauerle O, Younes M (1995) Role of ventilatory response to exercise in determining exercise capacity in COPD. J Appl Physiol 79:1870–1877

Benoit H, Busso T, Castells J, Denis C, Geyssant A (1995) Influence of hypoxic ventilatory response on arterial O_2 saturation during maximal exercise in acute hypoxia. Eur J Appl Physiol 72:101–105

Bergh U, Kanstrup IL, Ekblom B (1976) Maximal oxygen uptake during exercise with various combinations of arm and leg work. J Appl Physiol 41:191–196

Bicudo JE, Longworth KE, Jones JH, Taylor CR, Hoppeler H (1996) Structural determinants of maximal O2 transport in muscles of exercising foxes. Respir Physiol 103:243–251

Blomqvist CG, Saltin B (1983) Cardiovascular adaptations to physical training. Annu Rev Physiol 45:169–189

Bohr C (1909) Über die spezifische Tätigkeit der Lungen bei der respiratorischen Gasaufnahme und ihr Verhalten zu der durch die Alveolarwand stattfindenden Gasdiffusion. Skand Arch Physiol 22:221–280

Bringard A, Pogliaghi S, Adami A, De Roia G, Lador F, Lucini D, Pizzinelli P, Capelli C, Ferretti G (2010) Cardiovascular determinants of maximal oxygen consumption in upright and supine posture at the end of prolonged bed rest in humans. Respir Physiol Neurobiol 172:53–62

Brodal P, Ingjer F, Hermansen L (1977) Capillary supply of skeletal muscle fibers in untrained and endurance-trained men. Am J Phys 232:H705–H712

Broxterman RM, Hoff J, Wagner PD, Richardson RS (2020) Determinants of the diminished exercise capacity in patients with chronic obstructive pulmonary disease: looking beyond the lungs. J Physiol Lond 598:599–610

Buick FJ, Gledhill N, Froese AB, Spriet LL, Meyers EC (1980) Effect of induced erythrocythemia on aerobic work capacity. J Appl Physiol 48:636–642

Capelli C, Antonutto G, Azabji-Kenfack M, Cautero M, Lador F, Moia C, Tam E, Ferretti G (2006) Factors determining the time course of $\dot{V}O_2^{max}$ decay during bedrest: implications for $\dot{V}O_2^{max}$ limitation. Eur J Appl Physiol 98:152–160

Cardus J, Marrades RM, Roca J, Barberà JA, Diaz O, Masclans JR, Rodriguez-Roisin R, Wagner PD (1998) Effect of FIO_2 on leg $\dot{V}O_2^{max}$ during cycle ergometry in sedentary subjects. Med Sci Sports Exerc 30:697–703

Celsing F, Svedenhag J, Pihlstedt P, Ekblom B (1987) Effects of anaemia and stepwise-induced polycythaemia on maximal aerobic power in individuals with high and low hemoglobin concentrations. Acta Physiol Scand 129:47–54

Cerretelli P (1976) Limiting factors to oxygen transport on Mount Everest. J Appl Physiol 40:658–667

Cerretelli P (1980) Gas exchange at high altitude. In: West JB (ed) Pulmonary gas exchange, vol II. Academic Press, New York, pp 97–147

Cerretelli P, di Prampero PE (1987) Gas exchange in exercise. In: Farhi LE, Tenney SM (eds) Handbook of physiology. The respiratory system III. Vol 4: Gas Exchange. The American Physiological Society, Bethesda MD, pp 297–339

Cerretelli P, Hoppeler H (1996) Morphologic and metabolic response to chronic hypoxia. In: Fregly MJ, Blatteis CM (eds) Handbook of physiology. Environmental physiology, vol II. Oxford University Press, New York, sect. 4, pp 1155–1181

Cerretelli P, Margaria R (1961) Maximum oxygen consumption at altitude. Int Z Angew Physiol 18:460–464

Chilibeck PD, Paterson DH, Cunningham DA, Taylor AW, Noble EG (1997) Muscle capillarization O_2 diffusion distance, and $V'O_2$ kinetics in old and young individuals. J Appl Physiol 82:63–69

Clausen JP (1977) Effect of physical training on cardiovascular adjustments to exercise in man. Physiol Rev 57:779–815

Convertino VA, Hung J, Goldwater DJ, Debusk RF (1982) Cardiovascular responses to exercise in middle age men after 10 days of bed-rest. Circulation 65:134–140

Costill DL, Daniels J, Evans W, Fink W, Krahenbuhl G, Saltin B (1976) Skeletal muscle enzymes and fiber composition in male and female track athletes. J Appl Physiol 40:149–154

Davies CTM, Sargeant AJ (1974) Effects of training on the physiological responses of one- and two-leg work. J Appl Physiol 38:377–381

Dempsey JA, Hanson PG, Henderson KS (1984) Exercise-induced arterial hypoxaemia in healthy human subjects at sea level. J Physiol 355:161–175

Dempsey JA, Wagner PD (1999) Exercise–induced arterial hypoxemia. J Appl Physiol 87:1997–2006

di Prampero PE (1985) Metabolic and circulatory limitations to $\dot{V}O_2^{max}$ at the whole animal level. J Exp Biol 115:319–331

di Prampero PE (2003) Factors limiting maximal performance in humans. Eur J Appl Physiol 90: 420–429

di Prampero PE, Ferretti G (1990) Factors limiting maximal oxygen consumption in humans. Respir Physiol 80:113–128

Downey AE, Chenoweth LM, Townsend DK, Ranum JD, Ferguson CS, Harms CA (2007) Effects of inspiratory muscle training on exercise responses in normoxia and hypoxia. Respir Physiol Neurobiol 156:137–146

Ekblom B (1969) The effect of physical training on oxygen transport system in man. Acta Physiol Scand Suppl 328:1–45

Ekblom B (1986) Factors determining maximal aerobic power. Acta Physiol Scand Suppl 556:15–19

Ekblom B, Åstrand PO, Saltin B, Stenberg J, Wallström B (1968) Effect of training on circulatory response to exercise. J Appl Physiol 24:518–528

Ekblom B, Huot R (1972) Response to submaximal and maximal exercise at different levels of carboxyhemoglobin. Acta Physiol Scand 86:474–482

Ekblom B, Huot R, Stein EM, Thorstensson AT (1975) Effect of changes in arterial oxygen content on circulation and physical performance. J Appl Physiol 39:71–75

Ekblom B, Wilson G, Åstrand PO (1976) Central circulation during exercise after venesection and reinfusion of red blood cells. J Appl Physiol 40:379–383

Esposito F, Ferretti G (1997) The effects of breathing he–O_2 mixtures on maximal oxygen consumption in normoxic and hypoxic men. J Physiol 503:215–221

Esposito F, Limonta E, Alberti G, Veicsteinas A, Ferretti G (2010a) Effect of respiratory muscle training on maximum aerobic power in normoxia and hypoxia. Respir Physiol Neurobiol 170: 268–272

Esposito F, Mathieu-Costello O, Shabetai R, Wagner PD, Richardson RS (2010b) Limited maximal exercise capacity in patients with chronic heart failure: partitioning the contributors. J Am Coll Cardiol 55:1945–1954

Fagraeus L, Karlsson J, Linnarsson D, Saltin B (1973) Oxygen uptake during maximal work at lowered and raised ambient air pressure. Acta Physiol Scand 87:411–421

Ferretti G (1990) On maximal oxygen consumption in hypoxic humans. Experientia 46:1188–1194

Ferretti G (2003) Limiting factors to oxygen transport on Mount Everest 30 years after: a critique of Paolo Cerretelli's contribution to the study of altitude physiology. Eur J Appl Physiol 90:344–350

Ferretti G (2014) Maximal oxygen consumption in healthy humans: theories and facts. Eur J Appl Physiol 114:2007–2036

Ferretti G (2015) Energetics of muscular exercise. Springer, Heidelberg

Ferretti G, Antonutto G, Denis C, Hoppeler H, Minetti AE, Narici MV, Desplanches D (1997a) The interplay of central and peripheral factors in limiting maximal O_2 consumption in man after prolonged bed rest. J Physiol Lond 501:677–686

Ferretti G, Capelli C (2009) Maximal O_2 consumption: effects of gravity withdrawal and resumption. Respir Physiol Neurobiol 169:S50–S54

Ferretti G, di Prampero PE (1995) Factors limiting maximal O_2 consumption: effects of acute changes in ventilation. Respir Physiol 99:259–271

Ferretti G, Moia C, Thomet J, Kayser B (1997b) The decrease of maximal oxygen consumption during hypoxia in man: a mirror image of the oxygen equilibrium curve. J Physiol 498:231–237

Gavin TP, Derchak PA, Stager JM (1998) Ventilation's role in the decline in $\dot{V}O_2^{max}$ and SaO_2 in acute hypoxic exercise. Med Sci Sports Exerc 30:195–199

Gayeski TE, Honig CR (1986) O_2 gradients from sarcolemma to cell interior in red muscle at maximal $\dot{V}O_2$. Am J Phys 251:H789–H799

Gea J, Agustí A, Roca J (2013) Pathophysiology of muscle dysfunction in COPD. J Appl Physiol 114:1222–1234

Giesbrecht GG, Puddy A, Ahmed M, Younes M, Anthonisen NR (1991) Exercise endurance and arterial desaturation in normobaric hypoxia with increased chemosensitivity. J Appl Physiol 70: 1770–1774

Gollnick PD, Armstrong RB, Saubert CW IV, Piehl K, Saltin B (1972) Enzyme activity and fiber composition in skeletal muscle of trained and untrained men. J Appl Physiol 33:312–319

Gosker HR, Hesselink MK, Duimel H, Ward KA, Schols AM (2007) Reduced mitochondrial density in the vastus lateralis muscle of patients with COPD. Eur Respir J 30:73–79

Grataloup O, Prieur F, Busso T, Castells J, Favier FB, Denis C, Benoit H (2005) Effect of hyperoxia on maximal O_2 uptake in exercise-induced arterial hypoxaemic subjects. Eur J Appl Physiol 94: 641–645

Greenleaf JE, Bernauer EM, Ertl AC, Trowbridge TS, Wade CE (1989) Work capacity during 30 days of bed rest with isometric and isotonic exercise. J Appl Physiol 67:1820–1826

Haldane JS, Priestley JG (1935) Respiration. Oxford University Press, Oxford UK

Hanson P (1994) Exercise testing and training in patients with chronic heart failure. Med Sci Sports Exerc 26:527–537

Harms CA, McClaran SR, Nickele GA, Pegelow DF, Nelson WB, Dempsey JA (2000) Effect of exercise-induced arterial O_2 desaturation on $\dot{V}O_2^{max}$ in women. Med Sci Sports Exerc 32:1101–1108

Henriksson J (1977) Training induced adaptation of skeletal muscle and metabolism during submaximal exercise. J Physiol 270:661–675

Henriquez A, Schrijen F, Poincelot F, Gimenez M, Polu JM, Sadoul P (1986) Maximal oxygen consumption and pulmonary circulation in patients with chronic bronchitis. Eur J Clin Investig 16:526–530

Herbst R (1928) Der Gasstoffwechsel als Mass der körperlichen Leistungsfähigkeit. I. Mitteilung: die Bestimmung des Sauerstoffaufnahmevermögens bein Gesunden. Deut Arch Klin Med 162: 33–50

Hill AV, Lupton H (1923) Muscular exercise, lactic acid, and the supply and utilization of oxygen. Q J Med 16:135–171

Holloszy JO, Coyle EF (1984) Adaptations of skeletal muscle to endurance exercise and their metabolic consequences. J Appl Physiol 56:831–838

Hoppeler H, Howald H, Conley K, Lindstedt SL, Claassen H, Vock P, Weibel ER (1985) Endurance training in humans: aerobic capacity and structure of skeletal muscle. J Appl Physiol 59:320–327

Hoppeler H, Kleinert E, Schlegel C, Claassen H, Howald H, Kayar SR, Cerretelli P (1990) Morphological adaptations of human skeletal muscle to chronic hypoxia. Int J Sports Med 11 (suppl 1):S3–S9

Howald H (1982) Training-induced morphological and functional changes in skeletal muscle. Int J Sports Med 3:1–12

Howald H, Hoppeler H, Claassen H, Mathieu O, Straub R (1985) Influences of endurance training on the ultrastructural composition of the different muscle fibre types in humans. Pflügers Arch 403:369–376

Ingjer F (1979) Effects of endurance training on muscle fiber ATP-ase activity, capillary supply and mitochondrial content in man. J Physiol Lond 294:419–432

Kaijser L (1970) Limiting factors for aerobic muscle performance. Acta Physiol Scand Suppl 346: 1–96

Kashihara H, Haruna Y, Suzuki Y, Kawakubo K, Takenaka K, Bonde-Petersen F, Gunji A (1994) Effects of mild supine exercise during 20 days bed rest on maximal oxygen uptake rate in young humans. Acta Physiol Scand Suppl 616:19–26

Kayser B, Hoppeler H, Claassen H, Cerretelli P (1991) Muscle structure and performance capacity of Himalayan Sherpas. J Appl Physiol 70:1938–1942

Koistinen P, Takala T, Martikkala V, Leppaluoto J (1995) Aerobic fitness influences the response of maximal oxygen uptake and lactate threshold in acute hypobaric hypoxia. Int J Sports Med 16: 78–81

Lacour JR, Flandrois R (1977) Le rôle du métabolisme aérobie dans l'exercice intense de longue durée. J Physiol Paris 73:89–130

Levine BD (2008) $\dot{V}O_2^{max}$: what do we know, and what do we still need to know? J Physiol 586: 25–34

Levine BD, Lane LD, Watenpaugh DE, Gaffney FA, Buckey JC, Blomqvist CG (1996) Maximal exercise performance after adaptation to microgravity. J Appl Physiol 81:686–694

Lindstedt SL, Wells DJ, Jones JR, Hoppeler H, Thronson HA (1988) Limitations to aerobic performance in mammals: interaction of structure and demand. Int J Sports Med 9:210–217

Marconi C, Marzorati M, Grassi B, Basnyat B, Colombini A, Kayser B, Cerretelli P (2004) Second generation Tibetan lowlanders acclimatize to high altitude more quickly than Caucasians. J Physiol 556:661–671

Margaria R, Aghemo P, Rovelli E (1965a) Indirect determination of maximal O_2 consumption in man. J Appl Physiol 20:1070–1073

Margaria R, Camporesi E, Aghemo P, Sassi G (1972) The effect of O_2 breathing on maximal aerobic power. Pflügers Arch 336:225–235

Margaria R, Cerretelli P, Marchi S, Rossi L (1961) Maximum exercise in oxygen. Int Z Angew Physiol 18:465–467

Margaria R, Mangili F, Cuttica F, Cerretelli P (1965b) The kinetics of oxygen consumption at the onset of muscular exercise in man. Ergonomics 8:49–54

Mathur RS, Revill SM, Vara DD, Walton R, Morgan MD (1995) Comparison of peak oxygen consumption during cycle and treadmill exercise in severe chronic obstructive pulmonary disease. Thorax 50:829–833

Mitchell JH, Blomqvist CG (1971) Maximal oxygen uptake. New Engl J Med 284:1018–1022

Oelz O, Howald H, Di Prampero PE, Hoppeler H, Claassen H, Jenni R, Bühlmann A, Ferretti G, Brückner JC, Veicsteinas A, Gussoni M, Cerretelli P (1986) Physiological profile of world class high altitude climbers. J Appl Physiol 60:1734–1742

Ogawa T, Calbet JAL, Honda Y, Fuji N, Nishiyasu T (2010) The effects of breathing a helium–oxygen gas mixture on maximal pulmonary ventilation and maximal oxygen consumption during exercise in acute moderate hypobaric hypoxia. Eur J Appl Physiol 110:853–861

Ogawa T, Hayashi K, Ichinose M, Nishiyasu T (2007) Relationship between rest ventilatory chemosensitivity and maximal oxygen uptake in moderate hypobaric hypoxia. J Appl Physiol 103:1221–1226

Piiper J, Meyer M, Scheid P (1984) Dual role of diffusion in tissue gas exchange: blood-tissue equilibration and diffusion shunt. Respir Physiol 56:131–144

Piiper J, Scheid P (1981) Model for capillary-alveolar equilibration with special reference to O_2 uptake in hypoxia. Respir Physiol 46:193–208

Pirnay F, Dujardin J, Deroanne R, Petit JM (1971) Muscular exercise during intoxication by carbon monoxide. J Appl Physiol 31:573–575

Powers SK, Lawler J, Dempsey JA, Dodd S, Landry G (1989) Effects of incomplete pulmonary gas exchange on $\dot{V}O_2^{max}$. J Appl Physiol 66:2491–2495

Punzal PA, Ries AL, Kaplan RM, Prewitt LM (1991) Maximum intensity exercise training in patients with chronic obstructive pulmonary disease. Chest 100:618–623

Rahn H, Fenn WO (1955) A graphical analysis of the respiratory gas exchange: the O_2–CO_2 diagram. American Physiological Society, Washington DC

Rasmussen J, Hanel B, Diamant B, Secher NH (1991) Muscle mass effect on arterial desaturation after maximal exercise. Med Sci Sports Exerc 23:1349–1352

Reynafarjee B (1962) Myoglobin content and enzymatic activity of muscle and altitude adaptation. J Appl Physiol 17:301–305

Ried-Larsen M, Aarts HM, Joyner JM (2017) Effects of strict prolonged bed rest on cardiorespiratory fitness: systematic review and meta-analysis. J Appl Physiol 123:790–799

Roca J, Hogan MC, Story D, Bebout DE, Haab P, Gonzalez R, Ueno O, Wagner PD (1989) Evidence for tissue diffusion limitation of $\dot{V}O_2^{max}$ in normal humans. J Appl Physiol 67:291–299

Rowell LB (1974) Human cardiovascular adjustments to exercise and thermal stress. Physiol Rev 54:75–159

Rowell LB, Saltin B, Kiens B, Christensen NJ (1986) Is peak quadriceps blood flow in humans even higher during exercise with hypoxemia? Am J Phys 251:H1038–H1044

Saltin B (1977) The interplay between peripheral and central factors in the adaptive response to exercise and training. Ann N Y Acad Sci 301:224–231

Saltin B, Blomqvist CG, Mitchell RC, Johnson RL, Wildenthal K, Chapman CB (1968) Response to exercise after bed rest and after training. Circulation 38(suppl. 7):1–78

Saltin B, Nazar K, Costill DL, Stein E, Jansson E, Essén B, Gollnick PD (1976) The nature of the training response: peripheral and central adaptations to one-legged exercise. Acta Physiol Scand 96:289–305

Saltin B, Rowell LB (1980) Functional adaptations to physical activity and inactivity. Fed Proc 39: 1506–1513

Saltin B, Strange S (1992) Maximal oxygen uptake: "old" and "new" arguments for a cardiovascular limitation. Med Sci Sports Exerc 24:30–37

Scheuer J, Tipton CM (1977) Cardiovascular adaptations to physical training. Annu Rev Physiol 39:221–251

Seay D, Michaud A, Couillard A, Côté CH, Mador MJ, LeBlanc P, Jobin J, Maltais F (2005) Contractile fatigue, muscle morphometry, and blood lactate in chronic obstructive pulmonary disease. Am J Respir Crit Care Med 171:1109–1115

Secher N, Ruberg-Larsen H, Binkhorst RA, Bonde-Petersen F (1974) Maximal oxygen uptake during arm cranking and combined arm plus leg exercise. J Appl Physiol 36:515–518

Servera E, Gimenez M (1984) $\dot{V}O_2^{max}$ during progressive and constant bicycle exercise in patients with chronic obstructive lung disease Respiration 45: 197–206

Sheel AW, Koehle MS, Guenette JA, Foster GE, Sporer BC, Diep TT, McKenzie DC (2006) Human ventilatory responsiveness to hypoxia is unrelated to maximal aerobic capacity. J Appl Physiol 100:1204–1209

Shephard RJ (1969) A non-linear solution of the oxygen conductance equation: applications to performance at sea level and at an altitude of 7,350 ft. Int Z Angew Physiol 27:212–225

Shibata S, Perhonen M, Levine BD (2010) Supine cycling plus volume loading prevent cardiovascular deconditioning during bed rest. J Appl Physiol 108:1177–1186

Skattebo Ø, Calbet JAL, Rud B, Capelli C, Hallén J (2020) Contribution of oxygen extraction fraction to maximal oxygen uptake in healthy young men. Acta Physiol 230:e13486

Smorawinski J, Nazar K, Kaciuba-Uscilko H, Kaminska E, Cybulski G et al (2001) Effects of 3-day bed rest on physiological responses to graded exercise in athletes and sedentary men. J Appl Physiol 91:249–257

Spriet LL, Gledhill N, Froese AB, Wilkes DL (1986) Effect of graded erythrocythemia on cardiovascular and metabolic responses to exercise. J Appl Physiol 61:1942–1948

Stremel RW, Convertino VA, Bernauer EM, Greenleaf JE (1976) Cardiorespiratory deconditioning with static and dynamic leg exercise during bed rest. J Appl Physiol 41:905–909

Sutton JR (1992) Limitations to maximal oxygen uptake. Sports Med 13:127–133

Taylor CR (1987) Structural and functional limits to oxidative metabolism: insights from scaling. Annu Rev Physiol 49:135–146

Taylor CR, Weibel ER (1981) Design of the mammalian respiratory system. I Problem and strategy. Respir Physiol 44:1–10

Tesch PA, Karlsson J (1985) Muscle fiber types and size in trained and untrained muscles of elite athletes. J Appl Physiol 59:1716–1720

Trappe T, Trappe S, Lee G, Widrick J, Fitts R, Costill D (2006) Cardiorespiratory responses to physical work during and following 17 days of bed rest and spaceflight. J Appl Physiol 100: 951–957

Turner DL, Hoppeler H, Noti C, Gurtner HP, Gerber H, Schena F, Kayser B, Ferretti G (1993) Limitations to VO_2max in humans after blood retransfusion. Respir Physiol 92:329–341

Verzar F (1912) The gaseous metabolism of striated muscle in warm-blooded animals. J Physiol Lond 44:243–258

Vogel JA, Gleser MA (1972) Effect of carbon monoxide on oxygen transport during exercise. J Appl Physiol 32(234–239):287

Wagner PD (1992) Gas exchange and peripheral diffusion limitation. Med Sci Sports Exerc 24:54–58

Wagner PD (1993) Algebraic analysis of the determinants of $\dot{V}O_2^{max}$. Respir Physiol 93:221–237

Wagner PD (1996a) Determinants of maximal oxygen transport and utilization. Annu Rev Physiol 58:21–50

Wagner PD (1996b) A theoretical analysis of factors determining $\dot{V}O_2^{max}$ at sea level and altitude. Respir Physiol 106:329–343

Wagner PD (2012) Muscle intracellular oxygenation during exercise: optimization for oxygen transport, metabolism, and adaptive change. Eur J Appl Physiol 112:1–8

Wagner PD (2015) A re-analysis of the 1968 Saltin et al. "Bedrest" paper. Scand J Med Sci Sports 22(suppl 4):83–87

Wagner PD (2017) Operation Everest II and the 1978 Habeler/Messner ascent of Everest without bottled O_2: what might they have in common? J Appl Physiol 123:1682–1688

Wagner PD, Erickson BK, Seaman J, Kubo K, Hiraga A, Kai M, Yamaya Y (1996) Effects of altered FIO_2 on maximum $\dot{V}O_2^{max}$ in the horse. Respir Physiol 105:123–134

Wagner PD, Gillespie JR, Landgren GL, Fedde MR, Jones BW, DeBowes RM, Pieschl RL, Erickson HH (1989) Mechanism of exercise-induced hypoxemia in horses. J Appl Physiol 66: 1227–1233

Wehrlin JP, Hallén J (2006) Linear decrease in $\dot{V}O_2^{max}$ and performance with increasing altitude in endurance athletes. Eur J Appl Physiol 96:404–412

Weibel ER (1987) Scaling of structural and functional variables in the respiratory system. Annu Rev Physiol 49:147–159

Welch HG, Pedersen PK (1981) Measurement of metabolic rate in hyperoxia. J Appl Physiol 51: 725–731

West JB (1983) Climbing Mt. Everest without oxygen: an analysis of maximal exercise during extreme hypoxia. Respir Physiol 52:265–279

Williams JH, Powers SK, Stuart MK (1986) Hemoglobin desaturation in highly trained athletes during heavy exercise. Med Sci Sports Exerc 18:168–173

Woodson RD, Wills RE, Lenfant C (1978) Effect of acute and established anemia on O_2 transport at rest, submaximal and maximal work. J Appl Physiol 44:36–43

Zumstein A, Mathieu O, Howald H, Hoppeler H (1983) Morphometric analysis of the capillary supply in skeletal muscles of trained and untrained subjects. Its limitations in muscle biopsies. Pflugers Arch 397:277–283

Chapter 8
Respiratory Mechanics

Giuseppe Miserocchi

Abstract A rising interest focusing on respiratory mechanics developed around 1950, in particular the link with muscular energetics was quite attractive; this led to the estimate of the mechanical work of breathing and the corresponding energy cost in order to match the metabolic requirement. A full analysis of the statics of the respiratory system was performed thanks to the refined technique of esophageal recording. The concept of mechanical stability and interdependence was developed considering the differences in alveolar size. Moreover, the relationship between pattern of breathing and ventilation was analyzed. Further, pleural space mechanics, lubrication, and fluid dynamics were described. A model was developed concerning the role of pleural lymphatics to maintain a sub-atmospheric pressure in the pleural cavity. The effect of changing shape of the chest by changing the effect of gravity or increasing the pressure during water immersion was another theme of research. Other topics in applied respiratory physiology included the development of a method to estimate the mechanical properties of the lung during mechanical ventilation, the development of tests of pulmonary function to estimate the efficiency of the respiratory muscles and air flow limitation in the small airways leading to the concept of "closing volume."

G. Miserocchi (✉)
University of Milano Bicocca, Milan, Italy
e-mail: Giuseppe.Miserocchi@unimib.it

G. Ferretti (ed.), *Exercise, Respiratory and Environmental Physiology*, Perspectives in Physiology, https://doi.org/10.1007/978-3-031-19197-8_8

The human thorax by Michelangelo Buonarroti. Airways branching has been added to highlight chest wall-mechanical interaction. Interpretation by Giuseppe Miserocchi

8.1 Introduction

Chapters 8 and 9 present the work devoted to various aspects of the respiratory function performed by three generations of researchers who, in their early years, attended the Istituto di Fisiologia Umana of the University of Milano. Below the name of Rodolfo Margaria, we report here, ordered by birth date, the names of the researchers, who brought their contribution, as well as that of their own research groups, to developing knowledge in respiratory physiology and pathophysiology in Italy and abroad. Results are presented under several headings to include the wide span of interconnected research interests concerning the respiratory function.

Rodolfo Margaria, Head of the *Istituto di Fisiologia Umana* from 1938 up to 1977
Eloisa Milla, University of Milano, Italy
Giacomo Meschia, University of Colorado, Denver, USA
Ivo Setnikar, University of Milano, Italy
Emilio Agostoni, University of Milano, Italy
Joseph Milic-Emili, McGill University, Montreal, Canada
Giovanni A. Cavagna, University of Milano, Italy
Giuseppe Sant'Ambrogio, University of Texas, Medical Branch, Galveston, USA
Edgardo D'Angelo, University of Milano, Italy
Giuseppe Miserocchi, McGill University, University of Milano Bicocca
Enrico Mario Camporesi, University of South Florida, Tampa, USA
Jacopo Mortola, McGill University, Montreal, Canada
Luciano Zocchi, University of Milano, Italy
Daniela Negrini, University of Insubria, Varese, Italy.

8.2 The Link of Respiratory Physiology to Muscle Energetics

The interest to respiratory physiology rose high in the USA since 1940 and a considerable amount of staminal papers were published on this topic (Rahn et al. 1946; Otis and Proctor 1948; Otis et al. 1947, 1950; Farhi et al. 1957).

This interest crossed the ocean and was caught by Rodolfo Margaria and his pupils. Eloisa Milla (Fig. 8.1), the only lady of the group, who was in her forties at the time, drew the interest of young researchers to developing knowledge in respiratory physiology, in particular mechanics (Milla et al. 1952). Results from this preliminary study led to interpret the shape of the volume–pressure curve of the lung as being due to an elastic structure contained in a semi-rigid net (Setnikar and Meschia 1954; Setnikar 1955).

The link of respiratory physiology to muscular energetics was obviously quite attractive for people "breathing" at the Istituto di Fisiologia Umana of University of Milano: in fact, a key issue was the estimate of the work of breathing required to match the metabolic requirement, in particular when the oxygen demand was increased. Margaria went to estimate the mechanical work of breathing (Margaria et al. 1960). The brilliant conclusion of the study was that the maximum pulmonary ventilation useful to match the external work is attained when the energy cost of breathing, due to any additional unit of air ventilated (dW_{re}/dV), equals the additional energy provided by the same change in ventilation (dW_{tot}/dV), i.e. when $dW_{re}/dV = dW_{tot}/dV$. At the maximal values of ventilation attained during exercise, the mechanical work of breathing amounts to about 100–120 cal min^{-1} (7.0–8.4 W).

Emilio Agostoni (1929–2021) visited Rochester-Buffalo in the late 1950s and early 1960s, and went in contact with Wallace Fenn (1893–1971), who had strong relations with Margaria, and Hermann Rahn (1912–1990), with whom he published

Fig. 8.1 Rodolfo Margaria and Eloisa Milla around 1954. With the kind permission of Giuseppe Miserocchi

papers on respiratory mechanics and the role of respiratory muscles. Visiting Buffalo continued over the years to include Cerretelli, di Prampero, Veicsteinas, Ferretti, Capelli, Zamparo, amongst others. Joseph Milic-Emili (1931–2022), a brilliant student who graduated in Milano, following the suggestion of Jere Mead (1920–2009) from Harvard, Boston, based on previous work (Mead and Whittenberger 1953; Mead et al. 1955), brought a mighty contribution to respiratory mechanics by developing the technique allowing to derive pleural pressure from esophageal recording (Petit and Milic-Emili 1958; Milic-Emili and Petit 1959; Milic-Emili et al. 1964a), an idea actually proposed 70 years before by Luciani (1879). This tool gave rise to a myriad of papers dealing with lung functional evaluation in health and disease. Milic-Emili, together with Peter Macklem (1931–2011), also pupil of Jere Mead, brought their knowledge to McGill University (Montreal) and created an internationally renowned center for respiratory mechanics. Giuseppe Miserocchi joined the group at McGill in the 1970s for about 4 years to develop a new field of research, namely the functional link between lung mechanics and fluid balance. Miserocchi was appointed as an Associate Professor at McGill University, a position, which he left on returning to University Milano Bicocca, and was subsequently taken by Jacopo Mortola, another Milano's pupil.

8.3 Pressure, Volumes, and Mechanical Stability of the Alveoli

The Swiss physiologist Fritz Rohrer (1888–1926) developed a thorough and quantitative analysis of respiratory mechanics (Rohrer 1916). He considered the elastic forces required to move the respiratory system and those resulting from gas flow resistance and frictional resistance within the lung parenchyma and the surrounding tissues. He also considered of negligible magnitude, in physiological conditions, the forces relating to acceleration of mass. Under conditions of steady volume, only elastic forces would be present; he could then partition the elastic properties of lungs and chest wall and the resulting elastic properties of the whole respiratory system.

About 40 years later, a full analysis of the statics of respiratory system was performed by Agostoni and Mead (1964). Figure 8.2 presents the volume–pressure curves for the lung, the chest wall and for the overall respiratory system referring to upright and supine posture. All pressures are reported as difference relative to atmospheric.

The mechanical coupling between the chest wall and the lung was defined considering that the overall elastic recoil pressure of the respiratory system (P_{rs}) is given by the sum of the elastic recoil of the lung (P_L) plus that of the chest wall (P_{cw}), thus:

$$P_L + P_{cw} = P_{rs} \tag{8.1}$$

The graphs allow estimating which alveolar pressures (P_{alv}) are required to move the respiratory system based on the relationship:

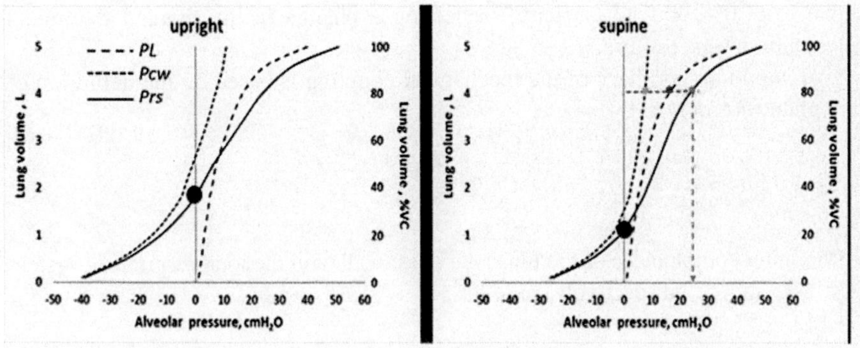

Fig. 8.2 Pressure–volume curves of the lung (P_L), of the chest wall (P_{cw}), and of the respiratory system (P_{rs}) in upright (left) and supine (right) posture. The resting point of the respiratory system is indicated by a black dot in both graphs. On the right, the partitioning between P_L and P_{cw} is shown at a lung volume of 4 L for an alveolar pressure, P_{alv}, of ~25 cmH_2O. From Agostoni and Mead (1964)

$$P_{alv} = P_{rs} = P_L + P_{cw} \tag{8.2}$$

Partitioning between P_L and P_{cw} was made possible by inserting a further variable, namely intrapleural pressure (commonly referred to as P_{pl}), reflecting the reciprocal elastic interaction between the lung and the chest wall. P_{pl} was derived from the esophageal pressure (Milic-Emili et al. 1964a, b). Knowing P_{pl}, we can define P_L as:

$$P_L = P_{alv} - P_{pl} \tag{8.3}$$

and substituting in Eq. (8.2), one can derive P_{cw} as:

$$P_{cw} = P_{pl} \tag{8.4}$$

As an example, wishing to inflate the lungs of an anesthetized paralyzed subject up to 4 L (consider the graph on the right), the required alveolar pressure is ~25 cmH_2O (blue dashed line and dot). At this value of P_{alv}, the relative values of P_L and P_{cw} are indicated by the red and orange dots, respectively.

An important point in both graphs is the one indicated by a black dot, since, at this volume, one has:

$$P_{alv} = P_L + P_{cw} = P_{rs} = 0 \tag{8.5}$$

This point corresponds, on mechanical basis, to the resting point of the respiratory system (also named as functional respiratory capacity, FRC) as the inward pull of the lung is exactly balanced by the outward pull of the chest wall. The concept that at FRC the lung and the chest wall pull in opposite directions had already been made clear by Franciscus Donders (1818–1889) (Donders 1853). Note that the resting point in supine posture is lowered remarkably relative to upright posture (from ~35 down to ~20% of vital capacity), reflecting a change in the thoraco-abdominal interaction in this posture.

An important corollary of the mechanical coupling is based on the definition of compliance (*C*) given by:

$$C = \frac{\Delta V}{\Delta P} \tag{8.6}$$

Defining compliances of the lung, the chest wall, and the total respiratory system (C_L, C_{cw}, and C_{rs}, respectively), one has:

$$\frac{1}{C_{rs}} = \frac{1}{C_L} + \frac{1}{C_{cw}} \tag{8.7}$$

Hundreds of papers have been published since the fundamental article by Agostoni and Mead (1964) relying on the measurement of esophageal pressure.

The technique of esophageal recording has been widely used and scientific interests continued until recently, pivoting on the debate about advantages and possible drawbacks (Akoumianaki et al. 2014).

A lot of interest focused on the quantification of the so-called heterogeneity index of lung expansion, considered as a helpful index to assess different stages of lung disease. Intense research efforts aimed at developing precise imaging of lung volumes via either invasive or minimally invasive methods in diseased lungs (Coxson et al. 1995; Salito et al. 2011).

The problem of mechanical stability of alveoli was an important issue concerning the efficiency of ventilation. On mechanical basis, the total alveolar distending pressure (P_d) can be defined as

$$P_d = P_{el} + P_\gamma \tag{8.8}$$

where P_{el} is the elastic recoil pressure and P_γ is the pressure reflecting the surface tension. For mechanical stability, P_d should be the same for adjacent alveoli. However, it was difficult to understand how could the alveolar equilibrium being considered stable, given the fact that alveoli in parallel differ in size and both elastic and surface forces change on changing lung volume. The heterogeneity of the mechanical properties of the alveoli still represents a real problem, aiming to frame the difference in elastic and surface forces within a stable model of inter-dependent units (Bates and Suki 2008).

Miserocchi's group, in collaboration with the Department of Biomedical Engineering of the Politecnico di Milano, afforded this problem by developing an experimental approach to estimate the distribution of alveolar mechanical properties based on a morpho-functional analysis of transpleural alveolar imaging on changing lung volume. It was found that absolute alveolar compliance was significantly lower in smaller compared to larger alveoli. However, specific compliance, obtained by normalizing absolute compliance to alveolar volume, varied much less. It was concluded that the relative constancy of specific compliance might contribute to reduce inter-regional differences in parenchymal and surface forces in the lung by contributing to assure a uniform stretching in a model of mechanically inter-dependent alveoli (Salito et al. 2014).

Further, considering three different scale levels of lung imaging, from single alveoli, to small cluster of alveoli (10–20 alveoli) and to larger cluster of alveoli (30–40 alveoli), it was found that the heterogeneity of alveolar size, and of corresponding mechanical behavior, was homogenously distributed on increasing scale level, suggesting a substantially homogenous mechanical behavior by extrapolating to the whole lung (Mazzuca et al. 2014). A further study went to investigate the nature of alveolar mechanical interdependence after having purposefully disturbed the equilibrium condition by administering exogenous surfactant in physiological non-surfactant deprived conditions (Salito et al. 2015). Surfactant instillation increased up to ~50% the surface area of alveoli smaller than 20,000 μm^2, reflecting a lowering of surface tension due to local surfactant enrichment. Conversely, the surface area of adjacent alveoli greater than 20,000 μm^2, decreased by ~5%. Opposite changes in alveolar surface suggest a decrease in the tethering effect exerted by the smaller alveoli on the larger ones. Accordingly, surfactant delivery helps reduce the heterogeneity in alveolar mechanical interdependence. This point has been recently reconsidered, as in severe lung inflammation with surfactant deficiency, microatelectasis results in distortion and overdistension of adjacent alveoli. Exogenous surfactant delivery proved indeed to reduce alveolar stress heterogeneity (Knudsen et al. 2018), thus preventing the spreading of lung lesion on cyclic recruitment/de-recruitment of the alveoli (Bates et al. 2020).

A further interesting point was raised by Cavagna et al. (1967), who showed that alveoli can resist to atelectasis on decreasing lung volume; the finding led to hypothesize the existence of an intrinsic alveolar structural support preventing alveolar collapse.

8.4 The Function of the Respiratory Muscles

Agostoni extensively studied the role of the diaphragm, the main respiratory muscle. The contraction of the diaphragm lowers pleural pressure and thus gives rise to inspiration. Agostoni analyzed the thoraco-abdominal interaction by relating abdominal to thoracic pressures when the diaphragm contracts (Agostoni and Rahn 1960); he further provided a graphical analysis of the mechanical contribution of the abdominal and thoracic components during different kinds of breathing cycles (Agostoni 1961). Further, the relationship of the electrical activity of the diaphragm with the increase in transdiaphragmatic pressure was also described (Agostoni et al. 1960). Based on thoraco-abdominal functional interaction, Agostoni also analyzed the distribution of forces generating a deformation of the rib cage (Agostoni et al. 1966b). Agostoni also went to study the onset of diaphragmatic activity that spontaneously arises during breath holding on approaching the breaking point of the apnea (Agostoni 1963). The interesting hypothesis from this study was that the onset of the diaphragm activity corresponds indeed to the onset of the so-called desire to breath.

A further study aimed at evaluating the role of the shortening of the expiratory muscle as a limiting factor to expiratory flow; the conclusion was that air flow is limited by the rate with which the muscles are able to mobilize chemical potential energy for the performance of work (Agostoni and Fenn 1960). Another interesting study was to compare the mechanical features of the respiratory system among mammals (Agostoni et al. 1959). It was found that the lung-thorax compliance, normalized to unit body weight and to unit of vital capacity, varied among species, yet, the pressure required for a normal inspiration was similar. It was also found that animals having a lower respiratory frequency have a larger FRC. This led to hypothesize that this feature would allow smaller changes in the composition of the alveolar air when the breathing cycle is longer. The interesting point also was

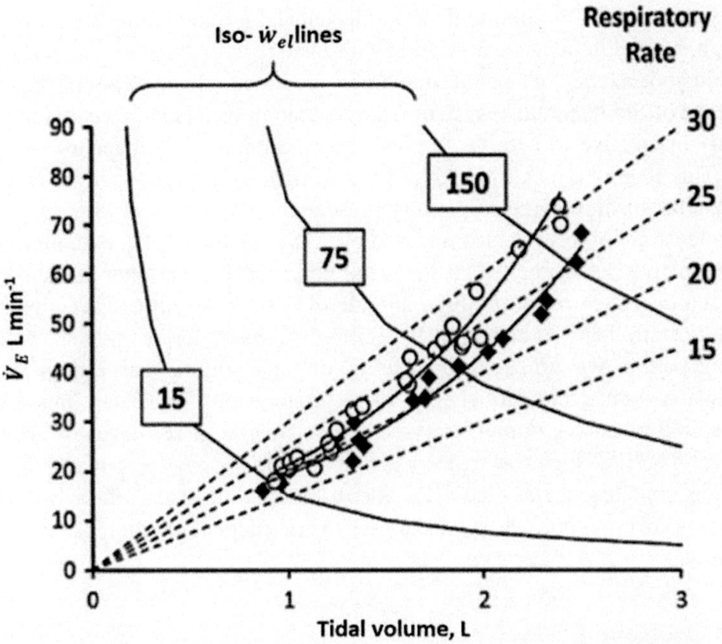

Fig. 8.3 Correlation between ventilation ($\dot{V}_E$) and tidal volume in one patient during incremental test before (closed symbols) and after (open symbols) training program. The figure reports iso-respiratory rate (dashed lines) and iso-inspiratory elastic power (continuous lines, $\dot{w}_{el}$). From Passoni et al. (2015)

that, from the relationship between work of breathing and breathing frequency, the frequency typical of each animal corresponded to the minimum work.

One year later, Mead took this point again and published a paper (Mead 1960) showing that there is indeed a particular frequency corresponding to a least costly work of breathing, stressing the point that inspiratory muscles work increases with tidal volume to the second power. Moreover, he demonstrated that the pattern of breathing is spontaneously chosen at rest and during exercise and hypothesized that the pattern of breathing would be controlled by neuronal afferents from lung receptors (a point dealt with in Sect. 9.3). A further analysis was carried in patients affected by metabolic syndrome (Passoni et al. 2015) undergoing a tailored exercise program.

Figure 8.3 shows that training (open symbols) induced a change in breathing pattern implying a larger contribution of respiratory rate (dashed lines), at the expense of tidal volume during exercise hyperventilation, resulting in a decrease of inspiratory power output (continuous lines) and in a decrease in the perceived fatigue index.

Further, an elegant study was published (Pengelly et al. 1971) on a specific functional feature of the respiratory machine by imposing sudden resistive loads to the system during the inspiratory activity. The mechanical interpretation of the results was based on the notion that, on mechanical ground, a constant flow source with high internal impedance is relatively insensitive to external loading. The results suggested indeed that, when the diaphragm contracts on inspiration, the internal impedance of the respiratory system is large enough to maintain ventilation, being relatively insensitive to external applied resistive loads that hinder the inspiratory action. This feature was attributed, at least in part, to the force–length and force–velocity relationships of the respiratory muscles.

An interesting point was also reported by Aliverti et al. (2011) by comparing the rate of perceived exertion (RPE) for respiratory and leg muscles to their specific power outputs at sea level and high altitude (4559 m). At altitude, ventilation was increased at any leg power output by $\sim$50%, however, for any given ventilation, breathing pattern was unchanged in terms of tidal volume, respiratory rate, and operational volumes and further RPE of respiratory muscles scaled uniquely with total respiratory power output, irrespectively of being at sea level or at altitude. Conversely, RPE of legs was exacerbated at altitude for any leg power output. The interesting conclusion was that RPE for respiratory muscles uniquely relates to respiratory power output, while RPE for legs varies depending on muscle metabolic conditions.

8.5 The Pressure Surrounding the Lung: Pleural Space Mechanics, Fluid Dynamics, and Lubrication

In 1900, Johannes Pieter Van der Brugh (1865–1944) developed a method to measure local intrapleural pressure from a small pneumothorax (Van der Brugh 1900) and in the same year, Aron (1900) measured pleural pressure in his compliant house servant with a liquid filled system: the latter measurements were more sub-atmospheric than the former. Conceptually, the measurement of Van der Brugh reflected the pressure generated by the opposing elastic recoil of the chest and the lung; conversely, the measurement of Aron referred to the pressure of the pleural fluid. Such difference represented the basis of a staminal work aimed at considering the mechanism responsible for keeping the lung expanded in the chest; in fact, Setnikar and colleagues went to explore the origin the sub-atmospheric pleural liquid pressure (Setnikar et al. 1955). Figure 8.4 shows the original graph that gave birth to a brilliant hypothesis (Setnikar et al. 1957): the two recordings from their paper refer to pleural liquid pressure measured at a high (*cannula superiore*—upper cannula) and at a low point (*cannula inferiore*—lower cannula) in the chest: pressure was definitely more negative at the upper cannula. Note that the liquid pressure became largely less sub-atmospheric at both levels following repeated injections of quite small amounts of saline. The conclusions of the study were that (1) a gravity dependent hydraulic gradient is acting on pleural fluid; (2) the pleural liquid pressure is more sub-atmospheric than what could be expected from the lung recoil pressures, as the injected volumes were too small to be explained by a

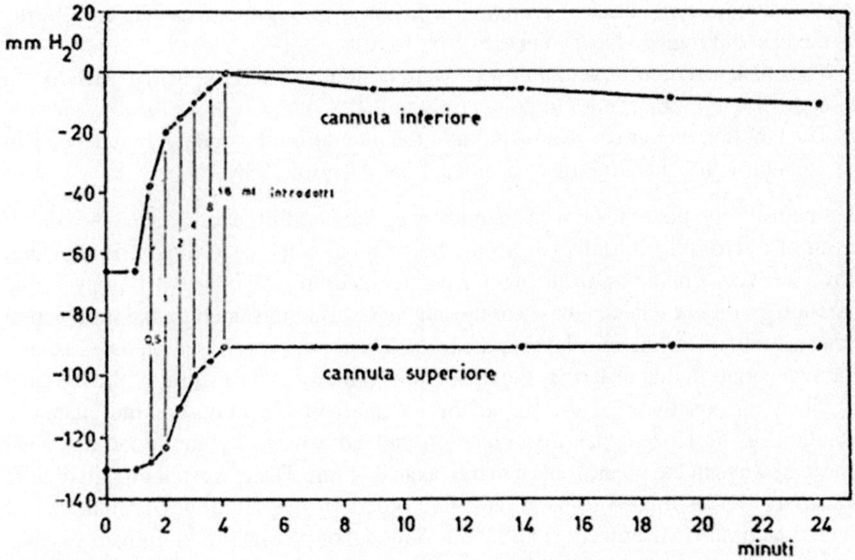

Fig. 8.4 Original recordings of intrapleural pressure. The remarkable change in pressure (becoming less sub-atmospheric) at two heights in the cavity after injection of tiny amounts of saline solution led to hypothesize that pleural liquid pressure cannot be explained by a corresponding change in lung elastic recoil. (*cannula superiore*—upper cannula), (*cannula inferiore*—lower cannula). From Setnikar et al. (1957)

change in the elastic recoil of the lung and the chest wall. The next step was to understand which mechanism could generate a rather sub-atmospheric pleural liquid pressure.

The hypothesis was put forward that the negative pressure would result from fluid absorption through the visceral pleura via a Starling dependent pressure gradient considering that plasma colloido-osmotic pressure exceeds pulmonary capillary pressure. Accordingly, an osmometer was devised to estimate such absorption pressure (Agostoni et al. 1957): the straightforward conclusion was that the negative pressure of the pleural fluid could indeed be attributed to a Starling dependent mechanism. This paper stated the important point that the resultant force of the reabsorption mechanism is higher than the elastic retraction force of the lung, accordingly, pleural liquid pressure is more sub-atmospheric that that generated by the opposite recoil of the lung and the chest wall. Thus, lung expansion in the chest is not the cause of a negative pleural liquid pressure but rather the consequence of the mechanism setting liquid pressure.

Some 40 years later, Miserocchi's group went to explore the role of lymphatics in the pleural fluid turnover that had not been considered at the time. The evidences suggesting this line of research were based on the following data:

- The ability of lymphatics to generate a sub-atmospheric pressure compatible with that found in the pleural space (Miserocchi et al. 1989)

- An incredible richness of lymphatics on the diaphragmatic surface and on the mediastinal region (Negrini et al. 1991, 1992)
- The existence of intrapleural flows toward regions rich in lymphatics (Miserocchi et al. 1984, 1986, 1988, 1992; Negrini et al. 1985)
- The fact that the rate of reabsorption of the pleural fluid is relatively independent of its protein concentration (Miserocchi and Negrini 1986).

Interestingly, pleural surfaces are quite irregular and further are coated by a dense mesh of microvilli protruding from the mesothelial cells; accordingly, intrapleural flows were considered as occurring in a porous medium down the existing hydraulic pressure gradients. The database for the analysis was represented by the intrapleural translocation of technetium-labeled albumin. The model was based on the Kozeny–Carman equation that describes the mean flow velocity as a function of the pressure gradient, the viscosity of fluid, and the mean hydraulic radius of the channels. Solution of the Kozeny–Carman equation yielded a mean hydraulic radius of the flow pathways in the pleural space in the range 2–4 μm. The corresponding hydraulic resistivity was estimated to be at least five orders of magnitude lower than that of interstitial tissue (Miserocchi et al. 1992). A model of pleural fluid turnover was also developed to describe the intrapleural fluid dynamics as detailed in Fig. 8.5. On the left, the polarization of filtration (white arrows), intrapleural flows (gray arrows), and drainage (black arrows) are shown (Miserocchi 2009). On the right, the features of the control of pleural liquid volume operated by lymphatics (Miserocchi et al. 1993).

The differential equations defining flows allowed to estimate that the "set point" of pleural liquid pressure under steady-state condition corresponds to a minimum volume of pleural fluid. Lymphatics act as an efficient negative feedback system to regulate pleural fluid dynamics through two mechanisms: (1) they can set a rather sub-atmospheric pressure thanks to their powerful draining action and (2) they can markedly increase draining flow to offset an increase in pleural liquid filtration thus avoiding an increase in liquid volume. Defining lymph flow as J_l, the lymphatic conductance K_l and pleural liquid pressure P_{liq}, the following equation holds:

$$\Delta J_l = K_l \, \Delta P_{liq} \tag{8.9}$$

where from the "gain" of the lymphatic control system can be defined as:

$$\frac{\Delta J_l}{\Delta P_{liq}} = K_l \tag{8.10}$$

As long as lymphatics can increase draining flow up to the maximum to face an increase in P_{liq}, the ratio $\Delta J_l/\Delta P_{liq}$ remains equal to the lymphatic conductance K_l that is largely exceeding the filtration coefficient of the pleurae. Accordingly, the "gain" of the lymphatic draining system appears quite high for the task to keep pleural liquid volume close to the minimum (Miserocchi et al. 1993; Miserocchi 2009).

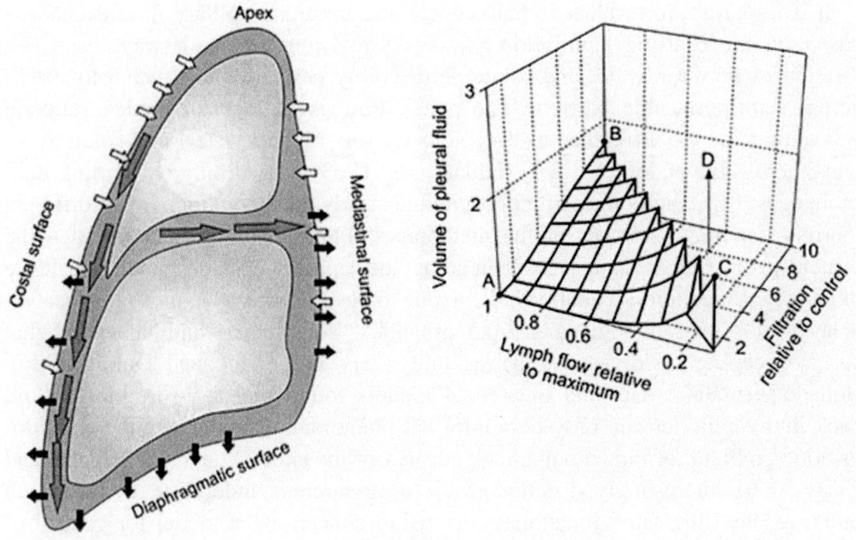

Fig. 8.5 Left. Polarization of pleural fluid filtration (white arrows), lymphatic drainage (black arrows), and intrapleural fluxes (gray arrows). From Miserocchi (2009). Right. Features of the control of pleural fluid volume (on the ordinate) operated by lymphatic drainage on increasing pleural fluid filtration. For a 10-fold increase in filtration rate, the volume of pleural fluid would not even double (point B, relative to point A) thanks to the increase in lymphatic flow. The chequered area delimited by A, B, C defines the extension of the efficiency of lymphatics to control pleural liquid volume. The increase in filtration rate coupled with a restriction of maximum lymphatic flow nullifies the control on pleural liquid volume (point D). From Miserocchi et al. (1993)

Figure 8.5 right shows a sensitivity analysis to describe the features of the lymphatic control. Point A corresponds to the physiological condition. The ordinate shows that pleural liquid volume would not even be doubled for a 10 times increase in filtration (point B). Lymphatic can still provide an efficient drainage even if their maximum draining capacity is limited to 1/10 of normal (point C). Outside the chequered area (point D) no lymphatic control is operating. A limitation in maximum lymph flow coupled with an increase in filtration rate lead to pleural effusion (Miserocchi 1997).

Evidence has been reported that reabsorption of pleural fluid may also occur via sodium–glucose cotransporters of mesothelial cells (Sironi et al. 2007, 2008; Agostoni and Zocchi 2007). Furthermore, some evidence for albumin transcytosis at pleural level has been provided (Bodega et al. 2002). In summary, three mechanisms co-operate to drain pleural fluid: a Starling dependent absorption into the pulmonary capillaries, lymphatic flow and an active absorption via sodium–glucose cotransporters. One shall note that the three outflow pathways are placed in parallel; accordingly, their relative contribution ought to be proportional to the corresponding conductance; in this respect, the conductance of lymphatics is by and large the highest of the three.

It is important to recall here that, concerning the transcapillary fluid exchanges based on the Starling equilibrium, a strong revision was underway, based on differences between water and solute permeability (Kedem and Katchalsky 1958) across semi-permeable barriers. The solute flow is in fact a complex function accounting for the difference in flow velocity for solvent (water) and solute (proteins) across the endothelial wall (Patlak et al. 1963). Based only on Starling fluid exchanges, it became quite difficult to define steady-state conditions with different morpho-functional features of the membranes. Further studies also revealed the difficulty for water exchanges to occur across the capillary wall, despite the existence of a hydraulic-colloido-osmotic gradient, due to the presence of an intact glycocalyx, a layer of glycosaminoglycans, ~0.5 μm thick, coating the luminal endothelial surface keeping microvascular permeability very low (Vink and Duling 1996). Both experimental data and theoretical models found that reabsorption of fluid back into capillaries can only be a transient phenomenon limited by the colloido-osmotic gradient forming along the glycocalyx of the endothelial clefts (Michel and Curry 1999). Interestingly, different groups of researchers, independently from each other, reached the same functional conclusion concerning a model for control of extravascular water, that is: capillary filtration depends on Starling forces, but this mechanism can hardly contribute to fluid reabsorption, the latter mainly occurring via lymphatics (Michel 1980; Levick 1991; Miserocchi et al. 1993).

Based on the fact that the opposing visceral and parietal pleura must touch each other and actually press one against the other, due to the negative pleural liquid pressure, a consideration was given to the friction and lubrication concerning the reciprocal movement of these surfaces. The hypothesis was then put forward that reducing friction in a cavity with negative pressure would be provided if the lubricant were directly adsorbed onto the pleural surfaces. Hills et al. (1982) postulated that such a system would conform to the concept of "boundary" lubrication, which, according to standard lubrication theory, is much more appropriate for slowly moving surfaces than hydrodynamic lubrication which is always associated with surfaces moving at high relative speeds.

Figure 8.6 is the original schema from the paper by Hills et al. (1982) for "boundary" lubrication, showing the adsorption of surfactant molecules to the mesothelium by means of a polar group at one end, leaving the hydrophobic tails at the other end orientated outward and free to engage those on the opposite surface. It is clear that charges of opposing sign are facing each other, thus avoiding real mechanical contact thus favoring sliding. At the time, the lubricating capability of the pleural extracts was tested with a standard industrial method of assessing boundary lubricants and the coefficient of kinetic friction (μ) was not directly measured but derived mathematically and estimated at about 0.35.

In a second paper (Hills and Brian 1992), electron microscopy revealed about 10 layers of surface active phospholipids stratified over the mesothelial membrane, so that the a graphite-like oligo-lamellar lubrication system was proposed representing an excellent anti-wear system.

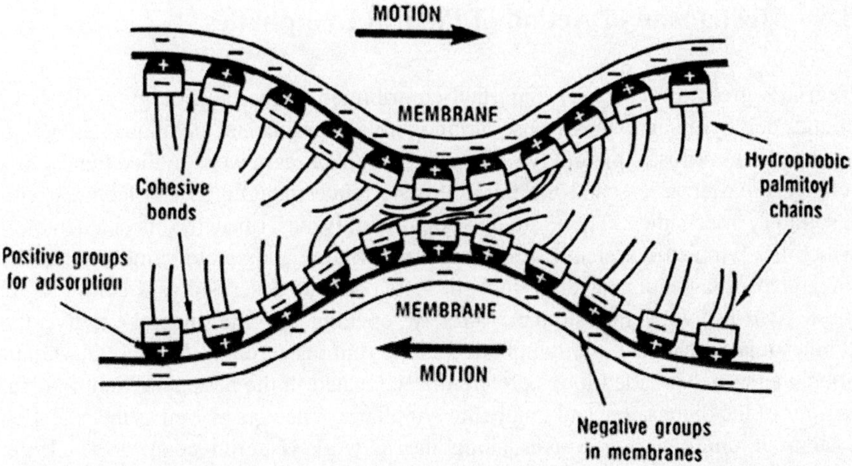

Fig. 8.6 Original model of "boundary" lubrication at pleural level proposed by Hills et al. (1982). Surfactant molecules adsorbed on the visceral and parietal mesothelial surfaces are facing through hydrophobic tails repulsing each other. From Miserocchi (2009)

Values of coefficient of friction μ were directly measured with an experimental set-up and this time ranged 0.05–0.06 fully consistent with boundary lubrication and about one order of magnitude lower than the previously estimated one. The problem of pleural lubrication was reconsidered by D'Angelo et al. (2004), who performed careful measurements of μ in vitro for rabbit's visceral pleura sliding against the parietal pleura. Oscillating at physiological velocities and amplitudes under physiological normal forces, μ averaged 0.02 with pleural liquid as lubricant.

A further interesting finding from this work was that gentle application of filter paper on pleural surface increased the coefficient of friction $\sim$10-fold irreversibly, suggesting alteration of the mesothelia: the simple message being the integrity of the mesothelium is a quite delicate affair! It is worth recalling that mesothelia display a thick layer of microvilli coated by the surfactant active molecules. In a further study, Pecchiari et al. (2016) found no correlation between μ and the features of the microvillar network (density, length, height). The author's interpretation was that the maximal lubricant effect is already reached with the amount of lubricants corresponding to the lowest microvilli density, the excess representing a "functional reserve" in case of increased consumption as, for example, during hyperventilation. In summary, pleural lubricant system appears well designed for its job. In fact, interference with the physiological sliding may represent a causative factor for potency to develop mesothelioma, in particular considering ultrafine amphibole fibers from asbestos (Miserocchi et al. 2008; Wylie et al. 2020).

8.6 Mechanism of Action of Pleural Lymphatics

Negrini's group provided an important contribution to the physiology of the lymphatics deploying an extensive use of innovative experimental techniques, imaging, and modeling. Mostly pleural lymphatics of the diaphragm were studied thanks to a relatively easier access and the possibility to relate their functional features with respiratory mechanics. These studies convincingly described the mechanisms by which the lymphatic system can control fluid volume and solute homeostasis. The hierarchal organization of the lymphatic network was described as a complex 3D mesh starting with the stomata directly opening on the pleural cavity, the submesothelial lacunae, the lymphatic vessels running through the muscular/tendinous fibers, the so-called loops preferentially located in the peripheral ventrolateral regions of the diaphragm, and finally the lymphatic collectors as well as the extended system of unidirectional valves along the network (Negrini et al. 1991, 1992; Grimaldi et al. 2006; Moriondo et al. 2008). Pressure gradients sustaining flows in the initial lymphatics were directly measured with the micropuncture technique and specifically related to the effect of tissue motion and muscle contraction (Negrini et al. 2004; Moriondo et al. 2005). The 3D arrangement of the diaphragmatic lymphatic network was considered as being finalized to efficiently exploit the stresses exerted by the contracting muscles on the surface of the lymphatic channels in order to promote lymph formation and its propulsion toward the deeper vessels (Moriondo et al. 2015). Knowledge of pressure gradients and estimated compliance values allowed full description of the mechanical features of the lymphatic walls (strain, stress, and elastic modulus) in relation of location of the channels in the network. A finite-element analysis (Moriondo et al. 2010) allowed characterization of three types of lymphatic conduits, as shown in Fig. 8.7: superficial channels with high compliance (Fig. 8.7a, left) which serve as distensible reservoirs for drained fluid; intermediate channels (Fig. 8.7b, left) subject to higher stress; deeper, stiffer lymphatics (Fig. 8.7c, left) with lower stress at the wall that guarantee a fast and

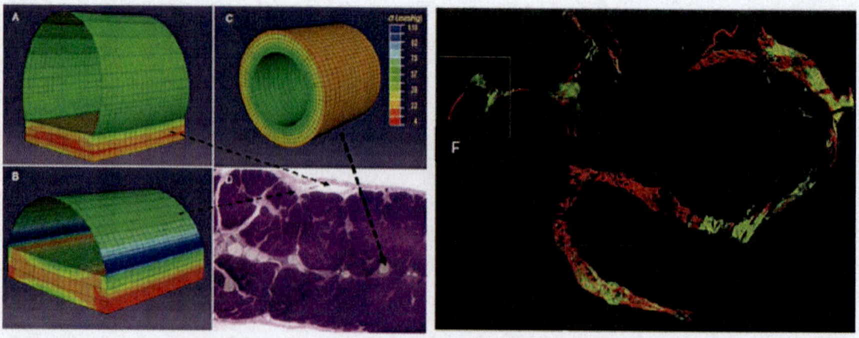

Fig. 8.7 (**a, b, c, d**): finite-element analysis for structural–mechanical characterization of diaphragmatic lymphatic channels (from Moriondo et al. 2010). (**e**) confocal image showing progression of FITC-dextran in a lymphatic loop. From Moriondo et al. (2013)

efficient transmission of tissue forces to the vessel lumen promoting lymph propulsion. Figure 8.7d shows the location of the lymphatic channels within the diaphragm. A novel and peculiar form of lymphatic network was also described and defined as "lymphatic loop" (Fig. 8.7e), whose function would be that of favoring lymph flow in regions where tissue movements are minimal as, for example, at the periphery of the diaphragm.

Lymph propulsion is also aided by spontaneous contractility. Negrini's work helped clarify the molecular machinery of muscle cells in the most peripheral diaphragmatic lymphatics which exhibit spontaneous contraction (Negrini et al. 2016). The ability to self-depolarize is related to sodium currents mediated by hyperpolarization-activated cyclic nucleotide-gated channels, similarly to what occurs in myocardial tissues. Other segments of the lymphatics are stretch-activated by the arrival of the fluid. Cross-correlation analysis also showed that the change in vessel diameter occurring in a spontaneously contracting site can also propagate to the stretch-activated segments. Hyposmolarity induces an early increase in pacemaker activity while, hyperosmotic conditions always decrease the frequency of lymphatic contraction (Solari et al. 2018). Further, a modulation was described in response to increasing temperature, due to the presence of specific temperature sensitive channels (Solari et al. 2017, 2020). It is noteworthy that in all of cases studied of lymphatic flow modulation, this occurred through a change in pacemaker activity, without any inotropic effect.

8.7 The Effect of Gravity on the Respiratory Function

The gravity dependence of several aspects of the respiratory function was the object of important and experimentally quite sophisticated approaches, particularly with reference to regional distribution in the lung of ventilation–perfusion. Further, no direct information were available concerning the air distribution in the alveolar compartment. Milic-Emili, relying on the use of Xenon 133, either delivered through inspiration or injected in the blood, detected the regional activity of Xenon 133 with scintillation counters placed on the chest. He was then able to provide a lucid and unsurpassed contribution to describe the regional subdivisions of lung volume as a function of gravity (Milic-Emili et al. 1966). A well-known figure from this paper, showing the gravity dependent regional distribution of lung volumes in the lung in head up posture, is shown in Fig. 8.8. The same experimental approach carried on obese patients revealed significant abnormality bearing a close relationship with the reduction in expiratory reserve volume due to the abnormal weight acting on the chest (Holley et al. 1967). A further contribution showed that pulmonary ventilation and perfusion were greater in the more dependent zones compared to the less dependent zones in all postures (Kaneko et al. 1966).

The gravity dependence of the ventilation and perfusion was confirmed by exposing subjects to increased positive acceleration on a human centrifuge. The

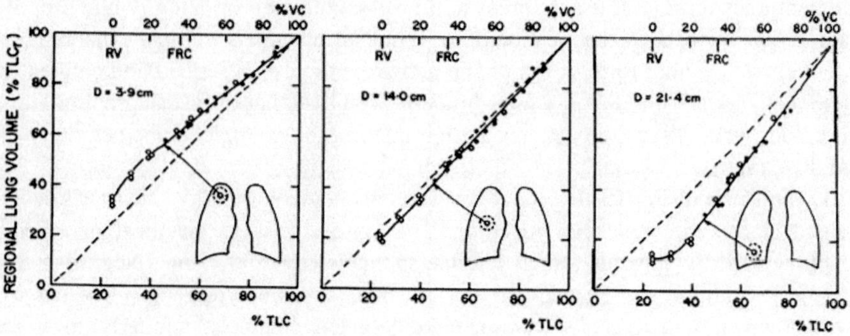

Fig. 8.8 Regional lung volume as a function of lung height. From Milic-Emili et al. (1966)

interesting hypothesis was also put forward that in weightlessness the regional lung volumes and ventilation distribution should be uniform (Bryan et al. 1966).

The effect of changing gravity on mechanical respiratory statics was also evaluated in a project funded to Bicocca University by the European Space Agency and Agenzia Spaziale Italiana (1995–2000). Results were obtained from parabolic flights providing a change in Gz from 1 to 1.8 on pull up and pull out, and 0 Gz at the top of parabola (Fig. 8.9). Results showed that, despite individual differences in respiratory elastic properties in microgravity condition, adding the load of the gravity factor brings the system close to its minimum compliance already at 1 Gz (Bettinelli et al. 2002a). Further, microgravity decreases the lung and the chest wall recoil pressures by removing the distortion of lung parenchyma and thorax induced by changing Gz. Interestingly, exposure to hyper-gravity does not greatly affect respiratory mechanics, confirming that mechanical distortion of the respiratory system is already close to maximum at 1 Gz (Bettinelli et al. 2002b).

Finally, the results revealed a remarkable feature in the mechanical arrangement of the respiratory system as the neuromuscular respiratory output is readjusted when deformations are abruptly induced or removed; accordingly, the ventilatory output is maintained with minor changes in inspiratory muscle work in face of considerable changes in configuration and mechanical properties.

Another peculiar point is that within the normal range of tidal volume, the changes in the lung and the chest wall elastic work are similar on changing the gravity vector, and since the two structures operate in opposite directions, the resultant change in respiratory work is relatively buffered (Dellacá et al. 2004).

8.8 Underwater Respiratory Physiology

This topic was afforded by Agostoni in 1966 (Agostoni et al. 1966a) who described the mechanical consequences of having a hydrostatic gradient acting on the respiratory system. Research on underwater physiology was continued by Camporesi who

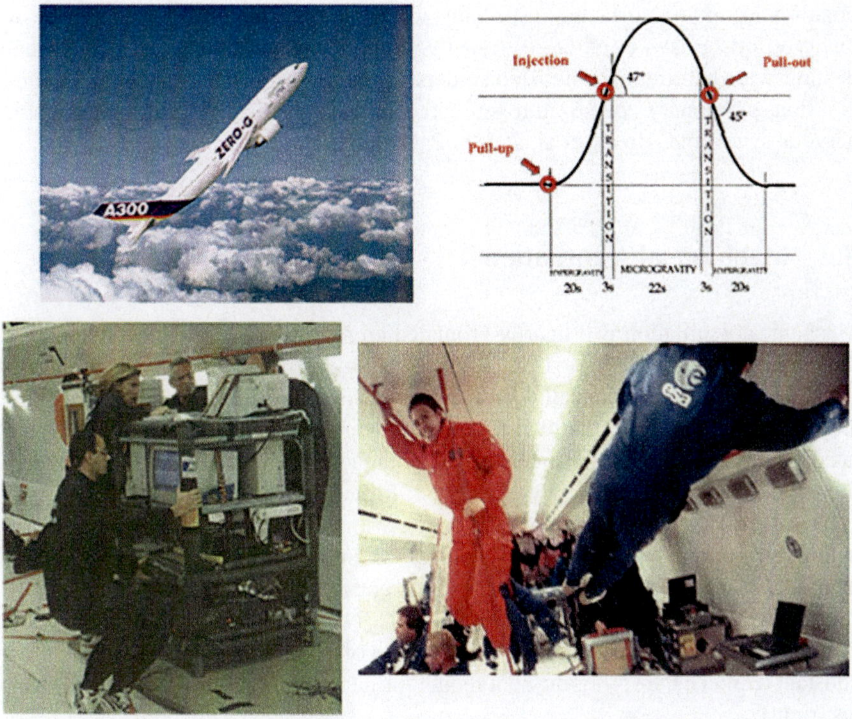

Fig. 8.9 The profile of the parabolic flight, the experimental set-up, and the floating at 0 Gz. With the kind permission of the European Space Agency

actually kept publishing on this topic. In 1984, Camporesi published an interesting paper dealing with work capacity at depth (Salzano et al. 1984). He found that moderate work was associated with alveolar hypoventilation, arterial hypercapnia, and higher levels of arterial lactate and signs of simultaneous respiratory and metabolic acidosis. Further, at variance with what occurs at surface, the onset of acidemia was not accompanied by an increase of ventilation. The point was made that an estimate of maximum work tolerance done at sea level may be an insufficient assessment of the physiological condition of a diver exposed to these high pressures. The pulmonary gas exchange in diving was the object of another study (Moon et al. 2009) that considered the effect of increased mechanical load due to immersion, the increased gas density, airways resistance, and work of breathing.

A further study estimated the effect of pre-breathing oxygen as being more effective to decrease air bubbles and platelet activation, to be considered therefore beneficial in reducing the development of decompression sickness (Bosco et al. 2010). Moreover, the first measurements were also obtained of blood gases at a depth of 40 m during a breath-hold descent in free-divers (Bosco et al. 2018a). Further, reconsidering a critical point in scuba diving, it was confirmed that the arterial partial pressure of oxygen reaches hazardously low values at the end of breath-hold upon resurfacing, especially when the dive requires considerable voluntary effort (Bosco et al. 2020). Finally, comprehensive reviews were published dealing with barotrauma, diving disorders (drowning, transient syncope), decompression, pulmonary edema, nitrogen narcosis, oxygen toxicity, and high pressure nervous syndrome (Bosco et al. 2018b; Paganini et al. 2022).

8.9 Mechanical Ventilation

Mechanical ventilation is primarily grounded on respiratory mechanics. It appeared then logic to transfer the knowledge for the assessment of lung mechanical properties from the awake to mechanically ventilated subjects. Milic-Emili pioneered this field and the point was made that two techniques would warrant investigation in ventilated adults and infants, the interrupter technique and the forced-oscillation technique (Gottfried et al. 1985; Sly et al. 1988; Milic-Emili et al. 1990).

D'Angelo's group, in close collaboration with Milic-Emili, provided a remarkable contribution to respiratory mechanics in anesthetized humans adopting the methods based on rapid occlusion during constant flow inflation. Figure 8.10 shows the original records by D'Angelo et al. (1989): the case of changing flow keeping lung volume constant (left), the case of changing volume keeping flow constant (center), the spring-dashpot model for interpretation of respiratory mechanics (right).

The respiratory system was modeled as a standard resistance ($R_{min,rs}$) in parallel with a standard elastance ($E_{st,rs}$) and includes spring-and-dashpot bodies that represent stress adaptation units. Distance between horizontal bars is analog of lung volume; tension between these bars is analog of pressure at airway opening. Based on the model, it was possible to partition the total resistance of the respiratory system into a standard resistance and a further resistive component that is an expression of time-dependent viscoelastic behavior of the thoracic tissues. In a further paper (D'Angelo et al. 1991), based on the same model, coefficients have been provided, representing a useful background for predicting the contribution of viscoelastic

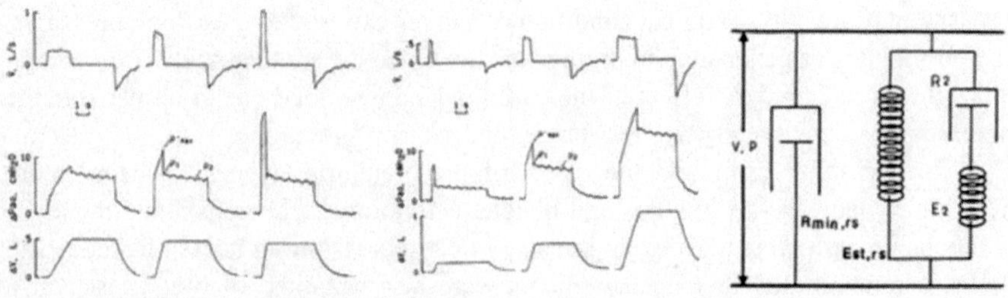

Fig. 8.10 Respiratory mechanics in anesthetized humans. Left: the case of changing flow keeping lung volume constant. Middle: the case of changing volume, keeping flow constant. Right: the spring-dashpot model for interpretation of respiratory mechanics. From D'Angelo et al. (1989)

mechanisms to frequency dependence of elastance and resistance of the lung and the chest wall and work of breathing.

The validity of the model was confirmed by studying the viscoelastic properties of the respiratory system with large tidal volumes reaching an end-inspiratory pressure up to 23 cmH_2O (D'Angelo et al. 2000). Respiratory mechanics was also studied maintaining an end-expiratory pressure (PEEP) in the range of 7–8 cmH_2O. The results showed that the applied PEEP caused a significant decrease in overall resistance, and in the lung and the chest wall compliance (D'Angelo et al. 1992). The problem of viscoelasticity, being responsible, in some cases, for most of the pressure dissipated during breathing, was reconsidered by Antonaglia et al. (2000), relying on the technique of rapid end-inspiratory airway occlusion. Direct measurement of the viscoelastic constants of the respiratory system was performed by an easier approach based on the analysis of only two breaths differing in a short or prolonged inspiratory time. This method was used in normal anesthetized paralyzed subjects and in patients with acute lung injury and proved to be of clinical utility, allowing performance of only a few bedside measurements. Another important contribution with major clinical implications was provided by Polese et al. (1999), showing that the size of the endotracheal tube and the inspiratory flow waveform must be taken into account to interpret the resistive work and thus the total work of breathing.

In a recent review article, mostly addressed to mechanical ventilation, a detailed analysis has been presented concerning factors that either favor or prevent lung lesion and the development of severe edema (Beretta et al. 2021). An important point was made concerning the repetitive lung recruitment and de-recruitment occurring during mechanical ventilation by increasing volume above 75–80% TLC as this generates excessive stretching of the septa determining an increase in permeability of the endothelial and epithelial barriers. The Starling pressure gradients across the endothelial and epithelial barriers have also been estimated in relation with lung distending pressures currently adopted. It has been suggested that severe edema develops down a relatively small driving pressure gradient acting over a large surface of damaged air–blood barrier.

8.10 Test of Pulmonary Function

8.10.1 The Efficiency of the Respiratory Action

This topic proved to be very attractive to respiratory physiologists reflecting indeed two basic problems: on one side, the fact that the available functional tests often lacked specificity, on the other, they were unable, and still are, to catch the early phase of progression of a disease.

Milic-Emili has always been very much interested in developing tests aiming to evaluate the efficiency of the respiratory action. He observed that the available methods at time were not capable of differentiating people, who will not breathe because of central or neuromuscular inadequacy, from those who cannot breathe because of mechanical abnormalities, such as airways obstruction and decreased compliance of the respiratory system. In 1972, his group devised a method to measure the so-called closing volume that is the lung volume at which the dependent lung zones cease to ventilate presumably as a result of airway closure (McCarthy et al. 1972). The measurement was done by a single-breath Argon bolus method for measuring the lung volume. It was concluded that measurement of the "closing volume" provides a test to detect small airway abnormalities much earlier than the conventional lung function tests.

Milic-Emili was also interested in the measurement of the pressure generated by the muscles during inspiration, by occluding the inspiratory line with a shutter; he considered the pressure measured at 0.1 s (known as $P_{0.1}$) after the onset of inspiration as an index of respiratory-center output. Using an animal experimental approach, he was able to correlate $P_{0.1}$ with the phrenic nerve discharge in response to increase in respiratory drive in hypoxia and hypercapnia (Grunstein et al. 1973). Based on these findings, he extended the technique to healthy and pneumopathic humans (Whitelaw et al. 1975; Milic-Emili et al. 1975; Aubier et al. 1980).

The idea underlying a measurement of central respiratory output was reconsidered 25 years later to propose a non-invasive estimate of the tension–time index of inspiratory muscles to include $P_{0.1}$; this time the technique was more complicated requiring the measurement of esophageal and gastric pressure (Hayot et al. 2000).

8.10.2 Flow Limitation

Milic-Emili and his group developed a method to estimate the entity of expiratory flow limitation, the main symptom of chronic obstructive pulmonary disease (COPD). The problem of an early diagnosis of developing COPD remains a difficult one, as flow limitation occurs in the smallest airways that in a healthy lung provide only about 5% of total respiratory flow resistance. Accordingly, relying on current diagnostic methods (forced expiratory maneuver, forced expiratory volume after 1 s, FEV1), flow limitation can be diagnosed only when a considerable share (~30%) of small airways close down and therefore lack any preventive approach.

A relatively simple method was proposed, consisting in applying negative pressure at the mouth during expiration. Patients in whom the pressure elicits an increase in flow throughout the expiration are not flow-limited; in contrast, patients that fail to respond with an increase in flow are flow-limited (Koulouris et al. 1997; Boczkowski et al. 1997). The same technique was applied to compare patients with restrictive and obstructive disorders (Baydur and Milic-Emili 1997; Murariu et al. 1998; Kosmas et al. 2004).

D'Angelo's group contributed to an interesting new analysis of the alveolar pressure–flow loops obtained during inspiration and expiration by using the body plethysmograph. On comparing healthy young and elderly people with COPD

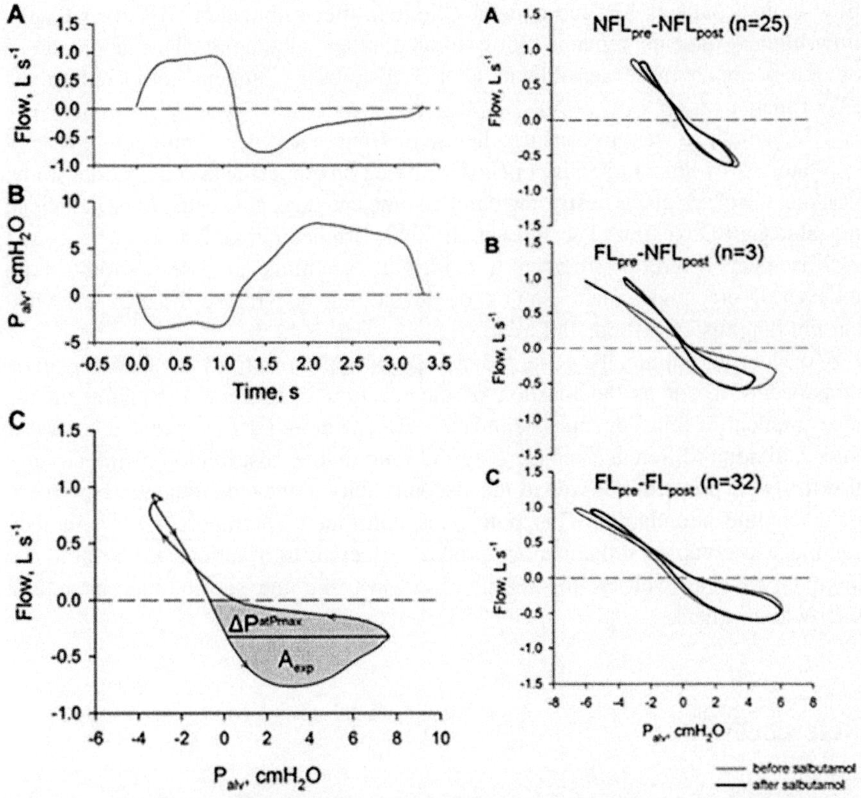

Fig. 8.11 Left: records of airflow and alveolar pressure (**a** and **b**, respectively) and flow–pressure diagram (**c**), showing the expiratory loop revealing the presence of expiratory flow limitation. Right: effect of salbutamol administration on the flow–pressure loops: no effect in normal subjects (**a**), COPD flow-limited patients becoming non-flow limited (**b**); COPD flow-limited patients remaining flow-limited. From Pecchiari et al. (2020)

patients, the results proved to contain potentially relevant information about the pathophysiology of COPD, reflecting changes in mechanical heterogeneity, recruitment/de-recruitment of lung units, and air trapping (Radovanovic et al. 2018). Expiratory flow limitation during tidal breathing is common to all COPD patients; however, Pecchiari et al. (2020) were able to grade the severity of the flow limitation by considering the shape of the relation between flow and volume recorded by a plethysmograph during spontaneous breathing. Their results are highlighted in Fig. 8.11. On the left, the records of airflow and alveolar pressure (A and B, respectively) are used to build the flow–pressure diagram (C) that on expiration (range of positive alveolar pressures) shows a characteristic loop revealing the presence of expiratory flow limitation. On the right, one can compare the flow–pressure loops before and after administration of a bronchodilator (salbutamol): in normal subjects (A) salbutamol had no effect; some COPD flow-limited patients became non-flow limited after salbutamol (B); other COPD flow-limited patients remained flow-limited after salbutamol. This approach is obviously much more sensible to COPD diagnosis treatment compared to the FEV1 maneuver.

D'Angelo also went to consider the use of Helium–oxygen ventilation in case of expiratory flow limitation. His conclusion, based on clinical data and a model study, was that the technique is costly and cumbersome and does not appear to be beneficial in stable COPD patients (Pecchiari et al. 2020; Brighenti et al. 2007).

Miserocchi's group provided a contribution aiming to assess subthreshold indexes of preclinical manifestation of developing respiratory disease, just what current diagnostic methods fail to do.

Considering asthma, the general diagnosis of hyper-reactivity remains in general non-specific as far as the location of the reaction is concerned. Relying on the measurement of impedenzometric indexes, Beretta et al. (2014) were able to assess inter-individual differences concerning the longitudinal distribution of the airways flow resistive properties as well as the viscoelasticity of the lung tissue in response to methacholine stimulation. This point was considered of importance as, in fact, asthma is a complex inflammatory process affecting to a various extent both the large airways as well as the terminal airways and the surrounding interstitial microenvironment.

References

Agostoni E (1961) A graphical analysis of thoraco-abdominal mechanics during the breathing cycle. J Appl Physiol 16:1055–1059

Agostoni E (1963) Diaphragm activity during breath holding: factors related to its onset. J Appl Physiol 18:30–36

Agostoni E, Fenn WO (1960) Velocity of muscle shortening as a limiting factor in respiratory air flow. J Appl Physiol 15:349–353

Agostoni E, Mead J (1964) Statics of the respiratory system. In: Fenn WO, Rahn H (eds) Handbook of physiology: respiration, Sec. 3, vol 1, American Physiological Society, Washington DC, pp 387–409

Agostoni E, Rahn H (1960) Abdominal and thoracic pressures at different lung volumes. J Appl Physiol 15:1087–1092

Agostoni E, Zocchi L (2007) Pleural liquid and its exchanges. Respir Physiol Neurobiol 159:311–323

Agostoni E, Taglietti A, Setnikar I (1957) Absorption force of the capillaries of the visceral pleura in determination of the intrapleural pressure. Am J Phys 191:277–282

Agostoni E, Thimm FF, Fenn WO (1959) Comparative features of the mechanics of breathing. J Appl Physiol 14:679–683

Agostoni E, Sant'Ambrogio G, Del Portillo Carrasco H (1960) Electromyography of the diaphragm in man and transdiaphragmatic pressure. J Appl Physiol 15:1093–1097

Agostoni E, Gurtner G, Torri G, Rahn H (1966a) Respiratory mechanics during submersion and negative-pressure breathing. J Appl Physiol 21:251–258

Agostoni E, Mognoni P, Torri G, Miserocchi G (1966b) Forces deforming the rib cage. Respir Physiol 2:105–117

Akoumianaki E, Maggiore SM, Valenza F, Bellani G, Jubran A, Loring SH, Pelosi P, Talmor D, Grasso S, Chiumello D, Guérin C, Patroniti N, Ranieri VM, Gattinoni L, Nava S, Terragni PP, Pesenti A, Tobin M, Mancebo J, Brochard L, PLUG Working Group (Acute Respiratory Failure Section of the European Society of Intensive Care Medicine) (2014) The application of esophageal pressure measurement in patients with respiratory failure. Am J Respir Crit Care Med 189:520–531

Aliverti A, Kayser B, Lo Mauro A, Quaranta M, Pompilio P, Dellacà RL, Ora J, Biasco L, Cavalleri L, Pomidori L, Cogo A, Pellegrino R, Miserocchi G (2011) Respiratory and leg muscles perceived exertion during exercise at altitude. Respir Physiol Neurobiol 177:162–168

Antonaglia V, Peratoner A, De Simoni L, Gullo A, Milic-Emili J, Zin WA (2000) Bedside assessment of respiratory viscoelastic properties in ventilated patients. Eur Respir J 16:302–308

Aron E (1900) Der intrapleurale Druch beim lebenden, gesunden Menschen. Arch Pathol Anat Physiol 160:226–234

Aubier M, Murciano D, Fournier M, Milic-Emili J, Pariente R, Derenne JP (1980) Central respiratory drive in acute respiratory failure of patients with chronic obstructive pulmonary disease. Am Rev Respir Dis 122:191–199

Bates JH, Suki B (2008) Assessment of peripheral lung mechanics. Respir Physiol Neurobiol 163: 54–63

Bates JHT, Gaver DP, Habashi NM, Nieman GF (2020) Atelectrauma versus volutrauma: a tale of two time-constants. Crit Care Explor 16:e0299

Baydur A, Milic-Emili J (1997) Expiratory flow limitation during spontaneous breathing: comparison of patients with restrictive and obstructive respiratory disorders. Chest 112:1017–1023

Beretta E, Tana F, Grasso GS, Bartesaghi M, Novelli L, Pesci A, Miserocchi G (2014) Regional differences in bronchial reactivity assessed by respiratory impedance. Respir Physiol Neurobiol 192:23–29

Beretta E, Romanò F, Sancini G, Grotberg JB, Nieman GF, Miserocchi G (2021) Pulmonary interstitial matrix and lung fluid balance from normal to the acutely injured lung. Front Physiol 12:781874

Bettinelli D, Kays C, Bailliart O, Capderou A, Techoueyres P, Lachaud JL, Vaïda P, Miserocchi G (2002a) Effect of gravity on chest wall mechanics. J Appl Physiol 92:709–716

Bettinelli D, Kays C, Bailliart O, Capderou A, Techoueyres P, Lachaud JL, Vaïda P, Miserocchi G (2002b) Effect of gravity and posture on lung mechanics. J Appl Physiol 93:2044–2052

Boczkowski J, Murciano D, Pichot MH, Ferretti A, Pariente R, Milic-Emili J (1997) Expiratory flow limitation in stable asthmatic patients during resting breathing. Am J Respir Crit Care Med 156:752–757

Bodega F, Zocchi L, Agostoni E (2002) Albumin transcytosis in mesothelium. Am J Phys 281:L3–L11

Bosco G, Yang ZJ, Di Tano G, Camporesi EM, Faralli F, Savini F, Landolfi A, Doria C, Fanò G (2010) Effect of in-water oxygen prebreathing at different depths on decompression-induced bubble formation and platelet activation. J Appl Physiol 108(5):1077–1083. https://doi.org/10.1152/japplphysiol.01058.2009

Bosco G, Rizzato A, Martani L, Schiavo S, Talamonti E, Garetto G, Paganini M, Camporesi EM, Moon RE (2018a) Arterial blood gas analysis in breath-hold divers at depth. Front Physiol 9. https://doi.org/10.3389/fphys.2018.01558

Bosco G, Rizzato A, Moon RE, Camporesi EM (2018b) Environmental physiology and diving medicine. Front Psychol 9:72. https://doi.org/10.3389/fpsyg.2018.00072

Bosco G, Paganini M, Rizzato A, Martani L, Garetto G, Lion J, Camporesi EM, Moon RE (2020) Arterial blood gases in divers at surface after prolonged breath-hold. Eur J Appl Physiol 120(2): 505–512. https://doi.org/10.1007/s00421-019-04296-2

Brighenti C, Barbini P, Gnudi G, Cevenini G, Pecchiari M, D'Angelo E (2007) Helium-oxygen ventilation in the presence of expiratory flow-limitation: a model study. Respir Physiol Neurobiol 157:326–334

Bryan AC, Milic-Emili J, Pengelly D (1966) Effect of gravity on the distribution of pulmonary ventilation. J Appl Physiol 21:778–784

Cavagna GA, Stemmler EJ, DuBois AB (1967) Alveolar resistance to atelectasis. J Appl Physiol 22: 441–452

Coxson HO, Mayo JR, Behzad H, Moore BJ, Verburgt L, Staples CA, Paré P, Hogg JC (1995) Measurement of lung expansion with computed tomography and comparison with quantitative histology. J Appl Physiol 79:1525–1530

D'Angelo E, Calderini E, Torri G, Robatto FM, Bono D, Milic-Emili J (1989) Respiratory mechanics in anesthetized paralyzed humans: effects of flow, volume, and time. J Appl Physiol 67:2556–2564

D'Angelo E, Robatto FM, Calderini E, Tavola M, Bono D, Torri G, Milic-Emili J (1991) Pulmonary and chest wall mechanics in anesthetized paralyzed humans. J Appl Physiol 70: 2602–2610

D'Angelo E, Calderini E, Tavola M, Bono D, Milic-Emili J (1992) Effect of PEEP on respiratory mechanics in anesthetized paralyzed humans. J Appl Physiol 73:1736–1742

D'Angelo E, Tavola M, Milic-Emili J (2000) Volume and time dependence of respiratory system mechanics in normal anaesthetized paralysed humans. Eur Respir J 16:665–672

D'Angelo E, Loring SH, Gioia ME, Pecchiari M, Moscheni C (2004) Friction and lubrication of pleural tissues. Respir Physiol Neurobiol 142:55–68

Dellacá RL, Bettinelli D, Kays C, Techoueyres P, Lachaud JL, Vaïda P, Miserocchi G (2004) Effect of changing the gravity vector on respiratory output and control. J Appl Physiol 97:1219–1226

Donders FC (1853) Beitrage zum Mechanismus der Respiration und Circulation im gesunden und kranken Zustande. Z Rat Med N F 3:287–319

Farhi L, Otis AB, Proctor DF (1957) Measurement of intrapleural pressure at different points in the chest of the dog. J Appl Physiol 10:15–18

Gottfried SB, Higgs BD, Rossi A, Carli F, Mengeot PM, Calverly PM, Zocchi L, Milic-Emili J (1985) Interrupter technique for measurement of respiratory mechanics in anesthetized humans. J Appl Physiol 59:647–652

Grimaldi A, Moriondo A, Sciacca L, Guidali ML, Tettamanti G, Negrini D (2006) Functional arrangement of rat diaphragmatic initial lymphatic network. Am J Phys 291:H876–H885

Grunstein MM, Younes M, Milic-Emili J (1973) Control of tidal volume and respiratory frequency in anesthetized cats. J Appl Physiol 35:463–476

Hayot M, Ramonatxo M, Matecki S, Milic-Emili J, Prefaut C (2000) Noninvasive assessment of inspiratory function during. J Respir Crit Care Med 162:2201–2207

Hills B, Brian A (1992) Graphite-like lubrication of mesothelium by oligolamellar pleural surfactant. J Appl Physiol 73:1034–1039

Hills B, Butler BD, Barrow E (1982) Boundary lubrication imparted by pleural surfactants and their identification. J Appl Physiol 53:463–469

Holley HS, Milic-Emili J, Becklake MR, Bates DV (1967) Regional distribution of pulmonary ventilation and perfusion in obesity. J Clin Invest 46:475–481

Kaneko K, Milic-Emili J, Dolovich MB, Dawson A, Bates DV (1966) Regional distribution of ventilation and perfusion as a function of body position. J Appl Physiol 21:767–777

Kedem O, Katchalsky A (1958) Thermodynamic analysis of the permeability of biological membranes to non-electrolytes. Biochim Biophys Acta 27:229–246

Knudsen L, Lopez-Rodriguez E, Berndt L, Steffen L, Ruppert C, Bates JHT, Ochs M, Smith BJ (2018) Alveolar micromechanics in bleomycin-induced lung injury. Am J Respir Cell Mol Biol 59:757–769

Kosmas EN, Milic-Emili J, Polychronaki A, Dimitroulis I, Retsou S, Gaga M, Koutsoukou A, Roussos C, Koulouris NG (2004) Exercise-induced flow limitation, dynamic hyperinflation and exercise capacity in patients with bronchial asthma. Eur Respir J 24:378–884

Koulouris NG, Dimopoulou I, Valta P, Finkelstein R, Cosio MG, Milic-Emili J (1997) Detection of expiratory flow limitation during exercise in COPD patients. J Appl Physiol 82:723–733

Levick JR (1991) Capillary filtration-absorption balance reconsidered in light of dynamic extravascular factors. Exp Physiol 76:825–857

Luciani L (1879) Della pressione intratoracica e intraddominale (1878). Arch Sci Med 2:177

Margaria R, Milic-Emili J, Petit JM, Cavagna G (1960) Mechanical work of breathing during muscular exercise. J Appl Physiol 15:354–358

Mazzuca E, Salito C, Rivolta I, Aliverti A, Miserocchi G (2014) From morphological heterogeneity at alveolar level to the overall mechanical lung behavior: an in vivo microscopic imaging study. Physiol Rep 7;2(2):e00221

McCarthy DS, Spencer R, Greene R, Milic-Emili J (1972) Measurement of "closing volume" as a simple and sensitive test for early detection of small airway disease. Am J Med 52:47–753

Mead J (1960) The control of respiratory frequency. Ann N Y Acad Sci 109:724–729

Mead J, Whittenberger JL (1953) Physical properties of human lungs measured during spontaneous respiration. J Appl Physiol 5:779–796

Mead J, McIlroy MB, Selverstone NJ, Krietem BC (1955) Measurement of intraesophageal pressure. J Appl Physiol 7:491

Michel CC (1980) Filtration coefficients and osmotic reflexion coefficients of the walls of single frog mesenteric capillaries. J Physiol Lond 309:341–355

Michel CC, Curry FE (1999) Microvascular permeability. Physiol Rev 79:703–761

Milic-Emili J, Petit JM (1959) Relationship between endoesophageal and intrathoracic pressure variations in dog. J Appl Physiol 14:535

Milic-Emili J, Mead J, Turner JM (1964a) Topography of esophageal pressure as a function of posture in man. J Appl Physiol 19:212–216

Milic-Emili J, Mead J, Turner JM, Glauser EM (1964b) Improved technique for estimating pleural pressure from esophageal balloons. J Appl Physiol 19:207–211

Milic-Emili J, Henderson JA, Dolovich MB, Trop D, Kaneko K (1966) Regional distribution of inspired gas in the lung. J Appl Physiol 21:749–759

Milic-Emili J, Whitelaw WA, Derenne JP (1975) New tests to assess lung function: occlusion pressure a simple measure of the respiratory center's output. N Engl J Med 293:1029–1030

Milic-Emili J, Robatto FM, Bates JH (1990) Respiratory mechanics in anaesthesia. Brit J Anaesth 65:4–12

Milla E, Meschia G, Setnikar I (1952) Sulle proprietà meccaniche del polmone. Arch Fisiol 52:1952

Miserocchi G (1997) Physiology and pathophysiology of pleural fluid turnover. Eur Respir J 10: 219–225

Miserocchi G (2009) Mechanisms controlling the volume of pleural fluid and extravascular lung water. Eur Respir Rev 18:244–252

Miserocchi G, Negrini D (1986) Contribution of Starling and lymphatic flows to pleural liquid exchanges in anesthetized rabbits. J Appl Physiol 61:325–330

Miserocchi G, Pistolesi M, Miniati M, Bellina CR, Negrini D, Giuntini C (1984) Pleural liquid pressure gradients and intrapleural distribution of injected bolus. J Appl Physiol 56:526–532

Miserocchi G, Negrini D, Para AF, Pistolesi M, Miniati M, Bellina R, Giuntini C (1986) Kinetics of the intrapleural distribution of a radioactive bolus. Respir Physiol 65:13–27

Miserocchi G, Negrini D, Pistolesi M, Bellina CR, Gilardi MC, Bettinardi V, Rossitto F (1988) Intrapleural liquid flow down a gravity-dependent hydraulic pressure gradient. J Appl Physiol 64:577–584

Miserocchi G, Negrini D, Mukenge S, Turconi P, Del Fabbro M (1989) Liquid drainage through the peritoneal diaphragmatic surface. J Appl Physiol 66:1579–1585

Miserocchi G, Venturoli D, Negrini D, Gilardi MC, Bellina R (1992) Intrapleural fluid movements described by a porous flow model. J Appl Physiol 73:2511–2516

Miserocchi G, Venturoli D, Negrini D, Del Fabbro M (1993) Model of pleural fluid turnover. J Appl Physiol 75:1798–1806

Miserocchi G, Sancini G, Mantegazza F, Chiappino G (2008) Translocation pathways for inhaled asbestos fibers. Environ Health 24(7):4. https://doi.org/10.1186/1476-069X-7-4

Moon RE, Cherry AD, Stolp BW, Camporesi EM (2009) Pulmonary gas exchange in diving. J Appl Physiol 106(2):668–677. https://doi.org/10.1152/japplphysiol.91104.2008

Moriondo A, Mukenge S, Negrini D (2005) Transmural pressure in rat initial subpleural lymphatics during spontaneous or mechanical ventilation. Am J Phys 289:H263–H269

Moriondo A, Bianchin F, Marcozzi C, Negrini D (2008) Kinetics of fluid flux in the rat diaphragmatic submesothelial lymphatic lacunae. Am J Phys 295:H1182–H1190

Moriondo A, Boschetti F, Bianchin F, Lattanzio S, Marcozzi C, Negrini D (2010) Tissue contribution to the mechanical features of diaphragmatic initial lymphatics. J Physiol 588:3957–3969

Moriondo A, Solari E, Marcozzi C, Negrini D (2013) Spontaneous activity in peripheral diaphragmatic lymphatic loops. Am J Phys 305:H987–H995

Moriondo A, Solari E, Marcozzi C, Negrini D (2015) Diaphragmatic lymphatic vessel behavior during local skeletal muscle contraction. Am J Phys 308:H193–H205

Murariu C, Ghezzo H, Milic-Emili J, Gautier H (1998) Exercise limitation in obstructive lung disease. Chest 114:965–968

Negrini D, Pistolesi M, Miniati M, Bellina R, Giuntini C, Miserocchi G (1985) Regional protein absorption rates from the pleural cavity in dogs. J Appl Physiol 58:2062–2067

Negrini D, Mukenge S, Del Fabbro M, Gonano C, Miserocchi G (1991) Distribution of diaphragmatic lymphatic stomata. J Appl Physiol 70(4):1544–1549

Negrini D, Del Fabbro M, Gonano C, Mukenge S, Miserocchi G (1992) Distribution of diaphragmatic lymphatic lacunae. J Appl Physiol 72:1166–1172

Negrini D, Moriondo A, Mukenge S (2004) Transmural pressure during cardiogenic oscillations in rodent diaphragmatic lymphatic vessels. Lymphat Res Biol 2:69–81

Negrini D, Marcozzi C, Solari E, Bossi E, Cinquetti R, Reguzzoni M, Moriondo A (2016) Hyperpolarization-activated cyclic nucleotide-gated channels in peripheral diaphragmatic lymphatics. Am J Phys 311:H892–H903

Otis AB, Proctor DF (1948) Measurement of alveolar pressure in human subjects. Am J Phys 152: 106–112

Otis AB, Proctor DF, Rahn H (1947) The measurement of alveolar pressure and the work of breathing. Fed Proc 6:173

Otis AB, Fenn WO, Rahn H (1950) Mechanics of breathing in man. J Appl Physiol 2:592–607

Paganini M, Moon RE, Boccalon N, Melloni GEM, Giacon TA, Camporesi EM, Bosco G (2022) Blood gas analyses in hyperbaric and underwater environments: a systematic review. J Appl Physiol 132(2):283–293. https://doi.org/10.1152/japplphysiol.00569.2021

Passoni E, Lania A, Adamo S, Grasso GS, Noè D, Miserocchi G, Beretta E (2015) Mild training program in metabolic syndrome improves the efficiency of the oxygen pathway. Respir Physiol Neurobiol 208:8–14

Patlak CS, Goldstein DA, Hoffman JF (1963) The flow of solute and solvent across a two-membrane system. J Theor Biol 5:426–442

Pecchiari M, Sartori P, Conte V, D'Angelo E, Moscheni C (2016) Friction and morphology of pleural mesothelia. Respir Physiol Neurobiol 220:17–24

Pecchiari M, Radovanovic D, Zilianti C, Saderi L, Sotgiu G, D'Angelo E, Santus P (2020) Tidal expiratory flow limitation induces expiratory looping of the alveolar pressure-flow relation in COPD patients. J Appl Physiol 129:75–83

Pengelly LD, Alderson AM, Milic-Emili J (1971) Mechanics of the diaphragm. J Appl Physiol 30: 797–805

Petit JM, Milic-Emili J (1958) Measurement of endoesophageal pressure. J Appl Physiol 13:481–485

Polese G, D'Angelo E, Rossi A, Milic-Emili J (1999) Effect of inspiratory flow waveform on work on endotracheal tubes: a model analysis. Intensive Care Med 25:503–507

Radovanovic D, Pecchiari M, Pirracchio F, Zilianti C, D'Angelo E, Santus P (2018) Plethysmographic loops: a window on the lung pathophysiology of COPD patients. Front Physiol 9:484

Rahn H, Otis AB, Chadwick LE, Fenn WO (1946) The pressure-volume diagram of the thorax and lung. Am J Phys 146:161–178

Rohrer F (1916) Der Zusammenhang der Atemkrafte und ihre Afhangigkeit vom Dehnungszustand der Atmungsorgane. Pflügers Arch 165:419–444

Salito C, Woods JC, Aliverti A (2011) Influence of CT reconstruction settings on extremely low attenuation values for specific gas volume calculation in severe emphysema. Acad Radiol 18: 1277–1284

Salito C, Aliverti A, Mazzuca E, Rivolta I, Miserocchi G (2014) Alveolar mechanics studied by in-vivo microscopy imaging through intact pleural space. Respir Physiol Neurobiol 202:44–49

Salito C, Aliverti A, Mazzuca E, Rivolta I, Miserocchi G (2015) The effect of exogenous surfactant on alveolar interdependence. Respir Physiol Neurobiol 210:7–13

Salzano JV, Camporesi EM, Stolp BW, Moon RE (1984) Physiological responses to exercise at 47 and 66 ATA. J Appl Physiol Respir Environ Exerc Physiol 57(4):1055–1068. https://doi.org/10.1152/jappl.1984.57.4.1055

Setnikar I (1955) Resistenza offerta dalle vie aeree al flusso gassoso durante i moti respiratori. Boll Soc It Biol Sper 31(Fasc 9-10):1–4

Setnikar I, Meschia G (1954) Entità e rendimento del lavoro respiratorio misurato sul cane. Arch Fisiol 54:231–244

Setnikar I, Agostoni E, Taglietti (1955) La pressione nella cavità pleurica in condizioni fisiologiche. Boll Soc Ital Biol Sper 31:1–3

Setnikar I, Agostoni E, Taglietti A (1957) Entità caratteristiche e origine della depressione pleurica. Arch Sci Biol 41:312–325

Sironi C, Bodega F, Porta C, Zocchi L, Agostoni E (2007) Expression of Na+−glucose cotransporter (SGLT1) in visceral and parietal mesothelium of rabbit pleura. Respir Physiol Neurobiol 159:68–75

Sironi C, Bodega F, Porta C, Monaco A, Zocchi L, Agostoni E (2008) Na+−glucose cotransporter is also expressed in mesothelium of species with thick visceral pleura. Respir Physiol Neurobiol 161:261–266

Sly PD, Brown KA, Bates JH, Spier S, Milic-Emili J (1988) Noninvasive determination of respiratory mechanics during mechanical ventilation of neonates: a review of current and future techniques. Pediatr Pulmonol 4:39–47

Solari E, Marcozzi C, Negrini D, Moriondo A (2017) Temperature-dependent modulation of regional lymphatic contraction frequency and flow. Am J Phys 313:H879–H889

Solari E, Marcozzi C, Negrini D, Moriondo A (2018) Fluid osmolarity acutely and differentially modulates lymphatic vessels intrinsic contractions and lymph flow. Front Physiol 9:871

Solari E, Marcozzi C, Bistoletti M, Baj A, Giaroni C, Negrini D, Moriondo A (2020) TRPV4 channels' dominant role in the temperature modulation of intrinsic contractility and lymph flow of rat diaphragmatic lymphatics. Am J Phys 319:H507–H518

Van der Brugh JP (1900) Über eine Methode zur Messung des intrapleuralen Druckes (1900). Pflügers' Arch 82:591–602

Vink H, Duling BR (1996) Identification of distinct luminal domains for macromolecules, erythrocytes, and leukocytes within mammalian capillaries. Circ Res 79:581–589

Whitelaw WA, Derenne JP, Milic-Emili J (1975) Occlusion pressure as a measure of respiratory center output in conscious man. Respir Physiol 23:181–199

Wylie AG, Korchevskiy A, Segrave AM, Duane A (2020) Modeling mesothelioma risk factors from dimensionality: mineralogical and epidemiological perspective. J Appl Toxicol 40:515–524

Chapter 9
The Air–Blood Barrier

Giuseppe Miserocchi

Abstract This chapter presents an integrated view of the various functions of the air–blood barrier considered as an interface for gas exchange. "Safety factors" are being described allowing to exert a strict control on the amount of the extravascular water that is kept at minimum in physiological conditions to favor oxygen uptake. The key role of the macromolecular structure network of the lung interstitial tissue is described to guarantee a minimal microvascular permeability; further, the role of precapillary vasoconstriction is emphasized to avoid an increase in alveolar-capillary pressure. The physiopathology of lung edema is described as a progressive failure of the "safety factors" in edemagenic conditions such as exercise and exposure to hypoxia. The kinetics of the oxygen diffusion/transport mechanism is discussed as a function of age (fetal, neonatal, adult), oxygen demand (rest, exercise), and development of lung edema. The inter-individual differences in the kinetics of oxygen uptake are being correlated with morpho-functional features of the air–blood barrier based on diffusion capacitance and extension of the capillary network. The chapter also reports data on the reflex response to pollutants, and results from an applied developmental project for drug uptake across the air–blood barrier via engineered nanoparticles.

G. Miserocchi (✉)
University of Milano Bicocca, Milano, Italy
e-mail: Giuseppe.Miserocchi@unimib.it

G. Ferretti (ed.), *Exercise, Respiratory and Environmental Physiology*, Perspectives in Physiology, https://doi.org/10.1007/978-3-031-19197-8_9

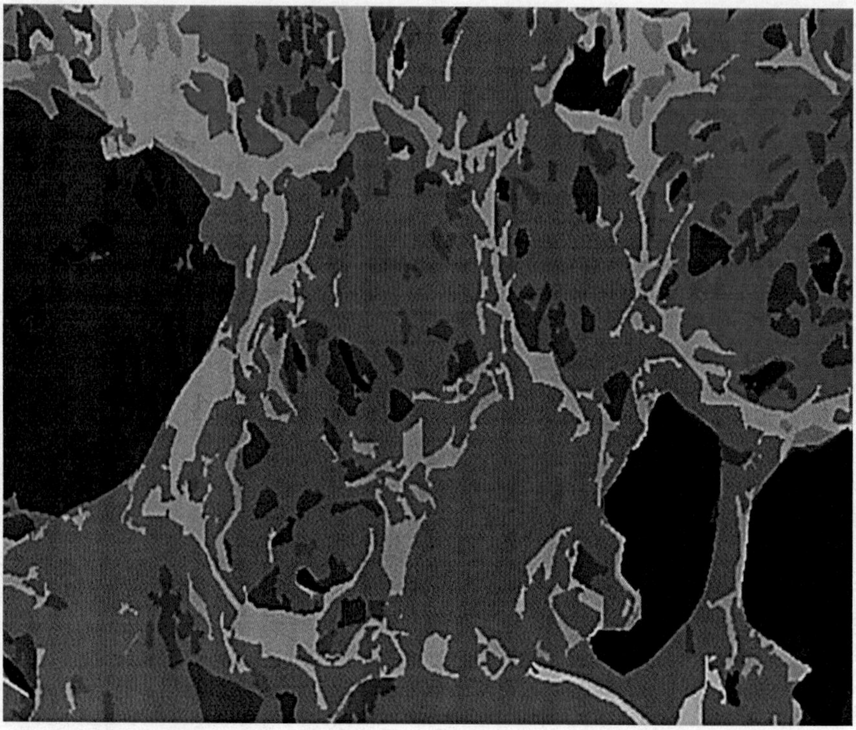

Septa and alveoli composing the air–blood barrier. Reinterpretation by Giuseppe Miserocchi of an original scanning by Ewald Weibel. With the kind permission of Giuseppe Miserocchi

9.1 Introduction

This chapter concerns the two other major branches of respiration physiology in which the School of Milano played a significant role in the advancement of knowledge, namely that of lung fluid balance and that of gas exchange across the air–blood barrier. It relies largely on the work carried out by Giuseppe Miserocchi, who started this line of investigation when he was with Jacopo Milic-Emili at McGill University, and by his pupil Daniela Negrini. Significant contributions by Emilio Agostoni are however to be acknowledged as well.

The understanding of microvascular fluid exchanges has an impact on the subject of the genesis of lung edema, a physiological problem of great clinical importance. The School of Milano generated significant knowledge on this subject as well.

9.2 Microvascular Fluid Exchanges in the Healthy Lung

The lung is a very efficient diffuser due to its enormous surface area and extremely thin air–blood barrier, averaging 0.2–0.3 mm; such thinness reflects the fact that the lung is a dry organ, meaning that its water content is kept at a minimum. By integrating experimental data of lung mechanics with microfluidics, Miserocchi's group developed a firm and, so far, unique theoretical framework to interpret the control of the extravascular water in lung placed in the biological context of its macromolecular interstitial structure. Figure 9.1 shows the complex macromolecular mesh at the level of the gas exchange units. This includes a network of essentially three types of molecules: (a) chemically stable and fairly resistant collagen fibers providing mechanical support, (b) elastic fibers providing elasticity, (c) a dense mesh of highly hydrophilic molecules, belonging to the proteoglycans family, located within the space left free by collagen and elastic fibers.

On the whole, the interstitial compartment is a crowd assembly of heterogeneous molecules behaving as a porous medium, through which water can circulate in tortuous *voids*; water flow depends on the permeability of the medium and of the endothelial barrier delimiting the interstitial compartment, namely the endothelium and the epithelium. In physiological conditions, proteoglycans coating the endothelial and epithelial barriers keep permeability to water very low and the amount of extravascular water is kept at minimum (Miserocchi and Rivolta 2012). A key mechanical variable to model lung fluid balance was the hydraulic pressure in the lung interstitial compartment (P_{ip}) that was measured by developing the trans-pleural micropuncture technique, whose details are shown in Fig. 9.2. In lungs physiologically expanded in the chest wall, P_{ip} was found rather sub-atmospheric, averaging – 10 cmH_2O (Miserocchi et al. 1990).

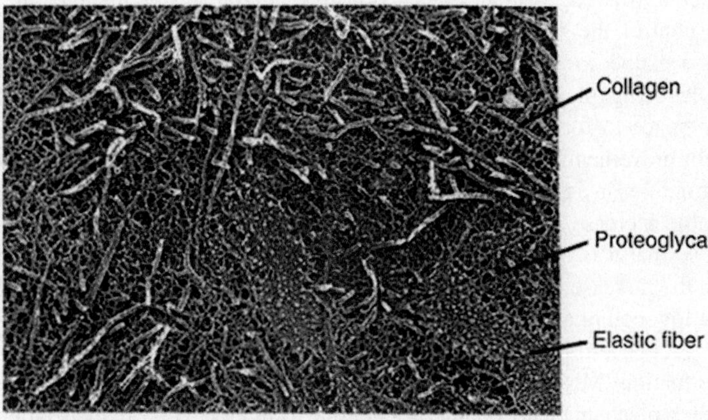

Fig. 9.1 The dense network of the macromolecular organization of the pulmonary interstitium. Redrawn from Miserocchi et al. 1990

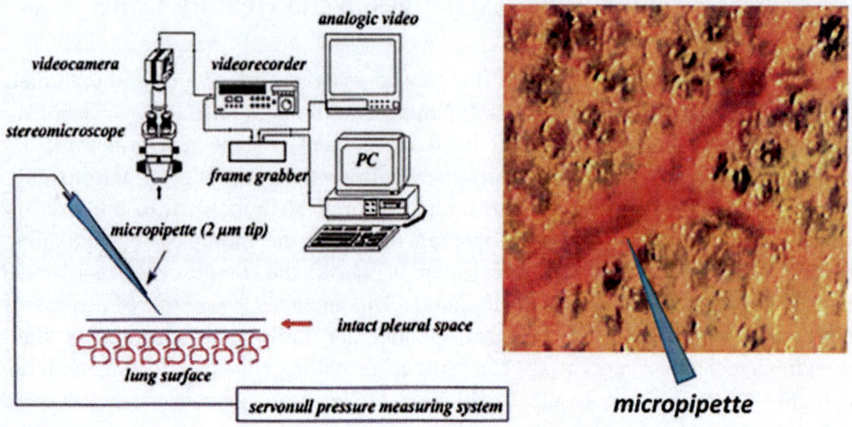

Fig. 9.2 Technique of trans-pleural micropuncture of the lung physiologically expanded in the chest. On the right imaging of superficial lung micro vessels and peri-microvascular interstitial compartment where hydraulic pressure was measured. Redrawn from Miserocchi et al. 1990

Since this value implies capillary filtration down an estimated Starling pressure gradient, it was reasoned that a sub-atmospheric pressure would result from the balance between fluid filtration through a high resistance system (capillary wall) and a powerful drainage through lymphatics (Miserocchi et al. 1990).

9.3 Pathophysiology of Lung Edema

Mild edemagenic conditions were experimentally created to test the resistance of the lung to edema (Miserocchi et al. 1993; Conforti et al. 2002). Two main features were found to control the volume of the extravascular water and allowing minimal variations relative to the steady state value. The first concerns proteoglycans (a key role being played by hyaluronan), which can bind excess water in the interstitial space to form gel-like structures that cause a restrictive sieve to water and protein movements. The other is represented by the greater steric hindrance of gel-like structures in a rigid (high elastance) macromolecular scaffold, which leads to a remarkable increase in *Pip* up to about 5 cmH_2O. This pressure buffers further filtration, so that it is called tissue *safety factor* against development of edema. At this stage, the increase in lung extravascular water is kept within 10%, thanks to the extremely low compliance of the interstitial compartment (Miserocchi et al. 1993). As microvascular filtration is blocked, no increase in lymphatic flow is required to face this situation (Miserocchi 2009).

A further mechanism to counteract edema formation is to reduce blood flow through vasoconstriction of the precapillary arterioles, by diverting blood away from regions at risk of developing edema (Negrini et al. 2001; Rivolta et al. 2011a;

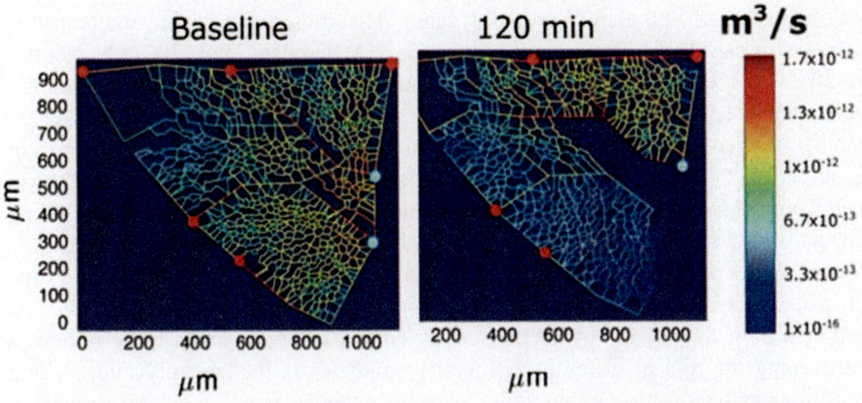

Fig. 9.3 Modelling the decrease in blood flow (shift towards blue) in the alveolar-capillary districts from baseline to 120 min of exposure to hypoxia, a well-known edemagenic condition From Mazzuca et al. 2019

Mazzuca et al. 2019). Mazzuca et al. (2016) modeled the decrease in microvascular filtration related to the decrease in blood flow in edemagenic conditions. Figure 9.3 shows the decrease in blood flow (shift towards blue) in the alveolar-capillary districts from baseline up to 120 min of exposure to hypoxia, a well-known edemagenic condition (Mazzuca et al. 2019). The importance of this finding actually provides the simplest explanation of the well-known fact that pulmonary hypertension on hypoxia exposure is the consequence of a marked precapillary vasoconstriction whose significance is that of limiting microvascular filtration preventing lung edema.

It is worth recalling here that a similar theoretical micro-fluidic framework model applies to the lung interstitial compartment and to the pleural space (see Chap. 8, Sect. 8.5). In fact, in both compartments, a sub-atmospheric pressure assures a steady state equilibrium with a minimum amount of water. The considerable undertaking aiming to integrate interstitial fluid mechanics with tissue mechanics as well as the scaling of parameters and variables for the biophysical modeling have been recently validated and mentioned in the book *Biofluid mechanics* by Grotberg (2021).

The integrated mechanical and fluid dynamic analysis of lung water balance allowed to decipher the critical event leading to the development of edema, namely: the loss of integrity of the macromolecular proteoglycan assembly that weakens and ends up nullifying the *safety factor*. The recovery of fragments of proteoglycans from the interstitial compartments varies depending on the specific condition studied that differently involves either the proteoglycans of the capillary wall (controlling permeability) or of the interstitial compartment (proving mechanical resistance). Further, data suggest that there is a critical threshold of damage beyond which two events contribute to trigger the uncontrolled, tumultuous, and dramatic phase leading to severe edema (so-called *accelerated phase of lung edema*): an increase in tissue compliance leading to a sharp drop of interstitial pressure to zero (due to fragmentation of matrix proteoglycans) and an increase in microvascular permeability (due to fragmentation of endothelial proteoglycans). The consequence is an uncontrolled increase in capillary filtration and protein leak (Miserocchi et al. 2001a,b; Negrini et al. 2006, 2008). Several conditions lead to this event; proteases, commonly activated in severe inflammation processes (Miserocchi et al. 1999; Passi et al. 1998), oxygen radicals being produced in hypoxia and hyperoxia (Miserocchi et al. 1999, 2001a,b); remarkable alveolar over distension as during mechanical ventilation (Miserocchi et al. 1991; Moriondo et al. 2007; Passi et al. 1998). Lung edema is also a severe complication of post-thoracic surgery, whose pathogenesis is rooted on the over distension of the resected lung in the chest wall imposed by the post-operative draining system (Miserocchi et al. 2010; Salito et al. 2016, 2014). The problem of pathophysiology of pulmonary edema has been extensively reviewed concerning the role of different ventilatory strategies in the severely injured lung focusing on factors that either favor or hinder the progression of the disease, a problem raising remarkable controversies during the SARS-CoV-2 pandemic (Beretta et al. 2021).

Routine respiratory mechanics is not a reliable clinical tool for the early detection of lung injury. In fact, it has been suggested to rely on determination of viscoelastic properties of the lung. Marked regional structural inhomogeneity in viscoelastic properties have been reported as a consequence of mechanical ventilation (D'Angelo et al. 2002). Further, the estimate of impedenzometric indexes has been proposed in order to derive impedance (Z_{rs}), resistance (R_{rs}) and reactance (X_{rs}) by oscillating the respiratory system at different frequencies (f), as long as:

$$Z_{rs}(f) = R_{rs}(f) + X_{rs}(f) \tag{9.1}$$

It was found that X_{rs} shifted significantly to more negative value for a 10% increase in extravascular water; accordingly, this method proved to be sensitive enough to detect the early phase of developing edema (Dellacà et al. 2008). The use of impedenzometric indexes has been recently re-considered to estimate the mechanical properties of the respiratory system in COPD patients and decompensated heart failure (Terraneo et al. 2021). One cannot ignore that the lung is well equipped to signal the risk of developing edema; in fact, an afferent vagal input from the so-called lung J receptors, placed in the interstitial space (Paintal 1969) signals the increase in interstitial pressure that in awake subjects evokes rapid and shallow breathing (Fig. 9.4).

A recent report by D'Angelo et al. (2020) offered the chance to estimate the extension of validity of the safety factor when facing an increase in mechanical stress related to the increase in lung distension by increasing tidal volume during mechanical ventilation. Exceeding 60% of lung distension fairly compromised the safety factor leading to an exponential increase in lung water.

The recovery from edema critically depends upon the re-deposition of the interstitial matrix. The initial steps of this phase were extensively described (Palestini et al. 2002, 2003; Daffara et al. 2004; Botto et al. 2006, 2008) and the signaling-transduction through the activation of signal platforms on the surface of endothelial

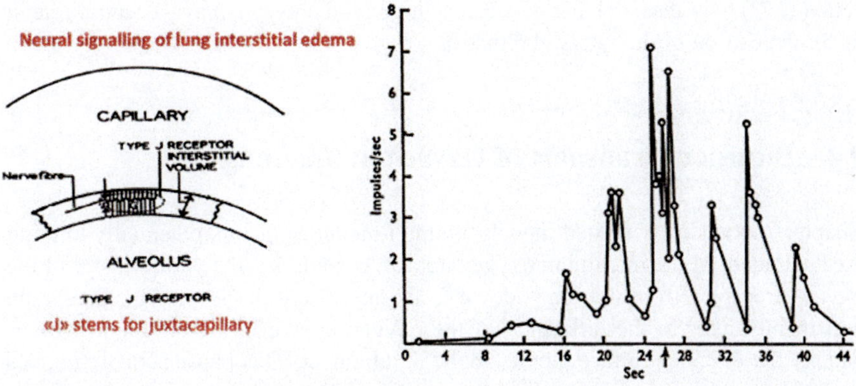

Fig. 9.4 In awake subjects, developing interstitial lung edema causes increased vagal afferents from J receptors placed in the lung interstitial space. Discharge from these receptors evokes rapid shallow breathing. Modified after Paintal 1969

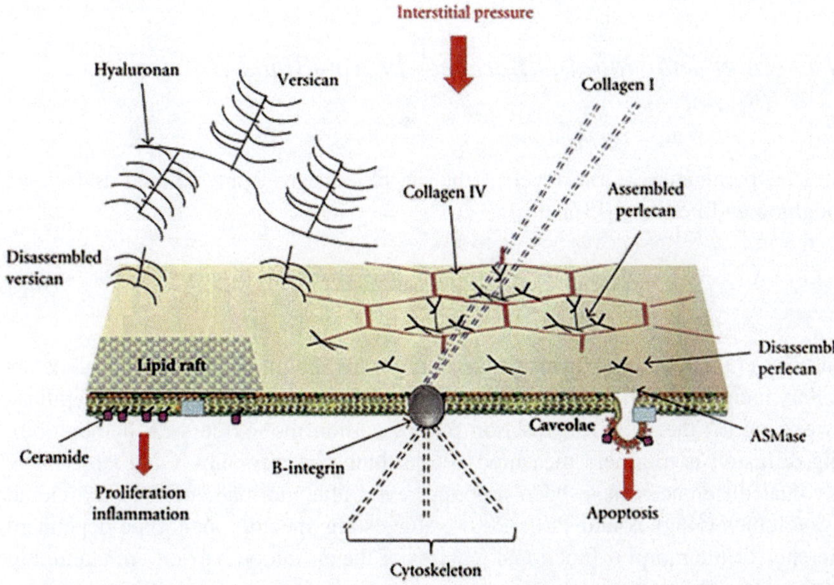

Fig. 9.5 Possible model for lung cellular response to an increase in extravascular water caused by proteoglycan fragmentation in order to trigger matrix re-deposition. Signaling platforms are the so-called "lipid rafts" and caveolae of the plasma membrane. From Palestini et al. 2011

cells evoked by changes in interstitial pressure is shown schematically in Fig. 9.5 (Palestini et al. 2011).

Specific gene expression related with the pro-inflammatory cytokines as well as to matrix re-deposition (Sabbadini et al. 2003) were also reported. Further, Passi et al. (1999) hypothesized that versican, a large proteoglycan, plays a central role in the re-deposition of the interstitial matrix.

9.4 Diffusion-Transport of Oxygen in the Lung

Chapter 7 extensively treated models aiming to identify factors potentially limiting the attainment of the maximum oxygen uptake, considering the various steps of the oxygen transport from the lungs down to the mitochondrial level. This paragraph will mainly refer to the efficiency of the oxygen uptake in the air–blood barrier, namely the first step to face the metabolic requirement. This phase actually implies two functionally related events: (1) oxygen diffusion across the alveolar membrane and (2) oxygen binding to hemoglobin in the blood flowing in the pulmonary capillaries. Accordingly, one is dealing with a coupled diffusion/perfusion mechanism.

9.4.1 Inter-Individual Differences in Air–Blood Barrier Phenotype

Diffusion parameters at the level of the air–blood barrier are defined as follows (Roughton and Forster 1957):

$$\frac{1}{D_LCO} = \frac{1}{D_m} + \frac{1}{\theta V_c} \tag{9.2}$$

being D_LCO the total lung diffusive capacity for carbon monoxide, D_m the alveolar-capillary membrane diffusing capacity, V_c the blood volume in the alveolar-capillary network, and θ the chemical reaction rate of carbon monoxide with hemoglobin. Lung diffusion is routinely measured at total lung capacity (TLC) and large inter-individual differences have been reported, even after normalization to individual lung volumes (Hughes and Pride 2001), suggesting specific phenotype dependent difference of the morpho-functional features of the air–blood barrier. An attempt to decipher such differences came from measurements of the diffusion parameters at different lung volumes from functional residual capacity (FRC) up to TLC (Miserocchi et al. 2008a). The interpretation of the data based on a numerical simulation allowed identification of two phenotypes concerning the morpho-functional features of the air–blood barrier that are characterized by the $\frac{V_c}{D_m}$ ratio, the distribution of which was found to be normal. It was proposed that a phenotype with low $\frac{V_c}{D_m}$ on the left tail of the distribution curve is characterized by a less developed capillary network and a relatively high number of small alveoli, while a phenotype with high $\frac{V_c}{D_m}$, towards the right tail, has a more extended capillary network and a lower number of larger alveoli.

9.4.2 Modeling the Alveolar-Capillary Equilibration

The next step was to envisage an estimate of the inter-individual differences in terms of efficiency of the diffusion/perfusion alveolar coupling, based on the model developed by Johannes Piiper (1924–2012) and Peter Scheid in Göttingen (Piiper and Scheid 1981). This model was grounded on the principle of mass conservation whereby, under steady state conditions, the mass transfer of diffused oxygen equals the uptake by the mixed venous blood affluent to the lung. The increase of the oxygen partial pressure along the capillary to reach a final equilibration value (L_{eq}) is exponential, with an exponent given by the ratio of diffusive capacitance for oxygen (D_LO_2) to perfusive capacitance, given by the product of cardiac output ($\dot{Q}$) times the oxygen binding capacity of hemoglobin (β_O):

$$L_{eq} = \frac{P_AO_2 - P_aO_2}{P_AO_2 - P_{\bar{v}}O_2} = e^{-\frac{D_LO_2}{\dot{Q}\,\beta_O}} \tag{9.3}$$

where P_AO_2, P_aO_2 and $P_{\bar{v}}O_2$ are the partial oxygen pressure in the alveoli and at the arterial (in fact, at the capillary end) and mixed venous sides of the lung capillary, respectively. In case $P_AO_2 = P_aO_2$ one has $L_{eq} = 0$, the case of maximum efficiency; on the other hand, if no oxygenation occurs, the numerator is equal to the denominator so that $L_{eq} = 1$ (the case of a shunt). Despite the fact that it is critically difficult to have reliable values of P_AO_2, P_aO_2 and $P_{\bar{v}}O_2$, it appears useful to rely on the ratio between diffusive and perfusive conductances for oxygen that can be measured. This approach was used to quantitate the inter-individual differences in alveolar-capillary oxygen equilibration by partitioning between a limitation due either to diffusive capacitance or to perfusive capacitance (Beretta et al. 2019). Equation 9.3 was also rewritten as:

$$L_{eq} = e^{-\frac{t_t}{\tau}} \tag{9.4}$$

where the transit time of the blood in pulmonary capillaries (t_t) is defined as

$$t_t = \frac{V_c}{\dot{Q}} \tag{9.5}$$

In Eq. (9.4), τ is the time constant of the equilibration process, which is given by

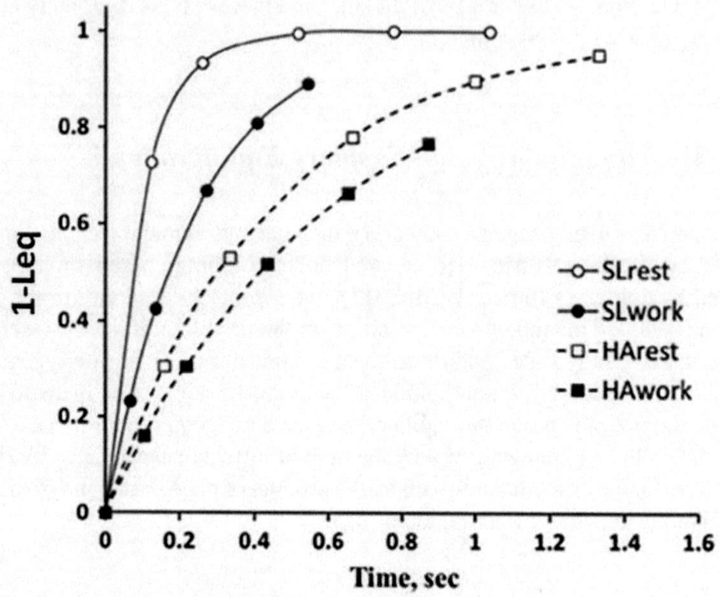

Fig. 9.6 Kinetics of alveolar-capillary oxygen equilibration in one subject in four different conditions of oxygen demand/supply. Slowing of the kinetics (increase in time constant) coupled with the shortening of blood transit time in the pulmonary capillaries, concur to prevent the complete alveolar-capillary equilibration. From Beretta et al. 2019

$$\tau = \frac{\beta_o V_c}{D_L O_2} \tag{9.6}$$

The analysis proved to be sensible enough to highlight inter-individual differences in alveolo-capillary equilibration among healthy subjects on increasing the oxygen demand.

The results showed that during work in hypoxia, the most demanding oxygen condition, $D_L O_2$ actually increased, and accordingly no diffusion limitation did occur. Conversely, the loss of efficiency of the equilibration process (increase in L_{eq}) strongly correlated with perfusion limitation reflecting both the shortening of t_t and the increase in τ: the former reflects the increase in $\dot{Q}$ and therefore the velocity of blood flow, the latter the fact that the increase in βo in hypoxia exceeds that of $D_L O_2$.

Figure 9.6 clearly shows that a slowing of the equilibration kinetics coupled with the shortening of transit time, prevents the complete alveolar-capillary equilibration (for simplicity, the ordinate is expressed as $1 - L_{eq}$).

Venous admixture, which includes the mismatch of ventilation to blood perfusion ($\dot{V}_A/\dot{Q}$) (Domino et al. 1993) and shunt (Lovering et al. 2006, 2008; Stickland et al. 2004) are well-known causes leading to inefficient alveolar-capillary oxygen equilibration. We present here a new reading of a shunt-like effect.

9.4.3 Alveolar Phenotype and Proneness to Develop Lung Edema

Thus, differences in air–blood barrier phenotype impact on the efficiency of gas exchange as well as on the control of extravascular lung water in edemagenic conditions. A phenotype with lower $\frac{V_c}{D_L O_2}$, implying a greater number of alveoli of smaller size (Miserocchi et al. 2008a), appears to be more efficient to guarantee gas exchange, as actually predicted by a theoretical morpho-functional model (Sapoval et al. 2002), and also appears more resistant to the development of lung edema.

Obviously, the same L_{eq} value is obtained regardless of the exponent being used, either $\frac{D_L O_2}{\dot{Q}\,\beta o}$ or $\frac{t_t}{\tau}$. These exponents provide different but complementary information concerning the functional response to the increase in oxygen demand in hypoxia. $\frac{D_L O_2}{\dot{Q}\,\beta o}$ is essentially the ratio between diffusion to perfusion capacitance. An decrease in 1-L_{eq} is expected on increasing oxygen demand, as the ratio $\frac{D_L O_2}{\dot{Q}\,\beta o}$ has to decrease, because perfusion capacitance increases about sixfold more than diffusion capacitance. In subjects where the increase in $D_L O_2$ is less pronounced, a greater increase in $\dot{Q}$ occurs, aiming to maintain oxygen delivery (lowering of $\frac{D_L O_2}{\dot{Q}\,\beta o}$). In these same subjects, there is a decrease of $\frac{t_t}{\tau}$ due to vasoconstriction and increase in cardiac output.

Further studies confirmed the inter-individual variability of the alveolar-capillary phenotype concerning in particular the lung vasoactive response to hypoxia (Beretta et al. 2017a,b; Bartesaghi et al. 2014). The modeling of alveolar-capillary equilibration in the specific case of hypoxia exposure is extensively treated in Chap. 10, Sect. 10.8.

A general interest concerning the efficiency of the oxygen pathway was also considered by the research group at Milano Bicocca, by extending the study from healthy people (Ferri et al. 2012, 2018) to a wide range of pathological states. These include the metabolic syndrome (Passoni et al. 2015), obesity (Salvadori et al. 1999), children recovering from lymphoblastic leukemia (Lanfranconi et al. 2014), and patients with amyotrophic lateral sclerosis (Lanfranconi et al. 2017a,b; Ferri et al. 2019).

The proneness to develop lung edema is known to vary greatly among subjects. This is well known, considering the exposure to hypobaric hypoxia (high altitude) that allowed to partition subjects in two groups: sensitive to HAPE (high-altitude pulmonary edema) and on the opposite Non-sensitive to HAPE (Busch et al. 2001; Dehnert et al. 2006; Eichstaedt et al. 2020a,b; Pham et al. 2012; Richalet et al. 2012; Richalet et al. 2021). One can therefore hypothesize inter-individual differences in the control of lung fluid balance. Finding reasons for these differences remained elusive for a long time. A line of research from Milano Bicocca afforded the problem by relating differences in control of lung fluid balance to individual air–blood barrier morpho-functional features (Miserocchi et al. 2008a). The latter could be related to individual values of the $\frac{V_c}{D_m}$ factor, the ratio between the membrane diffusion capacitance D_m (an index of extension of the alveolar surface and its diffusive properties) and capillary blood volume V_c (an index of extension of the alveolar-capillary network). The lung vasoactive response was estimated for subjects having either low or high $\frac{V_c}{D_m}$,ratio in edemagenic conditions, namely work and hypoxia exposure. Work is edemagenic due to increased lung blood flow (Hodges et al. 2007; McKenzie et al. 2005), and hypoxia is a well-known potent factor causing an increase in microvascular permeability to water and solutes (Hansen et al. 1994; Dehler et al. 2006).

The critical point was to verify the response of subjects with elevated $\frac{V_c}{D_m}$ ratio, who have a higher inborn extension of the alveolar-capillary network and a corresponding higher surface area for capillary filtration. It was found that, on exposure to edemagenic conditions, remarkable de-recruitment of pulmonary capillaries was found in subjects with high $\frac{V_c}{D_m}$, whereas minor de-recruitment, or actually some recruitment was documented in subjects with low $\frac{V_c}{D_m}$ (Bartesaghi et al. 2014; Beretta et al. 2017a,b). It was hypothesized that precapillary vasoconstriction in the high $\frac{V_c}{D_m}$ subjects represents a functional response aimed at limiting microvascular filtration, to prevent/attenuate edema formation reflecting a greater proneness to develop lung edema (Miserocchi et al. 2022). Notably, subjects being more prone to develop lung edema in hypoxia have a greater increase in pulmonary arterial pressure, in line with precapillary vasoconstriction. The clinical overlap of high-altitude pulmonary edema and pulmonary arterial hypertension was recently discussed in terms of genetic background (Eichstaedt et al. 2020a,b; Sharma et al. 2014). The hypothesis of a less efficient control of lung fluid balance in subjects with high $\frac{V_c}{D_m}$ was strengthened by the estimate of the mechanical properties of the respiratory system determined with the forced frequency oscillation technique. Relative to sea level at rest, the respiratory reactance of subjects with high $\frac{V_c}{D_m}$ decreased to a greater extent in hypoxia. Moreover, a fourfold increase in frequency dependence of respiratory resistance was found (Bartesaghi et al. 2014). Both results are considered indexes of greater perturbation of lung fluid balance (Dellacà et al. 2008).

In summary, defining the $\frac{V_c}{D_m}$ ratio through ambulatory pneumological evaluation may turn useful in order to define the patient's proneness to develop edema before an acute severe disease occurs (Miserocchi et al. 2022). There is a potential interest to consider people reaching high altitudes on trekking expeditions being exposed to the risk of HAPE. A clinical relevance may also be considered as cardio-pulmonary disorders as well as conditions of decrease in vascular bed (lung resection, thrombosis) are at risk of developing lung edema.

9.5 Fetal and Neonatal Respiration Physiology

Giacomo Meschia provided an impressive amount of work on the placenta considered as the "*baby's lung*"; Fig. 9.7 shows a schema of fetal circulation. Meschia's early studies considered the oxygen dissociation curves of maternal, fetal, and newborn blood (Meschia et al. 1961).

Further studies estimated the transplacental diffusion of inert molecules (thus not metabolized by the placenta nor bound in maternal and fetal blood) as a useful step towards the analysis of the exchange of respiratory gases between mother and fetus. The results suggested that the diffusion processes were either limited by blood flow or by placental permeability (Meschia et al. 1967a,b).

Concerning trans-placentar oxygen fluxes Meschia made estimates at different gestation age of: (1) oxygen partial pressure (PO_2) in blood samples drawn from chronic catheters in anesthetized ewes and their fetuses, (2) the rate of oxygen uptake by the umbilical circulation and (3) the diffusing capacity of the placenta for oxygen (Meschia et al. 1965). The results of the study showed that the quantity of oxygen

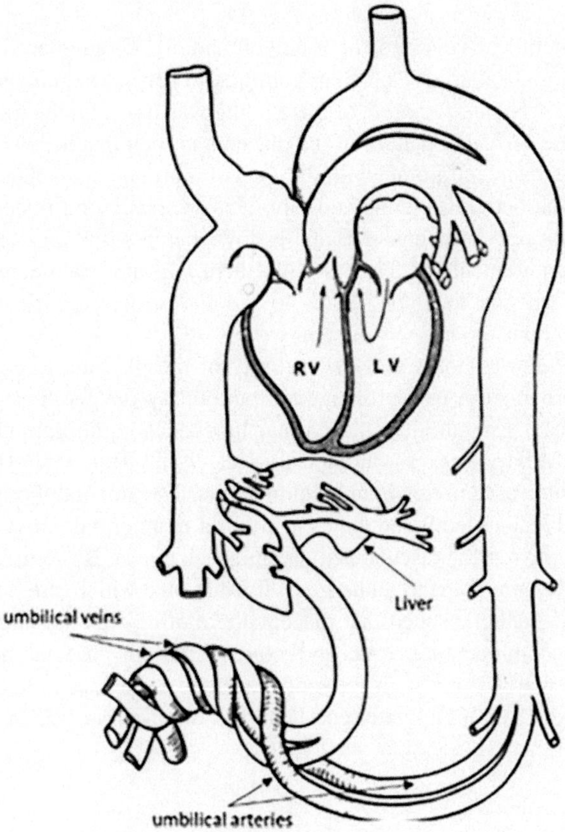

Fig. 9.7 Schema of fetal circulation. Arterial blood is carried via umbilical veins from the placenta, reaching the liver and the fetal venous circulation. Venous fetal blood returns to the placenta via umbilical arteries. Modified after Meschia 2011

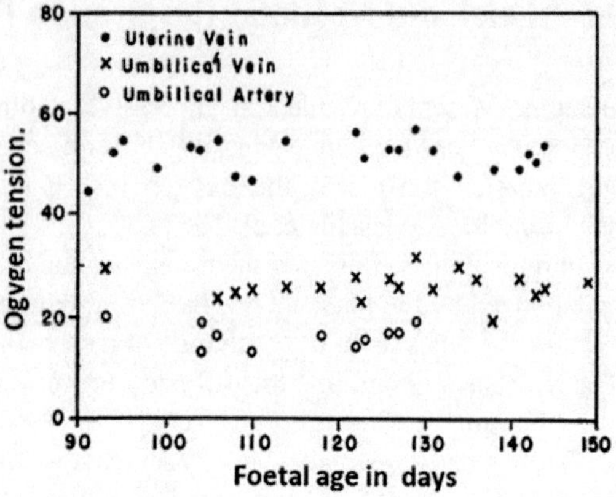

Fig. 9.8 PO_2 gradient from uterine artery to umbilical vein and between umbilical vein and umbilical artery. From Meschia et al. 1965

taken up by the umbilical blood and the diffusing capacity of the placenta for oxygen both increase with the fetal weight, the effect being attributed to the increase of the surface area of the trophoblast. Moreover, the associated maternal and fetal capillaries grow as gestation advances in the last trimester. The critical result from the above paper, as shown in Fig. 9.8, is the large PO_2 gradient from uterine artery to umbilical vein and the relatively small PO_2 gradient between umbilical vein and umbilical artery, a feature common to several mammals.

The interpretation of this finding was based on the following considerations: first, the perfusion pattern of the placenta is such that the PO_2 of umbilical venous blood tends to equilibrate with the PO_2 of uterine venous blood; secondly, PO_2 equilibration between uterine and umbilical venous blood is not attained because placental oxygen diffusing capacity is low, being however also relative low the oxygen requirements of fetus; thirdly, there is a moderate degree of shunting and uneven perfusion that contributes to the PO_2 difference between uterine and umbilical venous blood.

It was noted that the inability of the placenta to equilibrate maternal and fetal blood PO_2, puts placenta and pulmonary oxygen transport in sharp contrast as, in resting conditions, the normal lung does equilibrate alveolar and pulmonary end capillary PO_2, as outlined in Sect. 9.4.2. This contrast is likely to reflect a basic difference in functional requirements. The amount of oxygen that diffuses across the alveolar membrane varies many fold in a normal individual, as the individual shifts from rest to muscular activity and back to rest. By contrast, at any given fetal age, the oxygen demand undergoes fluctuations which are extremely small. Therefore, Meschia reasoned that placental evolution was not under the selective pressure of accommodating large and rapid variations in fetal oxygen flux. It was further remarked that high rates of placental perfusion, high oxygen affinity of fetal red cells and high relative perfusion of the tissues compensate for a relative inefficient placental exchange system (compared to the lung), due to the fact that fetal arterial blood is formed from a mixture between umbilical blood and venous maternal blood. Yet, under normal physiological conditions, fetal oxygenation provides an oxygen supply approximately twice the minimum level to sustain fetal life (Meschia 1985).

The point was also made concerning the model of gas exchange, either based on counter-current or on concurrent placental perfusion: the latter model appears more adequate to explain the experimental results (Meschia 1979).

A comparison was also done concerning the fetus oxygen carrying capacity, considering the difference in oxygen affinity (Wilkening et al. 1988). The conclusion was that difference in oxygen affinity was fully compensated for by differences in arterial oxygen saturation, by the perfusion rate of the uterus, and by the degree of PO_2 placental equilibration (Wilkening et al. 1988). The normal range of human umbilical venous PO_2 at term ranges from about 30 to 40 mmHg. Thus, the human placenta imposes normally on the term fetus an umbilical venous PO_2, which is in the 25–35 mmHg range, and fetal adaptation to this low PO_2, is a combination of increased oxygen affinity and increased hemoglobin concentration (*Everest* in utero, as Barcroft maintained!). Meschia (2011) also reported that alveolar ventilation increases in pregnancy, leading to a decrease in arterial carbon dioxide partial pressure ($PaCO_2$) to about 30 mmHg; the interpretation being that this prevents an increase in fetal PCO_2 that is always higher than the maternal one. Another important finding was that placenta has high metabolism and high rate of glucose utilization, but the bulk of the amino acids is transferred to the fetus (Meschia 2011).

Mechanical aspects of the first breath were studied by Agostoni et al. (1958). Experimental studies in guinea pig, cat, and goat showed that, on the first inspiration, a marked resistance ought to be overcome, due to (1) the high viscosity of the liquid present in the airways, (2) the surface tension of the air-liquid interphase, and (3) the fact that the expansion of the thorax is accomplished only by the inspiratory muscles and is not appreciably facilitated, as in the adult, by the tendency of the thorax to expand due to elastic forces.

Respiratory mechanics was also studied by Agostoni (1959) in dogs of different age from 1–3 up to 30 days of life and in the adult. In the newborn, the lung-thorax compliance, normalized to vital capacity, is larger than in the adult. It is however smaller during the first few days of life over the tidal volume range, due to still unexpanded alveoli. The compliance of the thorax is high at birth and decreases then progressively. The compliance of the lung increases after some days of life becoming similar to that of the adult. The FRC increases significantly during growth due to progressive recruitment of alveoli and airways closure on expiration (self-PEEP).

Most of Mortola's work was oriented towards respiratory physiology in newborn mammals, including inter-species comparative aspects and zoology. It was shown that the stiffening of the respiratory system gives more stability of the infant's breathing apparatus, providing a more ready volume response for any given change in driving pressure, at the cost however of a higher work of breathing (Fisher et al. 1982). Concerning the pattern of breathing, it is markedly variable immediately after birth, while from 10 to 90 min to a few days of life the tidal volume increases and respiratory rate decreases with no changes in minute ventilation; further, infants born vaginally behave similarly to those delivered by cesarean (Fisher et al. 1982). It was also remarked that a high chest wall/lung compliance ratio is an obligatory structural requirement in the newborn. In fact, a flexible chest wall is an essential prerequisite for the delivery through the birth canal. The goal to establish an FRC is mainly achieved through closure of the glottis and/or inspiratory muscle activity during expiration (Mortola 1983a).

An interesting study in newborn mammals of different size, from mice to piglets (Mortola 1983b) showed that the pattern of ventilation allowed coping for the metabolic requirement, as well as with the constraints of the mechanical properties of the newborn respiratory system. Considering respiratory energetics, the authors concluded that the breathing pattern adopted by newborn is the one that produces the adequate alveolar ventilation with minimal cost (as in the adult!). In a further study on infants (Mortola 1984), the data suggested that during tidal breathing the post-inspiratory muscle activity can substantially contribute to the increase in FRC although in most cases the final portion of expiration occurred by total muscle relaxation as in adults. In some infants, however, a braking mechanism, probably of laryngeal origin, further decreases the expiratory flow and may contribute to maintain a lung volume above FRC (Mortola et al. 1984).

In kittens, the frequency response of lung compliance, together with the relatively greater rate of stress relaxation, suggests that viscoelasticity contributes in newborn more than in adults to the dynamic stiffening of the lung (Sullivan and Mortola 1985). Mortola and Saetta (1987) proposed a method for measuring the respiratory system compliance in spontaneously-breathing infants, which circumvents the potential problems introduced by the breath-by-breath oscillations in the end-expiratory level (Mortola et al. 1993).

Mortola et al. (1999) also reported an interesting note referring to the newborn of a marsupial mouse, *Sminthopsis douglasi* (Fig. 9.9), a case where gas exchange in the first 5 days of life occurs primarily through the skin, which is hairless and highly vascularized.

The problem of the response to hypoxia was also considered (Mortola 1999). The most immediate response to acute hypoxia in newborn mammals is hyperventilation, like in the adult, accompanied however by a reduction in metabolic rate. This response relates to inhibition of thermogenesis in all its forms, shivering, non-shivering, and behavioral. Another study hypothesized that the circadian rhythms modifications of the metabolism and of body temperature found in newborn on exposure to hypoxia could also be at the origin of sleep disturbances in cardio-respiratory patients and at high altitude (Mortola and Seifert 2000). This hypothesis was confirmed by extending studies to humans (Bosco et al. 2003).

Studies performed on infants born at high altitude (Mortola et al. 1992) showed that the breathing pattern was deeper and slower at high altitude, the ventilatory oxygen extraction was higher (+50%) in the highlanders. Further, the ability to use a greater fraction of the inspired oxygen at high altitude was attributed to functional and structural alterations stimulated by fetal hypoxia.

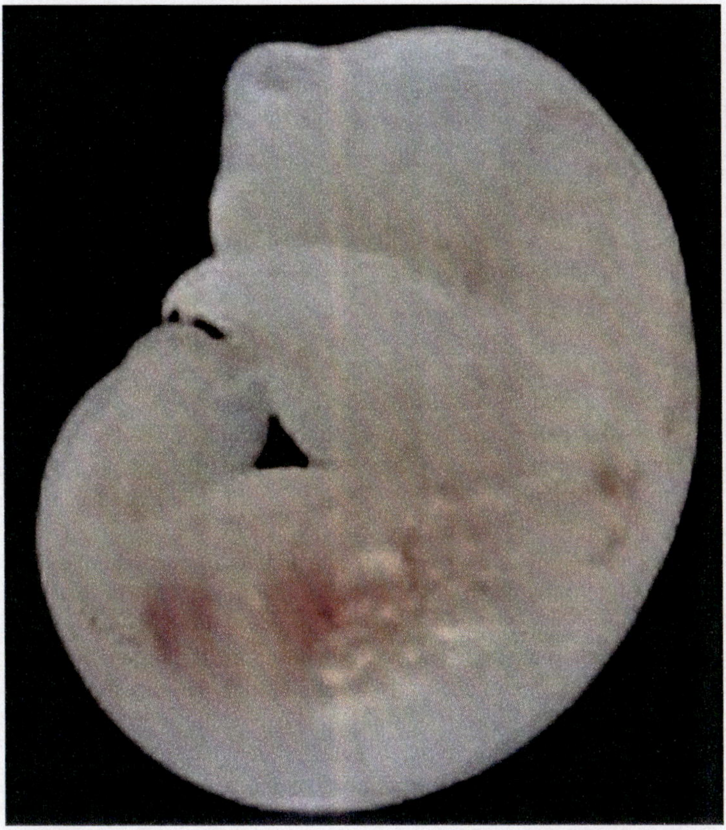

Fig. 9.9 Newborn from Australian marsupial mouse Sminthopsis douglasi, about 4-mm long, weighing 17 mg, where oxygen uptake occurs through the skin. Lungs are visible through the skin as air sacs. From Mortola et al. 1999

Identification of the mechanism of clearing lung interstitial water after birth contributed significantly to neonatal physiology. This mechanism implies two steps: the first one was elucidated by demonstrating an active sodium-dependent water transport by alveolar cells; the second, implying a water flow from the lung interstitium to the pulmonary capillaries, was still obscure. Relying on the trans-pleural micropuncture technique of the lung (Miserocchi et al. 1990), P_{ip} was measured in cesarean-delivered term rabbits at birth and at different postnatal times up to 16 days of life (Miserocchi et al. 1994). It was shown that P_{ip} was 0 cmH_2O, at birth and then increased remarkably peaking at about 2 h at 7 cmH_2O. The hypothesis was put forward that the increase in P_{ip} at 2 h reflects alveolar fluid reabsoprtion into a low compliant interstitial compartment; this increase in pressure would in turn provide a Starling-dependent fluid absorption into the pulmonary capillaries. A low compliance of the interstitial compartment in term newborn reflects the existence of a relatively mature extracellular matrix

(Miserocchi et al. 1993; Negrini et al. 1996). Wet/dry lung weight ratios decreased progressively indicating that most of water clearance occurred in the first postnatal hours in phase with the increase in P_{ip}. Interestingly, over time P_{ip} became sub-atmospheric, reaching a value of about −6 cmH_2O at 16 days of life, approaching the physiological value of −10 cmH_2O found in the adult (Miserocchi et al. 1993).

A similar study, carried out on premature newborn rabbits, showed that in the non-ventilated lung regions, P_{ip} remained at 0 cmH_2O, reflecting both a lack of active alveolar fluid absorption and/or a high compliance of the interstitial compartment; the second case can be explained with an immature mesh of the macromolecular matrix. In the ventilated regions of premature newborn P_{ip} increased but not as much as in the term newborn, and correspondingly, lung water clearance (as from the wet/dry ratio) was slowed down (Miserocchi et al. 1995). Surfactant administration, despite favoring alveolar expansion, failed to accelerate water clearance from the lung in premature human babies (Cattarossi et al. 2010); this finding indirectly stressed the important point that a physiological fluid balance in the preterm newborn can only be reached once a mature extracellular matrix is formed.

Interest was also addressed to a poorly documented phenomenon, namely the spontaneous contractions of the airways occurring from the early to late pseudoglandular stage of fetal life. It was found that, in fetal pig's lungs, the contractions propagated proximal to distal generated an intraluminal pressure (measured by micropuncture) in the range of 2–3 cmH_2O, higher than the pressure of the mesenchyme, which was close to zero (Schittny and Miserocchi 2000). It was proposed that the positive intraluminal pressure generated by the smooth muscles might contribute to act as a stimulus to promote alveolar growth into the surrounding mesenchyme.

9.6 The Air–Blood Barrier as Interface

9.6.1 Pollutants and Respiratory Reflexes

Sensory afferents from the lungs and their impact on the pattern of breathing was the major contribution from Sant'Ambrogio's work. The respiratory reflexes acting on the diaphragm and inspiratory intercostal muscle were described (Sant'Ambrogio and Widdicombe 1965) as well as the contribution of the diaphragm and of the other inspiratory muscles to the increase in lung volume during inspiration and to the force generated on inspiration (Mognoni et al. 1969).

Based on recording from vagal single fibers, Sant'Ambrogio was able to localize vagal receptors in the airways (Miserocchi et al. 1973; Bartlett Jr et al. 1976a). Figure 9.10 shows the frequency of discharge of a single vagal fiber from a stretch receptor in the lung on increasing lung volume (as indicated by lowering intra-esophageal pressure). The intrapulmonary localization of the receptor was defined following selective occlusion of bronchi, in this case localization was in the upper

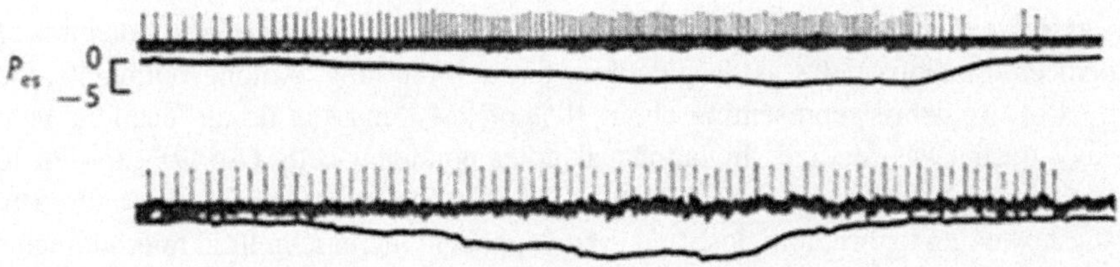

Fig. 9.10 Frequency of discharge of single vagal fiber from a stretch receptor. From Miserocchi et al. 1973

left bronchus: in fact, when the expansion of the bronchus was prevented, the receptor did not discharge.

Transduction properties of stretch receptor as well as their relationship with the mechanics of the trachea were also studied; it was found that receptors responded more to transverse than to longitudinal stretching of the wall of the trachea.

Furthermore, during pressure oscillations, receptors had a higher discharge frequency during the ascending phase of the pressure (simulating an inspiration) compared to a steady pressure (Bartlett Jr et al. 1976b). The contraction of airway smooth muscle brought about a remarkable increase in receptor discharge frequency (Bartlett Jr et al. 1976b; Mortola and Sant'Ambrogio 1979). Irritant receptors were also localized mostly in the upper segment of the trachea (Mortola et al. 1975). The general message from these studies is that stretch and irritant receptors convey, along vagal afferent fibers, information relative to the transmural pressure, the state of the bronchial smooth muscles and the presence of irritants on the mucosa.

Interestingly, Sant'Ambrogio and his group provided evidence for a potential vagus-mediated inspiratory-initiating effect from lung irritant receptors (Davies et al. 1981). Sant'Ambrogio gave also an exhaustive account on the role of vagal receptors to elicit cough and bronchoconstriction, which he considered as reflexes aiming at protecting the lung from inspired noxious agents. Both reflexes can be evoked from the larynx and the tracheobronchial tree. Within the airways, certain sites appear to be particularly sensitive to stimulation of cough (larynx and points of proximal airway branching), whereas bronchoconstriction can be triggered from the whole of the tracheobronchial tree. In the tracheobronchial tree and in the lung, slowly-adapting stretch receptors and irritant rapidly-adapting receptors have opposing effects on airway smooth muscle tone, the former mediating bronchodilation and the latter bronchoconstriction (Karlsson et al. 1988).

Sant'Ambrogio also went to debug the complex reflex response referred to as pulmonary chemoreflex, which includes the triad of bradycardia, hypotension, and apnea, and he concluded that vagal afferents from the lung would play an important role in this reflex (Sant'Ambrogio1996). This was the last scientific contribution from Giuseppe Sant'Ambrogio, who died prematurely.

The lung is by nature exposed to air pollutants with potential toxicity. Miserocchi's group at Università Milano Bicocca was involved in projects aiming to assess either the toxicity of nanoparticles as air pollutants or the potential of engineered nanoparticles as drug carriers through the lung. Among pollutants, the effect of tire debris representing about 10% of PM_{10} mass at urban locations with heavy traffic was studied. In human alveolar epithelial cells (A549) exposed to increasing extract concentrations (10, 60, and 75 mg ml^{-1}) for 24, 48, and 72 h, data showed up to threefold dose and time-dependent increase in lipid microdomains that represent sensible protein signaling platforms acting on the plasma membrane. In parallel, time- and dose-dependent toxic effects were documented at 48 h of exposure. Further, the increased leak of the cytosolic enzyme lactate dehydrogenase suggested a damage to plasma membrane representing a prenecrotic sign (Beretta et al. 2007). The effect of direct tire particle (in the range of 10 μm) instillation caused extensive inflammatory tissue infiltration, while 2.5 μm particles caused lysis of the alveolar barrier. Inflammatory cellular profiles in the broncho-alveolar lavage fluid showed dose-dependent responses compatible with strong cytotoxic effects (Mantecca et al. 2009, 2010). Further, for the first time, the translocation of inhaled asbestos fibers was modeled based on pulmonary and pleuro-pulmonary interstitial fluid dynamics (Miserocchi et al. 2008b).

The results of this study are briefly summarized here as they represent the only model so far available to explain the asbestos distribution in the body. Fibers can pass the alveolar barrier and reach the lung interstitium via the paracellular route down a water flow combined with active sodium absorption and to a hydraulic pressure gradient (lung interstitial pressure is sub-atmospheric). Fibers are dragged by pulmonary lymph flow (primary translocation), wherefrom they can reach the bloodstream and subsequently distribute to the whole body (secondary translocation). Secondary translocation to the pleural space may occur via the physiological route of pleural fluid formation across the parietal pleura; fibers accumulation in parietal pleura stomata (black spots) reflects indeed the role of parietal lymphatics in draining pleural fluid. Fibers translocation is a slow process developing over decades of life; it is aided by high biopersistence, by inflammation-induced increase in permeability, by low steric hindrance and by fibers motion pattern at incredibly low Reynolds number (~0.005). Ultrafine fibers (length < 5 μm, diameter < 0.25 μm) can travel larger distances due to low steric hindrance (in mesothelioma about 90% of fibers are ultrafine).

9.6.2 Drug Delivery Via Nanoparticles

Nanoparticles engineering and delivery have been the object of considerable scientific interest at Milano Bicocca and were supported by several European funding. Lungs and brain were considered as potential route for nanoparticles-mediated drug delivery, due to similar biophysical an morphological features of the air–blood barrier and of the blood-brain barrier. The alveolar surface may represent an important pathway for inhalation drug delivery, and is actually already routinely used to treat allergic, genetic, infective, or chronic diseases of the respiratory system. Only recently, though, the delivery to the lung has been considered for administering drugs not only for local pathology, but also for systemic diseases. The pulmonary route is attractive, as it is non-invasive and might allow drug absorption from a large highly vascularized surface area. This allows drugs reaching circulation avoiding the first-pass metabolism, which actually represents the main disadvantage of oral administration route. Concerning the blood-brain barrier, the point was made to try to overcome its limited penetration by functionalizing the surface of nanoparticles to achieve a site-specific targeting, and to allow prolonged release of the payload (Dal Magro et al. 2017). An attempt was made to use the intranasal route for nanoparticle-mediated brain delivery of oxcarbazepine, an anti-epileptic drug (Musumeci et al. 2018). Further, the problem of delivering drugs to prevent/treat Alzheimer disease was also afforded in transgenic Alzheimer mice model (Mancini et al. 2016). An attempt was also made to functionalize liposomes, aiming to overcome the permeability of the blood-brain barrier (Re et al. 2011; Salvati et al. 2013). Considerable interest focused on solid lipid nanoparticles (SLNs) that range in size between 50 and 1000 nm, and allow loading of hydrophobic drugs for potential time-controlled release.

The biophysical aspects of nanoparticle cellular uptake was a related interesting issue; this was made possible by developing a particle-tracking algorithm to optimize for low signal/noise images with a minimum set of requirements on the target size and with no a priori knowledge of the type of motion (Villa et al. 2010). This allowed exploring the feature of the intracellular traffic of nanoparticles (Rivolta et al. 2011c; Orlando et al. 2013). The uptake model allowed understanding a so far unexplained feature of the intracellular traffic, namely the accumulation of particulate in the perinuclear region. In fact, unlike the membrane crossing, this phenomenon occurs against a diffusion gradient. This process raises a potential interest for the biomedical use of delivery of drugs or genetic material close to the nuclear region. Figure 9.11 freezes an image of the time-dependent dye intracellular distribution with accumulation in the perinuclear region (Rivolta et al. 2011b).

The biophysical model coupled the initial diffusion-dependent phenomenon across the plasma membrane with the existence of an intracellular force field to drive the dye in the perinuclear region; this led to hypothesize that the force field relates to a hydraulic pressure gradient developing between the regions close to the plasma membrane and the perinuclear ones. The treatment with cytoD was found to affect the perinuclear accumulation indicating a major role of the cytoskeleton in sustaining the force field and the migration process.

The concept of "*latent toxicity*" was also developed (Panariti et al. 2012) based on the evidence that nanostructures not only passively interact with cells but also actively engage and mediate the molecular processes that are essential for regulating cell functions. Thus, the critical point was made that perturbed activities may or may not, late in time, reveal as stressful for the cells.

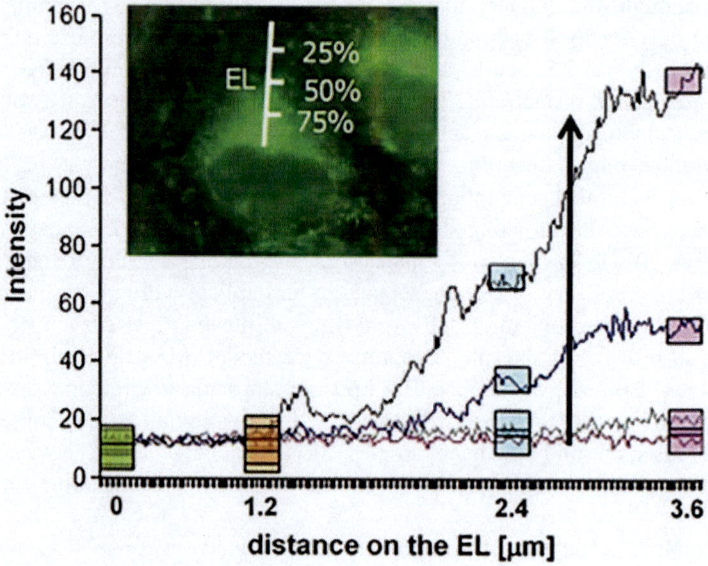

Fig. 9.11 The time-dependent intracellular distribution of solid lipid nanoparticles from the plasma membrane to the perinuclear region. From Rivolta et al. 2011b

References

Agostoni E (1959) Volume-pressure relationships of the thorax and lung in the newborn. J Appl Physiol 14:909–913

Agostoni E, Taglietti A, Agostoni AF, Setnikar I (1958) Mechanical aspects of the first breath. J Appl Physiol 13:344–348

Bartesaghi M, Beretta E, Pollastri L, Scotti V, Mandolesi G, Lanfranconi F, Miserocchi G (2014) Inter-individual differences in control of alveolar capillary blood volume in exercise and hypoxia. Respir Physiol Neurobiol 190:96–104

Bartlett D Jr, Jeffery P, Sant'Ambrogio G, Wise JC (1976a) Location of stretch receptors in the trachea and bronchi of the dog. J Physiol Lond 258:409–420

Bartlett D Jr, Sant'Ambrogio G, Wise JC (1976b) Transduction properties of tracheal stretch receptors. J Physiol Lond 258:421–432

Beretta E, Grasso GS, Forcaia G, Sancini G, Miserocchi G (2019) Differences in alveolo-capillary equilibration in healthy subjects on facing O_2 demand. Sci Rep 9:16693

Beretta E, Gualtieri M, Botto L, Palestini P, Miserocchi G, Camatini M (2007) Organic extract of tire debris causes localized damage in the plasma membrane of human lung epithelial cells. Toxicol Lett 173:191–200

Beretta E, Lanfranconi F, Grasso GS, Bartesaghi M, Alemayehu HK, Pratali L, Catuzzo B, Giardini G, Miserocchi G (2017a) Air blood barrier phenotype correlates with alveolo-capillary O2 equilibration in hypobaric hypoxia. Respir Physiol Neurobiol 246:53–58

Beretta E, Lanfranconi F, Grasso GS, Bartesaghi M, Alemayehu HK, Miserocchi G (2017b) Reappraisal of DLCO adjustment to interpret the adaptive response of the air-blood barrier to hypoxia. Respir Physiol Neurobiol 238:59–65

Beretta E, Romanò F, Sancini G, Grotberg JB, Nieman GF, Miserocchi G (2021) Pulmonary interstitial matrix and lung fluid balance from normal to the acutely injured lung. Front Physiol 12:781874

Bosco G, Ionadi A, Panico S, Faralli F, Gagliardi R, Data P, Mortola JP (2003) Effects of hypoxia on the circadian patterns in men. High Alt Med Biol 4:305–318

Botto L, Beretta E, Bulbarelli A, Rivolta I, Lettiero B, Leone BG, Miserocchi G, Palestini P (2008) Hypoxia-induced modifications in plasma membranes and lipid microdomains in A549 cells and primary human alveolar cells. J Cell Biochem 105:503–513

Botto L, Beretta E, Daffara R, Miserocchi G, Palestini P (2006) Biochemical and morphological changes in endothelial cells in response to hypoxic interstitial edema. Respir Res 7:7

Busch T, Bärtsch P, Pappert D, Grünig E, Hildebrandt W, Elser H, Falke KJ, Swenson ER (2001) Hypoxia decreases exhaled nitric oxide in mountaineers susceptible to high-altitude pulmonary edema. Am J Respir Crit Care Med 163:368–373

Cattarossi L, Copetti R, Poskurica B, Miserocchi G (2010) Surfactant administration for neonatal respiratory distress does not improve lung interstitial fluid clearance: echographic and experimental evidence. J Perinat Med 38:557–563

Conforti E, Fenoglio C, Bernocchi G, Bruschi O, Miserocchi G (2002) Morpho-functional analysis of lung tissue in mild interstitial edema. Am J Phys 282:L766–L774

D'Angelo E, Pecchiari M, Baraggia P, Saetta M, Balestro E, Milic-Emili J (2002) Low-volume ventilation causes peripheral airway injury and increased airway resistance in normal rabbits. J Appl Physiol 92:949–956

D'Angelo E, Koutsoukou A, Della Valle P, Gentile G, Pecchiari M (2020) The development of various forms of lung injury with increasing tidal volume in normal rats. Respir Physiol Neurobiol 274:103369

Daffara R, Botto L, Beretta E, Conforti E, Faini A, Palestini P, Miserocchi G (2004) Endothelial cells as early sensors of pulmonary interstitial edema. J Appl Physiol 97:1575–1583

Dal Magro R, Ornaghi F, Cambianica I, Beretta S, Re F, Musicanti C, Rigolio R, Donzelli E, Canta A, Ballarini E, Cavaletti G, Gasco P, Sancini G (2017) ApoE-modified solid lipid nanoparticles: a feasible strategy to cross the blood-brain barrier. J Control Release 249:103–110

Davies A, Sant'Ambrogio F, Sant'Ambrogio G (1981) Onset of inspiration in rabbit during artificial ventilation. J Physiol Lond 318:17–23

Dehler M, Zessin E, Bärtsch P, Mairbäurl H (2006) Hypoxia causes permeability edema in the constant-pressure perfused rat lung. Eur Respir J 27:600–606

Dehnert C, Risse F, Ley S, Kuder TA, Buhmann R, Puderbach M, Menold E, Mereles D, Kauczor HU, Bärtsch P, Fink C (2006) Magnetic resonance imaging of uneven pulmonary perfusion in hypoxia in humans. Am J Respir Crit Care Med 174:1132–1138

Dellacà RL, Zannin E, Sancini G, Rivolta I, Leone BE, Pedotti A, Miserocchi G (2008) Changes in the mechanical properties of the respiratory system during the development of interstitial lung edema. Respir Res 9:51

Domino KB, Eisenstein BL, Tran T, Hlastala MP (1993) Increased pulmonary perfusion worsens ventilation-perfusion matching. Anesthesiology 79:817–826

Eichstaedt CA, Benjamin N, Grünig E (2020a) Genetics of pulmonary hypertension and high-altitude pulmonary edema. J Appl Physiol 128:1432–1438

Eichstaedt CA, Mairbäurl H, Song J, Benjamin N, Fischer C, Dehnert C, Schommer K, Berger MM, Bärtsch P, Grünig E, Hinderhofer K (2020b) Genetic predisposition to high-altitude pulmonary edema. High Alt Med Biol 21:28–36

Ferri A, Adamo S, La Torre A, Marzorati M, Bishop DJ, Miserocchi G (2012) Determinants of performance in 1,500-m runners. Eur J Appl Physiol 2012(112):3033–3043

Ferri A, Lanfranconi F, Corna G, Bonazzi R, Marchese S, Magnoni A, Tremolizzo L (2019) Tailored exercise training counteracts muscle disuse and attenuates reductions in physical function in individuals with amyotrophic lateral sclerosis. Front Physiol 26(10):1537

Ferri A, Panariti A, Miserocchi G, Rocchetti M, Buoli Comani G, Rivolta I, Bishop DJ (2018) Tissue specificity of mitochondrial adaptations in rats after 4 weeks of normobaric hypoxia. Eur J Appl Physiol 118:1641–1652

Fisher JT, Mortola JP, Smith JB, Fox GS, Weeks S (1982) Respiration in newborns: development of the control of breathing. Am Rev Respir Dis 25:650–657

Grotberg JM (2021). Biofluid mechanics.. Cambridge Texts in Biomedical Engineering
Hansen JM, Olsen NV, Feldt-Rasmussen B, Kanstrup IL, Déchaux M, Dubray C, Richalet P (1994) Albuminuria and overall capillary permeability of albumin in acute altitude hypoxia. J Appl Physiol 76:1922–1927
Hodges AN, Sheel AW, Mayo JR, McKenzie DC (2007) Human lung density is not altered following normoxic and hypoxic moderate-intensity exercise: implications for transient edema. J Appl Physiol 103:111–118
Hughes JM, Pride NB (2001) In defence of the carbon monoxide transfer coefficient Kco (TL/VA). Eur Respir J 17:168–174
Karlsson JA, Sant'Ambrogio G, Widdicombe J (1988) Afferent neural pathways in cough and reflex bronchoconstriction. J Appl Physiol 65:1007–1023
Lanfranconi F, Ferri A, Corna G, Bonazzi R, Lunetta C, Silani V, Riva N, Rigamonti A, Maggiani A, Ferrarese C, Tremolizzo L (2017a) Inefficient skeletal muscle oxidative function flanks impaired motor neuron recruitment in amyotrophic lateral sclerosis during exercise. Sci Rep 7(7):2951
Lanfranconi F, Pollastri L, Corna G, Bartesaghi M, Novarina M, Ferri A, Miserocchi G (2017b) The elusive path of brain tissue oxygenation and cerebral perfusion in harness hang syncope in mountain climbers. High Alt Med Biol 18:363–371
Lanfranconi F, Pollastri L, Ferri A, Fraschini D, Masera G, Miserocchi G (2014) Near infrared spectroscopy (NIRS) as a new non-invasive tool to detect oxidative skeletal muscle impairment in children survived to acute lymphoblastic leukaemia. PLoS One 9:e99282
Lovering AT, Romer LM, Haverkamp HC, Pegelow DF, Hokanson JS, Eldridge MW (2008) Intrapulmonary shunting and pulmonary gas exchange during normoxic and hypoxic exercise in healthy humans. J Appl Physiol 104:1418–1425
Lovering AT, Stickland MK, Eldridge MW (2006) Intrapulmonary shunt during normoxic and hypoxic exercise in healthy humans. Adv Exp Med Biol 588:31–45
Mancini S, Minniti S, Gregori M, Sancini G, Cagnotto A, Couraud PO, Ordóñez-Gutiérrez L, Wandosell F, Salmona M, Re F (2016) The hunt for brain Aβ oligomers by peripherally circulating multi-functional nanoparticles: potential therapeutic approach for Alzheimer disease. Nanomedicine 12:43–52
Mantecca P, Farina F, Moschini E, Gallinotti D, Gualtieri M, Rohr A, Sancini G, Palestini P, Camatini M (2010) Comparative acute lung inflammation induced by atmospheric PM and size-fractionated tire particles. Toxicol Lett 198:244–254
Mantecca P, Sancini G, Moschini E, Farina F, Gualtieri M, Rohr A, Miserocchi G, Palestini P, Camatini M (2009) Lung toxicity induced by intratracheal instillation of size-fractionated tire particles. Toxicol Lett 189:206–214
Mazzuca E, Aliverti A, Miserocchi G (2016) Computational micro-scale model of control of extravascular water and capillary perfusion in the air blood barrier. J Theor Biol 400:42–51
Mazzuca E, Aliverti A, Miserocchi G (2019) Understanding vasomotion of lung microcirculation by in vivo imaging. J Imaging 5:22
McKenzie DC, O'Hare TJ, Mayo J (2005) The effect of sustained heavy exercise on the development of pulmonary edema in trained male cyclists. Respir Physiol Neurobiol 145:209–218
Meschia G (1979) Supply of oxygen to the fetus. J Reprod Med 23:160–165
Meschia G (1985) Safety margin of foetal oxygenation. J Reprod Med 30:308–311
Meschia G (2011) Foetal oxygenation and maternal ventilation. Clin Chest Med 32:15–19
Meschia G, Battaglia FC, Bruns PD (1967a) Theoretical and experimental study of transplacental diffusion. J Appl Physiol 22:1171–1178
Meschia G, Cotter JR, Makowski EL, Barron DH (1967b) Simultaneous measurement of uterine and umbilical blood flows and oxygen uptakes. Quart J Exp Physiol 52:1–18
Meschia G, Cotter JR, Caoimhghin S, Breathnach CS, Barron DH (1965) The diffusibility of oxygen across the sheep placenta. Q J Exp Physiol Cogn Med Sci 50:466–480

Meschia G, Hellegers A, Blechner JN, Wolkoff AS, Barron DH (1961) A comparison of the oxygen dissociation curves of the bloods of maternal, foetal and newborn sheep at various pHs. Q J Exp Physiol Cogn Med Sci 46:95–100
Miserocchi G (2009) Mechanisms controlling the volume of pleural fluid and extravascular lung water. Eur Respir Rev 18:244–252
Miserocchi G, Beretta E, Rivolta I (2010) Respiratory mechanics and fluid dynamics after lung resection surgery. Thorac Surg Clin 20:345–357
Miserocchi G, Beretta E, Rivolta I, Bartesaghi E (2022) Role of the air-blood barrier phenotype in lung oxygen uptakeand control of extravascular water. Front Physiol
Miserocchi G, Messinesi G, Tana F, Passoni E, Adamo S, Romano R, Beretta E (2008a) Mechanisms behind inter-individual differences in lung diffusing capacity. Eur J Appl Physiol 102: 561–568
Miserocchi G, Mortola J, Sant'Ambrogio G (1973) Localization of pulmonary stretch receptors in the airways of the dog. J Physiol Lond 235:775–782
Miserocchi G, Negrini D, Del Fabbro M, Venturoli D (1993) Pulmonary interstitial pressure in intact in situ lung: transition to interstitial edema. J Appl Physiol 74:1171–1177
Miserocchi G, Negrini D, Gonano C (1990) Direct measurement of interstitial pulmonary pressure in in-situ lung with intact pleural space. J Appl Physiol 69:2168–2174
Miserocchi G, Negrini D, Gonano C (1991) Parenchymal stress affects interstitial and pleural pressures in in situ lung. J Appl Physiol 71:1967–1972
Miserocchi G, Negrini D, Passi A, De Luca G (2001a) Development of lung edema: interstitial fluid dynamics and molecular structure. News Physiol Sci 16:66–71
Miserocchi G, Passi A, Albertini R, Negrini D, De Luca G (1999) Interstitial pressure and proteoglycan degradation in hydraulic- and elastase-induced lung edema. Chest 116:31S
Miserocchi G, Passi A, Negrini D, Del Fabbro M, De Luca GC (2001b) Pulmonary interstitial pressure and tissue matrix structure in acute hypoxia. Am J Physiol Lung 280:L881–L887
Miserocchi G, Poskurica BH, Del Fabbro M (1994) Pulmonary interstitial pressure in anesthetized paralyzed newborn rabbits. J Appl Physiol 77:2260–2268
Miserocchi G, Poskurica BH, del Fabbro M, Crisafulli B (1995) Pulmonary interstitial pressure in premature rabbits. Respir Physiol 102:239–249
Miserocchi G, Rivolta I (2012) Mechanistic considerations on the development of lung edema: vascular, perivascular and molecular aspects from early stage to tissue and vascular remodeling stage. Current Resp Med Rev 8:82–89
Miserocchi G, Sancini G, Mantegazza F, Chiappino G (2008b) Translocation pathways for inhaled asbestos fibers. Environ Health 24(7):4
Mognoni P, Saibene F, Sant'Ambrogio G (1969) Contribution of the diaphragm and the other inspiratory muscles to different levels of tidal volume and static inspiratory effort in the rabbit. J Physiol 202:517–534
Moriondo A, Pelosi P, Passi A, Viola M, Marcozzi C, Severgnini P, Ottani V, Quaranta M, Negrini D (2007) Proteoglycan fragmentation and respiratory mechanics in mechanically ventilated healthy rats. J Appl Physiol 103:747–756
Mortola JP (1983a) Some functional mechanical implications of the structural design of the respiratory system in newborn mammals. Am Rev Respir Dis 128:S69–S72
Mortola JP (1983b) Comparative aspects of the dynamics of breathing in newborn mammals. J Appl Physiol 54:1229–1235
Mortola JP (1984) Breathing pattern in newborns. J Appl Physiol 56:1533–1540
Mortola JP (1999) How newborn mammals cope with hypoxia. Respir Physiol 116:95–103
Mortola JP, Frappell PB, Frappell DE, Villena-Cabrera N, Villena-Cabrera M, Peña F (1992) Ventilation and gaseous metabolism in infants born at high altitude, and their responses to hyperoxia. Am Rev Respir Dis 146:1206–1209
Mortola JP, Frappell PB, Woolley PA (1999) Breathing through skin in a newborn mammal. Nature 397:660

Mortola JP, Hemmings G, Matsuoka T, Saiki C, Fox G (1993) Referencing lung volume for measurements of respiratory system compliance in infants. Pediatr Pulmonol 16:248–253
Mortola JP, Milic-Emili J, Noworaj A, Smith B, Fox G, Weeks S (1984) Muscle pressure and flow during expiration in infants. Am Rev Respir Dis 129:49–53
Mortola JP, Saetta M (1987) Measurements of respiratory mechanics in the newborn: a simple approach. Pediatr Pulmonol 3:123–130
Mortola JP, Sant'Ambrogio G (1979) Mechanics of the trachea and behaviour of its slowly adapting stretch receptors. J Physiol Lond 286:577–590
Mortola J, Sant'Ambrogio G, Clement MG (1975) Localization of irritant receptors in the airways of the dog. Respir Physiol 24:107–114
Mortola JP, Seifert EL (2000) Hypoxic depression of circadian rhythms in adult rats. J Appl Physiol 88:365–368
Musumeci T, Serapide MF, Pellitteri R, Dalpiaz A, Ferraro L, Dal Magro R, Bonaccorso A, Carbone C, Veiga F, Sancini G, Puglisi G (2018) Oxcarbazepine free or loaded PLGA nanoparticles as effective intranasal approach to control epileptic seizures in rodents. Eur J Pharm Biopharm 133:309–320
Negrini D, Candiani A, Boschetti F, Crisafulli B, Del Fabbro M, Bettinelli D, Miserocchi G (2001) Pulmonary microvascular and perivascular interstitial geometry during development of mild hydraulic edema. Am J Phys 281:L1464–L1471
Negrini D, Passi A, de Luca G, Miserocchi G (1996) Pulmonary interstitial pressure and proteoglycans during development of pulmonary edema. Am J Phys 270:H2000–H2007
Negrini D, Passi A, Moriondo A (2008) The role of proteoglycans in pulmonary edema development. Intensive Care Med 34:610–618
Negrini D, Tenstad O, Passi A, Wiig H (2006) Differential degradation of matrix proteoglycans and edema development in rabbit lung. Am J Phys 290:L470–L477
Orlando A, Re F, Sesana S, Rivolta I, Panariti A, Brambilla D, Nicolas J, Couvreur P, Andrieux K, Masserini M, Cazzaniga E (2013) Effect of nanoparticles binding beta-amyloid peptide on nitric oxide production by cultured endothelial cells and macrophages. Int J Nanomedicine 8:1335–1347
Paintal AS (1969) Mechanism of stimulation of type J pulmonary receptors. J Physiol Lond 203: 511–532
Palestini P, Botto L, Rivolta I, Miserocchi G (2011) Remodelling of membrane rafts expression in lung cells as an early sign of mechanotransduction signalling in pulmonary edema J Lipids:695369
Palestini P, Calvi C, Conforti E, Botto L, Fenoglio C, Miserocchi G (2002) Composition, biophysical properties and morphometry of plasma membranes in pulmonary interstitial edema. Am J Phys 282:L1382–L1390
Palestini P, Calvi C, Conforti E, Daffara R, Botto L, Miserocchi G (2003) Compositional changes in lipid microdomains of air-blood barrier plasma membranes in pulmonary interstitial edema. J Appl Physiol 95:1446–1452
Panariti A, Miserocchi G, Rivolta I (2012) The effect of nanoparticle uptake on cellular behavior: disrupting or enabling functions? Nanotechnol Sci Appl 5:87–100
Passi A, Negrini D, Albertini R, De Luca G, Miserocchi G (1998) Involvement of lung interstitial proteoglycans in development of hydraulic- and elastase-induced edema. Am J Phys 275:L631–L635
Passi A, Negrini D, Albertini R, Miserocchi G, De Luca G (1999) The sensitivity of versican from rabbit lung to gelatinase a (MMP-2) and B (MMP-9) and its involvement in the development of hydraulic lung edema. FEBS Lett 456:93–96
Passoni E, Lania A, Adamo S, Grasso GS, Noè D, Miserocchi G, Beretta E (2015) Mild training program in metabolic syndrome improves the efficiency of the oxygen pathway. Respir Physiol Neurobiol 208:8–14

Pham I, Würzner G, Richalet JP, Peyrard S, Azizi M (2012) Bosentan effects in hypoxic pulmonary vasoconstriction: preliminary study in subjects with or without high altitude pulmonary edema-history. Pulm Circ 2:28–33

Piiper J, Scheid P (1981) Model for capillary-alveolar equilibration with special reference to O_2 uptake in hypoxia. Respir Physiol 46:193–208

Re F, Cambianica I, Sesana S, Salvati E, Cagnotto A, Salmona M, Couraud PO, Moghimi SM, Masserini M, Sancini G (2011) Functionalization with ApoE-derived peptides enhances the interaction with brain capillary endothelial cells of nanoliposomes binding amyloid-beta peptide. J Biotechnol 156:341–346

Richalet JP, Larmignat P, Poitrine E, Letournel M, Canouï-Poitrine F (2012) Physiological risk factors for severe high-altitude illness: a prospective cohort study. Am J Respir Crit Care Med 185:192–198

Richalet JP, Pillard FLE, Moal D, Rivière D, Oriol P, Poussel M, Chenuel B, Doutreleau S, Vergès S, Demanez S et al (2021) Validation of a score for the detection of subjects with high risk for severe high-altitude illness. Med Sci Sports Exerc 53:1294–1302

Rivolta I, Lucchini V, Rocchetti M, Kolar F, Palazzo F, Zaza A, Miserocchi G (2011a) Interstitial pressure and lung edema in chronic hypoxia. Eur Respir J 37:943–949

Rivolta I, Panariti A, Collini M, Lettiero B, D'Alfonso L, Sironi L, Miserocchi G, Chirico G (2011b) A biophysical model of intracellular distribution and perinuclear accumulation of particulate matter. Biophys Chem 158:134–140

Rivolta I, Panariti A, Lettiero B, Sesana S, Gasco P, Gasco MR, Masserini M, Miserocchi G (2011c) Cellular uptake of coumarin-6 as a model drug loaded in solid lipid nanoparticles. J Physiol Pharmacol 62(1):45–53

Roughton FJ, Forster RE (1957) Relative importance of diffusion and chemical reaction rates in determining rate of exchange of gases in the human lung with special reference to true diffusing capacity of pulmonary membrane and volume of blood in the lung capillaries. J Appl Physiol 11:290–302

Sabbadini M, Barisani D, Conforti E, Marozzi A, Ginelli E, Miserocchi G, Meneveri R (2003) Gene expression analysis in interstitial lung edema induced by saline infusion. Biochim Biophys Acta 1638:149–156

Salito C, Bovio D, Mazzuca E, Rivolta I, Brunelli A, Aliverti A, Miserocchi G (2014) Experimental model to evaluate the effect of hydrothorax and lobar resection on lung compliance. Eur J Cardiothorac Surg 45:489–495. discussion 495

Salito C, Bovio D, Orsetti G, Salati M, Brunelli A, Aliverti A, Miserocchi G (2016) Effect **of** lung resection **on** pleuro- pulmonary mechanics **and** fluid balance. Respir Physiol Neurobiol 221: 35–40

Salvadori A, Fanari P, Fontana M, Buontempi L, Saezza A, Baudo S, Miserocchi G, Longhini E (1999) Oxygen uptake and cardiac performance in obese and normal subjects during exercise. Respiration 66:25–33

Salvati E, Re F, Sesana S, Cambianica I, Sancini G, Masserini M, Gregori M (2013) Liposomes functionalized to overcome the blood-brain barrier and to target amyloid-β peptide: the chemical design affects the permeability across an in vitro model. Int J Nanomedicine 8:1749–1758

Sant'Ambrogio G (1996) Pulmonary chemoreflex elicited by intravenous injection of lactic acid in anesthetized rats. J Appl Physiol 81:2347–2348

Sant'Ambrogio G, Widdicombe JG (1965) Respiratory reflexes acting on the diaphragm and inspiratory intercostal muscle of the rabbit. J Physiol Lond 180:766–779

Sapoval B, Filoche M, Weibel ER (2002) Smaller is better–but not too small: a physical scale for the design of the mammalian pulmonary acinus. Proc Natl Acad Sci U S A 99:10411–10416

Schittny JC, Miserocchi G (2000) Sparrow MP. Spontaneous peristaltic airway contractions propel lung liquid through the bronchial tree of intact and foetal lung explants Am J Respir Cell Mol Biol 23:11–18

Sharma M, Singh SB, Sarkar S (2014) Genome wide expression analysis suggests perturbation of vascular homeostasis during high altitude pulmonary edema. PLoS One 9:e85902

Stickland MK, Welsh RC, Haykowsky MJ, Petersen SR, Anderson WD, Taylor DA, Bouffard M, Jones RL (2004) Intra-pulmonary shunt and pulmonary gas exchange during exercise in humans. J Physiol 561:321–329

Sullivan KJ, Mortola JP (1985) Dynamic lung compliance in newborn and adult cats. J Appl Physiol 60:743–750

Terraneo S, Rinaldo RF, Sferrazza Papa GF, Ribolla F, Gulotta C, Maugeri L, Gatti E, Centanni S, Di Marco F (2021) Distinct mechanical properties of the respiratory system evaluated by forced oscillation technique in acute exacerbation of COPD and acute decompensated heart failure. Diagnostics 11:554

Villa CE, Caccia M, Sironi L, D'Alfonso L, Collini M, Rivolta I, Miserocchi G, Gorletta T, Zanoni I, Granucci F, Chirico G (2010) Accumulative difference image protocol for particle tracking in fluorescence microscopy tested in mouse lymphonodes. PLoS One 5(8):e12216

Wilkening RB, Molina RD, Meschia G (1988) Placental oxygen transport in sheep with different hemoglobin types. Am J Phys 254:R585–R589

Chapter 10
A School Goes to Altitude

Guido Ferretti and Giuseppe Miserocchi

Abstract This chapter concerns the contribution of the School of Milano to the study of human responses to altitude exposure. Paolo Cerretelli's contributions are firstly described. The studies performed during the expedition to Mount Kanjut-Sar, Karakorum, in 1957, and during the Italian Expedition to Mount Everest in 1973 are reported. This expedition generated a highly celebrated article, on the factors limiting oxygen transport on Mount Everest. The subsequent remarkable work on structural and functional muscle adaptation to altitude, performed along the Cerretelli–Hoppeler (Bern) axis is described. Then, Reinhold Messner attained the summit of Mount Everest without supplementary oxygen. Several projects devoted to understand Messner's achievement (AMREE, by John West; the Messner's study, set up by Oswald Oelz with Hoppeler and Cerretelli; Operation Everest II) are analyzed. Finally, Cerretelli's work at the new Italian laboratory close to the Everest basecamp (the Pyramid) is reported. Secondly, the studies coordinated by Giuseppe Miserocchi concerning the complex interaction between lung diffusion, alveolar-capillary blood volume, and lung water balance to affect the kinetics of alveolar-capillary equilibration on increasing oxygen demand is developed. Further, inter-individual differences in the proneness to develop pulmonary edema at altitude are related to specific morpho-functional features of the alveolar-capillary network.

G. Ferretti (✉)
Department of Molecular and Translational Medicine, University of Brescia, Brescia, Italy
e-mail: Guido.Ferretti@unibs.it

G. Miserocchi
University of Milano Bicocca, Milan, Italy
e-mail: Giuseppe.Miserocchi@unimib.it

G. Ferretti (ed.), *Exercise, Respiratory and Environmental Physiology*, Perspectives in Physiology, https://doi.org/10.1007/978-3-031-19197-8_10

Mount Everest, on the way in the Khumbu Valley. Pastel by Giuseppe Miserocchi after a picture, which he took in 2011. With the kind permission of Giuseppe Miserocchi

10.1 Introduction

Chapters 10–12 illustrate the contributions of the School of Milano to our understanding of human adaptation to special environments: specifically and respectively, altitude, microgravity, and diving. Although this is not the core of the scientific activity of the School, special environments inevitably attracted many of us. Paolo Cerretelli and Giuseppe Miserocchi were particularly active at altitude; Pietro Enrico di Prampero, Guido Ferretti, and Carlo Capelli in microgravity and space physiology; Guido Ferretti in breath-hold diving. Significant contributions came also from Giovanni Cavagna, Bruno Grassi, Claudio Marconi, Arsenio Veicsteinas (1944–2017), Guglielmo Antonutto, and Paola Zamparo.

This chapter is devoted to altitude. This is a long-lasting story, which started well before Margaria. Paul Bert (1833–1886) carried out the first systematic studies of the effects of hypoxia on the human body in his pneumatic chamber in Paris (Bert 1878). As discussed in detail in Chap. 1, Angelo Mosso (1846–1910) promoted several studies on human responses and adaptation to altitude exposure, along the track traced by Bert. These were carried out in two of the most historical altitude laboratories worldwide: the Capanna Regina Margherita (m. 4560 above sea level) and the *Istituto Scientifico Internazionale*, at the Col d'Olen (m. 2900 above sea level), both on the Mounte Rosa massif, in the Pennine Alps (Cogo et al. 2000). Mosso's book entitled *La Fisiologia dell'Uomo sulle Alpi* (The Physiology of Man on the Alps), in which he summarized his main results, has become one of the most classic textbooks on altitude physiology (Mosso 1897). Nathan Zuntz (1847–1920), a close friend to Mosso, and his coworkers made also wide use of the same facilities for their altitude studies (Zuntz et al. 1906).

Mosso's pupil and successor Amedeo Herlitzka (1872–1949) followed in part the tracks of Mosso and made wide use of the permanent laboratory at Col d'Olen. However, his interest in altitude was subsidiary to his interest in aviation medicine, in so far as his aim was to understand the responses of humans to hypoxia, since aircrafts were not pressurized in those days (see Chap. 1).

After World War II, Rodolfo Margaria considered altitude an interesting subsidiary field of research, which could be useful to understand some phenomena related to the energetics of muscular exercise, especially concerning oxygen delivery to working muscles. He was generally not directly involved himself, but he favored his pupils, especially Paolo Cerretelli. In 1954, Ardito Desio (1897–2001), a Professor of Geology at the University of Milano, organized the famous Italian expedition to K2, in Karakorum. The expedition was successful, since it ended with the first men, Achille Compagnoni (1914–2009) and Lino Lacedelli (1925–2009), posing their feet on the summit of the second highest mountain on Earth, 1 year after the conquest of Everest by Sir Edmund Hillary (1919–2008) and the Sherpa Tenzing Norgay (1914–1986). Yet it generated also stirred long-lasting polemics, due to the night that Walter Bonatti (1930–2011), perhaps the best, strongest Italian climber of those times, and the Sherpa porter accompanying him, had to spend unsheltered above 8000 meters, just before the conquest of the summit. In fact, they were unable to reach camp IX before night, because the camp was set in a different position from that agreed upon. All the participants in the expedition underwent functional evaluation and exercise testing in the Institute of Physiology in Milano, then already chaired by Margaria.

Compagnoni was an alpine guide in Valtournenche, Aosta Valley, on the Italian side of the Matterhorn. Margaria knew him well, because he attended Valtournenche regularly, being the home village of his mother. Shortly before the K2 expedition, Compagnoni had the venture of guiding young Guido Monzino (1928–1988) to the summit of the Matterhorn. Monzino would then have become a peculiar figure of climber and explorer. He came from one of the most influential families of the industrial bourgeoisie of Milano. His father Franco founded and owned one of the biggest and most successful department store chain in Italy, so Guido was destined to business. In fact, he became a chief executive officer of the family company for a few years, but he resigned in 1966, when the company was sold.

Monzino was a romantic aesthete. He was fascinated by the enchanting spectacle of mountains and wide landscapes. His doom was exploration, not business. Climbing the Matterhorn was a kind of revelation to him. His dream became to reach farther and farther destinations, prompted not only by a romantic spirit of

Fig. 10.1 Left. Prince Luigi Amedeo, Duke of the Abruzzi (middle), with two of his guides, César Ollier and Joseph Petigax, in Karakorum in 1906. Right. Portrait of Guido Monzino taken during the 1959 expedition to Kanjut-Sar, Karakorum. With the kind permission of Club Alpino Italiano

adventure, but also by the wish of reviving Italian tradition of great explorative expeditions, along the groove traced by Prince Luigi Amedeo di Savoia, Duke of the Abruzzi (1873–1933), who became famous for his polar, African, and Himalayan expeditions. Monzino organized his first expedition in 1955: the savannah crossing from Dakar, Senegal to Abidjan, Ivory Coast. The following year he started his enduring collaboration with Valtournenche guides. Then, his acquaintance with Compagnoni brought him in touch with Margaria, and this is where the present story, in fact an adventure, started. Pictures of Monzino and of the Duke of Abruzzi appear in Fig. 10.1.

Obviously enough, the School of Milano contributed only a tiny, yet scientifically significant part of the overall history of altitude studies. Those interested in it can be referred to specific, analytic books and reviews (Luks et al. 2021; Ward et al. 2000; West 1998). Generally speaking, two main attitudes to the study of human adaptation to altitude can be identified in the scientific community. On the one side, we find the synthetic approach of those who consider altitude exposure a tool to understand basic physiological issues. On the other side, there is the analytic approach of those who consider understanding altitude adaptation an aim *in se*, and thus investigate and describe how specific physiological functions are modified during prolonged altitude sojourns or during permanent life at altitude. It goes without saying that all the excursions of the School of Milano in altitude research followed the former approach. Finally, we remind that the history of altitude physiology and medicine includes also related chapters like nutrition (Stellingwerff et al. 2019), sleep (Bloch et al. 2015), genomics (Storz 2021; Semenza 2020), training (Lundby and Robach 2016; Saunders et al. 2009), and high-altitude pulmonary or cerebral edema (Luks et al. 2017; Swenson and Bärtsch 2012). Thanks to the activity of Giuseppe Miserocchi, the School of Milano provided also significant contributions to the understanding of the pathophysiology of high-altitude pulmonary edema.

10.2 The Kanjut-Sar Expedition

Monzino contacted Margaria in 1958, when he was preparing his first expedition to Karakorum, aimed at attaining the summit of Mount Kanjut-Sar (7760 m above sea level), which is near the K2, and was still unconquered at that time. A picture of the participants in the expedition is reproduced as Fig. 10.2.

Monzino was looking for a medical researcher apt to join such an expedition. Margaria charged Paolo Cerretelli of that task, who was then a young medical assistant at the Institute of Physiology in Milano. When he was told that he would

Fig. 10.2 The participants in the 1959 expedition to Kanjut-Sar at base camp. From left, back row: Marcello Carrel, Camillo Pellissier, Cap. Khalid, Guido Monzino (chief of the expedition), Jean Bich, Piero Nava; middle row: Paolo Cerretelli, Lino Tamone, Pierino Pession, Leonardo Carrel, Pacifico Pession, Lorenzo Marimonti; first row: high-altitude porters. With the kind permission of Club Alpino Italiano

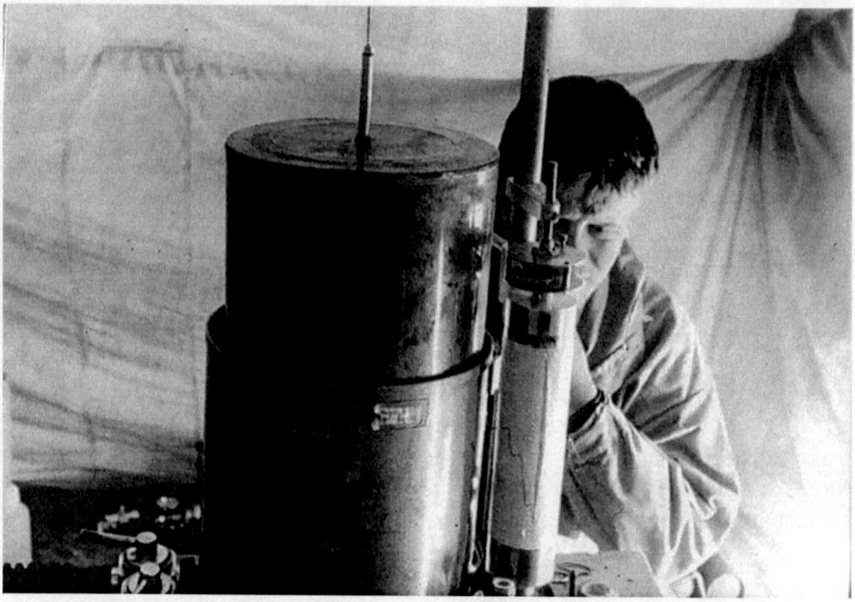

Fig. 10.3 Paolo Cerretelli behind a Tissot spirometer inside his tent at the base camp of Mt. Kanjut-Sar, 1959. With the kind permission of Club Alpino Italiano

have been sent as a medical researcher after a mountaineering expedition to Kashmir, Cerretelli felt unprepared: he was neither a climber nor a trekker.

Often unique occasions come unexpectedly, take it or leave it, and, notwithstanding his worries, Cerretelli took his. So he found himself in Cervinia (Aosta Valley, Italy, near Valtournenche) for 3 days, where the equipment for the expedition was to be tested, and where he was entrusted to a group of Valtournenche guides. Monzino instructed them to shake him off and show him the brilliant emotions of life at altitude. They trekked to the summit of Mt. Breithorn (4178 m above sea level, between the Matterhorn and the Mount Rose). On summit, they gave him hickory skis and told him that he was to descend with those skis under his feet. Good to him, Cerretelli was at least an average leisure skier, so he could accomplish his task somehow (Cerretelli, personal communication). After animate discussions, he was finally admitted as an expedition medical researcher and practitioner: Monzino needed one, after all. As a consequence, in 1959, he found himself at the Kanjut-Sar base camp, living in a tent with a Tissot spirometer, a Haldane apparatus, a 40-cm high step, and a couple of high pressure oxygen cylinders (Fig. 10.3).

The expedition was a success. Most important to Monzino, Camillo Pelissier (1924–1966), an alpine guide from Valtournenche (Fig. 10.4), attained the summit (7760 m above sea level) on July 10th, 1959. Although overlooked by the press (after all, Kanjut-Sar was less 8000 m tall…), this was an extraordinary climbing enterprise. First of all, Pelissier did it without supplementary oxygen supply: an approach that only the great Hermann Bühl (1924–1957) had dared to take until

Fig. 10.4 Camillo Pellissier upon return to camp III, after his conquest of the Kanjut-Sar summit. With the kind permission of Club Alpino Italiano

then. Secondly, he did it alone, because his companion Jean Bich (1916–2010) had to remain at camp VI for an initial frostbite at his toes (Monzino et al. 1960). Finally, climbing Kanjut-Sar is a technically difficult enterprise; suffice to say that the ascent was repeated only once, by a Japanese expedition in 1981.

Most important to us, Cerretelli succeeded in performing his first experiments at altitude and in carrying back the results. Three remarkable papers originated from the Kanjut-Sar expedition. The first (Cerretelli 1959) reports an analysis of the ventilatory response of climbers to chronic hypoxia. He used the technique of chemical denervation of peripheral chemoreceptors that Dejours et al. (1959) had just published. In fact, Dejours himself made the protocol available to Margaria to that purpose. In practice, the subjects, who were breathing ambient air, were suddenly administered four breaths of pure oxygen. The subsequent evolution of expired ventilation ($\dot{V}_E$) was followed on the spirometer. The results demonstrated that the $\dot{V}_E$ reduction induced by oxygen breathing was greater in chronic hypoxia than in normoxia. The same technique was resumed 25 years later to study the ventilatory response to hypoxia of climbers who had reached the summit of the highest peaks on Earth without supplementary oxygen (Oelz et al. 1986).

In the second paper, alveolar gas composition before and after altitude acclimation was measured (Cerretelli 1961). The data confirmed the predictions made by Fenn et al. (1946). In the third article (Cerretelli and Margaria 1961), the decrease of maximal oxygen consumption ($\dot{V}O_2^{max}$) was shown to be the same in chronic as in acute hypoxia. We discuss this issue more in detail in Chap. 7. Cerretelli (2013) told that he measured also vital capacity in 11 subjects after 60 days of exposure between 5000 and 7500 m, finding a reduction by 12.5%. However, we reckoned no publications on this issue by him.

After the Kanjut-Sar expedition, Cerretelli did not practice altitude physiology until the 1973 Italian Expedition to Mount Everest. He kept being acquainted with Monzino, but their relationship encompassed other geographic interests, such as Greenland and Kenya. Nevertheless, Cerretelli performed some experiments in acute hypoxia. He demonstrated (Cerretelli 1967) that the decrease of $\dot{V}O_2^{max}$ is accompanied by an equivalent decrease of the workload at which blood lactate accumulation appears. He also showed that the maximal blood lactate concentration is the same in acute hypoxia as in normoxia, at variance with what occurs in chronic hypoxia (Dill et al. 1931; Edwards 1936). He also argued that the fall of $\dot{V}O_2^{max}$ in acute hypoxia supports the hypothesis of central (cardiopulmonary) limitation of $\dot{V}O_2^{max}$ in humans (Cerretelli et al. 1967).

10.3 The Italian Expedition to Mount Everest

10.3.1 The Expedition

In 1972, Guido Monzino set up, and in 1973 realized, in collaboration with the Italian Army, a huge (65 members, 54 soldiers, and 11 civilians) expedition, the aim of which was to have the first Italian climbers posing their feet on the summit of Mount Everest. The Army provided transportation, logistics, telecommunications, the base camp settlement, and electric power supply. Nine transport aircrafts were used to carry more than 100 tons of equipment and goods, included two helicopters. The caravan ascending the Khumbu Valley toward the Everest base camp consisted of 150 individuals, the 65 Italians plus 85 Sherpas, some 2000 porters, and hundreds of yaks. The base camp was inaugurated on March 20th, 1973, at an altitude of 5350 m above sea level. It was a kind of village, with 60 tents, several heated lodgings, a canteen, a bar, a post office, an infirmary. The higher camps were then installed (Monzino 1976).

The two helicopters were used for transporting equipment along the Khumbu Valley and eventually to the base camp. That was the first time helicopters took off at altitudes higher than 4500 m above sea level. Cerretelli (2013) revealed that, while approaching and setting the base camp, pilots tried to raise the take-off altitude, mainly for technical purposes. One of the helicopters, which had been modified for extreme take-off and landing on ice, was lost without casualties in an attempt at landing at the advanced base camp (6300 m), to evacuate a Sherpa suffering of acute pulmonary edema. The second helicopter succeeded in landing, rescued the Sherpa and evacuated the three crewmembers who survived the accident of the first helicopter.

Despite bad weather and strong snowfalls, the expedition was a remarkable climbing success. Two tents could be installed at camp VI, at 8513 m above sea level, wherefrom five climbers (firstly Rinaldo Carrel and Mirko Minuzzo from Valtournenche, then, 2 days later, Virginio Epis, Claudio Benedetti, and Fabrizio Innamorati, who were enrolled in the alpine troops) and three Sherpas (Lakpa Tensing and Sambu Tamang on the first occasion, Sonan Gyaltzen on the second) reached the summit between May 5th and May 7th, 1973.

This was the last traditional Himalayan expedition, organized with large means and a large number of participants. Alpinism in the Himalayas and Karakorum was changing. Reinhold Messner had already reached the summit of Nanga Parbat without supplementary oxygen, in pure alpine style, following Hermann Bühl's example, after a dramatic ascent, which was marked by the death of his brother, hit and killed by an avalanche on the way back. Five years only separate the successful climb by Carrel and Minuzzo from the pivotal enterprise by Messner and Peter Habeler, who, on May 8th 1978, were the first to reach the summit of Mount Everest without supplementary oxygen. This was a real turning point, opening a new era in the history of alpinism and prompting new approaches to the study of extreme altitude adaptation.

10.3.2 Paolo Cerretelli at the Expedition

Paolo Cerretelli, who had just been appointed Professor of Physiology at the University of Milano, participated in the Italian expedition to Mount Everest as a medical scientist. He told the story of his participation and of the experiments that he carried out (Cerretelli 2013). We can thus make use of his words to give an account of what he did. The acronyms have been adapted to the present nomenclature.

> *My main task in the expedition was to develop my research project in the facility provided by the organization. I was assisted by a technician from my laboratory and, occasionally, by one of the three doctors participating in the expedition with professional assignments. A large number of porters and altitude Sherpas (70) were also recruited. I had obtained informed consent from all Italian members and a number of selected altitude Sherpas to collaborate as subjects for some of the experiments that were planned. The equipment was provided by the organization and later donated to my laboratory at the University of Milano. It included a special custom-built metabograph conceived for measurements of maximum oxygen consumption in a closed circuit, opposing low resistance up to very high ventilation rates, breathing ambient air or pure oxygen (Margaria et al.* 1959*), carbon dioxide and oxygen breath-by-breath meters, an earpiece oximeter coupled to amplifiers, and a recorder, battery-powered electrocardiographs,* PO_2, PCO_2, *and pH microelectrodes, a spectrophotometer, and the basic equipment for a standard laboratory of clinical biochemistry (see pictures). The laboratory was moderately heated (Fig. 10.5).*
>
> *Several variables were recorded in selected groups of Caucasians, before and after 8 weeks of continuous (5000–7000 m) or intermittent (2–6 h daily, between 2800 and 7000 m) altitude exposure, and in Sherpas operating at or above the base camp. Among these were red blood cell count, blood hemoglobin concentration, hematocrit;* $\dot{V}O_2^{max}$, *maximal expired ventilation* ($\dot{V}_E^{max}$), *maximal heart rate* (f_H^{max}) *and arterial oxygen*

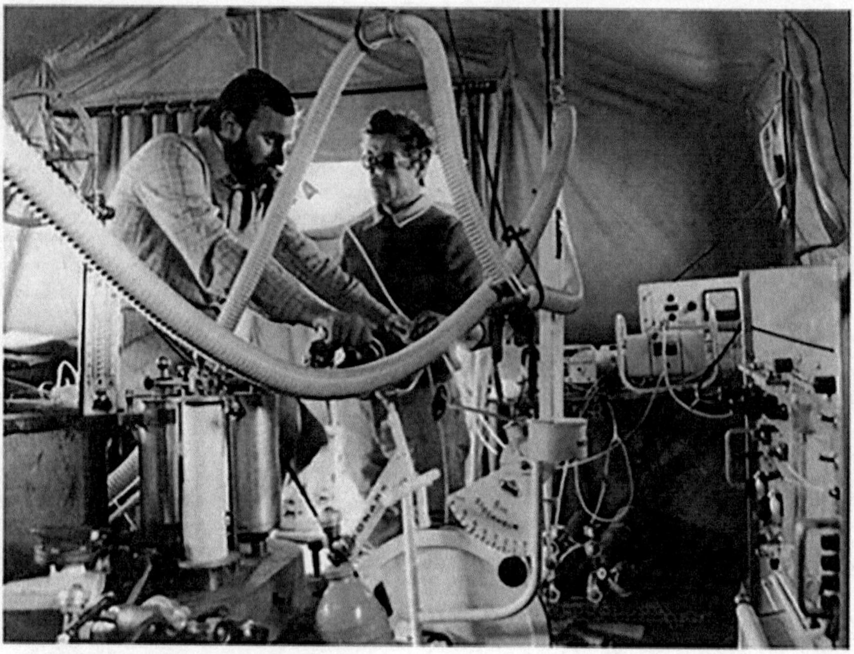

Fig. 10.5 Physiology laboratory at the base camp of the 1973 Mount Everest expedition. A bicycle ergometer, spirometers, gas analyzers, and electronic recorders are visible. With the kind permission of Paolo Cerretelli

saturation (S_aO_2); resting and exercise cardiac output ($\dot{Q}$); $\dot{V}O_2^{max}$ changes when administering hypobaric oxygen or normoxic air–oxygen mixtures to breathe to acclimatized Caucasians with increased hematocrit at the Mt. Everest base camp (Cerretelli 1976a*); and peak blood lactate concentration at exhaustion ($L\hat{a}_b$) and the kinetics of payment of the lactic oxygen debt (Cerretelli et al.* 1982*).*

The results of hematologic measurements confirmed essentially previous data obtained in similar conditions. An original finding concerning blood composition appears in Tables 1 and 2 (this data are reported in present Table 10.1): it confirms the different responses to chronic hypoxia of native Sherpas vs. acclimatized Caucasians, also showing an unexpected, peculiar reaction of the members of the helicopter crew. The latter, based at 2850 m, were regularly making use of supplemental oxygen when operating at above 4000 m, and used to spend several hours a day on the ground in the altitude range of 4500–6300 m. They were apparently very well acclimatized from the respiratory standpoint, as proven by the perfect altitude tolerance when stranded at above 6300 m, but were characterized by hematologic parameters similar to those of Sherpas. To the author's knowledge, these measurements have never been repeated.

Measurements of $\dot{Q}$ by the nitrogen–carbon dioxide rebreathing method (Cerretelli et al. 1966*) and of oxygen consumption have been carried out at the base camp (5350 m) in five subjects at rest and, in two of them, at increasing exercise loads on the bicycle ergometer and, compared with the sea level control values (Cerretelli* 1976b*).*

Resting $\dot{Q}$ values were found to be unchanged, as were the measurements obtained at moderate workloads. At about 90% of their $\dot{V}O_2^{max}$, the two investigated subjects attained $\dot{Q}$ levels of 19.0 and 20.7 L min^{-1}, respectively, versus sea level control data of 20.9 and

Table 10.1 The top and middle rows report the data contained in Cerretelli's Table 1, concerning data obtained on Italian subjects at sea level and at the two indicated altitudes. The bottom rows reports the data contained in Cerretelli's Table 2, concerning a comparison between Italians, either climbers, who underwent 8 weeks of continuous altitude exposure, or helicopter pilots, who underwent 5 weeks of intermittent altitude exposure, and Sherpas, who underwent 8 weeks of continuous altitude exposure. RBC, red blood cell count; [Hb], blood hemoglobin concentration; Hct, hematocrit; f_H^{max}, maximal heart rate; $\dot{V}O_2^{max}$, maximal oxygen consumption. This Table combines Tables 1 and 2 of Cerretelli (2013)

	Age (years)	Weight (kg)	RBC (μm^{-3})	[Hb] ($g\ L^{-1}$)
Sea level	29.1 ± 5.9	71.5 ± 9.7	4.70 ± 0.22	149.6 ± 8.6
Base camp (5350 m)		66.9 ± 8.7	6.65 ± 0.70	216.0 ± 24.2
Lukla (2850 m)		66.9 ± 8.7	6.65 ± 0.70	216.0 ± 24.2
	f_H^{max} (min^{-1})	$\dot{V}O_2^{max}$ ($L\ min^{-1}$)	$\dot{V}O_2^{max}$ ($ml\ kg^{-1}\ min^{-1}$)	
Sea level	187 ± 13	3.13 ± 0.29	44.3 ± 5.6	
Base camp (5350 m)	160 ± 9	2.36 ± 0.29	35.7 ± 6.2	
Lukla (2850 m)	168 ± 15	3.03 ± 0.25	45.8 ± 5.3	
	ss	Hct	RBC (μm^{-3})	[Hb] ($g\ L^{-1}$)
Italians (climbers)	Continuous	0.663 ± 0.043	7.03 ± 0.66	235.2 ± 16.4
Sherpas	Continuous	0.627 ± 0.047	5.02 ± 0.80	201.0 ± 12.6
Italians (pilots)	Intermittent	0.555 ± 0.024	5.02 ± 0.45	196.0 ± 13.2

23.7 L min^{-1}. $\dot{V}O_2^{max}$-extrapolated maximal $\dot{Q}$ ($\dot{Q}_{max}$) values were 10% lower than the control values at sea level, while f_H^{max} were 148 and 162 min^{-1}, respectively (vs. 160 and 178 at sea level). Thus, the stroke volume of the heart in the prevailing experimental conditions was essentially unchanged.

A major interest for me was the analysis of the factors limiting maximal exercise at altitude in acclimatized lowlanders. By the use of the metabograph, it was possible to show, right at the base camp ($P_B = 410$ mmHg), that these subjects ($n = 10$), when given 100% oxygen to breathe, could raise their $\dot{V}O_2^{max}$ from an average of 70% of the control sea level value only to 92% (Cerretelli 1976b*). Considering the limited reduction of $\dot{Q}_{max}$ and the 40% increase in blood hemoglobin concentration ([Hb]), to explain these results (also confirmed by indirect measurements of $\dot{V}O_2^{max}$ in a group of 13 subjects taken down by the helicopter to Lukla, 2850 m), two hypotheses were put forward, i.e., (a) that the release of oxygen in the muscles by peripheral capillaries could have been impaired by packing of erythrocytes due to extremely high Hct and/or (b) that the blood was shunted away from the muscles to reduce the load on the heart due to increased viscosity. Neither condition has been tested experimentally. However, more recent findings (Cerretelli et al.* 2009*) indicate that the reduced maximum rate of oxidations could have been mainly the result of muscle deterioration (reduction of tissue mass, decreased mitochondrial volume density, increase in the accumulation of lipofuscin, a marker of damage by reactive oxygen species, and enzymatic impairments of the muscle metabolic machinery).*

In the course of the expedition, I had the opportunity to evaluate in a large group of subjects, both acclimatized Caucasians and Sherpas, and in various experimental conditions, the apparent peak contribution to energy metabolism by anaerobic glycolysis. The Sherpas, for most of the time, were living at altitudes varying between 2500 and 3400 m, whereas during the expedition, they spent, like Caucasians, at least 8 weeks between 5000 and 7500 m. Both breathing ambient air or oxygen, resting lactate concentrations at altitude are approximately the same as those at sea level. By contrast, peak lactate concentration at exhaustion (work duration 3–5 min) appears to be considerably lower than at sea level in all investigated groups. These data were among the first after those of Dill et al. (1931) and Edwards (1936) confirming the existence of an apparent blunting of the maximum energy yield from anaerobic glycolysis at altitude. The latter phenomenon was later defined the "lactate paradox" by Hochachka (1988).

Cerretelli's account, although quite precise and complete, is nonetheless deceptive, because it remains confined to a relatively descriptive and superficial level, and thus recognizes less credits than his work deserves. His experiment on the effect on $\dot{V}O_2^{max}$ of having acclimatized subject, with elevated [Hb], breathing hypobaric hyperoxic mixtures at the base camp, thus reproducing the inspired oxygen pressure incurring at sea level, is among the most influential experiments ever carried out at altitude, for its tremendous and enduring onsequences.

10.3.3 Limiting Factors to Oxygen Transport on Mount Everest

Indeed, Cerretelli's participation in the Italian expedition to Mount Everest is memorable to the scientific community for essentially one paper (Cerretelli 1976a) on the factors limiting oxygen transport on Mount Everest, which we briefly summarize herewith, since it is poorly described in the above account of the scientific outcomes of the expedition.

Cerretelli was puzzled by the fact that the $\dot{V}O_2^{max}$ fall in chronic hypoxia was the same as in acute hypoxia (Cerretelli and Margaria 1961; Cerretelli et al. 1967; Dill and Adams 1971; Dill et al. 1966; Fagraeus et al. 1973; Pugh et al. 1964; Vogel et al. 1967, just to encompass the Kanjut-Sar 1959–Everest 1973 time window), because he thought it was in contradiction with the observation that altitude acclimatization implies an increase in [Hb]. Thus, he formulated the hypothesis that cardiovascular oxygen transport may not account for the entire $\dot{V}O_2^{max}$ change in chronic hypoxia, despite his own previous conclusions (Cerretelli et al. 1967). To test this hypothesis, he conceived and realized an experiment, wherein $\dot{V}O_2^{max}$ was measured on a man acclimatized to altitude, and thus polycythemic, who was breathing at the Everest base camp a gas mixture containing an inspired oxygen partial pressure (P_IO_2) equivalent to that existing at sea level (acute normoxia, P_IO_2= 150 mmHg). Of course, $\dot{V}O_2^{max}$ was measured also at sea level, before departure, and at altitude. He admitted two possible outcomes. If cardiovascular oxygen transport were indeed the factor limiting $\dot{V}O_2^{max}$ in chronic hypoxia, he would have found a $\dot{V}O_2^{max}$ higher than that measured at sea level before the expedition, and by a relative amount equivalent to the percent increase of arterial oxygen concentration (C_aO_2) due to the higher [Hb]. If conversely he would not have observed such a $\dot{V}O_2^{max}$ increment, then the limit to $\dot{V}O_2^{max}$ would be elsewhere than in cardiovascular oxygen transport.

The results are reported in Fig. 10.6. The combination of a higher [Hb] due to acclimatization with a lower S_aO_2 due to altitude, generated a slightly higher C_aO_2 in

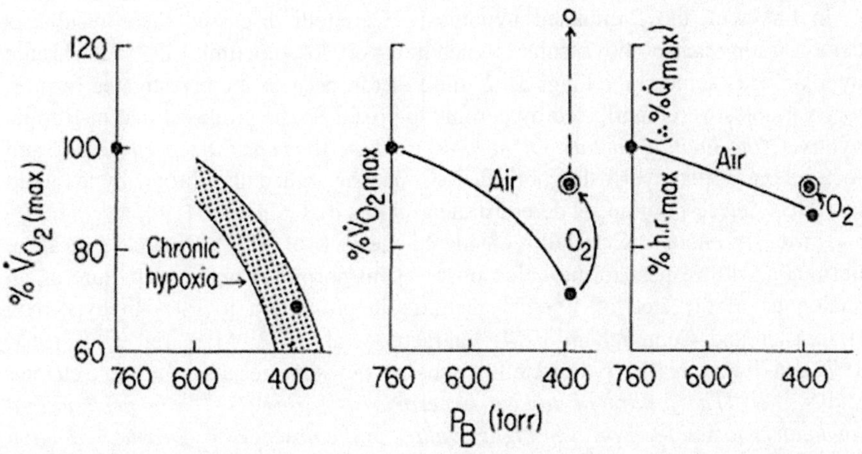

Fig. 10.6 Left panel: Maximal oxygen consumption ($\dot{V}O_2^{max}$), expressed relative to the value at sea level, set equal to 100%, as a function of barometric pressure (P_B). The filled dots refer to the values obtained in Milano before departure (normoxia) and at the Everest base camp after acclimatization (chronic hypoxia). The dotted area indicates the expected range for $\dot{V}O_2^{max}$ values in chronic hypoxia. Middle panel: Same as on left. A curve connects the two values in normoxia and in chronic hypoxia. Breathing the hyperoxic mixture at Everest base camp moves $\dot{V}O_2^{max}$ up to the circled dot, as indicated by the continuous bent arrow. The open dot shows the $\dot{V}O_2^{max}$ that Cerretelli expected to find after administration of the hyperoxic mixture, based on the observed increase in blood hemoglobin concentration. The dashed arrow indicates the apparent "lack" of $\dot{V}O_2^{max}$. Right panel: Maximal heart rate or cardiac output ($\dot{Q}_{max}$), expressed relative to the respective value at sea level, set equal to 100%, as a function of P_B. The values in normoxia and in chronic hypoxia (filled dots) are connected by a curve. Administration of the hyperoxic mixture at Everest base camp moves $\dot{Q}_{max}$ up to the circled dot only. From Ferretti (2003), redrawn after Cerretelli (1976a)

chronic hypoxia than in normoxia. Obviously enough, acute normoxia determined a huge increase in C_aO_2. The measured $\dot{V}O_2^{max}$ fall at altitude corresponded well to the expectations at the Everest base camp (Pugh et al. 1964). The drama exploded when the subjects were tested in acute normoxia: their $\dot{V}O_2^{max}$ went up, but not in proportion to the C_aO_2 increase; it rather remained definitely lower than observed in Milano, before the expedition. Although expressed only relative to the value in normoxia, it was clear that $\dot{Q}_{max}$ decreased in chronic hypoxia with respect to normoxia, before departure. Yet most important, when the subjects breathed the hyperoxic gas mixture, $\dot{Q}_{max}$ did not come back to the value observed in Milano, despite its slight increase. Interestingly enough, Pugh et al. (1964) reported an indirect finding along the same line: two of their subjects, while breathing oxygen at 150 mmHg at 5800 m after acclimatization (acute normoxia), could sustain a given supramaximal power for 4 min only, instead of 6 min in normoxia. This finding did not escape Cerretelli, who mentioned it in his paper (Cerretelli 1976a).

In line with the formulated hypothesis, Cerretelli discussed these results as demonstrating that cardiovascular oxygen transport does not limit $\dot{V}O_2^{max}$ in chronic hypoxia. He excluded the lungs as limiting factor, because the investigated respiratory variables were unaffected by chronic hypoxia. So, he proposed that in chronic hypoxia "*the limiting factors for aerobic work performance are peripheral*," and focused on capillary perfusion or diffusion, since he argued that chronic hypoxia did not affect cell respiration, as demonstrated on rats by Gold et al. (1973).

Curiously enough, Cerretelli considered the lack of $\dot{V}O_2^{max}$ increase in acute normoxia with respect to the value observed in normoxia before departure as an analogous of the lack of $\dot{V}O_2^{max}$ increase in proportion to P_IO_2 in hyperoxia (Bannister and Cunningham 1954; Fagraeus et al. 1973; Margaria et al. 1961, 1972). A few years later, Cerretelli discussed the same results in a book chapter and wrote: "*The failure of sudden hyperoxia to raise* $\dot{V}O_2{}^{max}$ *in acclimatized lowlanders to sea level or even higher values in the absence of a drastic reduction of maximal cardiac output* ($\dot{Q}_{max}$) *could be explained by a reduction of effective blood flow to the working muscles*" (Cerretelli 1982).

10.3.4 Critique of a Paper

In 1976, exercise physiologists were still searching the single limiting factor to $\dot{V}O_2^{max}$ (Ferretti 2014). As discussed in Chap. 7, two fields faced each other. On one side, we find those believing that the limits were imposed by central circulation (cardiovascular oxygen transport); on the other side, there were the supporters of peripheral muscular limitation. Indeed, the majority was on the former side (Åstrand 1952; Ekblom 1969; Holmgren and Åstrand 1966; Margaria et al. 1965; Mitchell and Blomqvist 1971; Ouellet et al. 1969; Rowell 1974; Saltin 1973, just to cite publications appearing before 1976). The contention of cardiovascular limitation of $\dot{V}O_2^{max}$ relied essentially on the findings that $\dot{V}O_2^{max}$ (i) is higher in endurance athletes than in non-athletes, (ii) is proportional to $\dot{Q}_{max}$, and (iii) increases with endurance exercise training (see Blomqvist and Saltin 1983; Ferretti 2014, Lacour and Flandrois 1977, for a summary of results). In the 1970s, the demonstration that $\dot{V}O_2^{max}$ increases after autologous blood reinfusion (Ekblom et al. 1972) in parallel with and in proportion to the induced increase in $\dot{Q}_{max}$ (Ekblom et al. 1976) gave further momentum to the cardiovascular limitation field. When he submitted his 1976 paper, Cerretelli was probably unaware of the latter article, published just 2 months earlier in the same journal, when his was already in press. No on-line first articles existed then. This may explain why he underscored the value of autologous blood reinfusion in his discussion.

It is fair to admit that Cerretelli's reasoning was logical, considering the knowledge and the theoretical vision of those times. To him, the fact that $\dot{V}O_2^{max}$ did not increase in acute normoxia above the level attained in Milano before departure (normoxia) necessarily implied that circulation was not the limiting factor, at least at altitude. Inevitably, he had to look elsewhere, and think of a possibly distal site of the single limiting factor: in fact, he excluded the lungs and concentrated on muscles. As possible limiting step, he focused on a limitation of muscle tissue perfusion and on changes in oxygen diffusion rates between capillaries and muscle fibers, but did not consider muscle oxidative capacity. He did so based on observations made on small animals (Gold et al. 1973), but, rather surprisingly, neglected what was then perhaps the most important work supporting a role for peripheral limitation. In fact, Reynafarjee (1962) had reported an increment of muscle myoglobin concentration and of oxidative enzyme activity in Andean altitude natives. Reynafarjee's paper was extremely influential in the 1960s and the 1970s. The belief that oxidative capacity had necessarily to increase in chronic hypoxia was so deeply rooted in the scientific community, that Peter Hochachka, a bright and imaginative zoologist and physiologist from Canada, constructed a keen interpretative hypothesis around this notion (Hochachka et al. 1983). Indeed, they stated that animals living permanently in chronic hypoxia had to cope with a major problem, namely "*how to maintain an acceptably high scope for aerobic metabolism in the face of the reduced oxygen availability of the atmosphere*." They postulated that this could be achieved only by enhancing the capacity for oxygen transport and the capacity for oxidative metabolism.

Cerretelli's paper just came out too early. A revolution in the way of looking at $\dot{V}O_2^{max}$ limitation took place in subsequent years. After Taylor and Weibel (1981) had resumed the oxygen conductance equation and had applied it to the condition of maximal exercise, multifactorial models of $\dot{V}O_2^{max}$ limitation were developed. The first and most comprehensive of these models was conceptually proposed by di Prampero (1985), algebraically formulated by di Prampero and Ferretti (1990), and later expanded to cover hypoxia (Ferretti and di Prampero 1995). At the same time, a concurrent model, with a different vision of the oxygen cascade, was constructed following the same basic concept (Wagner 1993). These models are analyzed in detail in Chap. 7. Suffice to remind here that the concept, according to which $\dot{V}O_2^{max}$ is not limited by a single factor (unifactorial theory), as believed some 50 years ago, but by the simultaneous action of multiple factors (multifactorial theory) has become a widely accepted notion.

In this new context, Ferretti (2003) reinterpreted Cerretelli's (1976a) results, attaining opposite conclusions to his. Let us use his words, with the sole precaution of adapting the symbols and equation numbers to those used in the present book.

*According to di Prampero and Ferretti (*1990*), the cardiovascular resistance to oxygen flow (R_Q) is equal to:*

$$R_Q = G_Q^{-1} = \left(\dot{Q}\,\beta_b\right)^{-1} \tag{10.1}$$

where G_Q is the equivalent conductance and β_b is the oxygen transport coefficient of blood. This in turn corresponds to the average slope of the oxygen equilibrium curve, as (Note that Eq. 10.2 is tantamount to Eq. 7.7.)*:*

$$\beta_b = \frac{(CaO_2 - C\overline{v}O_2)}{(PaO_2 - P\overline{v}O_2)} \tag{10.2}$$

*where C and P indicate the concentrations and pressures of oxygen, respectively, in arterial (a) and mixed venous ($\overline{v}$) blood. What Cerretelli did when he administered oxygen at 150 mmHg to his subjects at Everest base camp was actually to change three parameters pertaining to the oxygen conductance equation: (i) he increased the overall oxygen gradient, because he changed P_IO_2; (ii) he slightly increased $\dot{Q}_{max}$ (he measured this indeed); and (iii) he decreased the factor β_b. In fact, the increase in P_IO_2 brought about a subsequent increase in alveolar oxygen pressure (P_AO_2) and in P_aO_2, so that the arterial blood point was moved onto the flat part of the oxygen equilibrium curve. Since the mixed venous point was only slightly displaced, this implied a dramatic change in the average slope of the oxygen equilibrium curve, and thus in β_b. According to Eq. (10.1), this means a reduction of G_Q with consequent increase in R_Q. To sum up, the administration of oxygen at 150 mmHg did not induce a $\dot{V}O_2^{max}$ increase proportional to that in the overall oxygen gradient, because, due to the shape of the oxygen equilibrium curve, the latter increase was inevitably accompanied by a concomitant reinforcement of the cardiovascular resistance to oxygen flow. These are exactly the same reasons why $\dot{V}O_2^{max}$ does not increase in hyperoxia and why subjects with a high $\dot{V}O_2^{max}$ in normoxia, who are subjected to arterial oxygen desaturation (*Dempsey et al. 1984*), undergo a greater decrease in $\dot{V}O_2^{max}$ in hypoxia (*Ferretti et al. 1997*). Indeed Cerretelli's experiment provides nothing but a brilliant confirmation of the predominant role of cardiovascular oxygen transport in limiting $\dot{V}O_2^{max}$! Since $\dot{V}O_2^{max}$ did not come back to the level measured at Milano before the expedition, it is alike that the peripheral (muscular) resistances to oxygen flow factors, though smaller than R_Q, be larger than in normoxia. However, this by no means implies that the fractional limitation to $\dot{V}O_2^{max}$ in chronic hypoxia imposed by peripheral factors be increased.*

In conclusion, a reanalysis (Ferretti 2003) of a great paper (Cerretelli 1976a) based on unique data collected at the 1973 Italian expedition to Mount Everest, in which the author postulated that muscle blood flow is the factor that limits $\dot{V}O_2^{max}$ at altitude, demonstrated the main role of cardiovascular oxygen transport as a factor limiting $\dot{V}O_2^{max}$. These conclusions, by Cerretelli and by Ferretti, are opposite each in respect to the other. Nevertheless, and in spite of this, the outcomes of Cerretelli's paper were along a third, different track. Indeed it prompted a coherent series of experiments that unambiguously confuted Reynafarjee's hypothesis, demonstrating that the loss of muscle oxidative capacity is the main effect of chronic hypoxia on muscle structure and function, as we will describe more in detail in the next section. In sum, the story of Cerretelli's Mount Everest paper reveals a mixture of erroneous conclusions, yet coherent with the vision of that time, a belated reversal of that conclusion 25 years later, in the light of further knowledge and of a different cultural climate, and ensuing studies along a third track, independent of both conclusions, the original and the revisited. According to Ferretti (2003), this story prevents from thinking of a coherent linear evolution of scientific knowledge because of a logical series of hypotheses and refutations. It rather puts forward a great deal of serendipity, casualty, chance, and logical misdemeanor, which fits well with Feyerabend's anarchic theory of the evolution of scientific knowledge (Feyerabend 1975). Epistemology is wonderful, but life is a non-linear phenomenon.

10.3.5 The Consequences of a Paper

And so, it eventually happened that a paper showing the role of cardiovascular oxygen transport in limiting $\dot{V}O_2^{max}$ at altitude originated a lot of studies, which can now be considered classical indeed, on muscle structural and functional adaptation to chronic altitude exposure!

This was certainly not the original intention of Cerretelli. He was candidly convinced that muscle oxidative capacity had nothing to do with it. He pointed to muscle blood flow and capillary-fiber oxygen transfer indeed. However, nobody felt induced to study muscle blood flow in chronic hypoxia. Only Cerretelli and coworkers (Cerretelli et al. 1984) investigated muscle blood flow in a group of climbers upon return from a Himalayan expedition. Indeed, the only notable finding of that study was a slowing of the blood flow response at light exercise onset.

We neither know nor understand how it happened that all of a sudden people drew their attention toward muscle structural studies. Certainly, it was not a will of falsifying Reynafarjee's conclusions. The only apparent link between limiting factors to oxygen transport on Mount Everest and muscle structural studies might have been capillarity, which can be determined by histochemical or morphometric methods on muscle biopsy samples. Therefore, we cannot state whether the change of direction taken by Cerretelli in his altitude research in subsequent years was serendipitous or not. Yet it is a matter of fact that, in apparent contrast with his own conclusions, Cerretelli turned his focus to muscle morphometry.

In 1978, Cerretelli moved from Milano to Geneva. Then, he entered in touch with Ewald Weibel (1929–2019) and his pupil Hans Hoppeler from the Institute of Anatomy of the University of Bern, as well as with Hans Howald from the Swiss Federal School of Sport in Magglingen. Weibel was famous worldwide for the introduction of morphometric methods for electron microscope studies, together with a mathematician from Cuba, Domingo Gomez (1903–1978), and for the application of these methods to the quantitative analysis of alveolar morphology (Weibel 1963, 1973; Weibel and Gomez 1962). Weibel encouraged Hoppeler to initiate the morphometric study of human muscle (Hoppeler et al. 1973). Howald was a sport medicine specialist, whose research interests were in muscle biochemical adaptation to training (see, e.g., Howald 1982). When Cerretelli arrived at Geneva, Weibel was engaged, in collaboration with Dick Taylor (1939–1995) from Harvard, in a big study on the morphological characteristics of the mammalian respiratory system and on the structural limits to oxygen flow, which resulted in a major series of publications in Respiration Physiology (Taylor and Weibel 1981). Hoppeler was responsible for muscle morphometric studies, and Howald of muscle enzymatic activity studies.

The ensuing collaboration between Cerretelli and them led to several major studies on muscle structural and biochemical characteristics in chronic hypoxia. Elite climbers (Oelz et al. 1986), climbers before and after altitude acclimatization (Hoppeler et al. 1990; Howald et al. 1990), Sherpas (Kayser et al. 1991), Tibetans (Kayser et al. 1996), and Andean natives (Desplanches et al. 1996) were studied. Others also joined that research line, investigating the participants in Operation Everest II (Green et al. 1989; McDougall et al. 1991) or other acclimatized subjects (Poole and Mathieu-Costello 1989). All these studies coherently demonstrated the remarkable loss of muscle oxidative capacity either with altitude acclimatization, or in populations living permanently at altitude, whether by measuring muscle mitochondrial volume density or mass, or by determining muscle oxidative enzyme activities. These findings were opposite to what everybody expected. Furthermore, they were so univocal and coherent that they led to a clear-cut refutation of Reynafarjee's hypothesis and rejection of Hochachka's interpretative hypothesis (Hochachka et al. 1983).

Furthermore, all these studies demonstrated an increase in muscle capillary density. This did not come from capillary neoformation, as long as capillary volume is essentially unchanged. It rather came from a reduction of muscle fiber size. Notwithstanding its cause, this finding implies a shortening of muscle diffusion distances, thus reducing the peripheral resistance to oxygen flow and easing oxygen diffusion from capillaries to mitochondria. This tends to compensate for the loss of mitochondrial oxidative capacity, so that eventually the lumped peripheral resistance to oxygen flow (see Chap. 7) remains essentially unchanged in altitude-adapted humans with respect to non-acclimatized normoxic lowlanders.

Cerretelli and Hoppeler (1996) summarized these results in a splendid chapter of the Handbook of Physiology.

10.4 Messner on Top of Mount Everest Without Oxygen Bottles

On May 8th, 1978, Reinhold Messner and Peter Habeler reached the summit of Mount Everest in pure alpine style and without supplementary oxygen, as Pellissier did on Kanjut-Sar almost 20 years before. Whereas Pellissier's exploit made little surprise, after all Kanjut-Sar summit is definitely below 8000 m, Messner and Habeler shocked the entire physiological community: nobody indeed thought that it would have been possible to reach the top of Mount Everest without supplementary oxygen. At the time of the K2 expedition, Margaria did not take into account any hypothesis of refraining from using oxygen. All classical attempts to reach the 8000 m peaks were organized as big exploration expeditions. No trial succeeded before World War II. In the 1950s, all summiteers made use of supplementary

Fig. 10.7 Left: Portrait of Alexander Mitchell Kellas. Right: Kellas' grave in Kampa Dzong, Tibet. In public domain

oxygen, with only one yet very notable exception: Hermann Bühl on Nanga Parbat, who reached the summit alone, in alpine style and without oxygen supplementation, on July 3rd, 1953. Bühl opened a new way indeed, but Nanga Parbat is 700 m less than Everest above sea level.

Nevertheless, John West discovered that somebody predicted the possibility of reaching the summit of Mount Everest without supplementary oxygen indeed some 60 years before Messner achievement: Alexander Kellas (1868–1921), a Scottish chemist with interests in physiology, who also contributed deeply to our understanding of the geography of the Karakorum and the Himalayas area (West 1987, see Fig. 10.7). He traveled in Nepal and Sikkim, approaching Everest from East. Concerning the problem of climbing Mount Everest, he communicated a report to the Royal Geographical Society entitled "*A consideration of the possibility of ascending the loftier Himalaya*," which was published in the Geographical Journal (Kellas 1917).

Yet this was only a preliminary report. His considerations were further and more precisely detailed in an unpublished report, a copy of which West resumed from the Alpine Club archives in London. A detailed summary of this text, entitled "*A consideration of the possibility of ascending Mount Everest*," can be found in West (1987). The complete text has been reprinted in 2001 (Kellas 2001).

Here, we just highlight a few key points of physiological interest. In fact, Kellas (i) argued that hyperventilation induced by hypoxia would decrease alveolar carbon dioxide partial pressure (P_ACO_2) and increase P_AO_2; (ii) admitted the possibility of a left shift of the oxygen equilibrium curve at extreme altitude due to low P_ACO_2, yet corrected by a decrease of the alkaline reserve of blood; (iii) estimated, by means of an analogous of the alveolar air equation (FitzGerald 1913), P_AO_2= 23.6 mmHg, after assuming $P_B = 267$ mmHg and P_ACO_2= 18.6 mmHg on top; (iv) calculated

S_aO_2= 0.43[1]; (v[5]) discussed the potential effects of high-altitude polycythemia; (6) estimated $\dot{V}O_2^{max}$ at various altitudes, predicting a value of about 1 L min^{-1} on top, incredibly close to that measured on summiteer Chris Pizzo at the American Medical Research Expedition to Mount Everest (AMREE) (West et al. 1983a). He eventually concluded: *Mt. Everest could be ascended by a man of excellent physical and mental constitution in first rate training, without adventitious aids if the physical difficulties of the mountain are not too great, and with the use of oxygen even if the mountain can be classed as difficult from the climbing point of view.* Messner was probably unaware of Kellas' conclusion, nevertheless he inadvertently demonstrated, 60 years after, that Kellas was right.

Kellas fell into oblivion and, despite his acquaintance with Sir John Scott Haldane (1860–1936), one of the most eminent respiratory physiologists of those times, the physiological community remained unaware of him for long. Physiologists do not read the Geographical Journal, his main report remained in a drawer for many years. However, George Mallory (1886–1924), who more than any other human pursued the dream of climbing Mount Everest, and who was with Kellas in the preparatory 1920 expedition, trusted him. Kellas died in that expedition. Mallory took the flag, he tried to reach the summit in 1924 following Kellas' principles, he died on the mountain, we don't know whether before or after having reached the top. Nobody tried that way again before Messner's successful attempt. Hillary's conquest was with supplementary oxygen, and the chief of the expedition, Sir John Hunt (1910–1998), placed the final camp higher than suggested by Kellas, at the South Col. All repetitions after Hillary made use of oxygen, the 1973 Italian expedition as well. Then Messner came and revolutionized the game.

In the microcosm of high-altitude physiology, Messner achievement had essentially three more or less direct consequences. The first and more direct has to do with the School of Milano and represents the first collaboration between Cerretelli and Bern. It is the study on the functional characteristics of extreme climbers, which Messner himself was a subject of (Oelz et al. 1986). Cerretelli and Hoppeler were so struck by Messner achievement, that they expected these climbers having extraordinary physiological characteristics, like elite athletes. In fact, the results contradicted these expectations. Take, for instance, $\dot{V}O_2^{max}$, which is reported in Fig. 10.8: although higher than in sedentary individuals of equivalent age, it was by far lower than reported on top-level endurance athletes and corresponded to what we may expect in non-athletic well-trained individuals. The maximal anaerobic power was equal to that of sedentary controls. All subjects hyperventilated in hypoxia, but climbers had not an enhanced ventilatory response with respect to sedentary controls.

The most striking feature of the Discussion is the astonishment of the authors. The conclusion was, we quote: *The main features of a successful elite altitude*

[1]In fact, Kellas estimated lower P_AO_2, and thus higher P_ACO_2 and lower S_aO_2 than actually measured on top during the American Medical Research Expedition to Mount Everest almost 70 years later (West et al. 1983b).

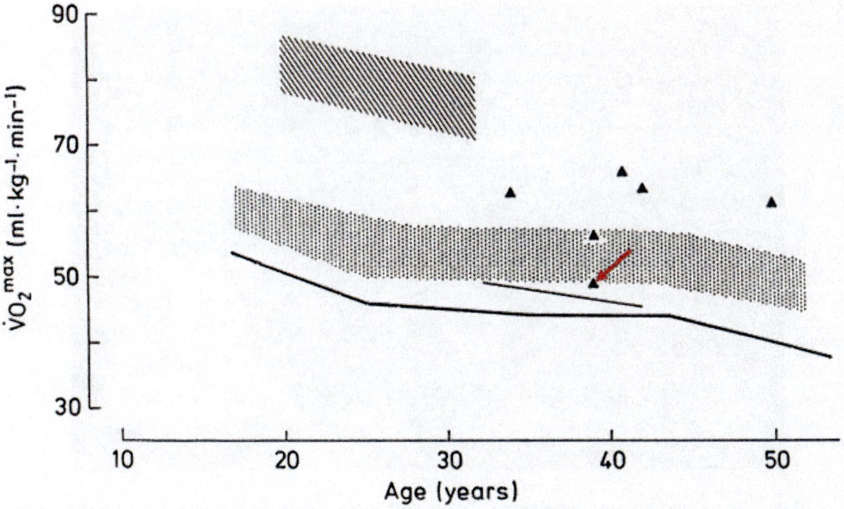

Fig. 10.8 Maximal oxygen consumption ($\dot{V}O_2^{max}$) as a function of age. Triangles: data on elite climbers (Oelz et al. 1986); dotted area: leisure marathon runners (di Prampero et al. 1986); thick line: sedentary controls (Åstrand and Christensen 1964); thin line, professional alpine guides (Cerretelli, unpublished data); hatched area, elite marathon runners (Costill et al. 1976). The red arrow indicates Messner's $\dot{V}O_2^{max}$. Modified after Oelz et al. (1986)

climber are, besides an optimal functional balance and an unusual skill, strong motivation and exceptional drive. Reinhold Messner, the first mountaineer to overcome the barrier of 8500 m without supplemental oxygen, really typifies these climbers. He is characterized by rather normal physiological features but by the obsessive need to be the first and the best, by "fair means" as he states it, i.e., *without the aid of oxygen in any phase of the ascent.* A quite peculiar, bizarre conclusion for a scientific paper, but a sincere testimony of the astonishment of the authors and of their inability to understand.

Notwithstanding the pioneering yet almost neglected study by Harrop (1919), the first modern account of arterial oxygen desaturation in individuals with high $\dot{V}O_2^{max}$ during maximal exercise at sea level (Dempsey et al. 1984) appeared 2 years before Oelz et al. (1986) published their paper. The multifactorial models of $\dot{V}O_2^{max}$ limitation were at their beginning (di Prampero 1985). More than 10 years after, Ferretti and di Prampero (1995) expanded the Prampero's multifactorial model to encompass hypoxia. Shortly afterward, Ferretti et al. (1997) demonstrated that athletes with high $\dot{V}O_2^{max}$ at sea level undergo a larger $\dot{V}O_2^{max}$ fall in hypoxia than sedentary individuals, who have a low $\dot{V}O_2^{max}$ at sea level. This is a consequence of exercise-induced arterial hypoxemia, which causes them operating close to the steep part of the oxygen equilibrium curve at maximal exercise already at sea level. Therefore, at extreme altitude, differences in $\dot{V}O_2^{max}$, which may be quite large at sea level, tend to disappear: indeed, there is no need of elevated $\dot{V}O_2^{max}$ values at sea level to attain the highest peaks on Earth!

Fig. 10.9 Chris Pizzo taking alveolar gas sampling from himself on top of Mount Everest. From West and Lahiri (1984), with the kind permission of the American Physiological Society

The second consequence was the AMREE. This expedition was organized by John West, a renowned Australian respiratory physiologist active for a long time in San Diego, California, whose worldwide fame largely relies on his small booklet of respiratory physiology for students. West has not the bright speculative mind of Hermann Rahn, Rodolfo Margaria, or Johannes Piiper. He refrains from theoretical analysis, had troubles with the cute analysis of ventilation–perfusion heterogeneity carried out in Rochester and in Buffalo, but has a strong practical attitude, which made him the ideal person for setting up the AMREE. Several physiologists and high-altitude clinical specialists joined to perform a wide set of measurements of a variety of physiological variables pertinent to altitude adaptation, aimed at demonstrating how was it possible to reach the summit of Everest without supplementary oxygen. The approach was descriptive, but the data collected were extremely important, highly influential, and quite unique.

A climbing practitioner, Chris Pizzo, attained the summit and performed alveolar air sampling on it (Fig. 10.9). The obtained value for P_ACO_2 of 7.5 mmHg demonstrated the extremely high degree of hyperventilation on the summit. P_AO_2 was 34 mmHg. Arterial blood composition was estimated, assuming $P_aCO_2 = P_ACO_2$: the resulting P_aO_2 turned out equal to 28 mmHg (West et al. 1983b). These estimates are coherent with the measured arterial blood gas data reported 25 years later (Grocott et al. 2009). Barometric pressure on that day (October 24th, 1981) was 253 mmHg, much higher than predicted from standard tables, whence a P_IO_2 of 43 mmHg was calculated (West et al. 1983c). Upon return to the highest camp, $\dot{V}O_2^{max}$ at an equivalent altitude to the summit was determined on summiteers. The famous 1.07 L min^{-1} value was then obtained (West et al. 1983a). Despite the increase of 2,3-diphosphoglycerate, the strong alkalosis shifted the oxygen equilibrium curve to the left (Winslow et al. 1984). This prevented an excessive fall of

P_aO_2. Yet the estimated P_aO_2 remained so low, that it lied anyway on the steeper part of the oxygen equilibrium curve. This implies diffusion limitation of alveolar-capillary oxygen transfer.

Of the third, Operation Everest II, it would be somehow out of scope to have it described here in detail. It was such an ambitious, huge, costly, demanding, comprehensive experiment, that it marked the history of altitude physiology in the late 1980s. Its aim was to simulate a Mount Everest ascension in hypobaric chamber and to perform as many measurements as possible concerning the response of the respiratory system in its broadest sense to chronic hypoxia exposure. Clinical, functional, chemical, and structural determinations were carried out. The results were reported in a long series of papers, characterized by the same initial title. An overview of the experiment can be found in Houston et al. (1987).

Several findings and predictions from AMREE were confirmed (Cymerman et al. 1989), but the reported alkalosis was less intense than on the field (Malconian et al. 1992). Many invasive measurements, unfeasible on the field, could be carried out (Reeves et al. 1987, 1990; Sutton et al. 1988; Wagner et al. 1987). Muscle structural studies provided results in line with those obtained by the Cerretelli–Hoppeler connection (Green et al. 1989; McDougall et al. 1991). Only researchers from the USA and from Canada were admitted. West was not, and he criticized the study deeply, demonstrating that the subjects were not fully acclimatized when they attained a degree of hypoxia equivalent to the summit of Mount Everest (West 1988). Yet the study remains an unequaled example of powerful cooperation among researchers of different cultural extraction, although far away from the scientific attitude of the School of Milano.

10.5 The Everest Pyramid

The permanent high-altitude laboratory named the Pyramid was established in 1990 just below the Everest base camp, at an altitude of 5050 m above sea level. Ardito Desio, the geologist who led the 1954 Italian expedition to K2, promoted its construction (Cogo et al. 2000). Although older than 90, Desio was still active and in fairly good shape. He eventually died in 2001, at age 104.

The original project, called EV-K2-CNR (Everest, K2, and the acronym of the Italian National Research Council), debuted in the mid-1980s. It was designed as a basic technological project. The first aim was to measure the altitude of Mount Everest and of K2 again with more modern tools, after the shattering claim, widely amplified by the press, that K2 might indeed be taller than Everest. Two teams of Italian extreme altitude climbers carried the necessary equipment to the summit of the two peaks, where eventually precise triangulation measures were completed. They confirmed the previous height measurements of both peaks, that is, 8846 m for Everest and 8611 m for K2.

In 1989, two Italian industrial sponsors furnished the glass and aluminum pyramidal structure, which was called the Pyramid, to host permanent scientific

Fig. 10.10 An evening picture of the Everest Pyramid. Mount Pumori is on background. With the kind permission of the National Research Council, Italy

laboratories. This donation gave a new allure and a wider perspective to the project. It is when Paolo Cerretelli, who was the Italian scientist active in altitude physiology, with the highest international reputation, was called upon to coordinate physiological studies in the new laboratory.

Initially, the Pyramid was to be installed in Tibet. In April 1989, the National Research Council signed a cooperation agreement with the Chinese Academy of Science. However, this project was abandoned after the political consequences of the repression of student's demonstrations in Tiananmen square in Beijing. So, it was finally decided to move the laboratory to Nepal. In 1990, the Pyramid was built in front of the south face of Mount Everest. A new cooperation agreement was established with the Royal Nepal Academy of Science and Technology before the official inauguration of the structure. Since then, the Pyramid hosts rooms for visitors and investigators, physiological and technological laboratories, a permanent meteorological and seismic station, and a telecommunication station. Power is supplied by solar panels and by a small hydroelectric central nearby. Further technical details on the Pyramid can be found in Cogo et al. (2000). Figure 10.10 shows a picture of it.

The availability of the Pyramid allowed scientists from many different areas to be involved in many physiological, clinical, physical, and technological projects. Focusing on what concerns us, and in particular the School of Milano, three lines of scientific activity, all marked by the prominent role played by Cerretelli, are briefly discussed herewith.

10.5.1 The Lactate Paradox

Dill et al. (1931) and Edwards (1936) were the first to report that the maximal blood lactate concentration ($\hat{La}_b$) attained upon exhaustion at the end of an incremental exercise test was reduced in chronic hypoxia. Almost all those who performed exercise studies at altitude, including Cerretelli, who reported the phenomenon on Caucasians and on Sherpas (Cerretelli et al. 1982), have invariably confirmed this finding. However, $\hat{La}_b$ is known to be unchanged in acute hypoxia (Asmussen et al. 1948; Cerretelli 1967; Cerretelli et al. 1982), suggesting that the phenomenon may be specific to altitude acclimatization. In agreement with this hypothesis, Lundby et al. (2000) demonstrated that subjects fully acclimatized to altitude, acutely brought back to normoxia by letting them breathe an oxygen-enriched mixture, have an $\hat{La}_b$ as low as in hypoxia. The same study showed that the reduction of $\hat{La}_b$ appears progressively during the acclimatization process. These low $\hat{La}_b$ values were considered "paradoxical" (West 1986), and the term "lactate paradox" was eventually created to define it (Hochachka 1988). Obviously enough, the term "paradox" in this context does not imply a logical paradox, but only an apparent contradiction with respect to the general expectations of those years.

The first project that Cerretelli set up at the Pyramid was on the lactate paradox. Several hypotheses were tested. He postulated that the lactate paradox could be due to reduced buffer capacity in chronic hypoxia (Cerretelli et al. 1982). To test this hypothesis on acclimatized subjects, $\hat{La}_b$ was measured before and after bicarbonate administration. No differences were found: $\hat{La}_b$ was reduced in both cases, a clear experimental evidence against the tested hypothesis (Kayser et al. 1993). Another hypothesis was that the maximal lactic power (Margaria et al. 1964) might be reduced in subjects acclimatized to altitude, and this was supported by the results (Grassi et al. 1995). The progressive reversibility of the lactate paradox upon return to sea level as a consequence of de-acclimatization was also demonstrated (Grassi et al. 1996). A lower $\hat{La}_b$ in chronic hypoxia was shown also after supramaximal exercise of 30 s, but not of 10 s duration (Grassi et al. 2001). On this basis, the hypothesis that exercise time to exhaustion may play some role in the lactate paradox was formulated.

All these results are coherent with the observation that there is a reduced net lactate release from contracting muscles in chronic hypoxia (Bender et al. 1989; Brooks et al. 1992, 1998). In fact, the ensemble of this data seem to support the hypothesis that the lactate paradox may reflect a reduction of proton transport outside the muscle fiber, perhaps due to a reduced activity of the lactate-proton co-transporters. This hypothesis was demonstrated to be essentially incorrect, since van Hall et al. (2001) showed that the lactate paradox disappears if the sojourn is prolonged beyond 6 weeks, suggesting that the lactate paradox may be a transient phenomenon. These authors had to admit the possibility of a progressive up-regulation of the lactate-proton co-transporter with time of acclimatization. In fact, Juel et al. (2003) found no changes in lactate-proton co-transporter proteins MCT1 and MCT4 in muscles with altitude acclimatization, and attributed

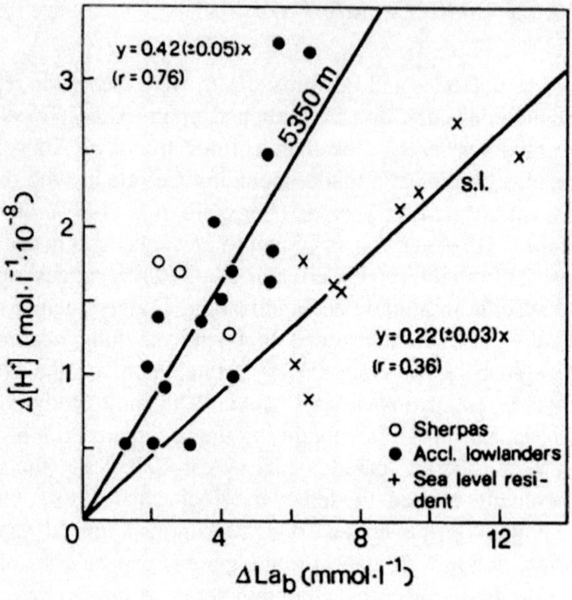

Fig. 10.11 Changes in blood [H$^+$] (Δ[H$^+$]) as a function of the increase in blood lactate concentration [ΔLa_b] in altitude natives and acclimatized lowlanders at 5350 m and sea-level residents. From Cerretelli and Samaja (2003)

altitude-induced changes in lactate flux across the sarcolemma to sarcolemmal carbonic anhydrase modulation of lactate-proton transporters.

The lactate-proton co-transporter hypothesis had the appreciation of Cerretelli. He admitted that during intense exercise muscles undergo important changes of the acid–base state, as a consequence of phosphocreatine hydrolysis and of the acceleration of glycolysis above and beyond the limits of aerobic metabolism (Cerretelli and Samaja 2003). Hydrogen ions flow across cell membranes, either by diffusion along concentration gradients or by facilitated diffusion thanks to various forms of facilitated transport, toward interstitial fluids and blood. Acid–base balance varies at altitude also by progressive excretion of bicarbonate, which reduces the buffering power of blood, and in particular of the cerebrospinal fluid. A demonstration of this state of affairs appears in Fig. 10.11.

Hochachka moved on a different track. He formulated a metabolic hypothesis, such that at maximal and supramaximal exercise, when lactate production exceeds lactate removal, there is a negative relationship between muscle phosphocreatine concentration and muscle proton or lactate concentration (Hochachka et al. 2002). According to this hypothesis, the lower the phosphorylation potential (see Chap. 2) at maximal exercise, the higher the lactate and proton concentrations. This hypothesis however proposes a correlation, not a mechanism, and forgets that the lactate paradox concerns blood, rather than muscle, lactate concentration. Cerretelli and Samaja (2003) interpreted it as the recognition of a tighter coupling of ATP demand and supply, which maximizes the energy yield per mole of oxidized fuel, and reduces the anaerobic capacity. This interpretation contains the implicit assumption that the stoichiometric relationship in glycolysis may change at altitude.

More recently, the occurrence of the lactate paradox was denied (van Hall et al. 2009). These authors found that both lowlanders and altitude natives had an elevated net leg lactate release, and arterial and muscle lactate concentrations during exercise in hypoxia. In the former group, these variables remained elevated during the entire acclimatization process. At exhaustion, however, high-altitude natives had higher muscle lactate and phosphate concentrations, and creatine-phosphocreatine than lowlanders. They concluded that lowlanders' muscles did not show a reduced capacity of producing lactate during acclimatization, which undermined the lactate paradox concept. The way they justified these results does not deserve a comment: suffice it to say that they claimed to have taken care of avoiding any muscle deteriorating condition in their study, at variance with previous studies. The lactate paradox fell into oblivion, yet it remains a curious, unresolved puzzle. Perhaps its understanding requires a more holistic approach than merely looking into muscle fibers.

10.5.2 *Effects of the Altitude History of Himalayan Ethnic Groups*

The main project that Cerretelli promoted at the Pyramid at the beginning of the twenty-first century was centered on the study of altitude adaptation in subjects characterized by different ethnic backgrounds, altitude exposure histories, and training conditions, during a standardized (approximately 4 weeks) sojourn at the Pyramid. According to him (Cerretelli 2013), this was the only comparative study carried out at such an altitude on homogeneous groups of selected subjects in identical environmental conditions, using the same protocols and experimental set-up.

An important finding from that project (Marconi et al. 2004) is that second-generation Tibetan lowlanders had a much smaller $\dot{V}O_2^{max}$ loss than untrained and trained Caucasians after the sojourn at the Pyramid. This was a consequence of the stronger ventilatory response to hypoxia in Tibetans than in the two latter groups. It was finally demonstrated that also in chronic hypoxia, as in acute hypoxia (Ferretti et al. 1997), the higher is the $\dot{V}O_2^{max}$ at sea level, the larger is the $\dot{V}O_2^{max}$ decrease at altitude. That paper has been discussed in Chap. 7 in terms of effects of arterial oxygen desaturation at maximal exercise and of position on which blood gases operate on the oxygen equilibrium curve.

Another interesting finding is that the remarkable work capacity of high-altitude natives highly relies on a lesser energy cost of pedaling on the cycle ergometer and of walking, and running on the treadmill (Marconi et al. 2005, 2006). The authors attributed the latter result to increased muscle myoglobin content, better antioxidant defense, and increased efficiency of oxidative phosphorylation.

10.5.3 The Links Between Energy Metabolism and Muscle Molecular Physiology

In more recent years, Cerretelli, who was always attracted and fascinated by technological developments, addressed his focus to new technologies allowing a deeper analysis of muscle molecular functional characteristics. These developments led to the identification of the hypoxia inducible factors, which are key regulator of cell hypoxic signaling (Semenza 2000, 2002; Wang et al. 1995). The Nobel Prize for physiology or medicine awarded to William Kaelin, Peter Radcliffe, and Gregg Semenza in 2019 for their contribution to our understanding of molecular oxygen sensing is the best recognition of the fundamental functional importance of hypoxia inducible factor 1 (HIF-1). This evolution opened a new scenario for a possible re-interpretation of adaptation to hypoxia. Cerretelli perceived this and acknowledged the importance of identifying also in humans the patterns of those proteins that are known to respond to hypoxia in isolated cells (Cerretelli and Gelfi 2011; Cerretelli et al. 2009). He and his coworkers, however, obtained contradictory results.

The molecular study of muscle energetics has made substantial steps forward, thanks to the development of advanced proteomic techniques. This allowed identification of a very large number of qualitative and quantitative protein changes induced by chronic external stimuli, such as altitude exposure. With this approach, it was possible to investigate, on approximately 500 protein spots, the patterns of muscle contractile and metabolic enzymatic proteins in the vastus lateralis and analyze the differences among altitude Tibetan natives, second-generation Tibetans born in Kathmandu, and Nepali lowlanders of Indian origin (Gelfi et al. 2004). Moreover, Viganò et al. (2008) investigated the adaptation patterns of muscle protein modulation during the first ten days of exposure to hypobaric hypoxia at the Capanna Margherita (4559 m). It appeared that a large number of proteins involved in iron transport, tricarboxylic acid cycle, oxidative phosphorylation, and response to oxidative stress were significantly decreased. By contrast, at the end of the sojourn, HIF-1α and pyruvate dehydrogenase kinase were at the pre-hypoxia levels, indicating that the long-term muscle adaptive response to hypoxia does not involve the stabilization of the key regulator of hypoxia signaling. In contrast, erythropoietin increases rapidly upon hypoxia exposure (Eckardt et al. 1989; Rodríguez et al. 2000). Since the connection of hypoxia inducible factors and erythropoietin production at altitude is solidly established since long (Jelkmann 2003; Lee and Percy 2011; Villafuerte 2015), this would suggest that stimulation of HIF-1 production prompts the erythropoietin increase only in the early phase of exposure to hypoxia.

10.6 The Lung as a Gas Exchanger

Another subject related to altitude adaptation, which attracted the interest of the School of Milano, concerns lung fluid balance and the gas exchange across the alveolar-capillary barrier, thanks in particular to the activity of Giuseppe Miserocchi, one of the last pupils of Margaria. In fact, Miserocchi took part as a young physiologist to the 1973 Italian Everest Expedition organized by Monzino (see above). At the time, it happened to him to take care of two cases of pulmonary edema above 6000 m, one was an Italian, the other a Nepalese. In both cases, he was stricken by the fact that, while severe symptoms of acute mountain sickness were present, lungs remained clean on auscultation for a reasonably long time until pulmonary edema acutely developed within minutes. This led to hypothesize that a mechanism preventing the development of edema is efficiently acting at the level of the air–blood barrier, and waning of this mechanism would explain the acute phase of lung flooding. Both individuals were rescued by helicopter.

After Miserocchi moved to the University of Milano Bicocca, his group devoted a considerable part of the research activity to frame the problem of lung fluid balance from normal to the acutely injured lung, in particular the proneness to develop lung edema (see recent review by Beretta et al. 2021).

The following research lines were developed:

1. Largely relying on experimental models, the mechanisms devoted to control of lung fluid balance were described.
2. Studies were then addressed to humans attempting to characterize the inter-individual differences in morpho-functional features of the air–blood barrier. These studies were stimulated by the large inter-individual variability reported for lung diffusion parameters as well as by the apparent different proneness to develop lung edema on exposure to high altitude.
3. In the same subjects studied under point 2, the adaptive response of the air–blood barrier was estimated in response to an increase in oxygen demand during exercise in hypoxia, which facilitates the development of lung edema.
4. Always in the same subjects, the inter-individual differences concerning the efficiency of the oxygen diffusion-transport in the air–blood barrier were explored.

10.7 Lung Fluid Balance from Normal to the Development of Edema

This section is only briefly recalled here, being central to the problem of lung edema; a detailed presentation of the experimental approach and results is provided in Chap. 9, Sect. 12. The very high surface area (*A*) of the air–blood barrier (~2500 $cm^2\ g^{-1}$) and its extreme thinness (~0.1 μm in its thinnest portion) serve the gas diffusion function. A minimum volume of extravascular water volume in the air–blood barrier is assured by the extremely low water permeability of the endothelial barrier (Miserocchi 2008). As a result, the air–blood barrier is very permeable to gases, but minimally permeable to water. The water content of the lung is well defined by the wet weight to dry weight ratio (*W/D*) that, in physiological conditions, is ~5. Trans-capillary and trans-epithelial water flows (J_v) are governed by the Starling law:

$$J_v = K_f[(P_1 - P_2) - \sigma(\pi_1 - \pi_2)] \tag{10.3}$$

where P and π are the hydraulic and the colloido-osmotic pressures across any two compartments, and K_f is the filtration coefficient. $K_f = L_p A$, where L_p is the water conductance. The term $[(P_1 - P_2) - \sigma(\pi_1 - \pi_2)]$ is the Starling pressure gradient generating flows, σ being the protein reflection coefficient that defines the selectivity of the barriers to plasma proteins.

The lung is normally well equipped to respond to edemagenic conditions, such as exercise and hypoxia, causing an increase in microvascular filtration. Exercise is edemagenic, as it causes increased lung blood flow and recruitment of the alveolar-capillary network (Hodges et al. 2007). Hypoxia is a potent factor leading to an increase in microvascular permeability to water and solutes (Hansen et al. 1994; Dehler et al. 2006).

Some features of the air–blood barrier are summarized here. The first one is the very low compliance of the interstitial compartment, ~0.5 ml $mmHg^{-1}$ 100 g^{-1} of wet weight (Miserocchi et al. 1993), reflecting the macromolecular organization of the proteoglycans component, which provides considerable rigidity (low compliance) of the lung interstitial compartment (Miserocchi et al. 1999; Negrini et al. 1996). Further, the proteoglycans coating the endothelial and epithelial barrier assure a very low permeability to water and solutes. A further interesting feature is that, in case of increase in microvascular filtration, a peculiar phenomenon occurs: water is captured by hyaluronan, a highly hydrophilic proteoglycan, to form gel; the increase in steric hindrance of the gel causes a remarkable increase in interstitial pressure, from ~−10 cmH_2O (physiological condition) up to ~+5 cmH_2O (Miserocchi et al. 1993). The increase in pressure provides the safety factor against edema formation (see Chap. 9), as it buffers further filtration and may actually favor some fluid reabsorption. As long as the safety factor is operating, water accumulation in the interstitial compartment is maintained within 10% of control value, so that *W/D* remains at ~5.5 (Miserocchi et al. 2001; Negrini et al. 2001). What deactivates the mechanism of protection against the development of edema? Inflammatory states, of either "sterile" (hypoxia, surgery, mechanical stretching) or infectious (viral, bacterial) type, may cause severe damage and disassembly to the native architecture of the proteoglycan family (Negrini et al. 1996; Passi et al. 1999; Miserocchi et al. 1999). The ensuing results are: (1) loss of low tissue compliance, increase of water and protein permeability, drop of interstitial pressure to zero, restoring of a filtration gradient. Modeling of Starling gradients and fluid exchanges indicates that the critical phase of developing severe edema pivots on reaching a threshold value of *W/D* (~6.5, Beretta et al. 2021). Above this threshold, fluid seeps out from capillaries and reaches the alveoli down a relatively small pressure gradient, acting however over a progressive increase in surface of damaged air–blood barrier. The time constant of this acute phase is surprisingly short, ranging ~4–6 min (Mazzuca et al. 2016; Parker and Townsley 2004). This suggests involvement of an extended surface of the air–blood barrier. Lymphatics can provide a passive negative-feedback control loop (Miserocchi 2008) and can actually increase flow in proportion to the increase in lung weight, but are only effective for $W/D < \sim 6.5$ (Mitzner and Sylvester 1986; Roselli et al. 1984).

10.8 The Air–Blood Barrier

10.8.1 Inter-individual Differences in Air–Blood Barrier Phenotype

As discussed in Chap. 9, diffusion parameters at the level of the air–blood barrier are defined as (Roughton and Forster 1957) (This equation is tantamount to Eq. (9.2), Chap. 9. It is repeated and discussed again here for reader's easiness.):

$$\frac{1}{D_L CO} = \frac{1}{D_m} + \frac{1}{\theta V_c} \tag{10.4}$$

being $D_L CO$ the total lung diffusive capacity for carbon monoxide, D_m the alveolar-capillary membrane diffusing capacity, V_c the blood volume in the alveolar-capillary network, and θ the chemical reaction rate of carbon monoxide with hemoglobin. Lung diffusion is routinely measured at total lung capacity (TLC) and large inter-individual differences have been reported even after normalization to individual lung volumes (Hughes and Pride 2001); these findings may suggest a specific phenotype-dependent difference of the morpho-functional features of the air–blood barrier. An attempt to decipher such differences came from measurements of the diffusion parameters at different lung volumes from functional residual capacity (FRC) up to TLC (Miserocchi et al. 2008).

Figure 10.12a shows that the increase in D_m remarkably differs among subjects on increasing lung volume. The highest D_m values at TLC were found in subjects displaying the highest increase in D_m on increasing lung volume. These differences have been interpreted considering that:

$$D_m \propto \frac{S_A}{d_A} \tag{10.5}$$

being S_A the overall surface of the air–blood barrier and d_A its thickness. The decrease in d_A on increasing lung volume was calculated as $\frac{1}{S_A}$, considering the

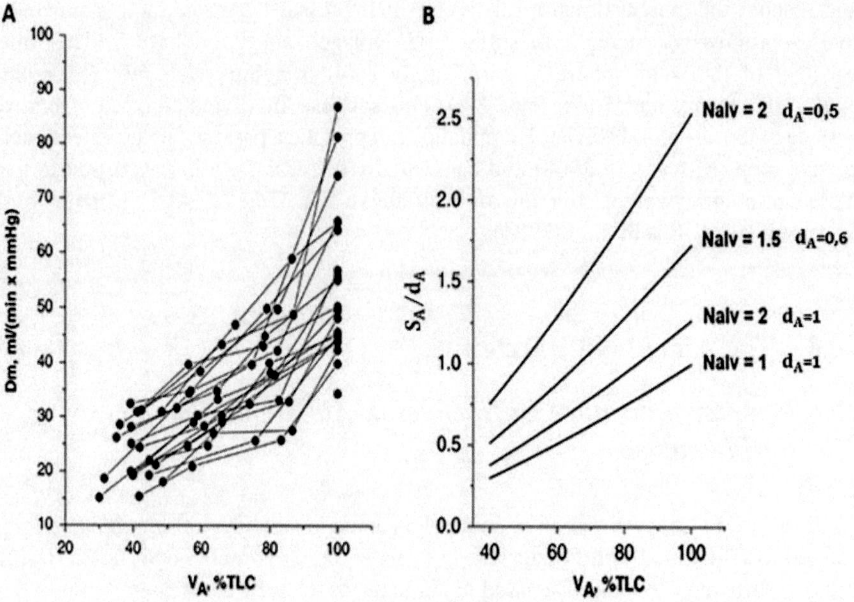

Fig. 10.12 (**a**) Individual plots of air–blood barrier diffusive capacity (D_m) vs. alveolar volume (V_A, % of TLC). (**b**) Numerical simulation to show the lung volume dependence of the ratio $\frac{S_A}{d_A}$ (surface to thickness ratio of the air–blood barrier) by modulating the number of alveoli (N_{alv}). From Miserocchi et al. (2008)

air–blood barrier as a lamina of constant volume (for details of the computational model refer to Miserocchi et al. 2008).

A numerical simulation (Fig. 10.12b) allowed an estimate of the volume dependence of $\frac{S_A}{d_A}$ by varying the number of alveoli (N_{alv}) and/or of d_A, expressed relative to that at FRC, in which N_{alv}, the alveolar volume (V_A), S_A, and d_A are set equal to unity. One can appreciate that a ~threefold difference in $\frac{S_A}{d_A}$ at FRC (say from 0.25 to 0.75) may justify a similar difference in D_m on increasing lung volume up to TLC. Accordingly, an inter-individual difference in N_{alv}, and in d_A can justify a corresponding difference in D_m. Note also that a lower number of alveoli implies a higher radius, compared to the case of a higher number of smaller alveoli.

Figure 10.13 shows that V_c (normalized to lung volume) decreases on increasing lung volume, as expected, due to the parenchymal squeezing of the pulmonary capillaries (Glazier et al. 1969; Koyama and Hildebrandt 1991; Mazzone et al. 1978). Again, large inter-individual differences were found: higher V_c values at FRC suggest a greater extension of the alveolar-capillary network; further, the higher the V_c value at FRC, the greater its decease on increasing lung volume.

It was concluded that the lung volume dependence of V_c and D_m may provide a hint to delineate the individual morpho-functional features of the air–blood barrier. In particular, the $\frac{V_c}{D_m}$ ratio was considered as being indicative of the geometry of the alveoli and the extension of the capillary network.

Fig. 10.13 Individual plots of V_c/V_A (ratio of pulmonary capillary blood volume normalized to lung volume) vs. lung volume (V_A, % of TLC). From Miserocchi et al. (2008)

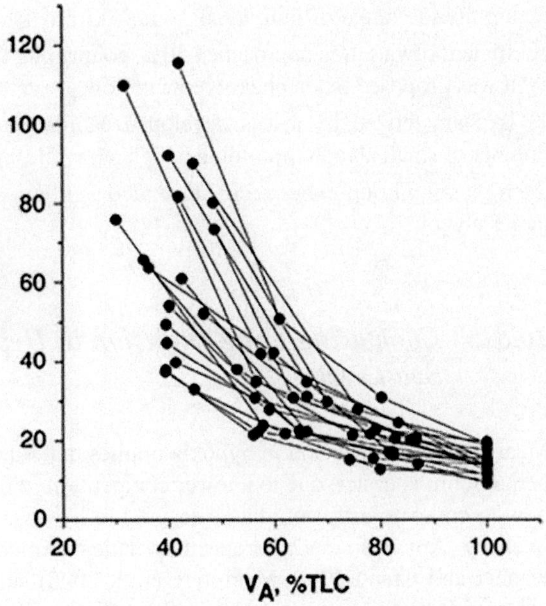

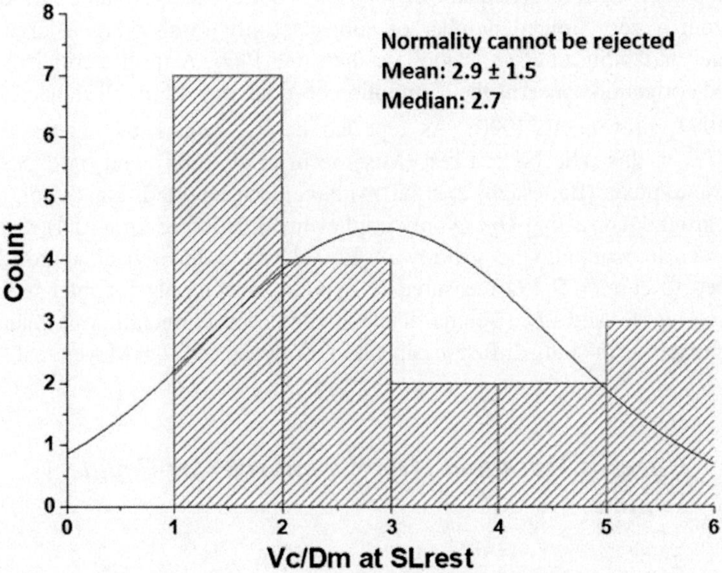

Fig. 10.14 Distribution of the $\frac{V_c}{D_m}$ ratio in the population studied. Data from Beretta et al. (2017b)

According to Fig. 10.14, the distribution of $\frac{V_c}{D_m}$ at 100% of TLC (the volume routinely chosen for pneumological evaluation) appears to be normal. Note that the coefficient of variation for repeated intra-subject measurements did not exceed 12%, facing a wide range of individual $\frac{V_c}{D_m}$ ratio (from 1 to 6), while on pooled data the coefficient of variation approached 50%, confirming the inter-individual differences.

It was proposed that a phenotype with low $\frac{V_c}{D_m}$ on the left tail in Fig. 10.14 (Group 1) is characterized by a less developed capillary network and a relatively high number of small alveoli, providing a high S_A, while a phenotype with high $\frac{V_c}{D_m}$ toward the right tail (Group 2) has a more extended capillary network and a lower number of larger alveoli.

10.8.2 Comparing Lung Diffusion in Hypoxia and at Sea Level

Measuring lung diffusion in hypoxia implies that binding of carbon monoxide with hemoglobin is easier, due to a lower competition with oxygen. The problem is then how to compare diffusion data measured in hypoxia with sea-level values. Guidelines by American and European societies (American Thoracic Society 1995; Kanner and Crapo 1986; MacIntyre et al. 2005) suggest correcting the measured value of D_LCO in hypoxia by accounting for the difference in P_AO_2. However, the regression coefficient of the equation proposed is quite low (0.36) and, further, data came from a very limited number of subjects (only 7) showing a large inter-individual variability of P_AO_2 values for the same P_IO_2. A further criticism to the proposed correction concerns the assumption of $\frac{V_c}{D_m}$ set to 1.42 for all subjects (Cotes et al. 1972; Frey et al. 1990). As reported before, remarkable inter-individual differences in this ratio both at rest (Miserocchi et al. 2008), and on exposure to hypobaric hypoxia (Bartesaghi et al. 2014) have been reported. Finally, on mathematical ground, correcting D_LCO only implies invalidation of Eq. (10.4).

In order to maintain the validity of Eq. (10.4), a theoretical approach was developed to correct D_LCO measured in hypoxia, accounting for inter-individual differences in V_c and D_m (Beretta et al. 2017a). The correction was mandatory wishing to derive the lung diffusion capacity for oxygen (D_LO_2) (Meyer et al. 1981).

10.8.3 Effect of Work and Hypoxia on Alveolar-Capillary Volume

Diffusion parameters have been measured at sea level at rest and after exposure to edemagenic lung conditions (Bartesaghi et al. 2014; Beretta et al. 2017b), namely work at sea level and in hypobaric hypoxia at rest (at 3269 m and 3850 m; $P_IO_2$107 mmHg and 90 mmHg, respectively) and during work (~60% of $\dot{V}O_2^{max}$). Figure 10.15 shows the relationship between the increase of D_LO_2 and that of $\dot{Q}$, both normalized to sea-level rest conditions. The increase in D_LO_2 favors oxygen

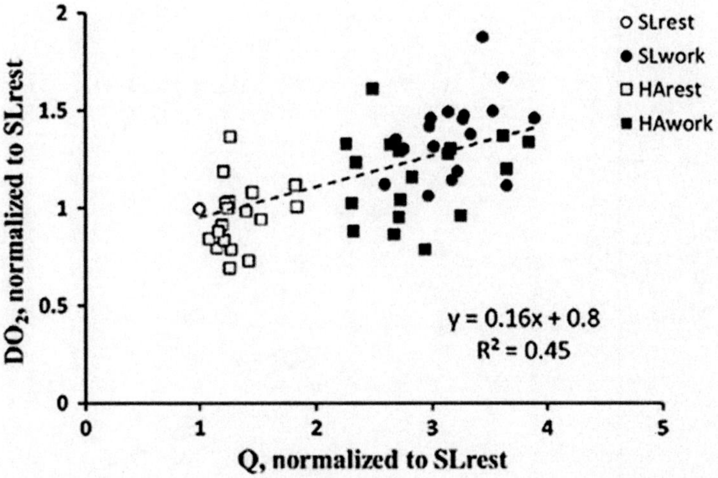

Fig. 10.15 Lung diffusive capacity for oxygen (D_LO_2) vs. cardiac output ($\dot{Q}$). Both variables are expressed as % of sea-level values at rest. From Beretta et al. (2019)

uptake, while the increase in $\dot{Q}$ favors oxygen delivery; note that the increase in $\dot{Q}$ is ~6 times greater than that in D_LO_2.

Given the differences in air–blood barrier phenotype, based on the $\frac{V_c}{D_m}$ ratio, a reasonable question was to estimate how edemagenic conditions would impact on pulmonary vasomotion in the capillary district. The question appeared justified, considering that subjects with high $\frac{V_c}{D_m}$ ratio (Group 2) would be more exposed to edemagenic conditions, being endowed with a more extended alveolar-capillary network and thus a greater overall capillary surface. Capillary recruitment has the advantage to favor gas diffusion by increasing S_A and the pool of hemoglobin to fix oxygen; on the other hand, the disadvantage is the increase in surface area available for microvascular filtration, a critical point in edema formation.

Subjects in Group 1 responded to edemagenic conditions by increasing V_c, suggesting recruitment of the capillary vascular network; conversely subjects of Group 2 did the opposite (Bartesaghi et al. 2014). Group 2 had a ~20% higher $\dot{V}O_2^{max}$ at sea level, but a greater decrease (−23%) relative to Group 1 (−19%) at 3269 m, the difference being significant. This is in line with the general concept that subjects with elevated $\dot{V}O_2^{max}$ at sea level have a greater fall of $\dot{V}O_2^{max}$ at altitude (Ferretti et al. 1997, see Sect. 10.3.4). An inverse correlation was found between changes in V_c and corresponding changes in D_m (Fig. 10.16). The simplest interpretation of this finding is that a volumetric increase in capillary blood (the case of Group1) is at the expense of alveolar volume, the opposite being true for Group 2.

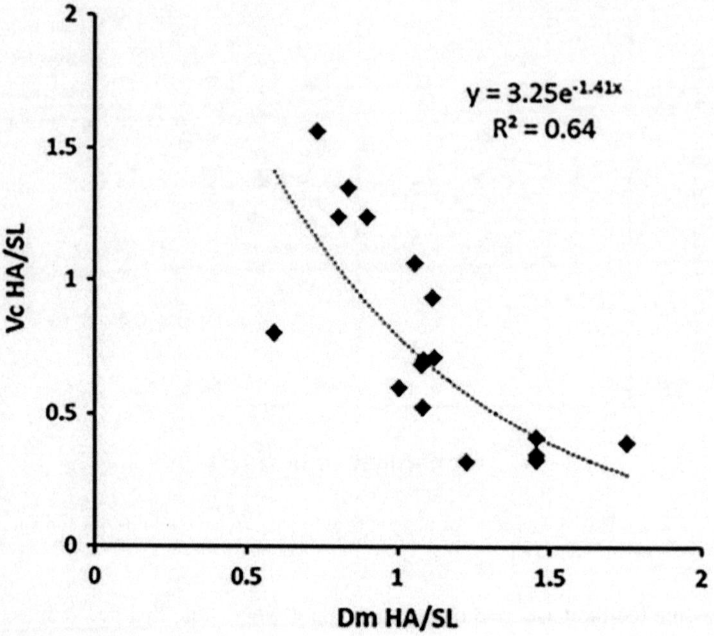

Fig. 10.16 Relationship between capillary blood volume (V_c) and membrane diffusion capacity of the air–blood barrier (D_m); both variables expressed as ratios to sea-level values at rest. From Beretta et al. (2017b)

10.8.4 Modeling the Alveolar-Capillary Equilibration

It is known that the alveolar-arterial difference in oxygen pressure ($P_AO_2 - P_aO_2$) is due to increase in work and on hypoxia exposure; what is the reason for this phenomenon? As already pointed out in Chap. 9, Johannes Piiper (1924–2012) and Peter Scheid proposed a sound model defining the alveolar-capillary equilibration across the air–blood barrier has been proposed based on a mass balance equation (Piiper and Scheid 1981). Setting oxygen uptake by the blood as $d\dot{M}O_2$ (molar rate of oxygen uptake) one has:

$$d\dot{M}O_2(x) = \dot{Q}\, dCO_2(x) \tag{10.6}$$

where from the corresponding increase in oxygen carried by the blood is given by the product of $\dot{Q}$ times the increasing concentration of oxygen in the blood, dCO_2, along the capillary length *(x)*.

The development of the theoretical model considers P_AO_2, P_aO_2, and $P_{\bar{v}}O_2$, and allows inclusion also of D_LO_2 and β_b (see Chap. 9, Sect. 8; Piiper and Scheid 1981; Ferretti et al. 2022). The equation defining the alveolar-capillary equilibration for oxygen (L_{eq}) at the exit from the pulmonary capillary (Chap. 9, Eq. 9.3) defines an exponential increase of oxygen uptake across the air–blood barrier to reach an equilibrium value at the exit from the pulmonary capillary. Equation (9.3) tells that, in case of perfect equilibration, $P_AO_2 - P_aO_2 = 0$ mmHg, so that $L_{eq} = 0$; conversely, in case of 100% shunt, the numerator is equal to the denominator, and thus $L_{eq} = 1$.

Aiming to characterize the kinetics of the equilibration process as a function of time (t) as blood flows along the capillary, one can write (Beretta et al. 2019):

$$d\dot{M}O_2(t) = \dot{Q}\, dCO_2(t) \tag{10.7}$$

Since the oxygen uptake along the capillary varies exponentially, the equilibrium reached at the exit from the capillary may be also expressed as:

$$L_{eq} = e^{-\frac{t_t}{\tau}} \tag{10.8}$$

where t_t is the blood transit time in the pulmonary capillary, estimated as the ratio of V_c to $\dot{Q}$. Note that t_t reflects both a local phenomenon relating to vasomotion concerning the lung capillary network and the increase in $\dot{Q}$.

Knowledge of t_t allows drawing the kinetics of oxygen equilibration as a function of time as:

$$L_{eq} = e^{-\frac{D_LO_2}{\dot{Q}\,\beta_o}\frac{t_t}{\tau}} \tag{10.9}$$

t being a fraction of t_t. The time constant τ of the equilibration process is defined as

$$\tau = \frac{\beta_o\, V_c}{D_LO_2} \tag{10.10}$$

Figure 10.17 shows two cases for subjects with either a high or low $\frac{V_c}{D_m}$ ratio (A and B, respectively).

For the graphical representation of the kinetics, the ordinate reports 1-L_{eq}; accordingly, in the case of full equilibration one has $1 - L_{eq} = 1$. Continuous lines show the time course of alveolar-capillary equilibration in normoxia at rest. The time course of the equilibration depends on the time constant τ. Note that at rest, τ in the subject with a high $\frac{V_c}{D_m}$ was longer, reflecting a greater V_c. In both subjects, t_t was long enough to allow complete equilibration. During work in hypoxia (3840 m, P_IO_2= 90 mmHg, dashed lines) in both subjects τ was increased (mostly reflecting a ~twofold increase in β_b), thus slowing down the kinetics of equilibration; further, $\frac{V_c}{D_m}$, decreased in A and increased in B. Changes in t_t reflect not only a local phenomenon (vasoconstriction in A; vasodilation in B), but also the increase in $\dot{Q}$. At a similar relative workload during work in hypoxia, $\dot{Q}$ (normalized to body weight) was ~50% greater in the subject with high $\frac{V_c}{D_m}$, compared to subject with low $\frac{V_c}{D_m}$.

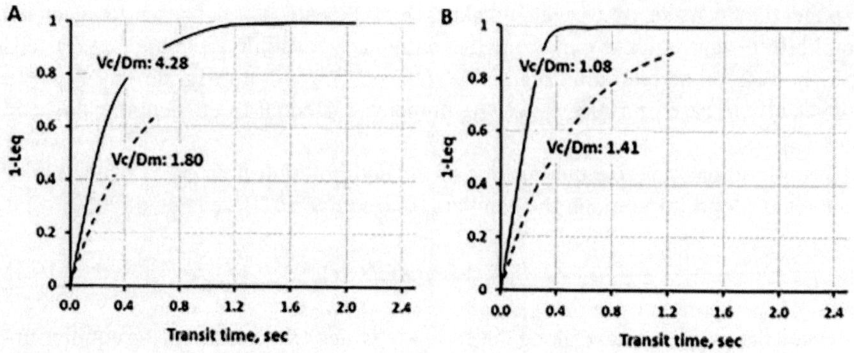

Fig. 10.17 Kinetics of alveolar-capillary equilibration for oxygen in two subjects at sea level at rest (continuous line) and during work in hypoxia (dashed line). For the graphical representation, the ordinate reports 1-L_{eq}. Data from one subject with high $\frac{V_c}{D_m}$ (panel A) and low $\frac{V_c}{D_m}$ (panel B) at rest, respectively. Data from Beretta et al. (2019)

In panel A, the remarkable shortening of t_t limited the equilibration at 0.6, while in panel B, due to a longer t_t, equilibration was only slightly decreased. Thus, during work performed in hypoxia, facing an average $P_AO_2 \sim 55$ mmHg, some individuals are able to still reach a satisfactory alveolar-capillary equilibration (Group 1), while, in other subjects (Group 2) this process may be limited by the decrease in t_t. The lack of equilibration increases exponentially for $t_t < \sim 1.5$ s (Beretta et al. 2017b).

The increase in blood velocity, consistent with the decrease in t_t, implies an increase in shear rate, a recognized cause of increase in endothelial microvascular and protein permeability (Kang et al. 2014; Lakshminarayanan et al. 2000; Sill et al. 1995). Figure 10.18 shows L_{eq} plotted vs. $1/t_t$ (an index of velocity). The relatively large dispersion of the data suggests that the net effect of increasing blood velocity on L_{eq} varies among subjects, being however definitely greater in subject with high $\frac{V_c}{D_m}$ (red dot in Fig. 10.18) as opposed to subject with low $\frac{V_c}{D_m}$ (blue dot in Fig. 10.18).

Figure 10.19 shows that, in healthy people in the most edemagenic conditions (work in hypoxia, $P_AO_2 \sim 55$ mmHg) the distribution of L_{eq} is normal. The positions of the two subjects referring to Fig. 10.18 (red and blue dots, respectively) reflect the remarkable inter-individual variability in the efficiency of alveolar-capillary equilibration, when shifting from low to high $\frac{V_c}{D_m}$.

10.9 Lung Water Balance and Phenotype

Miserocchi and his group aimed also at estimating potential changes in lung water balance in the subjects exposed to edemagenic conditions (Miserocchi et al. 2022). They relied on the measurement of the mechanical properties of the respiratory system by the forced oscillation technique (Bartesaghi et al. 2014) that proved

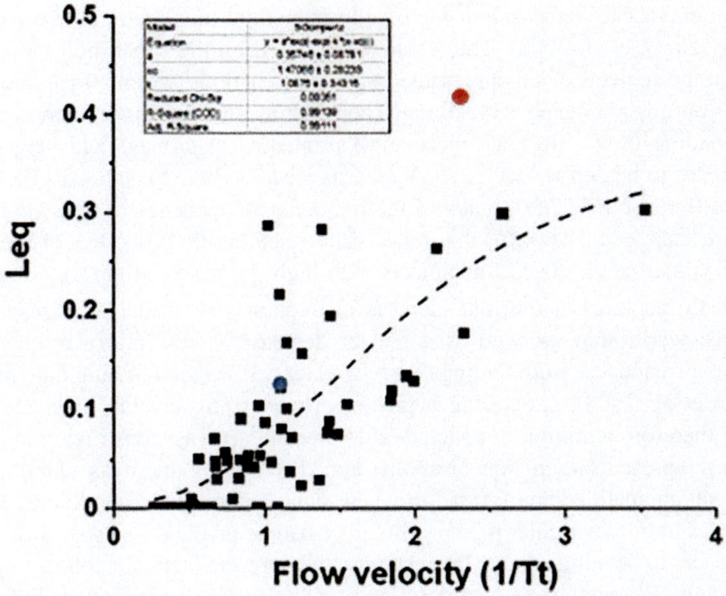

Fig. 10.18 Plot of L_{eq} vs. $1/t_t$ as index of blood flow velocity in the pulmonary capillaries. Data from Beretta et al. (2019)

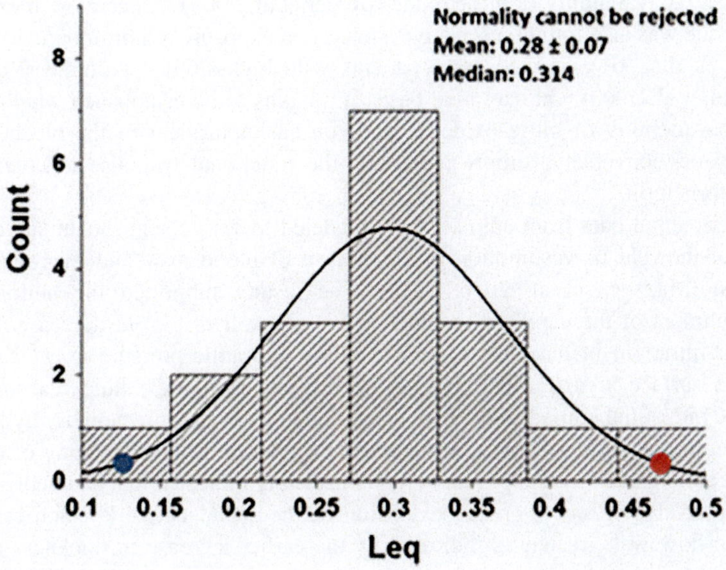

Fig. 10.19 Distribution of L_{eq} in the population studied during work in hypoxia. From Beretta et al. (2019) and Miserocchi et al. (2022)

sensitive enough to detect minor increase in interstitial lung water in experimental model (Dellacà et al. 2008). The technique is based on the superimposition at the mouth of the subjects of small pressure impulse oscillations with multiple frequency contents during spontaneous breathing. The resulting small changes in pressure and corresponding flow reflect the mechanical properties of airways and lung tissue. The indexes to be derived were: (1) the resistance at 5 Hz (*R5*) and at 20 Hz (*R20*), (2) the difference *R5* - *R20* (index of the frequency dependence of resistance), and (3) the reactance at 5 Hz (*X5*) that reflects the visco-elastic properties of the distal lung (Goldman et al. 2005). In subjects with high $\frac{V_c}{D_m}$ ratio (Group 2), comparing hypoxia to sea level, a fourfold increase in frequency dependence of respiratory resistance was found, as well as a greater decrease in respiratory reactance on decreasing frequency. Both findings reveal greater perturbation of lung fluid balance (Dellacà et al. 2008), suggesting a greater proneness to develop lung edema. It appears therefore tempting to conclude that precapillary vasoconstriction in Group 2 subjects represents a functional response aimed at limiting microvascular filtration, to prevent/attenuate edema formation when facing edemagenic conditions. Inborn differences in microvascular permeability may also be invoked to justify differences in proneness to develop edema. Interestingly, the opposite behavior concerning lung vasomotion in hypoxia was also described for the systemic circulation. Indeed, in mountaineers who are sensitive to high-altitude pulmonary edema, a decrease in forearm blood flow was found on hypoxia exposure, unlike non-sensitive subjects. This finding was attributed to impaired vascular endothelial function, due to decreased bioavailability of nitric oxide (Berger et al. 2005). A decrease in exhaled nitric oxide was also found in sensitive subjects on exposure to normobaric hypoxia (Busch et al. 2001), as well as in patients with high-altitude pulmonary edema (Duplain et al. 2000). On causative basis, it remains to be established whether the low bioavailability of nitric oxide depends on an impairment of the biochemical pathway or, conversely, simply represents the functional response to counteract edema formation.

Experimental data from animals were modeled to derive semi-quantitative estimates of the role of vasomotion in the control of blood flow and microvascular filtration (Mazzuca et al. 2016, 2019). The results supported the notion that de-recruitment of the capillary vascular bed is a powerful tool to decrease microvascular filtration: in fact this reduces capillary hydraulic pressure as well as the filtration surface area. Furthermore, in unperfused capillaries, fluid reabsorption from the interstitial compartment may occur due to a decrease in capillary hydraulic pressure, thus favoring recovery from edema. Modeling in-vivo imaging of development of lung edema on exposure to hypoxia clearly indicated that in alveolar units with larger alveoli and a greater extension of the septal network, microvascular filtration flow was greater, as indicated by the earlier increase in thickness of the interstitial space (Mazzuca et al. 2019). This can be appreciated in Fig. 10.20 showing a 2D image-based model of the decrease in capillary blood flow as change in color from yellow to blue in regions becoming edematous on exposure to hypoxia (12% oxygen, balanced nitrogen). This finding provides the simplest explanation of

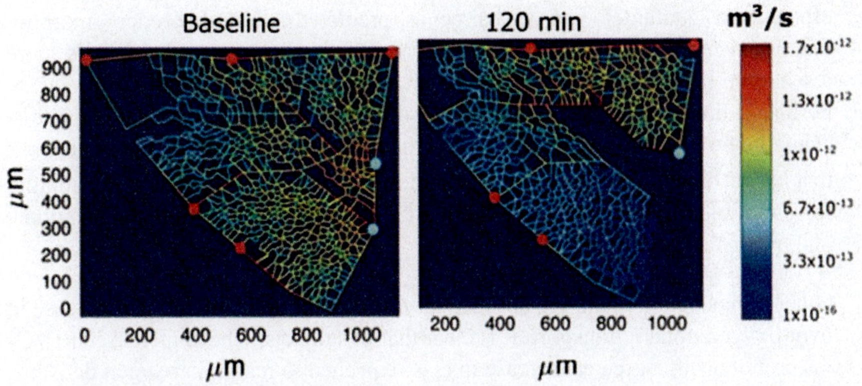

Fig. 10.20 Modeling the decrease in blood flow (shift toward blue) in the alveolar-capillary districts from baseline to 120 min of exposure to hypoxia, a well-known edemagenic condition. From Mazzuca et al. (2019)

the well-known fact that pulmonary hypertension on hypoxia exposure is the consequence of a marked precapillary vasoconstriction whose significance is that of limiting microvascular filtration preventing lung edema.

Were it possible to extend to humans these indications, they would lend support to the concept that subjects endowed with larger alveoli, the case of a high $\frac{V_c}{D_m}$ ratio, are more exposed to the risk of edema. The model also shows that flow limitation in the alveolar-capillary network shifts blood flow to alveolar corner vessels: although this prevents filtration, it represents a disadvantage to alveolar-capillary equilibration by generating a shunt-like effect.

10.10 Diffusion and Perfusion Limitation in the Air–Blood Barrier

Obviously, the same L_{eq} value is obtained considering either $\frac{D_L O_2}{\dot{Q}\beta_o}$ or $\frac{t_t}{\tau}$. These exponents provide, in fact, different but complementary related information. $\frac{D_L O_2}{\dot{Q}\beta_o}$ is the ratio between diffusion to perfusion capacity (ratio of two conductances indeed). On increasing oxygen demand, perfusion capacity increases ~sixfold more than diffusion capacity; as a consequence, a decrease in $\frac{D_L O_2}{\dot{Q}\beta_o}$ (and in 1 - L_{eq}) is expected.

In general, one could re-define the concepts of diffusion and perfusion limitation at the level of the air–blood as follows:

1. Diffusion limitation to oxygen flow may be restricted to air–blood barrier diffusivity, essentially reflecting the state of the lung extravascular water. A phenotype with lower $\frac{V_c}{D_m}$, implying a greater number of alveoli of smaller size, appears more

efficient to guarantee gas exchange as predicted by a theoretical morpho-functional model (Sapoval et al. 2002); further, indications are that it is more resistant to the development of lung edema.

2. Perfusion limitation to oxygen carried in the blood relates with t_t, that is the time available for gas equilibration. t_t reflects the flow velocity in the capillary network that in turn depends upon the extension of the capillary bed and on $\dot{Q}$. Accounting for $\frac{t_t}{\tau}$ allows definition of a new reading of a shunt-like effect. Again, a phenotype with low $\frac{V_c}{D_m}$ seems favored.

Other things being equal, the absence of vasoconstriction and a lower increase in $\dot{Q}$ favor oxygen uptake/transport. It is clear that in subjects, whose increase in D_LO_2 is less pronounced, a greater increase in $\dot{Q}$ is requested to maintain oxygen delivery: accordingly, in these same subjects, the decrease of $\frac{t_t}{\tau}$ reflects both vasoconstriction and the increase in $\dot{Q}$. Note that in hypoxia, the increase in τ, mostly reflecting the increase in β_b, becomes an important factor limiting oxygen uptake/transport.

The present data are of interest to people reaching high altitudes on trekking expeditions, who are exposed to the risk of high-altitude pulmonary edema. However, one cannot exclude a potential clinical relevance aiming to characterize cardiopulmonary disorders as well as conditions of decrease of the available pulmonary vascular bed (lung resection, thrombosis) that are risk factors for developing lung edema. Based on the correlation existing between the efficiency of the alveolar-capillary equilibration with the $\frac{V_c}{D_m}$ ratio, one can suggest the methods for its determination at ambulatory pneumological level:

- Estimate of $\frac{V_c}{D_m}$ at TLC, relying on the D_LNO/D_LCO technique (D_LNO being the lung diffusing capacity for nitrogen monoxide), at sea level, at rest, and on exercise.
- Estimate of t_t that requires the measurement of $\dot{Q}$ (by echocardiography with semi-recumbent set-up at rest and on exercise).
- A valid potentiation of the trial requires the same determinations on exposure to normobaric hypoxia.

10.11 Further Considerations on $\dot{V}O_2^{max}$ Limitation at Altitude

Within the concept of a multifactorial $\dot{V}O_2^{max}$ limitation (di Prampero and Ferretti 1990; Ferretti and di Prampero 1995), we propose here a further analysis concerning a lung mechanism, yet with an impact on the oxygen transport and use (cardiovascular and mitochondrial function, respectively). One can therefore attempt to provide a contribution to the understanding of highly debated questions relating with factors limiting the increase in $\dot{V}O_2^{max}$ that brought to conflictual interpretation.

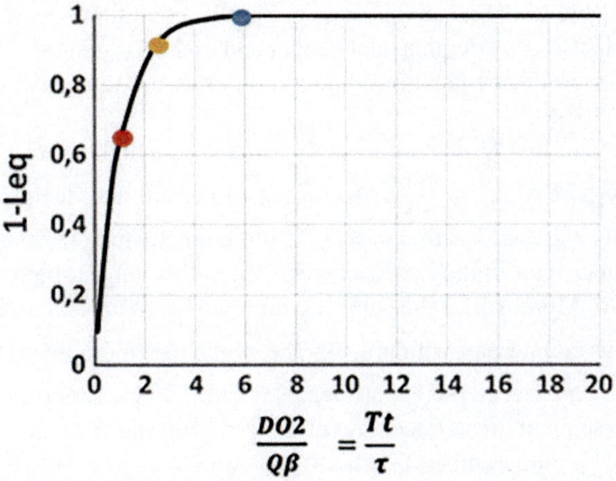

Fig. 10.21 Efficiency of alveolar-capillary equilibration ($1 - L_{eq}$) as a function of Piiper's equilibration deficit ($\frac{D_L O_2}{\dot{Q} \beta_o}$), equivalent to the ratio between capillary transit time (Tt, t_t in the text) and the time constant (τ) of the exponential equation. The red and blue points refer, respectively, to the subjects with high and low $\frac{V_c}{D_m}$ described in Fig. 10.16 (Beretta et al. 2019). The yellow point is an estimate for the subjects of Cerretelli (1976a). See text for further details

Let us come back to the celebrated experiment by Cerretelli at the 1973 Italian Expedition to Mount Everest, which is discussed at length in Sects. 10.3.3 and 10.3.4 (Cerretelli 1976a). We remind that Cerretelli, who was searching for a single factor of $\dot{V}O_2^{max}$ limitation, concluded his article by stating that the limiting factor was peripheral. He addressed his attention to a possible reduction of effective blood flow to the working muscles (Cerretelli 1982). Some twenty years later, Ferretti (2003) argued that Cerretelli (1976a) demonstrated the main role of cardiovascular oxygen transport as a factor limiting $\dot{V}O_2^{max}$. Further considerations on this issue can be made after the analysis of oxygen transfer at altitude using the model of Piiper and Scheid (1981) carried out by Beretta et al. (2019). These are summarized in Fig. 10.21, where the efficiency of alveolar-capillary equilibration ($1 - L_{eq}$) is plotted as a function of Piiper's equilibration deficit, id est the dimensionless $\frac{D_L O_2}{\dot{Q} \beta_b}$ ratio.

We highlight the sharp decrease in $1 - L_{eq}$ on decreasing $\frac{D_L O_2}{\dot{Q} \beta_o}$ and $\frac{t_t}{\tau}$ below 6 in Fig. 10.21. Note also that the subject represented by the red dot, who had high $\frac{V_c}{D_m}$ (Fig. 10.17a), had also clear signs of perturbation in lung fluid balance unlike the subject with low $\frac{V_c}{D_m}$ (Fig. 10.17b, blue dot in Fig. 10.21). No data are available to estimate the efficiency of the alveolar-capillary exchange in the subjects referred to by Cerretelli (1976a). One may propose that, other things being equal, a 10% loss of alveolar-equilibration (yellow point), coherent with a 10% decrease in cardiovascular oxygen transport, may well be justified by some degree of "interstitial edema" that did not wane on sudden exposure to normoxia. Neither CT scan nor ultrasound are, at present, able to diagnose this state before the condition becomes life threatening. One shall also consider that, at the other end of the oxygen cascade, a decrease in mitochondrial efficiency has been reported on exposure to high altitude (Murray and Horscroft 2016).

The subject with low $\frac{D_L O_2}{\dot{Q}\beta_o}$ and thus high $\frac{V_c}{D_m}$ had also a higher $\dot{V}O_2^{max}$ at sea level than his counterpart with low $\frac{D_L O_2}{\dot{Q}\beta_o}$. According to Ferretti and di Prampero (1995), this would imply a greater fraction of $\dot{V}O_2^{max}$ limitation by lung resistance to oxygen flow in the former than in the latter subject, Cerretelli's subject being intermediate between the two. Moreover, if the subject with low $\frac{D_L O_2}{\dot{Q}\beta_o}$ has also a higher $\dot{V}O_2^{max}$ at sea level than his counterpart with low $\frac{D_L O_2}{\dot{Q}\beta_o}$, he would also more likely be exposed to exercise-induced arterial hypoxemia (Dempsey et al. 1984; Dempsey and Wagner 1999), with consequent greater decrease of $\dot{V}O_2^{max}$ at altitude (see also Sect. 10.4 on the low $\dot{V}O_2^{max}$ of elite climbers).

Exercise-induced arterial hypoxemia is discussed at large in Chap. 7. Some further considerations are proposed here, with an impact on the onset of high-altitude pulmonary edema. The degree of ventilation/perfusion ($\dot{V}_A/\dot{Q}$) heterogeneity was found to correlate with a larger ventilatory response to exercise at sea level and at simulated altitude (Wagner et al. 1986), as well as with a higher $\dot{V}O_2^{max}$ (Hopkins et al. 1994). The last feature also characterizes subjects with high $\frac{V_c}{D_m}$, whose respiratory impedance data revealed a greater perturbation in lung fluid balance in hypoxia (Bartesaghi et al. 2014). Further, the increased $\dot{V}_A/\dot{Q}$ heterogeneity in response to exercise and hypoxia has been attributed to a perturbation of the fluid balance at the level of the air–blood barrier (Gale et al. 1985; Podolsky et al. 1996; Schaffartzik et al. 1992; Wagner et al. 1986). Moreover, a greater pulmonary hemodynamic response to exercise and hypoxia was found in subjects who had a previous history of high-altitude pulmonary edema (Eldridge et al. 1996), thus a greater proneness to develop lung edema. In summary, data suggest diffusion limitation in subjects having a phenotype characterized by a high $\frac{V_c}{D_m}$ due to some degree of edema developing in the air–blood barrier.

10.12 An Unsettled Scientific Controversy

In the previous paragraphs, the pathophysiology of lung edema has been discussed along the following functionally related steps, developed within the School of Milano:

- Lung edema develops as a consequence of an increase in microvascular permeability in edemagenic conditions (hypoxia is the specific case being considered).
- Precapillary vasoconstriction represents the adaptive response to prevent edema as it counteracts microvascular filtration.

- Pulmonary hypertension is the consequence of a downstream increase in flow resistance.
- In summary, from the physiological standpoint, one is leaning to interpret adaptive responses of pulmonary microcirculation as being protective to preserve the functional homeostasis, in this specific case avoiding edema.

A different point of view derived from the publication of a paper (West et al. 1991), showing that increasing transmural pulmonary capillary pressure ≥40 mmHg (experimental model in rabbits) caused capillary leak (concept of "stress failure" of pulmonary capillaries). Although it may appear reasonable that over-distension of an elastic tube causes a wall rupture, the paper by West et al. (1991) led to the development of the following, different paradigm: pulmonary hypertension developing on hypoxia exposure is not the consequence, but the cause of high-altitude pulmonary edema, in that it leads to increase in capillary pressure that, in turn, causes "stress failure" leading to edema (Maggiorini et al. 2001).

A complex and functionally fragmented evidence is being presented in the literature concerning both the response to hypoxia and the results from different approaches in the attempt to clarify the longitudinal distribution of vascular resistance from the arterial to the venous end. Thus, in general, this view is at variance with the physiological standpoint, as it invokes a self-destructing strategy of the lung in response to an edemagenic condition.

This alternative paradigm, however, presents several problematic issues. Firstly, it is a contradiction to state that high-altitude pulmonary edema is "*a high-permeability-type lung edema in the absence of inflammation, a concept first introduced under the term stress failure*" (Maggiorini et al. 2001). To be fair, one might state that, in presence of inflammation (hypoxia is a "sterile" form of inflammation, Andersson and Tracey 2011; Andersson et al. 2020), the increase in permeability (Dehler et al. 2006; Hansen et al. 1994; Kolliputi et al. 2010; Waxman and Kolliputi 2009) causes endothelial and epithelial lesions comparable to those caused by the "stress failure."

Maggiorini et al. (2001) consider high-altitude pulmonary edema as being characterized by exaggerated and unevenly distributed pulmonary vasoconstriction that results in significant increase in pulmonary artery and capillary pressure. What is not clear is why should pulmonary capillary pressure increase in face of "exaggerated" precapillary vasoconstriction. No model so far has been developed to confirm this fluid dynamic hypothesis. Heterogeneity in the regional pulmonary blood flow has been reported as a risk factor for lung edema (Hopkins and Levin 2006). An increase in precapillary resistance was demonstrated in parallel with pulmonary hypertension on hypoxia exposure (Parker et al. 1981), which protects indeed against the increase in capillary pressure.

Further, one can possibly cast some doubts concerning the estimate of capillary pressure derived from pulmonary artery occlusion pressure. Indeed, pulmonary capillary pressure was not found to bear a constant relationship to pulmonary artery occlusion pressure or wedge pressure (Cope et al. 1989). Moreover, the increase in capillary pressure cannot be accurately predicted by the pulmonary arterial or occlusion pressure (Cope et al. 1992). Finally, it is recognized that pulmonary artery occlusion pressure cannot be taken as a surrogate of pulmonary capillary pressure (Ganter et al. 2006).

How can then one explain the reported success of vasodilators to prevent/treat high-altitude pulmonary edema? One shall admit that vasodilators would decrease peripheral resistance by increasing capillary recruitment and this would inevitably cause an increase in microvascular filtration. In other words, vasodilators would vanish the anti-edemagenic role of vasoconstriction. One may comment that, since the distribution of edema is heterogeneous by nature, some control of lung fluid balance is only possible in regions where safety factors are still acting so that the Starling gradient still allows fluid reabsorption and nowhere else (see the recent model developed by Beretta et al. 2021). One should therefore hypothesize that in these regions the conditions of recovery should be lasting, despite the administration of a vasodilator. We are prone to consider that the reported exaggerated endothelin release in high-altitude pulmonary edema is an extreme form of defense against the development of edema rather than a biochemical disturbance (Sartori et al. 1999). One can certainly admit that the decrease in pulmonary artery pressure would alleviate the burden of the right ventricle.

10.13 The Harness Hang Syndrome

Several mountaineers, wearing harness and undergoing a fall, but remaining however hanging in orthostatic posture, were found dead on rescue, despite the absence of any trauma/injury. Thus, this specific case exposes the mountaineer to a severe risk of death. This condition, named as *harness hang syndrome*, may lead to unconsciousness, syncope, and to a chain of events leading to a lethal multivisceral hypoxia distress (Seddon 2002; Van Lieshout et al. 2003). Lanfranconi et al. (2017, 2019) investigated the impairments along the oxygen transport-utilization chain, with the aim of shading light on the pathophysiology of the harness hang syndrome (Lanfranconi et al. 2017, 2019).

Figure 10.22 shows the experimental model with a subject suspended on a tripod wearing the harness undergoing a motionless suspension test. Several cardiovascular and respiratory variables were continuously monitored, including tissue oxygenation at the level of the brain and of a skeletal muscle, relying on the near-infrared spectroscopy (NIRS) technique.

Subjects who did not develop syncope were able to activate a cardiovascular response to defend brain oxygen metabolic requirement (systolic and diastolic blood pressures were either unchanged or increased). Conversely, subjects who developed syncope (~20 min of suspension) had a remarkable decrease in systemic blood pressure, reflecting the failure of the vasomotor centers to trigger a cardiovascular adaptation. Results also clearly indicated that the most important and early pathophysiological parameter to monitor the occurrence of syncope was cerebral hypoxia as monitored by NIRS. Greater resistance to harness hang syndrome was found in

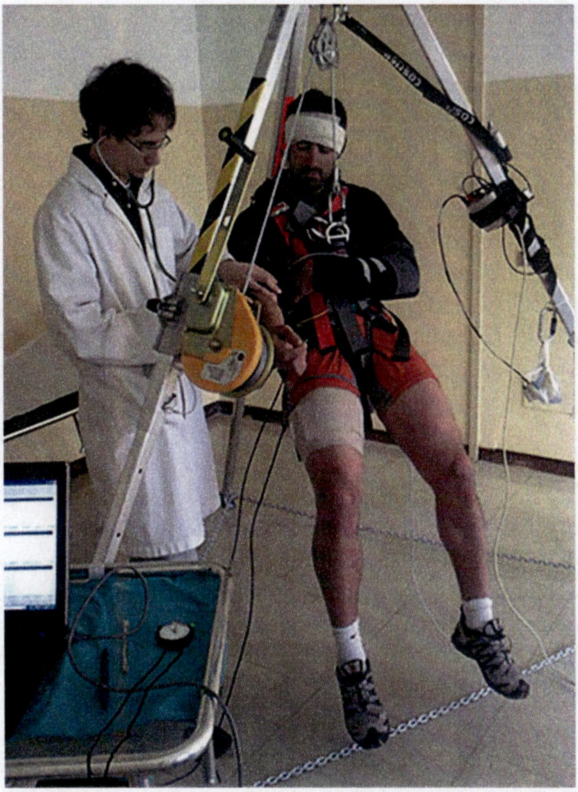

Fig. 10.22 Experimental set-up for suspension test. Bandages indicate the location of the NIRS probe. With the kind permission of Giuseppe Miserocchi

people responding with less marked fluctuations of both respiratory and cardiovascular reflex responses.

Unexpectedly, neither the long-term experience in using a harness nor higher aerobic fitness levels appear to protect against the development of harness hang syndrome. Given the relatively short time for developing this syndrome, the following recommendations were suggested: (1) climbers should avoid solo conditions; (2) harnesses should be designed to assure comfort, rapid rescue, and a body posture that does not result in more critical cerebral oxygen perfusion when losing consciousness; and (3) as to the immediate treatment of harness hang syndrome, keep the subject in supine position to prevent more severe cerebral hypoxia.

10.14 Conclusions

In conclusion, an Ariadne's thread can be perceived connecting the Pyramid to Angelo Mosso and his experiments on the Mount Rose. This thread has been unraveled from Mosso to Viale, Aggazzotti, and Herlitzka, then to Margaria and

Monzino, and Cerretelli and Miserocchi. Each of these men had a different practical, scientific, and psychological attitude to the study of the physiology of adaptation to altitude. Mosso was the pioneer who set the path, realizing the first permanent laboratory at altitude (current Istituto Mosso at Col d'Olen, Mount Rose, see Chap. 1), open to international collaboration. For Cerretelli, who was an adventurous man, the study of altitude played a central role in his scientific life. For Margaria, as for Miserocchi, it was a subsidiary field of activity providing some experimental opportunities for understanding or better focusing some aspects of their mainstream scientific activity. Margaria however, at variance with Miserocchi, did not involve himself directly, he rather encouraged and supported his pupils to participate in altitude research activities, namely in Monzino's expeditions. The case of Herlitzka was a peculiar one, as pointed out in Chap. 1: to him, altitude was the best environment for investigating the problem of hypoxia in aviation medicine, in a time when aircrafts were not pressurized yet and thus pilots were exposed to hypoxia during flights.

Along this Ariadne's thread, notwithstanding the different motivations of the involved scientists, and the various paths that they took, it eventually happened that the School of Milano contributed largely and significantly to the present knowledge of the physiology of human adaptation to high altitude. The contributions of the School of Milano to our understanding of the causes of $\dot{V}O_2^{max}$ decrease, of the lactate paradox, of altitude muscle deterioration, of the genesis of high-altitude pulmonary edema witness this engagement and deserve high consideration.

References

American Thoracic Society (1995) Single-breath carbon monoxide diffusing capacity (transfer factor) recommendations for a standard technique–1995 update. Am J Respir Crit Care Med 152:2185–2198

Andersson U, Tracey KJ (2011) HMGB1 is a therapeutic target for sterile inflammation and infection. Annu Rev Immunol 29:139–162

Andersson U, Ottestad W, Tracey KJ (2020) Extracellular HMGB1: a therapeutic target in severe pulmonary inflammation including COVID-19? Mol Med 26:42

Asmussen E, von Dobeln W, Nielsen M (1948) Blood lactate and oxygen debt after exhaustive work at different oxygen tensions. Acta Physiol Scand 15:57–62

Åstrand PO (1952) Experimental studies of physical working capacity in relation to sex and age. Munksgaard, Kopenhagen

Åstrand PO, Christensen EH (1964) Aerobic work capacity. In: Dickens F, Neil E, Widdas WF (eds) Oxygen in the animal organism. Pergamon Press, New York, pp 295–303

Bannister RG, Cunningham DJC (1954) The effects on respiration and performance during exercise of adding oxygen to the inspired air. J Physiol Lond 125:118–137

Bartesaghi M, Beretta E, Pollastri L, Scotti V, Mandolesi G, Lanfranconi F, Miserocchi G (2014) Inter-individual differences in control of alveolar capillary blood volume in exercise and hypoxia. Respir Physiol Neurobiol 190:96–104

Bender PR, Groves BM, McCullough RE, McCullough RG, Trad L, Young AJ, Cymerman A, Reeves JT (1989) Decreased exercise muscle lactate release after high altitude acclimatization. J Appl Physiol 67:1456–1462

Beretta E, Lanfranconi F, Grasso GS, Bartesaghi M, Alemayehu HK, Miserocchi G (2017a) Reappraisal of DLCO adjustment to interpret the adaptive response of the air-blood barrier to hypoxia. Respir Physiol Neurobiol 238:59–65

Beretta E, Lanfranconi F, Grasso GS, Bartesaghi M, Alemayehu HK, Pratali L, Catuzzo B, Giardini G, Miserocchi G (2017b) Air blood barrier phenotype correlates with alveolo-capillary O_2 equilibration in hypobaric hypoxia. Respir Physiol Neurobiol 246:53–58

Beretta E, Grasso GS, Forcaia G, Sancini G, Miserocchi G (2019) Differences in alveolo-capillary equilibration in healthy subjects on facing O_2 demand. Sci Rep 9:16693

Beretta EB, Romanò F, Sancini GA, Grotberg JB, Nieman GF, Miserocchi G (2021) Pulmonary interstitial matrix and lung fluid balance from normal to the acutely injured lung. Front Physiol 12. https://doi.org/10.3389/fphys.2021.781874

Berger MM, Hesse C, Dehnert C, Siedler H, Kleinbongard P, Bardenheuer HJ, Kelm M, Bärtsch P, Haefeli WE (2005) Hypoxia impairs systemic endothelial function in individuals prone to high-altitude pulmonary edema. Am J Respir Crit Care Med 172:763–767

Bert P (1878) La pression barométrique, recherches de physiologie expérimentale. Masson, Paris

Bloch KE, Buenzli JC, Latshang TD, Ulrich S (2015) Sleep at high altitude: guesses and facts. J Appl Physiol 119:1466–1480

Blomqvist CG, Saltin B (1983) Cardiovascular adaptations to physical training. Annu Rev Physiol 45:169–189

Brooks GA, Wolfel EE, Groves BM, Bender PR, Butterfield GE, Cymerman A, Mazzeo R, Sutton JR (1992) Muscle accounts for glucose disposal but not blood lactate appearance during exercise after acclimatization to 4,300 m. J Appl Physiol 72:2435–2445

Brooks GA, Wolfel EE, Butterfield GE, Cymerman A, Roberts AC, Mazzeo R, Reeves JT (1998) Poor relation between arterial [lactate] and leg net release during exercise at 4,300 m altitude. Am J Phys 275:R1192–R1201

Busch T, Bärtsch P, Pappert D, Grünig E, Hildebrandt W, Elser H, Falke KJ, Swenson ER (2001) Hypoxia decreases exhaled nitric oxide in mountaineers susceptible to high-altitude pulmonary edema. Am J Respir Crit Care Med 163:368–373

Cerretelli P (1959) Esistenza di una permanente stimolazione ipossica del centro respiratorio in individui acclimatati a quote di 5000-7500 m. s.l.m. (Himalaya) Atti Congr Int Med Aeron Spaziale:1–8

Cerretelli P (1961) Some aspects of the respiratory function in man acclimatized to high altitude (the Himalayas). Int Z Angew Physiol 18:386–392

Cerretelli P (1967) Lactacid O_2 debt in chronic and acute hypoxia. In: Margaria R (ed) Exercise at Altitude. Excerpta Medica, Amsterdam, pp 58–64

Cerretelli P (1976a) Limiting factors to oxygen transport on Mount Everest. J Appl Physiol 40:658–667

Cerretelli P (1976b) Metabolismo ossidativo ed anaerobico nel soggetto acclimatato all'altitudine. Minerva Aerospaziale 8:11–26

Cerretelli P (1982) O_2 breathing at altitude : effects on maximal performance. In: Brendel W, Zinc RA (eds) High altitude physiology and medicine. Springer, Berlin, pp 9–15

Cerretelli P (2013) Career perspective: Paolo Cerretelli. Extreme Physiol Med 2:13

Cerretelli P, Gelfi C (2011) Energy metabolism in hypoxia: reinterpreting some features of muscle physiology on molecular grounds. Eur J Appl Physiol 111:421–432

Cerretelli P, Hoppeler H (1996) Morphologic and metabolic response to chronic hypoxia. In: Fregly MJ, Blatteis CM (eds) Handbook of physiology. Environmental physiology. Oxford University Press, New York, sect. 4, vol II, pp 1155–1181

Cerretelli P, Margaria R (1961) Maximum oxygen consumption at altitude. Int Z Angew Physiol 18:460–464

Cerretelli P, Samaja M (2003) Acid-base balance at exercise in normoxia and in chronic hypoxia. Revisiting the "lactate paradox". Eur J Appl Physiol 90:431–448

Cerretelli P, Cruz JC, Farhi LE, Rahn H (1966) Determination of mixed venous O_2 and CO_2 tensions and cardiac output by a rebreathing method. Respir Physiol 1:258–264

Cerretelli P, Bordoni U, Debijadji R, Saracino F (1967) Respiratory and circulatory factors affecting the maximal aerobic power in hypoxia. Arch Fisiol 65:344–357

Cerretelli P, Veicsteinas A, Marconi C (1982) Anaerobic metabolism at high altitude : the lactacid mechanism. In: Brendel W, Zinc RA (eds) High altitude physiology and medicine. Springer, Berlin, pp 94–102

Cerretelli P, Marconi C, Dériaz O, Giezendanner D (1984) After effects of chronic hypoxia on cardiac output and muscle blood flow at rest and exercise. Eur J Appl Physiol 53:92–96

Cerretelli P, Marzorati M, Marconi C (2009) Muscle bioenergetics and metabolic control at altitude. High Alt Med Biol 10:165–174

Cogo A, Ponchia A, Pecchio O, Losano G, Cerretelli P (2000) Italian high altitude laboratories: past and present. High Alt Med Biol 1:137–147

Cope DK, Parker JC, Taylor MD, Houston M, Taylor AE (1989) Pulmonary capillary pressures during hypoxia and hypoxemia: experimental and clinical studies. Crit Care Med 17:853–857

Cope DK, Grimbert F, Downey JM, Taylor AE (1992) Pulmonary capillary pressure: a review. Crit Care Med 20:1043–1056

Costill DL, Fink WJ, Pollock ML (1976) Muscle fiber composition and enzyme activities of elite distance runners. Med Sci Sports 8:96–100

Cotes JE, Dabbs JM, Elwood PC, Hall AM, McDonald A, Saunders MJ (1972) Iron-deficiency anaemia: its effect on transfer factor for the lung (diffusion capacity) and ventilation and cardiac frequency during sub-maximal exercise. Clin Sci 42:325–335

Cymerman A, Reeves JT, Sutton JR, Rock PB, Groves BM, Malconian MK, Young PM, Wagner PD, Houston CS (1989) Operation Everest II: maximal oxygen uptake at extreme altitude. J Appl Physiol 66:2446–2453

Dehler M, Zessin E, Bärtsch P, Mairbäurl H (2006) Hypoxia causes permeability edema in the constant-pressure perfused rat lung. Eur Respir J 27:600–606

Dejours P, Girard F, Labrousse Y, Teillac A (1959) Etude de la régulation de la ventilation de repos chez l'homme en haute altitude. Rev Franc Etud Clin Biol 4:115–127

Dellacà RL, Zannin E, Sancini G, Rivolta I, Leone BE, Pedotti A, Miserocchi G (2008) Changes in the mechanical properties of the respiratory system during the development of interstitial lung edema. Respir Res 9:51

Dempsey JA, Wagner PD (1999) Exercise – induced arterial hypoxemia. J Appl Physiol 87:1997–2006

Dempsey JA, Hanson PG, Henderson KS (1984) Exercise-induced arterial hypoxaemia in healthy human subjects at sea level. J Physiol Lond 355:161–175

Desplanches D, Hoppeler H, Tüscher L, Mayet MH, Spielvogel H, Ferretti G, Kayser B, Leuenberger M, Grünefelder A, Favier R (1996) Muscle tissue adaptations of high altitude natives to training in chronic hypoxia or acute normoxia. J Appl Physiol 81:1946–1951

di Prampero PE (1985) Metabolic and circulatory limitations to VO_2max at the whole animal level. J Exp Biol 115:319–331

di Prampero PE, Ferretti G (1990) Factors limiting maximal oxygen consumption in humans. Respir Physiol 80:113–128

di Prampero PE, Atchou G, Brueckner J-C, Moia C (1986) The energetics of endurance running. Eur J Appl Physiol 55:259–266

Dill DB, Adams WC (1971) Maximal oxygen uptake at sea level and at 3,090 m altitude in high school champion runners. J Appl Physiol 30:854–859

Dill DB, Edwards HT, Folling A, Oberg SA, Pappenheimer AM, Talbot JH (1931) Adaptations of the organism to changes in oxygen pressure. J Physiol Lond 71:47–63

Dill DB, Myhre LG, Phillips EE, Brown DK (1966) Work capacity in acute exposures to altitude. J Appl Physiol 21:1168–1176

Duplain H, Sartori C, Lepori M, Egli M, Allemann Y, Nicod P, Scherrer U (2000) Exhaled nitric oxide in high-altitude pulmonary edema: role in the regulation of pulmonary vascular tone and evidence for a role against inflammation. Am J Respir Crit Care Med 162:221–224

Eckardt KU, Boutellier U, Kurtz A, Schopen M, Koller EA, Bauer C (1989) Rate of erythropoietin formation in humans in response to acute hypobaric hypoxia. J Appl Physiol 66:1785–1788
Edwards HT (1936) Lactic acid in rest and work at high altitude. Am J Phys 116:367–375
Ekblom B (1969) The effect of physical training on oxygen transport system in man. Acta Physiol Scand Suppl 328:1–45
Ekblom B, Goldbarg AN, Gullbring B (1972) Response to exercise after blood loss and reinfusion. J Appl Physiol 33:175–180
Ekblom B, Wilson G, Åstrand PO (1976) Central circulation during exercise after venesection and reinfusion of red blood cells. J Appl Physiol 40:379–383
Eldridge MV, Podolsky A, Richardson RS, Johnson DH, Knight DR, Johnson EC, Hopkins SR, Michimata H, Grassi B, Feiner J, Kurdak SS, Bickler PE, Wagner PD, Severinghaus JW (1996) Pulmonary hemodynamic response to exercise in subjects with prior high-altitude pulmonary edema. J Appl Physiol 81:911–921
Fagraeus L, Karlsson J, Linnarsson D, Saltin B (1973) Oxygen uptake during maximal work at lowered and raised ambient air pressures. Acta Physiol Scand 87:411–421
Fenn WO, Rahn H, Otis AB (1946) A theoretical study of the composition of the alveolar air at altitude. Am J Phys 146:637–653
Ferretti G (2003) Limiting factors to oxygen transport on Mount Everest 30 years after: a critique of Paolo Cerretelli's contribution to the study of altitude physiology. Eur J Appl Physiol 90:344–350
Ferretti G (2014) Maximal oxygen consumption in healthy humans: theories and facts. Eur J Appl Physiol 114:2007–2036
Ferretti G, di Prampero PE (1995) Factors limiting maximal O_2 consumption : effects of acute changes in ventilation. Respir Physiol 99:259–271
Ferretti G, Moia C, Thomet JM, Kayser B (1997) The decrease of maximal oxygen consumption during hypoxia in man : a mirror image of the oxygen equilibrium curve. J Physiol Lond 498: 231–237
Ferretti G, Fagoni N, Taboni A, Vinetti G, di Prampero PE (2022) A century of exercise physiology: key concepts on coupling respiratory oxygen flow to muscle energy demand during exercise. Eur J Appl Physiol 122:1317
Feyerabend P (1975) Against method. Verso, London
FitzGerald MP (1913) The changes in the breathing and the blood at various high altitudes. Philos Trans R Soc Lond 203:351–371
Frey TM, Crapo RO, Jensen RL, Kanner RE, Kass JE, Castriotta RJ, Mohsenifar Z (1990) Adjustment of DLCO for varying COHb, and alveolar PO_2 using a theoretical adjustment equation. Respir Physiol 81:303–311
Gale GE, Torre-Bueno JR, Moon RE, Saltzman HA, Wagner PD (1985) Ventilation-perfusion inequality in normal humans during exercise at sea level and simulated altitude. J Appl Physiol 58:978–988
Ganter CC, Jakob SM, Takala J (2006) Pulmonary capillary pressure. A review. Minerva Anestesiol 72:21–36
Gelfi C, De Palma S, Ripamonti M, Eberini I, Wait R, Bajracharya A, Marconi C, Schneider A, Hoppeler H, Cerretelli P (2004) New aspects of altitude adaptation in Tibetans: a proteomic approach. FASEB J 18:612–614
Glazier JB, Hughes JM, Maloney JE, West JB (1969) Measurements of capillary dimensions and blood volume in rapidly frozen lungs. J Appl Physiol 26:65–76
Gold AJ, Johnson TF, Costello LC (1973) Effects of altitude stress on mitochondrial function. Am J Phys 224:946–949
Goldman MD, Saadeh C, Ross D (2005) Clinical applications of forced oscillation to assess peripheral airway function. Respir Physiol Neurobiol 148:179–194
Grassi B, Ferretti G, Kayser B, Marzorati M, Colombini A, Marconi C, Cerretelli P (1995) Maximal rate of blood lactate accumulation during exercise at high altitude in humans. J Appl Physiol 79: 331–339

Grassi B, Marzorati M, Kayser B, Bordini M, Colombini A, Conti M, Marconi C, Cerretelli P (1996) Peak blood lactate and blood lactate vs. workload during acclimatization to 5050 m and in the deacclimatization. J Appl Physiol 80:685–692

Grassi B, Mognoni P, Marzorati M, Mattiotti S, Marconi C, Cerretelli P (2001) Power and peak blood lactate at 5050 m with 10 and 30 s "all out" cycling. Acta Physiol Scand 172:189–194

Green HJ, Sutton JR, Cymerman A, Young PM, Houston CS (1989) Operation Everest II: adaptations in human skeletal muscle. J Appl Physiol 66:2454–2461

Grocott MP, Martin DS, Levett DZ, McMorrow R, Windsor J, Montgomery HE (2009) Arterial blood gases and oxygen content in climbers on Mount Everest. New Engl J Med 360:140–149

Hansen JM, Olsen NV, Feldt-Rasmussen B, Kanstrup IL, Déchaux M, Dubray C, Richalet JP (1994) Albuminuria and overall capillary permeability of albumin in acute altitude hypoxia. J Appl Physiol 76:1922–1927

Harrop GA (1919) The oxygen and carbon dioxide content of arterial and of venous blood in normal individuals and in patients with anemia and heart disease. J Exp Med 30:241–257

Hochachka PW (1988) The lactate paradox : analysis of the underlying mechanisms. Ann Sports Med 4:184–188

Hochachka PW, Stanley C, Merkt J, Sumar-Kalinowski J (1983) Metabolic meaning of elevated levels of oxidative enzymes in high altitude adapted animals: an interpretive hypothesis. Respir Physiol 52:303–313

Hochachka PW, Beatty CL, Burelle Y, Trump ME, McKenzie DC, Matheson GO (2002) The lactate paradox in human high altitude physiological performance. News Physiol Sci 17:122–126

Hodges AN, Sheel AW, Mayo JR, McKenzie DC (2007) Human lung density is not altered following normoxic and hypoxic moderate-intensity exercise: implications for transient edema. J Appl Physiol 103:111–118

Holmgren A, Åstrand PO (1966) D_L and the dimensions and functional capacities of the O_2 transport system in humans. J Appl Physiol 21:1463–1470

Hopkins SR, Levin DL (2006) Heterogeneous pulmonary blood flow in response to hypoxia: a risk factor for high altitude pulmonary edema? Respir Physiol Neurobiol 151:217–228

Hopkins SR, McKenzie DC, Schoene RB, Glenny W, Robertson HT (1994) Pulmonary gas exchange during exercise in athletes. I. Ventilation—perfusion mismatch and diffusion limitation. J Appl Physiol 77:912–917

Hoppeler H, Lüthi P, Claassen E, Weibel ER, Howald H (1973) The ultrastructure of the normal human skeletal muscle. A morphometric analysis of untrained men, women, and well-trained orienteers. Pflügers Arch 334:217–232

Hoppeler H, Kleinert E, Schlegel C, Claassen H, Howald H, Cerretelli P (1990) Muscular exercise at high altitude. II. Morphological adaptation of skeletal muscle to chronic hypoxia. Int J Sports Med 11(Suppl 1):S3–S9

Houston CS, Sutton JR, Cymerman A, Reeves JT (1987) Operation Everest II: man at extreme altitude. J Appl Physiol 63:877–882

Howald H (1982) Training-induced morphological and functional changes in skeletal muscle. Int J Sports Med 3:1–12

Howald H, Pette D, Simoneau JA, Uber A, Hoppeler H, Cerretelli P (1990) Muscular exercise at high altitude. III. Effects of chronic hypoxia on muscle enzyme activity. Int J Sports Med 11 (Suppl 1):S10–S14

Hughes JM, Pride NB (2001) In defence of the carbon monoxide transfer coefficient Kco (TL/VA). Eur Respir J 17:168–174

Jelkmann W (2003) Erythropoietin. J Endocrinol Investig 26:832–837

Juel C, Lundby C, Sander M, Calbet JAL, van Hall G (2003) Human skeletal muscle and erythrocyte proteins involved in acid-base homeostasis: adaptations to chronic hypoxia. J Physiol Lond 548:639–648

Kang H, Cancel LM, Tarbell JM (2014) Effect of shear stress on water and LDL transport through cultured endothelial cell monolayers. Atherosclerosis 233:682–690

Kanner RE, Crapo RO (1986) The relationship between alveolar oxygen tension and the single-breath carbon monoxide diffusing capacity. Am Rev Respir Dis 133:676–678

Kayser B, Hoppeler H, Claassen H, Cerretelli P (1991) Muscle structure and performance capacity of Himalayan Sherpas. J Appl Physiol 70:1938–1942

Kayser B, Ferretti G, Grassi B, Binzoni T, Cerretelli P (1993) Maximal lactic capacity at high altitude. Effect of bicarbonate loading. J Appl Physiol 75:1070–1074

Kayser B, Hoppeler H, Desplanches D, Broers B, Marconi C, Cerretelli P (1996) Muscle ultrastructure and biochemistry of lowland Tibetans. J Appl Physiol 81:419–425

Kellas AM (1917) A consideration of the possibility of ascending the loftier Himalaya. Geogr J 49(26–47):1917

Kellas AM (2001) A consideration of the possibility of ascending Mount Everest. High Altitude Med Biol 2:431–461

Kolliputi N, Shaik NS, Waxman AB (2010) The inflammasome mediates hyperoxia-induced alveolar cell permeability. J Immunol 184:5819–5826

Koyama S, Hildebrandt J (1991) Air interface and elastic recoil affect vascular resistance in three zones of rabbit lungs. J Appl Physiol 70:2422–2431

Lacour JR, Flandrois R (1977) Le rôle du métabolisme aérobie dans l'exercice intense de longue durée. J Physiol Paris 73:89–130

Lakshminarayanan S, Gardner TW, Tarbell JM (2000) Effect of shear stress on the hydraulic conductivity of cultured bovine retinal microvascular endothelial cell monolayers. Curr Eye Res 21:944–951

Lanfranconi F, Pollastri L, Corna G, Bartesaghi M, Novarina M, Ferri A, Miserocchi GA (2017) The elusive path of brain tissue oxygenation and cerebral perfusion in harness hang syncope in mountain climbers. High Alt Med Biol 18:363–371

Lanfranconi F, Ferri A, Pollastri L, Bartesaghi M, Novarina M, De Vito G, Beretta E, Tremolizzo L (2019) Impact of hanging motionless in harness on respiratory and blood pressure reflex modulation in mountain climbers. High Alt Med Biol 20:122–132

Lee FS, Percy MJ (2011) The HIF pathway and erythrocytosis. Annu Rev Pathol 6:165–192

Luks AM, Swenson ER, Bärtsch P (2017) Acute high-altitude sickness. Eur Respir Rev 26:160096

Luks AM, Ainslie PN, Lawley JS, Roach RC, Simonson TS (2021) Ward Milledge and West's high altitude physiology and medicine. CRC Press, Boca Raton

Lundby C, Robach P (2016) Does 'altitude training' increase exercise performance in elite athletes? Exp Physiol 101:783–788

Lundby C, Saltin B, van Hall G (2000) The "lactate paradox", evidence for a transient change in the course of acclimatization to severe hypoxia in lowlanders. Acta Physiol Scand 170:265–269

MacIntyre N, Crapo RO, Viegi G, Johnson DC, van der Grinten CPM, Brusasco V, Burgos F, Casaburi R, Coates A, Enright P, Gustafsson P, Hankinson J, Jensen R, McKay R, Miller MR, Navajas D, Pedersen OF, Pellegrino R, Wanger J (2005) Standardization of the single breath determination of carbon monoxide uptake. Eur Respir J 26:720–735

Maggiorini M, Mélot C, Pierre S, Pfeiffer F, Greve I, Sartori C, Lepori M, Hauser M, Scherrer U, Naeije R (2001) High-altitude pulmonary edema is initially caused by an increase in capillary pressure. Circulation 103:2078–2083

Malconian MK, Rock PB, Reeves JT, Cymerman A, Houston CS (1992) Operation Everest II: gas tensions in expired air and arterial blood at extreme altitude. Aviat Space Environ Med 64:37–42

Marconi C, Marzorati M, Grassi B, Basnyat B, Colombini A, Kayser B, Cerretelli P (2004) Second generation Tibetan lowlanders acclimatize to high altitude more quickly than Caucasians. J Physiol Lond 556:661–671

Marconi C, Marzorati M, Sciuto D, Ferri A, Cerretelli P (2005) Economy of locomotion in high altitude Tibetan migrants exposed to normoxia. J Physiol Lond 569:667–675

Marconi C, Marzorati M, Cerretelli P (2006) Work capacity of permanent residents of high altitude. High Alt Med Biol 7:105–115

Margaria R, Galante E, Cerretelli P (1959) An efficient CO_2 absorber for experiments on metabolism. J Appl Physiol 14:1066–1068
Margaria R, Cerretelli P, Marchi S, Rossi L (1961) Maximum exercise in oxygen. Int Z Angew Physiol 18:465–467
Margaria R, Cerretelli P, Mangili F (1964) Balance and kinetics of anaerobic energy release durino strenuous exercise in man. J Appl Physiol 19:623–628
Margaria R, Mangili F, Cuttica F, Cerretelli P (1965) The kinetics of the oxygen consumption at the onset of muscular exercise in man. Ergonomics 8:49–54
Margaria R, Camporesi E, Aghemo P, Sassi G (1972) The effect of O_2 breathing on maximal aerobic power. Pfluegers Arch 336:225–235
Mazzone RW, Durand CM, West JB (1978) Electron microscopy of lung rapidly frozen under controlled physiological conditions. J Appl Physiol 45:325–333
Mazzuca E, Aliverti A, Miserocchi G (2016) Computational micro-scale model of control of extravascular water and capillary perfusion in the air blood barrier. J Theor Biol 400:42–51
Mazzuca E, Aliverti A, Miserocchi G (2019) Understanding vasomotion of lung microcirculation by in vivo imaging. Imaging 5:22
McDougall JD, Green HJ, Sutton JR, Coates G, Cymerman A, Young P, Houston CS (1991) Operation Everest II. Structural adaptations in skeletal muscle in response to extreme simulated altitude. Acta Physiol Scand 142:421–427
Meyer M, Scheid P, Riepl G, Wagner HJ, Piiper J (1981) Pulmonary diffusion capacities for O_2 and CO measured by a rebreathing technique. J Appl Physiol 51:1643–1650
Miserocchi G (2008) Mechanisms controlling the volume of pleural fluid and extravascular lung water. Eur Respir Rev 18:244–252
Miserocchi G, Negrini D, Del Fabbro M, Venturoli D (1993) Pulmonary interstitial pressure in intact in situ lung: transition to interstitial edema. J Appl Physiol 74:1171–1177
Miserocchi G, Passi A, Albertini R, Negrini D, De Luca G (1999) Interstitial pressure and proteoglycan degradation in hydraulic- and elastase-induced lung edema. Chest 116(1 Suppl):31S
Miserocchi G, Passi A, Negrini D, Del Fabbro M, De Luca G (2001) Pulmonary interstitial pressure and tissue matrix structure in acute hypoxia. Am J Phys 280:L881–L887
Miserocchi G, Messinesi G, Tana F, Passoni E, Adamo S, Romano R, Beretta E (2008) Mechanisms behind inter-individual differences in lung diffusing capacity. Eur J Appl Physiol 102:561–568
Miserocchi G, Beretta E, Rivolta I, Bartesaghi E (2022) Role of the air-blood barrier phenotype in lung oxygen uptakeand control of extravascular water. Front Physiol 13:811129
Mitchell JH, Blomqvist CG (1971) Maximal oxygen uptake. New Engl J Med 284:1018–1022
Mitzner W, Sylvester JT (1986) Lymph flow and lung weight in isolated sheep lungs. J Appl Physiol 61:1830–1835
Monzino G (1976) La spedizione italiana all'Everest. Mondadori, Verona
Monzino G, Meciani P, Marimonti L, Bich J, Carrel M, Pelissier C (1960) La spedizione G. M. '59 al Kanjut Sar. Rivista Mensile del Club Alpino Italiano 79:3–4
Mosso A (1897) La fisiologia dell'uomo sulle Alpi. Treves, Milano
Murray AJ, Horscroft JA (2016) Mitochondrial function at extreme high altitude. J Physiol Lond 594:1137–1149
Negrini D, Passi A, De Luca G, Miserocchi G (1996) Pulmonary interstitial pressure and proteoglycans during development of pulmonary edema. Am J Phys 270:H2000–H2007
Negrini D, Candiani A, Boschetti F, Crisafulli B, Del Fabbro M, Bettinelli D, Miserocchi G (2001) Pulmonary microvascular and perivascular interstitial geometry during development of mild hydraulic edema. Am J Phys 281:L1464–L1471
Oelz O, Howald H, di Prampero PE, Hoppeler H, Claassen H, Jenni R, Bühlmann A, Ferretti G, Brückner JC, Veicsteinas A, Gussoni M, Cerretelli P (1986) Physiological profile of world class high altitude climbers. J Appl Physiol 60:1734–1742
Ouellet Y, Poh SC, Becklake MR (1969) Circulatory factors limiting maximal aerobic exercise capacity. J Appl Physiol 27:874–880

Parker JC, Townsley MI (2004) Evaluation of lung injury in rats and mice. Am J Phys 286:L231–L246

Parker RE, Granger DN, Taylor AE (1981) Estimates of isogravimetric capillary pressures during alveolar hypoxia. Am J Phys 241:H732–H739

Passi A, Negrini D, Albertini R, Miserocchi G, De Luca G (1999) The sensitivity of versican from rabbit lung to gelatinase A (MMP-2) and B (MMP-9) and its involvement in the development of hydraulic lung edema. FEBS Lett 456:93–96

Piiper J, Scheid P (1981) Model for capillary-alveolar equilibration with special reference to $_{O2}$ uptake in hypoxia. Respir Physiol 46:193–208

Podolsky A, Eldridge MW, Richardson RS, Knight DR, Johnson EC, Hopkins SR, Johnson DH, Michimata H, Grassi B, Feiner J, Kurdak SS, Bickler PE, Severinghaus JW, Wagner PD (1996) Exercise-induced $\dot{V}_A/\dot{Q}$ inequality in subjects with prior high-altitude pulmonary edema. J Appl Physiol 81:922–932

Poole DC, Mathieu-Costello O (1989) Skeletal muscle capillary geometry: adaptation to chronic hypoxia. Respir Physiol 77:21–30

Pugh LGCE, Gill MB, Lahiri S, Milledge JS, Ward MP, West JB (1964) Muscular exercise at great altitudes. J Appl Physiol 19:431–440

Reeves JT, Groves BM, Sutton JR, Wagner PD, Cymerman A, Malconian MK, Rock PB, Young PM, Houston CS (1987) Operation Everest II: preservation of cardiac function at extreme altitude. J Appl Physiol 63:531–539

Reeves JT, Groves BM, Cymerman A, Sutton JR, Wagner PD, Turkevich D, Houston CS (1990) Operation Everest II: cardiac filling pressures during cycle exercise at sea level. Respir Physiol 80:147–154

Reynafarjee (1962) Myoglobin content and enzymatic activity of muscle and altitude adaptation. J Appl Physiol 17:301–305

Rodríguez FA, Ventura JL, Casas M, Casas H, Pagés T, Rama R, Ricart A, Palacios L, Viscor G (2000) Erythropoietin acute reaction and haematological adaptations to short, intermittent hypobaric hypoxia. Eur J Appl Physiol 82:170–177

Roselli RJ, Parker RE, Harris TR (1984) A model of unsteady-state transvascular fluid and protein transport in the lung. J Appl Physiol 56:1389–1402

Roughton F, Forster RE (1957) Relative importance of diffusion and chemical reaction rates in determining rate of exchange of gases in the human lung with special reference to true diffusing capacity of pulmonary membrane and volume of blood in the lung capillaries. J Appl Physiol 11:290–302

Rowell LB (1974) Human cardiovascular adjustments to exercise and thermal stress. Physiol Rev 54:75–159

Saltin B (1973) Oxygen transport by the circulatory system during exercise in man. In: Keul J (ed) Limiting factors of physical performance. Thieme, Stuttgart, pp 235–252

Sapoval B, Filoche M, Weibel ER (2002) Smaller is better–but not too small: a physical scale for the design of the mammalian pulmonary acinus. Proc Natl Acad Sci U S A 99:10411–10416

Sartori C, Vollenweider L, Löffler BM, Delabays A, Nicod P, Bärtsch P, Scherrer U (1999) Exaggerated endothelin release in high-altitude pulmonary edema. Circulation 99:2665–2668

Saunders PU, Pyne DB, Gore CJ (2009) Endurance training at altitude. High Alt Med Biol 10:135–148

Schaffartzik W, Poole DC, Derion T, Tsukimoto K, Hogan MC, Arcos JP, Bebout DE, Wagner PD (1992) $\dot{V}_A/\dot{Q}$ distribution during heavy exercise and recovery in humans: implications for pulmonary edema. J Appl Physiol 72:1657–1667

Seddon P (2002) Harness suspension: review and evaluation of existing information (2002). Health and Safety Executive (UK). Contract Research Report 451/2002. pp 3–42. www.hse.gov.uk/research/crr_htm//crr02451.ht

Semenza GL (2000) HIF-1: mediator of physiological and pathophysiological responses to hypoxia. J Appl Physiol 88:1474–1480

Semenza G (2002) Signal transduction to hypoxia-inducible factor 1. Biochem Pharmacol 64:993–998
Semenza GL (2020) The genomics and genetics of oxygen homeostasis. Annu Rev Genomics Hum Genet 21:183–204
Sill HW, Chang YS, Artman JR, Frangos JA, Hollis TM, Tarbell JM (1995) Shear stress increases hydraulic conductivity of cultured endothelial monolayers. Am J Phys 268:H535–H543
Stellingwerff T, Peeling P, Garvican-Lewis LA, Hall R, Koivisto AE, Heikura IA, Burke LM (2019) Nutrition and altitude: strategies to enhance adaptation, improve performance and maintain health: a narrative review. Sports Med 49(suppl 2):169–184
Storz JF (2021) High-altitude adaptation: mechanistic insights from integrated genomics and physiology. Mol Biol Evol 38:2677–2691
Sutton JR, Reeves JT, Wagner PD, Groves BM, Cymerman A, Malconian MK, Rock PB, Young PM, Walter SD, Houston CS (1988) Operation Everest II: oxygen transport during exercise at extreme simulated altitude. J Appl Physiol 64:1309–1321
Swenson ER, Bärtsch P (2012) High-altitude pulmonary edema. Compr Physiol 2:2753–2773
Taylor CR, Weibel ER (eds) (1981) Design of the mammalian respiratory system. Respir Physiol 44:1–164
van Hall G, Calbet JAL, Sondergaard H, Saltin B (2001) The re-establishment of the normal blood lactate response to exercise in humans after prolonged acclimatization to altitude. J Physiol Lond 536:963–975
van Hall G, Lundby C, Araoz M, Calbet JA, Sander M, Saltin B (2009) The lactate paradox revisited in lowlanders during acclimatization to 4100 m and in high-altitude natives. J Physiol Lond 587:1117–1129
Van Lieshout JJ, Wieling W, Karemaker JM, Secher NH (2003) Syncope, cerebral perfusion and oxygenation. J Appl Physiol 94:833–848
Viganò A, Ripamonti M, De Palma S, Capitanio D, Vasso M, Wait R, Lundby C, Cerretelli P, Gelfi C (2008) Proteins modulation in human skeletal muscle in the early phase of adaptation to hypobaric hypoxia. Proteomics 8:4668–4679
Villafuerte FC (2015) New genetic and physiological factors for excessive erythrocytosis and chronic mountain sickness. J Appl Physiol 119:1481–1486
Vogel JA, Hansen JE, Harris CW (1967) Cardiovascular responses in man during exhaustive work at sea level and high altitude. J Appl Physiol 23:531–539
Wagner PD (1993) Algebraic analysis of the determinants of VO_2max. Respir Physiol 93:221–237
Wagner PD, Gale GE, Moon RE, Torre-Bueno JR, Stolp BW, Saltzman HA (1986) Pulmonary gas exchange in humans exercising at sea level and simulated altitude. J Appl Physiol 61:260–270
Wagner PD, Sutton JR, Reeves JT, Cymerman A, Groves BM, Malconian MK (1987) Operation Everest II: pulmonary gas exchange during a simulated ascent of Mt. Everest. J Appl Physiol 63: 2348–2359
Wang GL, Jiang B-H, Rue EA, Semenza GL (1995) Hypoxia-inducible factor 1 is a basic-helix-loop-helix-PAS heterodimer regulated by cellular O_2 tension. Proc Natl Acad Sci 92:5510–5514
Ward MP, Milledge JS, West JB (2000) High altitude medicine and physiology. Arnold, London
Waxman AB, Kolliputi N (2009) IL-6 protects against hyperoxia-induced mitochondrial damage via Bcl-2-induced Bak interactions with mitofusins. Am J Respir Cell Mol Biol 41:385–396
Weibel ER (1963) Morphometry of the human lung. Springer, Berlin
Weibel ER (1973) Morphological basis of alveolar-capillary gas exchange. Physiol Rev 53:419–495
Weibel ER, Gomez DM (1962) Architecture of the human lung. Science 137:577–585
West JB (1986) Lactate during exercise at extreme altitude. Fed Proc 45:2953–2957
West JB (1987) Alexander M. Kellas and the physiological challenge of Mt. Everest. J Appl Physiol 63:3–11
West JB (1988) Rate of ventilatory acclimatization to extreme altitude. Respir Physiol 74:323–333
West JB (1998) High life: a history of high altitude physiology and medicine. Springer, New York

West JB, Lahiri S (eds) (1984) High altitude and man. Clinical Physiology Series, American Physiological Society

West JB, Boyer SJ, Graber DJ, Hackett PH, Maret KH, Milledge JS, Peters RM, Pizzo CJ, Samaja M, Sarnquist FH, Schoene RB, Winslow RM (1983a) Maximal exercise at extreme altitudes on Mount Everest. J Appl Physiol 55:688–702

West JB, Hackett PH, Maret KH, Milledge JS, Peters RM, Pizzo CJ, Winslow RM (1983b) Pulmonary gas exchange on the summit of Mount Everest. J Appl Physiol 55:678–687

West JB, Lahiri S, Maret KH, Peters RM, Pizzo CJ (1983c) Barometric pressure at extreme altitudes on Mount Everest: physiological significance. J Appl Physiol 54:1188–1194

West JB, Tsukimoto K, Mathieu-Costello O, Prediletto R (1991) Stress failure in pulmonary capillaries. J Appl Physiol 70:1731–1742

Winslow RM, Samaja M, West JB (1984) Red cell function at extreme altitude on Mount Everest. J Appl Physiol 56:109–116

Zuntz N, Löwy A, Müller F, Caspari W (1906) Höhenklima und Bergwanderungen in ihrer Wirkung auf den Menschen. Bong, Berlin

Chapter 11
A School Goes into Space

Pietro Enrico di Prampero and Guido Ferretti

Abstract This chapter summarizes the work carried out within the School of Milano in the physiology of space flight responses and adaptation. Only the studies during actual space flight and on the exploration of celestial bodies are accounted for. After a short historical account of the earliest studies, the pioneering work on locomotion on the Moon is described. Then the studies on the effects of prolonged space flight on maximal explosive power and the underlying mechanisms are analyzed. A physical countermeasure to cardiovascular and muscular deconditioning, the Twin Bike System, is described. A way is proposed and analyzed to increase gravity acceleration on the Moon artificially. The effects of gravity acceleration on the mechanical efficiency and the internal work during cycling are discussed, as well as the cardiovascular response to exercise on outer Planets.

P. E. di Prampero
Emeritus Professor of Physiology, University of Udine, Udine, Italy
e-mail: Pietro.Prampero@uniud.it

G. Ferretti (✉)
Department of Molecular and Translational Medicine, University of Brescia, Brescia, Italy
e-mail: Guido.Ferretti@unibs.it

G. Ferretti (ed.), *Exercise, Respiratory and Environmental Physiology*, Perspectives in Physiology, https://doi.org/10.1007/978-3-031-19197-8_11

Extra-vehicular activity during the joint Russia–ESA Euromir-95 prolonged space mission in the Russian Mir space station. On left, Yuri Gidzenko (crew commander); on right, ESA astronaut Thomas Reiter. With the kind permission of the European Space Agency

11.1 Introduction

In an extraordinary night of July 1969, for the first time in history, a man (Neil Armstrong, 1930–2012) put his feet on the lunar soil, and, while he was cautiously leaving the Lunar module, quietly sitting in our armchairs on Earth, we could observe on our TV screens his wobbling gait. At the Institute of Human Physiology of the University of Milano, these memorable images produced an enthusiastic emotional feeling. A colleague of ours, Giovanni Cavagna, and "the grand old master," Rodolfo Margaria, on the bases on their studies on the biomechanics of walking and running on Earth, had foreseen the main characteristics of human locomotion on the Moon (Margaria and Cavagna 1964; Margaria 1966), and the images arriving on the TV screen from ≈350,000 km afar qualitatively supported their predictions.

Yet there was more than this in that emotion. That was the apparent final outcome of a long story, which accompanied the evolution of the School of Milano, since Amedeo Herlitzka started aviation medicine in Italy during World War I (see Chap. 1), and of the sister School of Rochester—Buffalo, since Wallace Fenn obtained the first contract from the US Air Force during World War II. If we look at the history of human flight as a mixture of adventure, exploration, and conquest of new worlds, then a man walking (or skipping. . .) on the Moon can be considered as the ultimate, extreme achievement of that adventure.

After the conquest of the Moon, the exploration of outer Space took other tracks. The construction of long-lasting space stations orbiting around the Earth became a priority, and long-term space flights were organized, allowing the performance of several types of experiments in any field of science, including human physiology. The worldwide community of physiologists took this as an opportunity, and we owe much of our understanding of long-term human adaptation to microgravity to studies carried out in space stations. However, the possibility or performing physiological experiments in Space has manifest logistic and financial limits; therefore, micro-gravity simulation conditions have been exploited by Space Agencies, the most important being head-down tilt bed rest.

The aim of this Chapter is to discuss human adaptation to space flight, with special reference to the contribution of the School of Milano. Priority is given to space flight data, whereas we refer to bed rest experiments only in case of lack of data during space flight. We note that the effects of microgravity on maximal oxygen consumption and on respiratory mechanics are discussed, respectively, in Chaps. 7 and 8.

11.2 A Short Summary of Space Exploration

For long time, space exploration was confined to fantastic literature. Lucian of Samosata (~120 to ~185) deserves special mention under this respect, for his fantastic, untruthful description of a journey to the Moon in his *True Stories* ("Αληθῆ διηγήματα"), an English version of which was published by Francis Hickes (Hickes 1902). His greatest merit was that he overwhelmed the divine conception of the Moon, typical of Greek culture, and considered it just a physical body, similar to Earth, on which it was conceivable that a man could pose, one day, his feet: the psychological breakthrough of a journey to the Moon was attained.

But, how can we send a spacecraft to the Moon? In fact, the idea of the rocket came again from science fiction: after all, the space gun imagined by Jules Verne (1828–1905) in his *De la Terre à la Lune, trajet direct en 97 heures 20 min* (Verne 1865) to propel the "projectile" is a literary anticipation of the Saturn rocket propelling the Apollo capsule. Russian engineer Konstantin Tsiolkovsky (1857–1935, see Fig. 11.1), the real pioneer of astronautics, was the first to suggest the physical feasibility of a space rocket, although the technology of his times did not allow such an enterprise to be realized. Nevertheless, his work inspired the Soviet space program after World War II.

World War II determined a strong development of rocket technology for military purposes. Germany was particularly active in this domain, thanks to the impulse given by Wernher von Braun (1912–1977), who designed the well-known *V-2* bomb-carrying rockets. At the end of the war, he surrendered and delivered himself to the US forces (Fig. 11.2), to become, after some years in a prisoner camp, the leader of the American space project.

Shortly after the end of the war, the Soviet Union launched its own space program, under the impulse and the lead of Sergej Korolëv (1907–1966). While the Americans developed the program of supersonic aircrafts, the ultimate aim of which was to send an airplane into Earth orbit, Korolëv, an ancient pupil of the late Tsiolkovsky, opted for rocket carriers. His was the right choice, so the Soviet Union, on October 4th, 1957, preceded the USA in launching a satellite into orbit.

Those were the days of the so-called Cold War for world supremacy between the two main powers winners of World War II, characterized by different social regimes in competition between them. A strong competition started also for space supremacy. In 1958, President Eisenhower eventually charged the newly created National Aeronautics and Space Administration (NASA) to develop a human space flight

Fig. 11.1 Portrait of Konstantin Tsiolkovsky with his steel dirigibles in his garden, 1913. Picture in public domain

program. But the Soviets had an advantage and thus, thanks to the work of Korolëv, they were the first to send a man into orbit: Jurij Gagarin (1934–1967), who, on April 12th, 1961, carried out the first human orbital flight around the Earth, completing one orbit in 1 h and 48 min. Korolëv and Gagarin are portrayed in Fig. 11.3.

Gagarin tolerated the environmental conditions of his flight very well and returned in good shape. However, when German Titov (1935–2000) came back from the second space flight, completing four orbits on August 6th, 1961, he showed signs of vestibular troubles and of motion sickness, providing the first demonstration of physiological alterations by space flight.

After Gagarin's flight, President Kennedy set a new political goal: sending an American astronaut to the Moon. He charged NASA of the task, with unlimited funding. The first American in orbit, John Glenn (1921–2016), flew in 1962. Then the NASA launched the Gemini program, the aim of which was to develop and test the procedures for docking, re-entry, and extra-vehicular activities, propaedeutic to a

Fig. 11.2 Wernher von Braun with a group of German officers at Peenemünde, in 1941. With the kind permission of Wikimedia Commons and the German Federal Archive (Bundesarchiv Bild 146-1978-Anh.030-02)

flight to the Moon. Then came the Apollo project, by which humans left the Earth orbit, attained a lunar orbit, and eventually, on July 21st, 1969, Neil Armstrong and Edwin Buzz Aldrin posed their feet on the lunar soil for the first time (Fig. 11.4). That was undoubtedly a great political and propaganda success for the USA, perhaps the last time in which a Western democratic government dared to put the heart beyond the obstacle, setting a national adventure and pursuing it without hesitation and without narrow-minded calculations and weightings.

Korolëv also started conceiving a lunar program. The first extra-vehicular activity in orbit, performed by Alexei Leonov (1934–2019) on March 18th, 1965 during the Voskhod 2 flight, might already have been in that perspective. However, Korolëv died prematurely and unexpectedly by hemorrhage during a surgical intervention on January 14th, 1966. The Soviet lunar space program and the Soviet supremacy in space flight competition ended that day.

The completion of the Apollo program in 1972, which eventually led twelve astronauts to walk on the Moon, put an end to the heroic epoch of space conquest. The perspective had already changed after Korolëv's death. In 1967, the Soviet Union announced a wide project of orbit colonization. This project converted the space carriers of previous programs to the need of installing and maintaining permanent space stations. The first module of the newly conceived *Salijut* Space Station was launched in 1971. The station was completed within 1977. The *Sojuz* spacecraft played a great role in this shuttling activity. Conceived in the lunar flight

Fig. 11.3 Sergei Pavlovich Korolëv (right) and Jurij Alekseevich Gagarin at Star City in 1961. Picture in public domain

program, then converted into a human transporter to and from space stations, the *Sojuz* spacecraft is still in use after more than 50 years, and never stopped evolving ever since in parallel with the technological developments. It has been the most successful and performant spacecraft ever conceived, with only one deadly accident in its history.

Meanwhile, the Americans developed their own permanent station, the *Skylab*, the first module of which was launched on May 14th, 1973. But that was a short-term program, though successful. In fact, the American priority was different: to develop a reusable spacecraft, to be launched by a rocket, but landing by planar flight on an airport track, the *Space Shuttle*. Five Shuttle models were constructed, supporting more than a hundred flights in 30 years. The first Shuttle, Columbia, flew on April 12th, 1981, commanded by John Young (1930–2018), who was a veteran on lunar flights. The internal organization of the *Space Shuttles*, which could host up to seven crewmembers at a time, greatly improved the working conditions, allowing a variety of activities, from military to scientific, to be carried out. The flight duration, however, was short (a Shuttle flight lasted at most 18 days), so it was not suitable for long-term flights in preparation of eventual planetary explorations. After having served, together with the *Sojuz*, as a support carrier for the construction of the current International Space Station, the *Shuttle* program was dismissed in 2011 for budgetary reasons.

Fig. 11.4 Buzz Aldrin on the Moon, portrayed by Neil Armstrong. From Wikimedia Commons, in free domain, as agreed upon with Neil Armstrong

On the other side, the Soviets kept promoting permanent space stations. In 1985, they launched the *Mir* Station, a remarkable step forward in the field, in terms of space availability and crew comfort, which was kept active for 15 years. After the crumbling of the Soviet Union in 1991, the newly established Russian Space Agency, which inherited the entire installations, equipment, and staff of the Soviet space program, reoriented its activity toward international collaboration and became a privileged partner of the new International Space Station, the first component of which was launched in 1998. The station hosted its first long-term crew in 2000. Its ownership and use is established by intergovernmental treaties and agreements, involving the USA, Russia, the European Union, Japan, and Canada. The standard permanence of a crew in the Station has been fixed to six months, which is not of great use for the preparation of a human flight to Mars. This implies that the longest continuous sojourns in space are still those attained in the *Mir* Station, which still holds the record for the longest single human spaceflight, with Valerij Poljakov (1942–2022) spending 437 days and 18 h on the station between 1994 and 1995. The operational status of the International Space Station has been recently prolonged to 2030.

11.3 The Early Studies

After the reports of vestibular troubles on German Titov returning from his space flight in 1961, a number of clinical and physiological variables were monitored in all subsequent flights in the 1960s, including blood pressure, ECG, and body temperature. Upon return from *Mercury* and *Gemini* flights, weight loss, orthostatic intolerance, increased hemoglobin concentration were reported, notwithstanding the vestibular troubles, systematically confirmed. Yet the first discussion of motion sickness was published as early as in 1961. The author mentioned Gagarin's flight, but was not yet aware of Titov's report of vestibular troubles. Therefore, his discussion was purely conjectural, based on his own experience during parabolic flights (Johnson 1961). The evidence became rapidly so compelling, that the same author, less than five years later, stated that *we now have ample evidence for believing that, of all the various disturbing stimuli to which space travelers will be exposed, the disabling effects of angular and linear accelerations on the non-auditory labyrinth can reduce an astronaut's efficiency to such an extent as to jeopardize his very survival* (Johnson and Ireland 1966). And this, despite his conviction that it would be possible, with adequate preparation, to send a human to the Moon and have him/her coming back again in good shape, with at least one day available for lunar exploration.

Despite theoretical expectations of orthostatic intolerance occurrence, supported by water immersion experiments, no evidence of it came from tests upon return from the earliest space flights, perhaps because of their shortness. The first evidence of cardiovascular intolerance came from longer flights, especially from the 34-hr flight by Gordon Cooper (1927–2004) in the last *Mercury* mission (Di Giovanni and Chambers 1964). These authors paid attention to the observations from ECG recordings, showing several cases of arrhythmias. Most of the reported cases concerned supraventricular arrhythmias, but in one case, from a bed rest study, a subject, incidentally an athlete, underwent intermittent ventricular tachycardia, which reverted to normal sinus rhythm after quinidine administration.

Webb (1966) analyzed the reported data of body mass before and after space flight, indicating a consistent weight loss in all astronauts upon return, with a tendency toward a greater weight loss, the longer the flight duration. Since the weight loss occurred also in the shortest flights, and in most cases, it was recovered within 24 h after landing, he suggested that it was due primarily to water loss. Interestingly, he suggested that water loss might follow the activation of the Henry-Gauer reflex (Gauer and Henry 1963): stimulation of central venous volume receptors induced by blood shift in weightlessness would induce an increase in urine production due to inhibition of the effect of the anti-diuretic hormone on kidney water reabsorption. In fact, subsequent evidence demonstrated reduction of plasma volume upon return from longer *Gemini* missions (Fischer et al. 1967). This was expected to generate a small increase in hematocrit and in hemoglobin concentration, which however were not observed, because of an unexpected reduction of red blood cell mass upon return, which the authors attributed to a mild hemolysis of unknown cause (Fischer et al. 1967).

A great impulse to the study of life science in microgravity came from the development of permanent space stations, allowing prolonged sojourns in more comfortable conditions. The first reports on the development of muscle hypotrophy and of bone resorption during prolonged flights came from space stations in the 1970s. The first countermeasures against physiological alterations in microgravity were developed in those same years. The *Salijut* space station contained a cycle ergometer for the performance of aerobic exercise training, and a system for astronaut exposure to lower body negative pressure (LBNP), which was used for the prevention of orthostatic intolerance and hypovolemia. Moreover, the re-entry protocols included a procedure of forced rehydration by means of saline injection to restore blood volume, as a preventive treatment of orthostatic intolerance.

Further advances in countermeasures were introduced as long as the duration of flights progressively increased in the *Salijut* station. The increasing evidence that muscle hypotrophy aggravates with the progressive increase in flight duration led to the introduction of electrical muscle stimulation as a tool to limit this phenomenon. Improved and more powerful LBNP devices were made available. New anti-*G* suites were developed, to be used during landing. Specific prescriptions of physical training were applied during recovery after long-term flights.

Meanwhile, in the *Skylab*, several types of experiment, even at prolonged space flight times, could be carried out. Kinetic studies demonstrated that the alterations observed on the cardiovascular system in microgravity attained a steady state within four to six weeks from departure, and thus they were not such as to impair the astronauts' ability to work. Pharmacological treatment of motion sickness was implemented and tested. The bone loss was found to be accompanied by an increased urinary excretion of calcium, phosphates, and nitrates. This last was associated also to a progressive protein loss related to the development of muscle hypotrophy. A detailed report of the results obtained in the *Skylab* and *Salijut* missions can be found elsewhere (Nicogossian et al. 1996).

With the *Mir* Station, and even more with the International Space Station, the opportunities for physiological studies during long-term space flights increased remarkably, whereas the *Space Shuttle* program provided important occasions for experiments during short-term flights. Meanwhile, the European Space Agency started its programs of head-down tilt bed rest studies, in order to allow performance in simulated microgravity of experiments otherwise unfeasible in-flight. The launch of the *Mir* Station and the *Shuttle* program opened the contemporary epoch of the space exploration era, in which the performance of life science experiments in Space became systematically part of the civil space flight tasks. That is when the contemporary, deeper physiological knowledge of human adaptation to microgravity, and of the subsequent functional troubles upon re-exposure to gravity after re-entry from Space or after standing up again at the end of bed rest, was generated. That is when the School of Milano entered microgravity life science research consistently.

However, before focusing on the contributions of the last 30 years, we need to make a step back and discuss human locomotion on the Moon and on other planets, along the principles of the cute analysis that Margaria and Cavagna carried out before Armstrong and Aldrin stepped on the lunar soil. In fact, the two physiologists from Milano predicted in an amazingly correct way how astronauts would have moved on the Moon. These predictions were part of the enthusiasm that pervaded Milano's physiologists on watching the images transmitted from the Moon on July 21st, 1969.

11.4 Walking on the Moon or on Other Celestial Bodies

During natural walking at the optimal speed on Earth, the potential and kinetic energy changes at each stride are about equal, but of opposite sign (Chap. 5). As a consequence, the mechanical energy to move forward is essentially equal to the potential energy accumulated in the initial phase of the stride (ΔE_p):

$$\Delta E_p = M\, g\, h \tag{11.1}$$

where the mass M is the body mass, g is the acceleration of gravity, and the height h corresponds to the vertical displacement of the center of mass. In turn, the subsequent kinetic energy change (ΔE_k) is given by:

$$\Delta E_k = M\, v_m\, \Delta v \tag{11.2}$$

where Δv is the velocity change during the step, relative to the mean speed of locomotion (v_m). [1] At the optimal walking speed (v_{opt}), assuming: [2]

$$\Delta E_p = \Delta E_k \tag{11.3}$$

and since under these conditions $\Delta v = 0.2\, v_m$ and $v_m = v_{opt}$, from the right members of Eqs. 11.1 and 11.2, one obtains:

[1]Equation 11.2 can be derived as follows:

$$\Delta E_k = E_{k2} - E_{k1} = 0.5\, M(v_m + 0.5\, \Delta v)^2 - 0.5\, M(v_m - 0.5\, \Delta v)^2$$

where E_{k2} and E_{k1} are the kinetic energy values at the maximal and minimal velocity, respectively. Assuming further that Δv follows a symmetrical time course in respect to the mean velocity (v_m), rearranging and simplifying, one obtains Eq. 11.2.

[2]Equation 11.3 remains valid even knowing that the percent Recovery (see Chap. 5, Cavagna et al. 1976) during walking at v_{opt} is less than 1.0, because the missing energy is replaced by the mechanical energy generated by muscle contraction.

Table 11.1 Optimal walking speed ($m\ s^{-1}$ or $km\ h^{-1}$) on the indicated celestial bodies, whose gravity is expressed as a fraction of the Earth gravity ($g_{Earth} = 9.81\ m\ s^{-2}$)

Planet	Moon	Mars	Earth	Jupiter
g/g_{Earth}	0.16	0.38	1.00	2.53
($m\ s^{-1}$)	0.61	0.92	1.49	2.50
v_{opt}				
($km\ h^{-1}$)	2.19	3.31	5.36	9.00

$$M\ g\ h = M\ v_m\ \Delta v = M\ v_{opt}\ 0.2\ v_{opt} \tag{11.4}$$

Hence, solving Eq. 11.4 for v_{opt} yields:

$$v_{opt} = \sqrt{\frac{g\ h}{0.2}} = \sqrt{5\ g\ h} = 2.24\ \sqrt{g\ h} \tag{11.5}$$

It follows that, on the Moon, where g is about 1/6 of that on Earth (1.635 $m\ s^{-2}$), for any given value of h, the optimal speed will be about 40% of that observed on Earth, since $\sqrt{1/6} = 0.408$, and thus about 0.6 $m\ s^{-1}$ (2.2 $km\ h^{-1}$, Margaria 1975).

Equation 11.5 is formally and conceptually equal to Eq. 5.16, except for the fact that the lower limb length (L) is replaced by h. Obviously enough, h and L are directly proportional quantities.

At the end of the twentieth century, Cavagna et al. (1998) determined the energy cost of walking under low-gravity conditions, during two parabolic flights campaigns organized by the European Space Agency (ESA).[3] They investigated vertical and horizontal ground reaction forces on three subjects walking on force platforms fixed to cabin floor during a series of parabolas, such that the overall gravity vector acting along the subjects' head-to-feet axis was equal to the gravity acceleration on Mars (3.72 $m\ s^{-2}$). Under these conditions, the maximal % Recovery, calculated as described in Chap. 5 (Eq. 5.13), was rather close to that observed on Earth (≈65%). It occurred, however, at a lower speed (3.4, as compared to 5.4 $km\ h^{-1}$). It is interesting to note that the so determined v_{opt} is very close to that calculated from Eq. 11.5, setting h = 0.045 m (average vertical displacement within a step for an adult of normal body size and a lower limb length of 0.9 m). Indeed, since the gravity acceleration on Mars is about 38% of that on Earth, other things being equal, this equation yields the prediction that v_{opt} on Mars amounts to ≈60% of that observed on Earth ($\sqrt{0.38} = 0.61$).

These considerations show that the theory of dynamic similarity can be reasonably extended to include, beyond the different body dimensions (see Chap. 5), also gravity conditions different from those prevailing on Earth. The v_{opt} on different celestial bodies is indicated in Table 11.1. For a more detailed analysis, the reader is referred to Saibene and Minetti (2003).

[3]During these flights, throughout the duration of each parabola (about 25–30 s), the airplane and whatever it contains is subjected to pre-established gravity conditions, as given by the vectorial sum of gravity and the plane's centrifugal acceleration that results from an appropriately selected parabolic trajectory.

It goes without saying that all the considerations reported above for walking in "extra-terrestrial" conditions apply to freely moving subjects, thus neglecting the physical constraints (not least the thermoregulatory ones) brought about by the heavy space suits necessary on extra-terrestrial planets.

11.5 The School of Milano Enters the Microgravity Game

This brief overview on the effects of the prevailing acceleration of gravity on the biomechanics of walking does show the pioneering interest in and the original approach to "Space Science" by Rodolfo Margaria and his School at large. As such, it does not come as a great surprise that the number of studies devoted to these problems increased substantially with the advent of the "Space Exploration" era. Pietro Enrico di Prampero opened the way, as he started collaborating with the European Space Agency. He involved Carlo Capelli and Guglielmo Antonutto, who were then part of his newly established research group in Udine, Italy. He informed Guido Ferretti of the opportunities that started to be offered, with modest concurrence. At the same time, NASA selected a project proposed by Paolo Cerretelli. Then, a 30-years long collaboration started, involving an extensive participation to several bed rest and space flight programs. More specifically, many a study by several members of the Milano School, often in combination with other groups, were devoted to the effects of real or simulated (bed rest) microgravity on several aspects of the physiological responses to exercise.

In the next Sections, we review several studies carried out during flight and discuss some proposals for countermeasures against physiological deconditioning in prolonged space flights. The available space does not allow a complete analysis of the numerous studies carried out by scientists from the School of Milano during bed rest. We resignedly address our patient readers to the original publications, of which we quote here the most significant and the most related to the way of approaching physiology by the School of Milano (Adami et al. 2013; Bringard et al. 2010; Capelli et al. 2006, 2008, 2009; de Boer et al. 2008; Desplanches et al. 1998; di Prampero and Narici 2003; Ferretti et al. 1997, 1998, 2001, 2009; Milesi et al. 2000; Moriggi et al. 2010; Narici et al. 1997; Porcelli et al. 2010; Reeves et al. 2005; Rejc et al. 2015a, 2015b; Salanova et al. 2014; Salvadego et al. 2011, 2016, 2018; Zamparo et al. 2002a, 2002b). As already pointed out, we remind that the articles on maximal oxygen consumption limitation during and after bed rest are discussed at length in Chap. 7.

11.6 Maximal Explosive Power in Microgravity

The maximal explosive power (MEP) developed during all-out efforts of the lower limbs of about 0.3 s duration before and after space flight was assessed on a specially designed Multipurpose Ergometer Dynamometer (Fig. 11.5) (Zamparo et al. 1997).

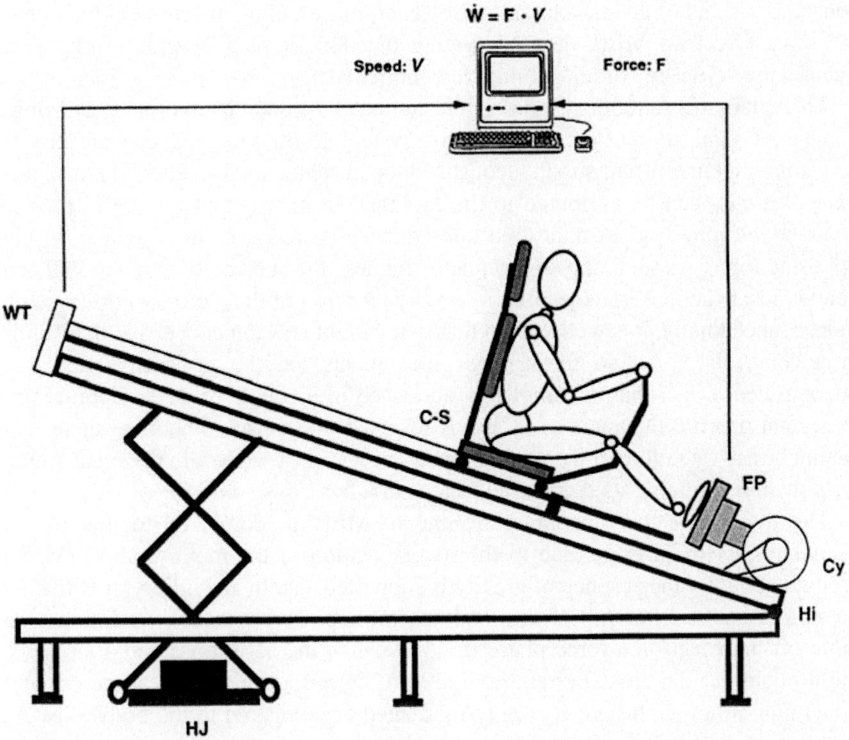

Fig. 11.5 The Multipurpose Ergometer Dynamometer (schematic). The carriage seat (C-S) is free to move on a rail that can be inclined to a preselected angle with respect to the horizontal. The subject, sitting on the seat, propels himself backward pushing on two force platforms (FP). The instantaneous product of the backward velocity of the seat, as obtained thanks to the wire tachometer (WT), and the corresponding overall force yields the instantaneous mechanical power. Note that, before the onset of the push the carriage-seat is supported by two blocks appropriately positioned on the rail, and that, at the top of the backward excursion, it is automatically blocked by a braking system. From Zamparo et al. 1997.

The experiments were performed on astronauts before and after space flights of up to 6 months duration on board of the *MIR* Space Station on the occasion of the 1994–1995 EUROMIR missions. The data were collected in Star City, near Moscow, Russia, shortly before flight departure and 2 days after return to Earth gravity. The results show that, after 6 months in microgravity, the MEP of the lower limbs was reduced to about 45%, whereas the muscle mass of the same limbs, as well as the maximal oxygen consumption, were reduced only to about 90% of the corresponding pre-flight values (Antonutto and di Prampero 2003; Antonutto et al. 1999; Zange et al. 1997).

It should be noted here that during the flight the astronauts were systematically exercising on specially designed ergometers for ≈2 h per day at moderate exercise intensity, a fact that might have reduced substantially the decline of maximal oxygen consumption and of the muscle mass. However, this training procedure did not have the same effect on MEP, thus suggesting that long-term microgravity exposure reduces the "efficacy" of all-out muscle contractions.

To explain this set of observations, the authors proposed that microgravity brings about a reorganization of the motor control system, leading to a reduced capacity of activating maximally and simultaneously all the available muscle fibers. In turn, this state of affairs can be attributed to the fact that, in microgravity, forceful muscle contractions may lead to undesired consequences because of the absence of any opposing force. As such, the environment "invites" the subject to favor smooth and gentle muscle actions, thus leading to a reorganization of the motor control system. Indeed, anecdotally, it is well known that space flight newcomers behave like "bulls in a china shop," in so far as their movements tend to be exaggerated and uncontrolled, a fact that is completely abolished in a couple of days. Upon return to normal gravity, this state of affairs is reversed: the muscle fibers are there, but cannot be activated forcefully and maximally as was the case before the space flight, the situation resuming its "terrestrial" status in a few days.

The hypothesis that the larger decline of MEP, as compared to that of the available muscle mass, be due to the reorganization of the motor control system brought about by the absence of gravity is supported also by the observation that, at the end of bed rest, the fall of maximal anaerobic power, as measured during high jump off both feet on a force platform, is less than the MEP observed after space flight of similar duration (Ferretti et al. 2001). Indeed, during bed rest the gravitational pull is not abolished; it is only reoriented as compared to the body axis. As such, no reorganization of the motor control system is required.

Finally, in animal muscles, long-term space flight leads to a shift of the fibers toward the fast type (Caiozzo et al. 1994, 1996; Widrick et al. 1999); as such, after return to normal gravity one would expect a larger (rather than smaller) MEP per unit muscle mass. This rules out any eventual changes of muscle fiber type composition to explain the experimentally observed fall of MEP, so lending further support to the role of the motor control system as *primum movens* of this state of affairs.

A possible additional explanation of the greater loss of force and MEP has been more recently proposed. In fact, the loss of contractile material and the reduced size of muscle fibers imply changes in muscle architecture susceptible to reduce both the maximal velocity of contraction and the maximal isometric force, so that also the maximal muscle power, and thus the MEP, are reduced (Kawakami et al. 2001; de Boer et al. 2008). These changes, however, are not such as to overthrow the role of motor control alterations, the occurrence of which has been clearly demonstrated upon return from space flight of various duration (Layne et al. 2001; Recktenwald et al. 1999) as well as at the end of prolonged bed rest (Buehring et al. 2011; Mulder et al. 2008; Yuan et al. 2018).

The fall of MEP during long-term space flight could be reduced, were it possible to install, where needed, appropriate ergometers allowing the astronauts to perform explosive efforts during their daily exercise routine (di Prampero and Antonutto 1996). It goes without saying that these explosive ergometers should be constructed

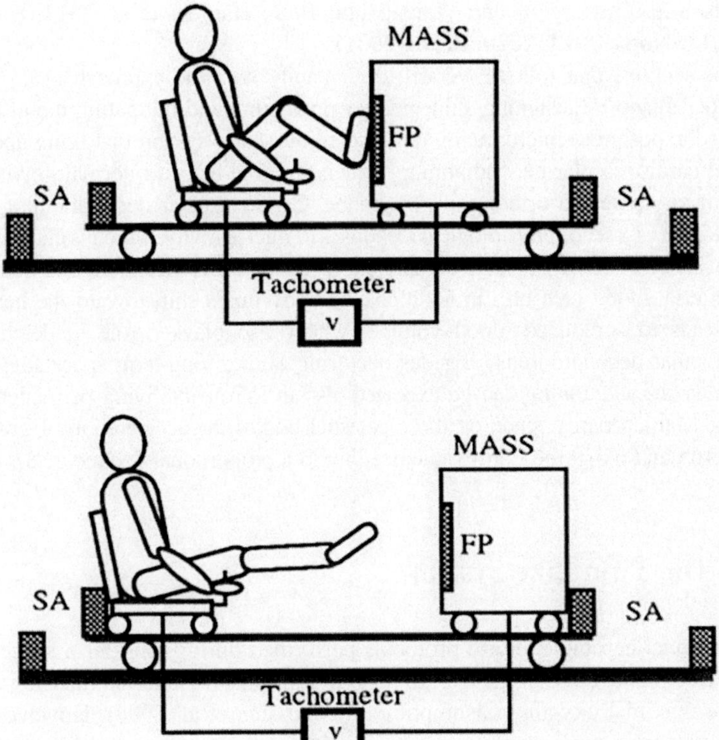

Fig. 11.6 Ergometer for maximal explosive power in space (schematic): *FP* force platforms, *SA* shock absorbers. The subject accelerates himself backward and an appropriate mass in the opposite sense. The product of force and relative velocity (v), as obtained thanks to the indicated tachometer and force platforms, yields the instantaneous mechanical power. The opposite motion of subject and mass is expected to minimize changes of the overall center of mass of the system, thus reducing unwanted mechanical side effects. Note also that, to avoid uploading unnecessary masses into space, the "mass" could be replaced by another subject. From di Prampero and Antonutto 1996

in such a way to avoid unwanted vibrations; a simplified model thereof is reported in Fig. 11.6.

11.7 Cardiovascular Deconditioning

As discussed in Sect. 11.3, it has been clear since the first days of the space flight era that exposure to microgravity leads to a series of consequences on the cardiovascular system, eventually leading to orthostatic intolerance upon return. Comprehensive analyses of the cardiovascular alterations induced by microgravity exposure appear in numerous review articles (Antonutto and di Prampero 2003; Hargens and

Richardson 2009; Hargens and Watenpaugh 1996; Hargens et al. 2013; Hughson et al. 2018; Norsk 2014, 2020; Zhang 2001).

In the sections that follow, we will deal mainly with the approach taken by the School of Milano in designing strategies for preventing and mitigating the effects of prolonged exposure to microgravity on the cardiovascular system that bring about the so-called cardiovascular deconditioning, which, upon re-entry to normal gravity conditions, leads to severe orthostatic intolerance. Cardiovascular deconditioning is the final outcome of a set of physiological responses to microgravity, which is characterized by the absence of the hydrostatic component (pressure difference along the body, ΔP) of the arterial blood pressure. In addition, the body fluids shift toward the head and thorax leads to a reduced blood volume, which also plays a role in determining cardiovascular deconditioning. Besides occurring during long-term space flight, cardiovascular deconditioning can be expected also in astronauts living permanently on Lunar or Martian bases, since on these celestial bodies the acceleration of gravity is about 0.165 and 0.379 the Earth value, leading to a proportional decline of ΔP.

11.8 The Twin Bike System

The traditional aerobic exercise protocols performed during long-term space flight prevent an excessive fall of the overall mass of the anti-gravitational muscles as well as of the maximal oxygen consumption (Nicogossian et al. 1996). However, they have no effects on cardiovascular deconditioning, which seems to be prevented only by artificial gravity (Clément and Pavy-Le Traon 2004; Hargens et al. 2013; Lackner and Di Zio 2000; Lazzer et al. 2011; Zhang 2013). Therefore, to prevent cardiovascular deconditioning, or at least to reduce its severity, several scenarios aimed at simulating gravity in space have been proposed; suffice it here to mention the Ergometer-Centrifuge (Greenleaf et al. 1997) and the Bed-Centrifuge (Cardus et al. 1991) (Fig. 11.7). The unifying principle underlying all these proposals is to have the astronauts positioned on a rotating system that generates a centrifugal force along their body axis in the head–feet direction, thus mimicking gravity.[4] All these systems, the fore-runner of which has been described in 1929 by Hermann Potočnik (1892–1929) in a very remarkable book that was published posthumous under the pseudonym of Noordung (Noordung 1929, see Fig. 11.8), have remained intellectual exercises so far.

To prevent or reduce the long-term effects of microgravity on the cardiovascular system, Antonutto et al. (1991) proposed to simulate gravity on a space vehicle by the use of two mechanically coupled counter rotating bicycles (Twin Bike System) which move along appropriately constructed structures. More specifically, the Twin Bike System consists of two bicycles moving at the very same speed, but in the

[4] Science fiction was once more ahead of reality in this respect. In the celebrated Stanley Kubrick movie *A Space Odyssey*, the authors imagined a rotating space craft, with a long spinning arm, such as to reproduce a gravitational environment on board.

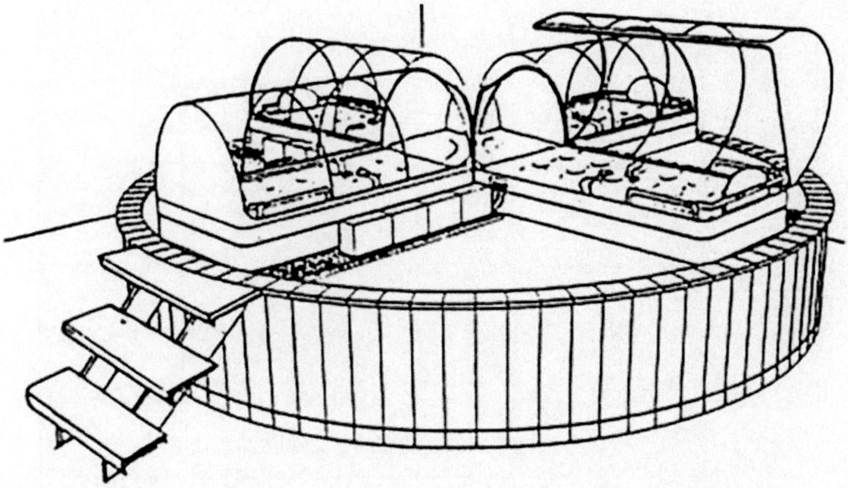

Fig. 11.7 Centrifuge platform with four beds and canopies. From Cardus et al. 1991

opposite direction, along the inner wall of a cylindrically shaped space module. Two adjustable masses, applied to an axle mounted in opposition to each bike, prevent the repetitive yaws that would otherwise occur when the two bicycles are on the same side of the space module (Fig. 11.9).

The wheels run on two parallel rails providing the necessary initial friction. The circular trajectories induce centrifugal acceleration vectors (a_c) oriented along the head-to-feet direction of each subject. Since a_c is given by the ratio of the square of the tangential velocity (v_t) to the radius of gyration (R_g):

$$a_c = \frac{{v_t}^2}{R_g} \tag{11.6}$$

the tangential and angular velocities yielding $a_c = 1$ g (= 9.81 m s^{-2}) can be calculated. For $R_g = 2$ m or 6 m, this is attained when $v_t = 4.43$ and 7.67 m s^{-1}, respectively.

The mechanical and metabolic powers yielding a given v_t and hence a given acceleration, depend on the air density (and hence barometric pressure P_B and temperature T) prevailing inside the space module, and to a lesser extent on the wheels rail friction. Thus, one can calculate the mechanical and metabolic powers yielding 1 g at the feet level, assuming a rolling resistance applying to a knobby tired bike on a concrete surface on Earth (see Chap. 6), and assuming for simplicity that the air in the space module remains still, in spite of the two counter rotating cyclists. For a 175 cm tall 70 kg body mass astronaut, assuming $P_B = 760$ mmHg (101.3 kPa), $T = 20$ °C (293 °K), they amount to 75 W and 1.2 L O_2 min^{-1}, for $R_g = 2$ m ($v_t = 4.43$ m s^{-1}). If, *ceteris paribus*, we have $R_g = 6$ m instead, and thus $v_t = 7.67$ m s^{-1}, the corresponding mechanical and metabolic powers would turn out

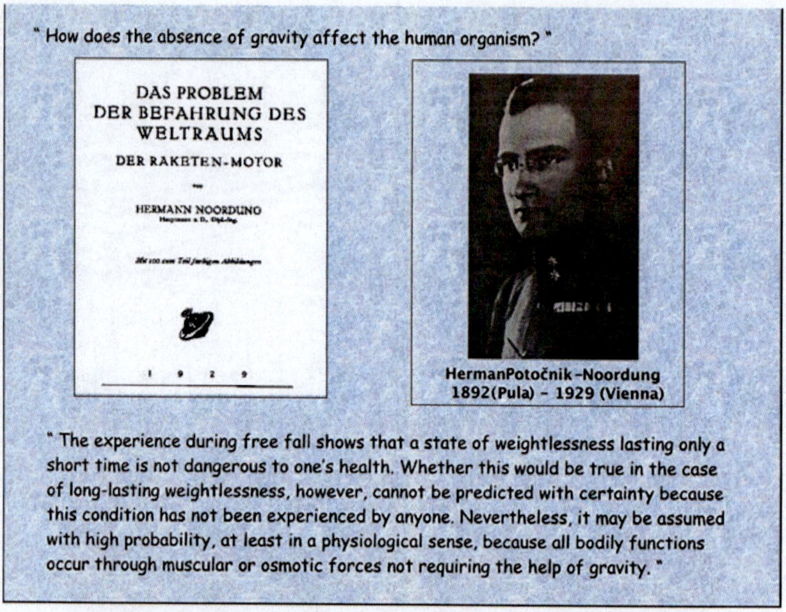

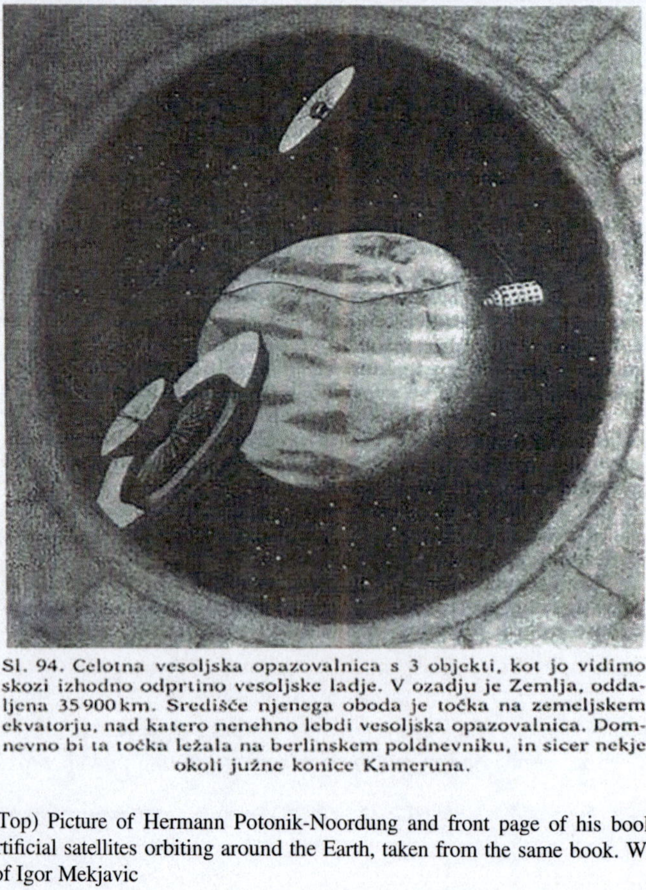

Sl. 94. Celotna vesoljska opazovalnica s 3 objekti, kot jo vidimo skozi izhodno odprtino vesoljske ladje. V ozadju je Zemlja, oddaljena 35 900 km. Središče njenega oboda je točka na zemeljskem ekvatorju, nad katero nenehno lebdi vesoljska opazovalnica. Domnevno bi ta točka ležala na berlinskem poldnevniku, in sicer nekje okoli južne konice Kameruna.

Fig. 11.8 (Top) Picture of Hermann Potonik-Noordung and front page of his book. (Bottom) Images of artificial satellites orbiting around the Earth, taken from the same book. With the kind permission of Igor Mekjavic

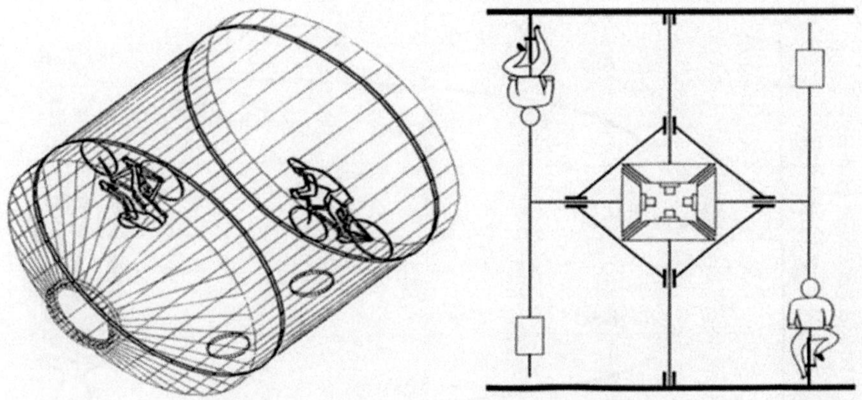

Fig. 11.9 (Left) Cyclists moving along the inner wall of a cylindrically shaped space module generate an acceleration vector mimicking gravity. (Right) Schematic view of the Twin Bike System. Thick lines indicate space module walls. The differential gear coupling the two cyclists is drawn on a larger scale. Adjustable masses are also shown. From Antonutto et al. 1991

240 W and 3.05 L O_2 min^{-1}. R_g is substantially smaller at the head than at the feet, so that a_c will also be smaller, the ratio of a_c at the feet to a_c at the head being 5 for $R_g =$ 2 m and 1.5 for $R_g = 6$ m.

The relationship between the arterial blood pressure at the feet and at the head as a function of R_g is shown in Fig. 11.10, for an astronaut 1.75 m tall riding a Twin Bike System at the speed yielding $a_c = 1$ g at the feet level. The pressure prevailing at any level in the circulatory system is given by the sum of the pressure generated by the heart plus (or minus) the hydrostatic component of pressure (ΔP), equal to the weight of the column of blood from the heart to the point in question. In turn, ΔP is given by:

$$\Delta P = \rho_b \, a_c \, h \tag{11.7}$$

where ρ_b is the blood density, and height h corresponds to the distance between the heart and the point in question. Since, according to Equation 11.6, a_c depends on R_g, the arterial pressures prevailing at any given point in the circulatory system at any given R_g and v_t can be easily calculated from simple geometrical considerations, provided that the pressure at the heart level is known.[5]

For a subject riding a conventional bike at a v_t yielding $a_c = 1$ g at the subject's feet, assuming a mean arterial pressure of 100 mmHg in the aortic bulb, the mean arterial pressure at the feet would be 160 mmHg for $R_g = 2$ m, and 175 mmHg for R_g = 6 m, the corresponding values at the head being 95 and 80 mmHg, respectively.

[5]For example, for $R_g = 2$ m, as is generally the case for a conventional space module, $a_c = 1$ g is attained at $v_t = 4.43$ m s^{-1}. The corresponding blood pressure values for our typical astronaut of 1.75 m stature amount to ≈160 and ≈95 mmHg, respectively at the feet and the head.

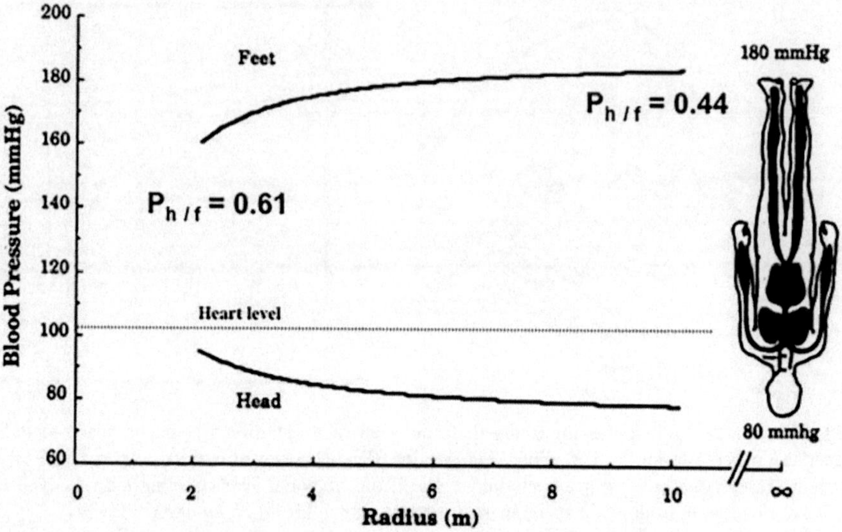

Fig. 11.10 Arterial blood pressure values at the head and feet levels for an astronaut of 1.75 m stature when riding the Twin Bike System at the speed yielding $a_c = 1$ g at the feet level, as a function of the radius of gyration of the space module. From di Prampero 2000

These same values become essentially equal to those expected in a standing healthy subject on Earth for $R_g \geq 10$ m.

When riding the Twin Bike System, slight head movements may lead to simultaneous rotations of the semi-circular canals about more than one axis. It is generally believed that this state of affairs and the resulting sensorial conflict is one of the major determinants of acute motion sickness (Nicogossian et al. 1996). To test this possibility, six healthy male subjects were tested on the human centrifuge at the Karolinska Institute in Stockholm (Antonutto et al. 1993). A cycle ergometer was fixed to one arm of the centrifuge, at a distance of 2.2 m from the rotation center, very close to the radius of a conventional space module. The cycle ergometer was inclined to 45°, so that the subject's head was closer to the rotation axis. The centrifuge rotation rate was set at 21 rounds per minute (0.35 Hz), yielding an angular velocity close to that required in the Twin Bike System to attain 1 g at the subject's feet. This resulted in a horizontal outward acceleration vector of 1 g at the inner ear level. The sum of this vector plus the Earth gravity resulted in a vector of 1.41 g applied to the subject's inner ear and aligned along his body axis.

The experiment consisted of 20 min pedaling at 50 W (pedal frequency = 1 Hz) during centrifuge rotation. The subject was asked to keep the head still or to move it according to a protocol involving various degrees of rolling, pitching, or yawing with eyes open or closed. The subjects were interviewed during the centrifuge runs and asked to rate their acute motion sickness symptoms, according to the diagnostic categorization proposed by Lackner and Graybiel (1986). In short, only one subject out of six suffered a mild level of acute motion sickness (score 3, out of a maximum of 16), corresponding to a subjective definition of moderate malaise. The symptoms worsened with eyes open and disappeared rapidly after the end of the run.

These data demonstrate that the discomfort deriving from the rotating environment necessary to generate artificial gravity is reasonably low and well tolerated. Thus, the Twin Bike System may indeed prove to be a useful tool for maintaining the physical fitness and cardiovascular conditioning of astronauts. In addition, at variance with other systems which have been proposed for mimicking gravity, the Twin Bike System: (i) does not need any external power, being operated by the subjects themselves, and (ii) by combining exercise and simulated gravity ought to prevent at one and the same time muscle atrophy, bone demineralization, and cardiovascular deconditioning.

The ESA has recently attempted to generate a similar gravitational countermeasure by developing a short-arm centrifuge. Two models have been constructed for ground utilization and located in bed rest facilities at Toulouse, France, and Cologne, Germany. Before its use as a countermeasure, no studies were carried out to investigate cardiovascular responses to centrifugation yielding, for instance, $a_c = 1$ g at the level of the heart or of the feet. And this, despite the fact that the very short arm of the centrifuge (the subject is placed parallel to the radius of gyration, with the head closest to the center of rotation and the feet at most at 1.9 m from the center of rotation) (Rittweger et al. 2015) implies a remarkable a_c gradient appears from the head to the feet indeed. Instead, in spite of the lack of a preliminary physiological evaluation, it was immediately tested as a countermeasure in short-term bed rest studies (Rittweger et al. 2015). Neuromuscular and bone functions were investigated, with ambiguous results. Cardiovascular effects were not studied. The doubt arises that the chosen bed rest duration was too short. Nevertheless, 90-min centrifugation (a_c= 0.8 g in men, and 0.6 g in women at the heart level) as compared to 90-min resting supine on the centrifuge bed without spinning, was claimed to improve orthostatic intolerance (Goswami et al. 2015b), with cerebrovascular and cardiovascular responses similar to standing (Goswami et al. 2015a, 0.75 g at heart level).

The effects of varying the a_c value at the feet by moving the rotation axis position from above the head to the heart was later investigated (Laing et al. 2020). This is not exactly what one would dream of, because a rotation axis position at the heart level generates two opposite gravitational gradients. Notwithstanding this confounding factor, that was the first time in which cardiovascular variables were studied during short-arm centrifugation. The data supported the hypothesis that displacing the rotation axis position toward the heart in such a way as to maintain a fix a_c at the feet, reduced the orthostatic challenge, leading to improved *g*-tolerance. Yet the effects of peripheral blood shift remained ambiguous, to the point that the authors themselves suggested that further investigations are required to understand the problem better. Further, the centrifuge was used as a countermeasure in a 60-day duration bed rest study. The occurrence of mastoid effusions was investigated, with no positive effects of centrifugation on it (Lecheler et al. 2021). Moreover, lumbar muscle atrophy and increased relative intramuscular lipid concentration were not counteracted by daily administration of artificial gravity exposure (De Martino et al.

2021). To the best of our knowledge, no articles on cardiovascular deconditioning have been published from that bed rest so far. Nine years after the inauguration of the first centrifuge, we are still at the starting point. Incidentally, the short-arm centrifuge will not be implemented in the International Space Station for technical reasons, as easily predictable. These things typically happen when other priorities instead of scientific and technological considerations prevail. Scientists and technicians should define the lines, along which the subject of countermeasures for the prevention of functional deterioration during space flight should be treated.

To date, the Twin Bike System, although its development has never been supported by Space Agencies, still represents, on theoretical grounds, the most promising and convincing idea as a potential countermeasure protecting astronauts from cardiovascular deconditioning and orthostatic intolerance upon re-entry.

11.9 Artificial Gravity on the Moon or Mars

On the Moon and on Mars, ΔP is reduced in direct proportion with the prevailing a_g, amounting to 16.5 and 38.0% of that on Earth, respectively (see Table 11.1). Hence, the average arterial pressure at the level of the carotid bodies or of the dorsal artery of the foot of astronauts standing on the Moon or on Mars, for an average mean pressure of 100 mmHg generated by the left ventricle in the aortic bulbus, will turn out as indicated in Table 11.2.

The reduced acceleration of gravity on the Moon and on Mars, and hence the corresponding fall of the hydrostatic component of the arterial blood pressure reported in Table 11.2 may bring about a reduced efficacy of arterial baroreflexes and hence determine cardiovascular deconditioning (Brunelli 1993). However, astronauts pedaling on appropriately designed circular or elliptic tracks, may maintain at one and the same time an appropriate muscular and circulatory status (di Prampero et al. 2009). Indeed, when moving along the curvilinear sections of the track, the astronauts will generate a centrifugal acceleration that depends on v_t and on R_g, according to Eq. 11.6. Since a_c is applied horizontally outward, the vectorial sum (g') of a_c and a_g can be easily calculated geometrically:

Table 11.2 Acceleration of gravity (g, m s^{-2}), average arterial pressures at the level of the carotid and dorsal foot arteries (mmHg), together with the ratio thereof are reported for a subject standing on the indicated celestial bodies

	g (m/s^{-2})	Carotid (mmHg)	Dorsal foot (mmHg)	Ratio
MOON	1.62	97	117	0.83
MARS	3.73	92	145	0.63
EARTH	9.81	80	200	0.40

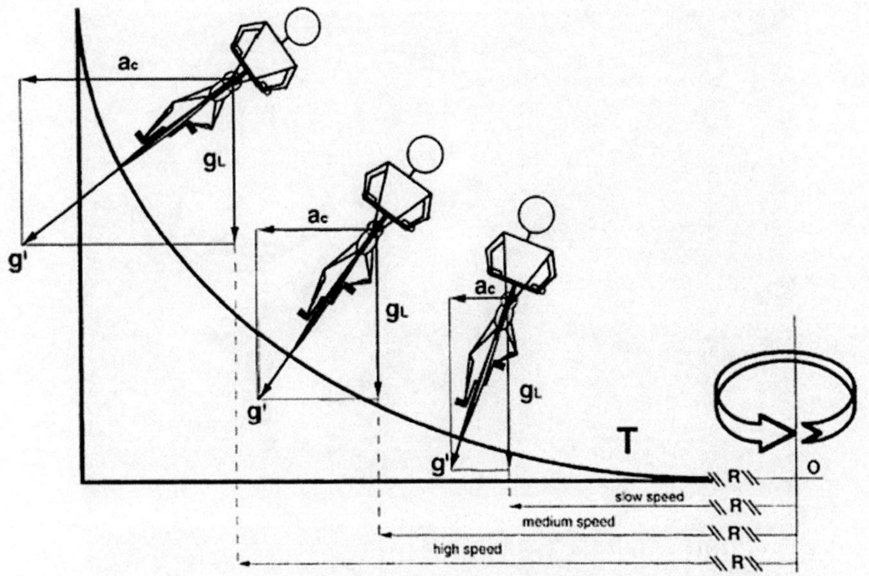

Fig. 11.11 On the curvilinear sections of the track, the acceleration vector acting along the head to feet axis (g'), coinciding with a line joining the center of mass of the system and the point of contact between wheels and terrain, can be easily calculated provided that the velocity in respect of the terrain, the radius of gyration (R in the Figure, R_g in the text), as well as the gravity acceleration on the celestial body in question are known (Eqs. 11.6 and 11.8). From di Prampero 2000.

$$g' = \sqrt{{a_c}^2 + {a_g}^2} \tag{11.8}$$

The geometric representation of the acting forces is shown in Fig. 11.11.

This line of thinking prompted di Prampero et al. (2009) to propose the construction of appropriate circular tracks around the Lunar or Martian bases. Based on the dimensions of the lunar station as described by Grandl (2007), the authors proposed to construct a circular track tunnel with a radius of 25 m positioned around the Lunar base itself and easily accessible from it (Fig. 11.12).

When cycling on flat terrain, at constant speed in the absence of wind, the overall mechanical power ($\dot{w}$) generated by the cyclist, as from Eq. 6.4, is equal to:

$$\dot{w} = v\left(F_{n-a} + k_w v^2\right) = v F_{n-a} + k_w v^3 \tag{11.9}$$

where v is the velocity relative to the air. $F_{n\text{-}a}$ is the non-aerodynamic component of the opposing force (force counteracting the rolling resistance), so that the product $v F_{n-a}$ indicates the mechanical power necessary to overcome the rolling resistance, whereas k_w is a proportionality constant (for details on the meaning of k_w, see Eq. 5.5, Chap. 5), so that the product $k_w v^3$ corresponds to the mechanical power dissipated against the air resistance.

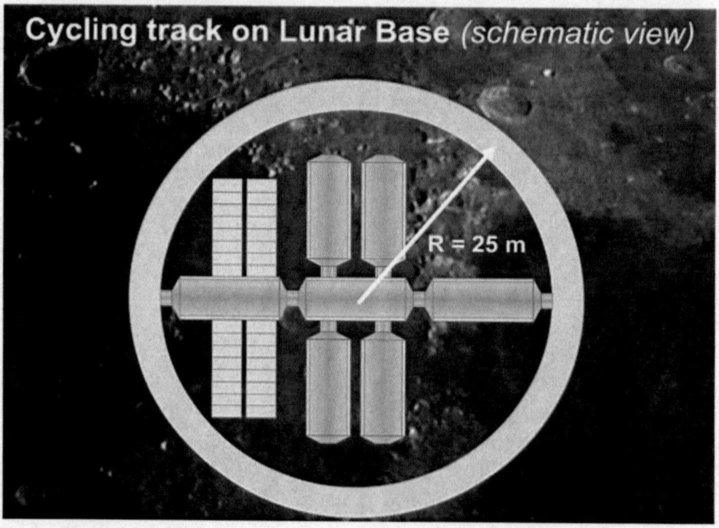

Fig. 11.12 (Top) Lunar base as proposed by Grandl (2007). (Bottom) Grandl's Lunar base surrounded by a circular "Track-Tunnel." Modified after di Prampero et al. 2009

On the one hand, $F_{n\text{-}a}$ is a function of the overall weight (cyclist + bike) and depends mainly on the friction of the tires with the terrain and, to a rather minor extent, on the friction of the transmission chain and of the axles of the pedals and wheels. On the other hand, the constant k_w is a function of the area projected on the frontal plain by the cyclist and bike, and depends on the air density and the drag coefficient. The two constants in question are rather well known, thanks to the numerous studies on the biomechanics of cycling, to a large extent carried out within the School of Milano and analyzed in Chap. 6. For a subject with a 70 kg body mass

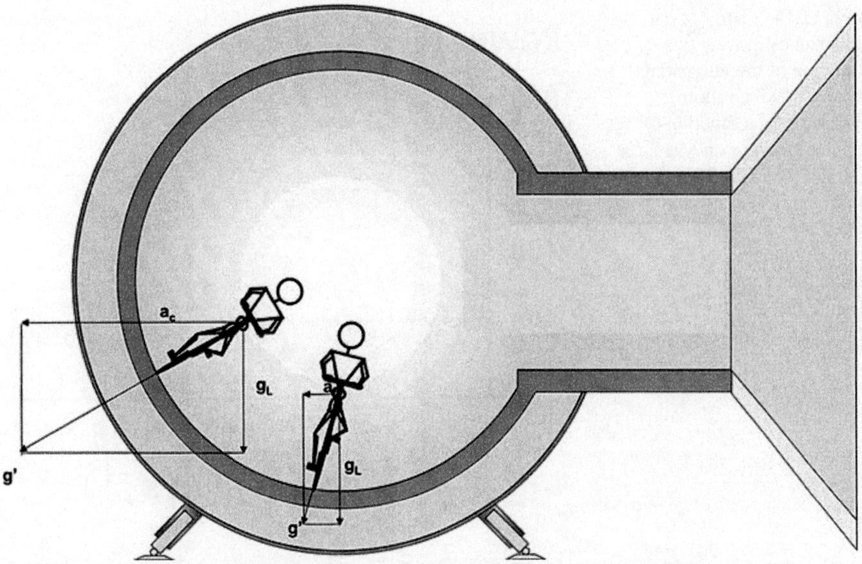

Fig. 11.13 Cross section of a Lunar Track-Tunnel (schematic). The centrifugal acceleration (a_c) is equal to the ratio between the square of the tangential speed (v_t) and the radius of gyration (R_g). The acceleration vector acting along the head-to-feet axis (g') is given by: $g' = \sqrt{a_c{}^2 + a_g{}^2}$ (Eq. 11.8), where a_g is the gravity acceleration. Therefore, on the Moon ($a_g = 1.62$ m s^{-2}), for $R_g = 25$ m, a_c and hence g' can be easily calculated for any v_t value. Modified after di Prampero et al. 2009

and 1.75 m stature, riding a standard racing bicycle in fully dropped posture on Earth at sea level ($P_B = 760$ mmHg) and at an ambient temperature of 20 °C, we obtain $F_{n\text{-}a} \approx 2.8$ J m^{-1} and $k_w = 0.19$ N s^2 m^{-2} (di Prampero 2000). For further details on the biomechanics of cycling, see Chap. 6, Sect. 6.3.

Knowledge of $F_{n\text{-}a}$ and of k_w makes it possible to calculate the evolution of $\dot{w}$ as a function of v, and hence to estimate, for any give R_g, the power required to generate the vector g' acting along the direction head—feet, i.e., the simulated gravity vector. This is reported in Fig. 11.13 for $R_g = 25$ m on the Moon or on Mars, in air at 760 mmHg and 20 °C.

Figure 11.14 shows the relationships between either $\dot{w}$ (top panel) or g' (bottom panel), and v_t. The $\dot{w}$ values reported in this figure, have been calculated for an overall mass (cyclist plus bike) of 85 kg on Earth, breathing air at a pressure of 760 mmHg. A man pedaling on Earth has a rolling coefficient that varies with the bike and tires characteristics and with the pressure of inflation of the tires, as shown in Table 6.1. The rolling resistance $F_{n\text{–}a}$ for any given cyclist can be obtained multiplying these rolling coefficients by the overall mass and by the gravity acceleration. However, $F_{n\text{–}a}$ is a function of the actual weight, which in the Lunar Track Tunnel corresponds to the product of the overall mass times the ongoing g'. As a consequence, for any given v_t and R_g, the $F_{n\text{–}a}$ value must be obtained multiplying the value applying on Earth times the ratio $g'/9.81$. As indicated above, this ratio can be calculated after Eqs. 11.6 and 11.8. Hence, for an overall mass of 85 kg, and for v_t

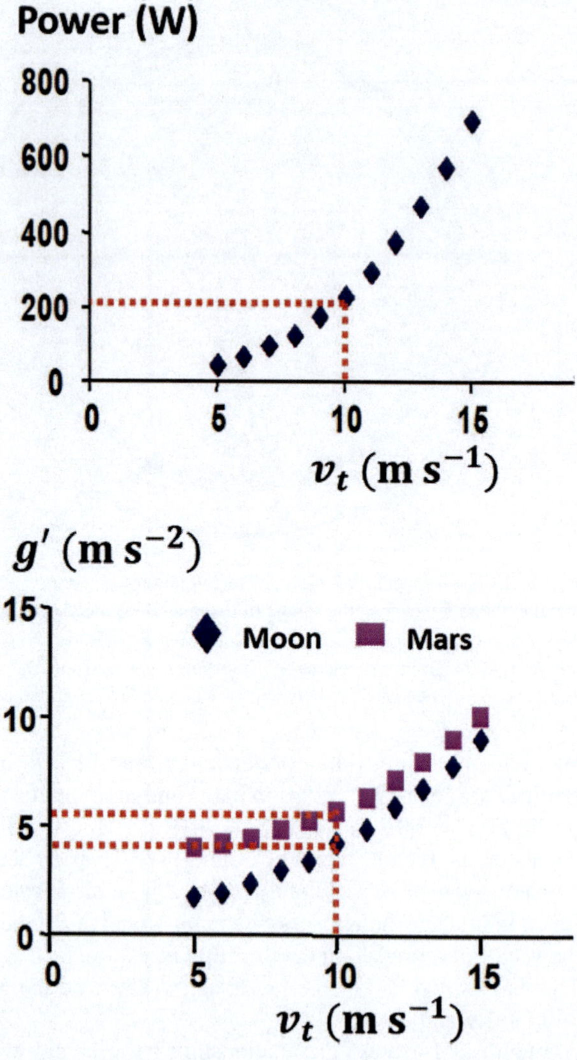

Fig. 11.14 (Top) Mechanical power as a function of the tangential velocity (v_t). (Bottom) Vector g' as a function of v_t, on the Moon or on Mars, for a track radius of 25 m in air (barometric pressure = 760 mmHg, temperature = 20 °C). Equations appears on the Figure. Modified after di Prampero et al., 2009

= 0, we have $F_{n\text{-}a} = 0.46$ J m^{-1} on the Moon and 1.06 J m^{-1} on Mars, which increases to 2.6 J m^{-1} on the Moon and to 2.8 J m^{-1} on Mars at $v_t = 15$ m s^{-1}.

Fig. 11.14 shows also that the speed attained with $\dot{w} = 200$ W is about 10 m s^{-1}, thus generating (for $R_g = 25$ m) a g' value of about 5 m s^{-2}, with small differences between Moon and Mars, due to the different gravity prevailing on these celestial bodies. However, as concerns permanently manned Lunar bases, in order to reduce the overall energy requirements of the base itself, present day proposals foresee an overall atmospheric prevailing pressure of ≈350 mmHg with an oxygen fraction increased to about 35%, to avoid hypoxic conditions for the astronauts living permanently there. Therefore, the air density within the lunar base (track tunnels included) will be substantially reduced, thus leading to an equal reduction of the aerodynamic constant k_w. Under these conditions, the v_t attained with $\dot{w} = 200$ W will be larger than reported in Fig. 11.14 ($\approx$12.3 m s^{-1}), thus leading to an increase of g' to about 6.3 m s^{-2}, i.e. about 65% of the Earth gravitational acceleration.

It should finally be pointed out that, in view of the mechanical efficiency of cycling ($\approx$0.25), the $\dot{w}$ of 200 W considered above implies a metabolic power about four times larger, corresponding to an oxygen consumption of about 2.3 L min^{-1} above resting, substantially lower than the maximal oxygen consumption of young active non-athletic subjects. As such, the effort to attain a substantial increase of g' will be well tolerated, and within the average daily exercise dose required to avoid muscular deconditioning. In any case, Fig. 11.14 shows that also for $\dot{w}$ values below 200 W, g' is substantially larger than the gravity values prevailing on the Moon and on Mars.

In conclusion, the track tunnels appear to be useful tools to avoid cardiovascular deconditioning and maintain the astronauts' physical fitness on permanently manned bases on the Moon or on Mars. Future studies should investigate the doses of g' that are necessary to avoid cardiovascular deconditioning during prolonged exposure to the Lunar or Martian acceleration of gravity, both in terms of intensity with respect to the Earth acceleration of gravity, and in terms of duration (per day) and frequency (per week) of the exposure. It goes without saying that further studies on the characteristics of the described system, engineering constraints, and manufacturing costs, are needed, in view of its possible installation on a Lunar base.

11.10 Mechanical Efficiency and Internal Power During Cycling

The rate at which chemical energy is transformed into mechanical work and heat during exercise is defined as the metabolic power of exercise ($\dot{E}$). If η is constant, since:

$$\eta = \frac{\Delta \dot{w}}{\Delta \dot{E}} \tag{11.10}$$

the relationship between $\dot{E}$ and $\dot{w}$ is linear, with slope equal to $1/\eta$. However, $\dot{E}$ is not entirely used for external power development. A resting individual does not perform mechanical work to move his/her body, yet he/she has a positive $\dot{E}$ (resting metabolic power, $\dot{E}_r$, about 1.4 W kg^{-1} on average), sustaining resting ventilation and blood flow, smooth muscle contraction and molecule displacement against concentration gradients, plus the isometric muscle contractions that sustain a standing or sitting body. Moreover, during exercise, a fraction of $\dot{E}$ ($\dot{E}_i$) is used to sustain the internal mechanical power, corresponding to the power that is not done to move along the direction of movement on a bicycle or to overcome the brake resistance on a cycle ergometer. During cycling at constant pedaling frequency (f_p), $\dot{E}_i$ is a constant. Therefore, the linear relationship between $\dot{E}$ and $\dot{w}$ must have a positive y-intercept, $\dot{E}_0$, equal to $\dot{E}_r + \dot{E}_i$.

This is indeed so during dynamic light exercise on a cycle ergometer, the most common type of exercise in a laboratory of integrative physiology nowadays. The linear relationships between metabolic and mechanical power obtained at various a_g values are depicted in Fig. 11.15. These relationships take the following algebraic form:

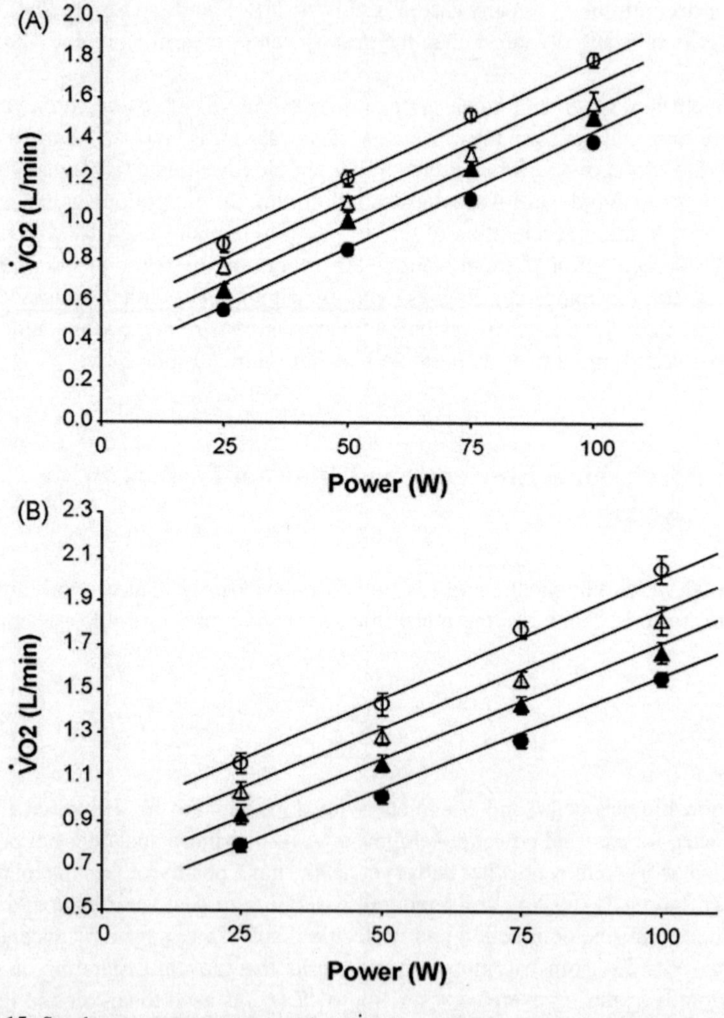

Fig. 11.15 Steady-state oxygen consumption ($\dot{V}O_2$) as a function of the mechanical power at a pedaling frequency (f_p) of 1 Hz (panel **a**) and 1.5 Hz (panel **b**). Data are given as mean and standard error. Each line and symbol corresponds to a given gravity acceleration (● = 1.0 G; ▲ = 1.5 G; Δ = 2.0 G; ○ = 2.5 G). At each f_p, the lines have the same slope, but are displaced upward as the gravity acceleration is increased. From Bonjour et al. 2010

$$\dot{E} = \dot{E}_0 + \frac{\dot{w}}{\eta} \quad (11.11)$$

The reciprocal of the slope of the fitting lines in Fig. 11.15, corresponding to η in Eq. 11.11, is about 0.25, which corresponds pretty well to the mechanical efficiency of concentric muscle contractions, and which is sometimes called the delta-efficiency of exercise (Gaesser and Brooks 1975).

Since $\dot{E}_0 = \dot{E}_r + \dot{E}_i$, Eq. 11.11 can be rewritten as follows:

$$\dot{E} - \dot{E}_r = \dot{E}_i + \frac{\dot{w}}{\eta} \quad (11.12)$$

indicating that constant $\dot{E}_i$ is independent of $\dot{w}$.

Francescato et al. (1995) provided empirical demonstration that $\dot{E}_i$ is a power function of f_p. Theoretical analyses of $\dot{E}_i$ during cycling suggest that $\dot{E}_i$ varies with the third power of f_p(Minetti 2011; Vinetti et al. 2022; Zamparo et al. 2002a, 2002b):

$$\dot{E}_i = k_i f_p^{\,3} \quad (11.13)$$

where k_i is a proportionality constant depending on the lower limb mass, on a_g (see Fig. 11.16) and on the square of the pedal diameter. Equation 11.13 conveniently

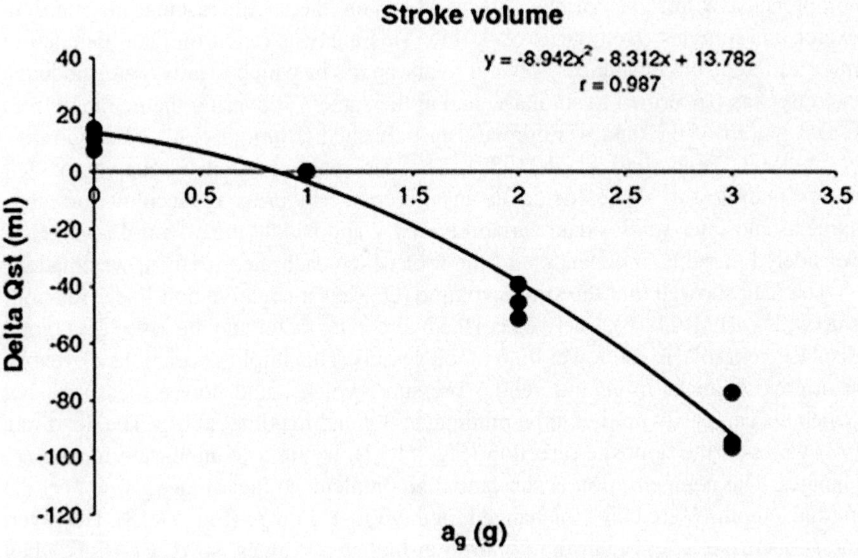

Fig. 11.16 Stroke volume (Q_{st}), expressed as the difference (Delta) with respect to the corresponding values observed at 1 g, as a function of gravity acceleration (a_g). The regression line assumes a quadratic model, and is calculated through all the individual values. Each point is the mean value at each power. From Bonjour et al. 2011

describes data in air on Earth (Bonjour et al. 2010; Formenti et al. 2015; Francescato et al. 1995; Martin et al. 2002; Vinetti et al. 2022).

The fact that constant k_i depends on a_g implies that $\dot{E}_i$ ought to be lower in microgravity than on Earth. This prediction was confirmed by the data that Girardis et al. (1999) obtained during the EUROMIR long-term space mission, and was better focused by their theoretical demonstration that, *ceteris paribus*, $\dot{E}_i$ (and thus k_i) is directly proportional to a_g. This theoretical statement was later supported by the data obtained in the long-arm human centrifuge at Karolinska Institute in the a_g range from 1.0 to 2.5 g (Bonjour et al. 2010).

A direct consequence of these effects of a_g on $\dot{E}_i$ is that the ability to perform exercise improves on celestial bodies smaller than the Earth, but deteriorates on celestial bodies bigger than the Earth. The former case might have an impact on a future human flight to Mars, as it could attenuate the impact of muscle atrophy (Fitts et al. 2001; di Prampero and Narici 2003) on exercise capacity.

11.11 Cardiovascular Response to Exercise on Outer Planets

The characteristics of the relationship between a_g and $\dot{E}_i$ and the associated vertical displacement of the relationship between $\dot{E}$ and $\dot{w}$ during moderate exercise, in which the oxygen consumption ($\dot{V}O_2$) is equal to $\dot{E}$, have set the basis for a comprehensive analysis of the effects of a_g on the cardiovascular response to exercise in humans (Bonjour et al. 2011). This analysis relied on data obtained in three different circumstances, several years apart, in which steady-state moderate exercise was performed at similar $\dot{w}$ and at the same f_p, covering the a_g range from 0 to 3 g. Part of the data were previously published (Bonjour et al. 2010; Girardis et al. 1999; Pendergast et al. 1987). The data were presented relative to their corresponding 1 g value (delta) at any given $\dot{w}$, in order to account for inter-subjects and inter-studies data variability at 1 g and isolate the effect of a_g on each considered variable, independent of the $\dot{w}$ at which each measurement was made.

The data showed that the stroke volume (Q_{st}) has a negative non-linear relationship with a_g (Fig. 11.16), likely due to the hydrostatic factor and the associated blood displacement toward the lower limb blood vessels. This implies variations of venous return and changes in central venous pressure, which could not be measured, but which we can easily predict to be minimal at 3 g and maximal at 0 g. The heart rate (f_H) varies in the opposite direction (Fig. 11.17), trying to compensate for the Q_{st} changes. The compensation is substantially complete, so that at any given $\dot{V}O_2$ the cardiac output ($\dot{Q}$) remains unchanged, in the 0-to-1 g range (Fig. 11.18). However, the f_H increase is less than the Q_{st} drop at higher a_g values, so that $\dot{Q}$ falls when $a_g > 1$ g. This looks positive for exercise performance on the Moon and on Mars, on which $a_g < 1$ g.

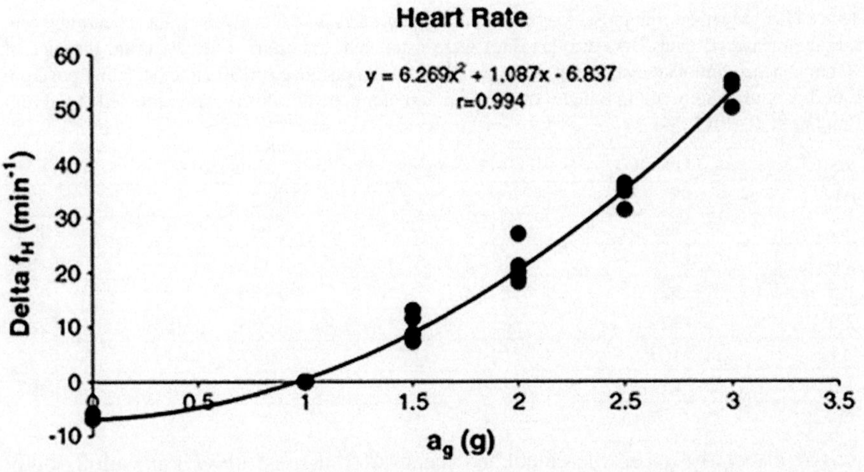

Fig. 11.17 Heart rate (f_H), expressed as the difference (Delta) with respect to the corresponding values observed at 1 g, as a function of gravity acceleration (a_g). The regression line assumes a quadratic model, and is calculated through all individual values. Each point is the mean value at each power. From Bonjour et al. 2011

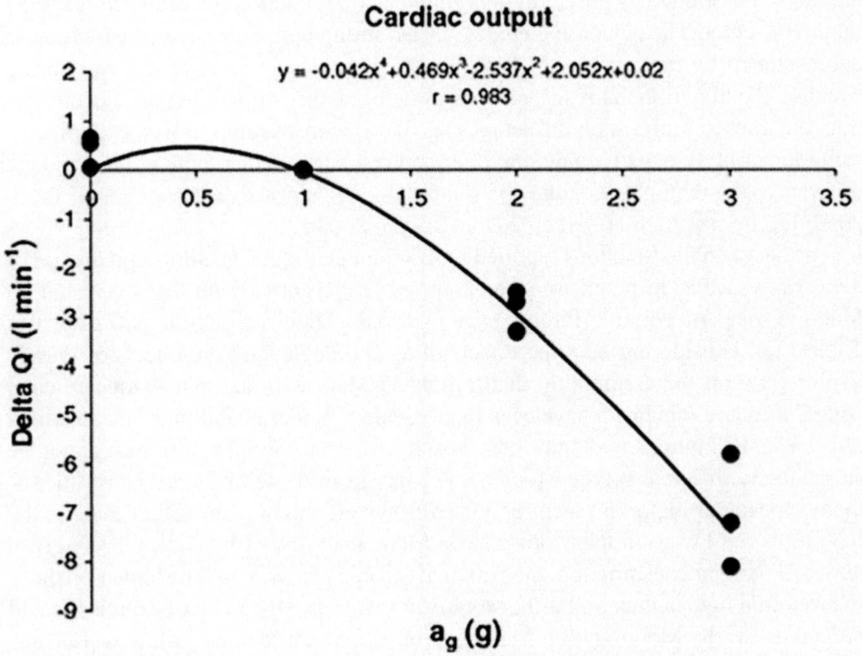

Fig. 11.18 Cardiac output ($\dot{Q}$), expressed as the difference (Delta) with respect to the corresponding values observed at 1 g, as a function of gravity acceleration (a_g). Since $\dot{Q}$ is the product of stroke volume times heart rate, the regression line assumes a fourth power function, which was calculated through all individual values. Each point is the mean value at each power. From Bonjour et al. 2011

Table 11.3 Mean arterial pressure at heart, carotid, buttocks, and feet levels. Data at heart level are average measured data. Data at other sites were estimated, assuming a heart–carotid distance of 30 cm, a heart–buttocks distance of 40 cm, and heart–feet distance of 130 cm in standing position. Blood pressure values are in mmHg; the acceleration along the Gz axis (a_g) is expressed in g. From Bonjour et al. 2011

a_g	Heart	Carotid	Buttocks	Feet
0.0	82	82	82	82
0.5	84	73	99	132
1.0	87	65	117	183
1.5	92	58	136	235
2.0	97	53	156	288
2.5	103	48	177	342
3.0	111	44	199	397

The effects of a_g on the calculated mean arterial pressure ($\overline{P}$) at various body heights in vertical posture are reported in Table 11.3. The $\overline{P}$ at the level of the heart increases with a_g. Such an increase is necessary to prevent an excessive pressure fall in the carotid artery, and is likely mediated by sympathetic activation through inhibition of cardiopulmonary and arterial baroreceptors (Goodman et al. 2000; Strempel et al. 2008). The reduction of central blood volume may inhibit the former, the fall of carotid artery pressure, implying tension reduction in baroreceptors, may inhibit the latter. The response includes at least strong peripheral vasoconstriction, as demonstrated by the increase in total peripheral resistance both at rest and during exercise. Total peripheral resistance is a systemic index of peripheral vasoconstriction, but does not inform on the actual sites of vasoconstriction. In fact, experiments using lower body negative pressure after bed rest suggest that there are remarkable differences in regional vascular resistance among various districts (Edgell et al. 2012; Fischer et al. 2007; Arbeille et al. 2008a, 2008b).

The data and the functions reported by Bonjour et al. (2011) allow a prediction of the cardiovascular response to exercise on a cycle ergometer on the Moon and on Mars, in view of possible future space missions. These predictions are shown in Table 11.4, considering only the effects of a_g and neglecting the effects of muscle hypotrophy, on the assumption that a flight to Mars will take place after efficient countermeasures against it have been implemented. In these conditions, the maximal oxygen consumption, and thus the capacity of performing work, would not be substantially different on the Moon or on Mars from those on Earth. Nevertheless, there appears an important feature, related to the effects of a_g on $\dot{E}_i$: at each $\dot{w}$, the $\dot{V}O_2$ turns out lower on the Moon and on Mars, than on Earth. Thus, for any given maximal oxygen consumption, the maximal aerobic power would be higher on those two celestial bodies than on Earth. Moreover, the gross efficiency of exercise would be greater on the Moon and on Mars than on Earth, particularly at low workloads, due to the effect of a lower $\dot{w}_i$. On the opposite side, there would be a planet mass limitation in the ability to perform exercise on outer planets. Extrapolation of the f_H versus a_g relationship to the maximal f_H (see Fig. 11.17) indicates that, for a human with a standard resting f_H on Earth of 70 min^{-1}, the maximal f_H would be attained

Table 11.4 Metabolic and cardiovascular response to exercise on the Moon and on Mars. The values on Earth represent the overall mean at 1 g from the three experiments reported by Bonjour et al. (2011), weighted for subject's number. The values on Moon and Mars are estimated after Bonjour et al. (2011). At each power, the reported values were obtained in steady state. $\dot{V}O_2$, oxygen consumption; f_H, heart rate, $\dot{Q}$, cardiac output, Q_{st}, stroke volume, $\overline{P}$, mean arterial pressure

Power (W)		50	75	100
$\dot{V}O_2$ (L min^{-1})	On Earth	0.87	1.12	1.42
	On Moon	0.74	0.99	1.29
	On Mars	0.78	1.03	1.33
f_H (min^{-1})	On Earth	93	102	112
	On Moon	87	96	105
	On Mars	88	97	106
$\dot{Q}$ (L min^{-1})	On Earth	11.3	12.4	14.5
	On Moon	11.6	12.7	14.8
	On Mars	11.8	12.9	14.9
Q_{st} (mL)	On Earth	121	122	130
	On Moon	133	134	142
	On Mars	130	131	139
$\overline{P}$ (mmHg)	On Earth	87	93	94
	On Moon	84	91	91
	On Mars	85	92	92

already at rest on planets with masses yielding an a_g of 4.0–4.5 g, depending on astronaut's age. On such planets, no exercise would be possible, at most, *ceteris paribus*, one could stand. However, *cetera non sunt paria.* In fact, Fig. 11.18 shows that, at the prevailing a_g on Jupiter (2.53 g), the heart would pump only some 2 L of blood per minute.

Assuming a $\overline{P}$ of 100 mmHg at the level of the heart (several factors act in opposite directions on the heart, see above), the corresponding $\overline{P}$ at the level of the carotid bifurcation would be about 50 mmHg. In fact it might even be less, if we take into account also the very long journey in microgravity to attain Jupiter: the pressure of the blood flowing to the brain would be less than that compatible with consciousness (Adami et al. 2013). Therefore, no human would be able to stand conscious on Jupiter, as on other planets of such masses or greater: apart from their dense gaseous surfaces, no man could ever be sent to Jupiter or Saturn.

11.12 Conclusions

It appears from this Chapter, that the School of Milano has delivered a considerable intellectual and experimental effort in conceiving efficient countermeasures for and collecting experimental data during space flight and hypergravity exposure in long-arm centrifuges. Yet this effort *in se* is not superior to the effort put by other investigators and other physiological Schools in studying life science in Space. What distinguishes the School of Milano is its proneness to theoretical analysis in support of experimental data, stemming from the cultural *milieu* within which all of us have been working since Margaria's days. This attitude led to conceive and formulate physiological models that could profitably find applications in the countermeasure payload of future space flights and outer celestial body colonization. Nevertheless, this attitude, although capable of satisfying our intellectual pleasure deeply, did not translate into actual applications currently in use or approved for future space missions. The reasons of this are multiple and complex, and have to do, on one side, with our psychology, on the other, with the current organization of space activity in our societies. The School of Milano had a greater practical impact on altitude activities, through Monzino's expeditions and the Everest pyramid, than in Space activities. Yet, ideas and models are there and remain available for further exploitation.

References

Adami A, Pizzinelli P, Bringard A, Capelli C, Malacarne M, Lucini D, Simunič B, Pišot R, Ferretti G (2013) Cardiovascular re-adjustments and baroreflex response during clinical reambulation procedure at the end of 35-day bed rest in humans. Appl Physiol Nutr Metab 38:673–680

Antonutto G, Capelli C, di Prampero PE (1991) Pedalling in space as a countermeasure to microgravity deconditioning. Microgravity Q 1:93–101

Antonutto G, di Prampero PE (2003) Cardiovascular deconditioning in microgravity: some possible countermeasures. Eur J Appl Physiol 90:283–291

Antonutto G, Linnarsson D, di Prampero PE (1993) On Earth evaluation of neurovestibular tolerance to centrifuge simulated artificial gravity in humans. The physiologist 36. Supplement 1:S85–S87

Antonutto G, Capelli C, Girardis M, Zamparo P, diPrampero PE (1999) Effects of microgravity on maximal power of lower limbs during very short efforts in humans. J Appl Physiol 86:85–92

Arbeille P, Kerbeci P, Mattar L, Shoemaker JK, Hughson RL (2008a) WISE-2005. Tibial and gastrocnemius vein, and calf tissue response to LBNP after a 60 day bedrest with and without countermeasures. J Appl Physiol 104:938–943

Arbeille P, Kerbeci P, Mattar L, Shoemaker JK, Hughson RL (2008b) Insufficient flow reduction during LBNP in both splanchnic and lower limb areas is associated with orthostatic intolerance after bedrest. Am J Phys 295:H1846–H1854

Bonjour J, Bringard A, Antonutto G, Capelli C, Linnarsson D, Pendergast DR, Ferretti G (2011) Effects of acceleration in the Gz axis on human cardiopulmonary responses to exercise. Eur J Appl Physiol 111:2907–2917

Bonjour J, Capelli C, Antonutto G, Calza S, Tam E, Linnarsson D, Ferretti G (2010) Determinants of oxygen consumption during exercise on cycle ergometer: the effects of gravity acceleration. Respir Physiol Neurobiol 171:128–134

Bringard A, Pogliaghi S, Adami A, De Roia G, Lador F, Lucini D, Pizzinelli P, Capelli C, Ferretti G (2010) Cardiovascular determinants of maximal oxygen consumption in upright and supine posture. Respir Physiol Neurobiol 172:53–62

Brunelli F (1993) Fundamentals of space life sciences. In: Churchill SE (ed) Human factors and habitability issues for the Design of Space Habitats. Krieger, Melbourne, Australia, pp 221–246

Buehring B, Belavy DL, Michaelis I, Gast U, Felsenberg D, Rittweger J (2011) Changes in lower extremity muscle function after 56 days of bed rest. J Appl Physiol 111:87–94

Caiozzo VJ, Baker MJ, Herrick RE, Tao M, Baldwin KM (1994) Effect of spaceflight on skeletal muscle: mechanical properties and myosin isoform content of a slow muscle. J Appl Physiol 76: 1764–1773

Caiozzo VJ, Haddad F, Baker MJ, Herrick RE, Prietto N, Baldwin KM (1996) Microgravity-induced transformations of myosin isoforms and contractile properties of skeletal muscle. J Appl Physiol 81:123–132

Capelli C, Adami A, Antonutto G, Cautero M, Tam E (2009) Oxygen deficits and oxygen delivery kinetics during submaximal intensity exercise in humans after 14 days of head-down tilt-bed rest. Eur J Appl Physiol 107:51–59

Capelli C, Antonutto G, Kenfack MA, Cautero M, Lador F, Moia C, Tam E, Ferretti G (2006) Factors determining the time course of $\dot{V}O_2^{max}$ decay during bedrest: implications for $\dot{V}O_2^{max}$ limitation. Eur J Appl Physiol 98:152–160

Capelli C, Antonutto G, Cautero M, Tam E, Ferretti G (2008) Metabolic and cardiovascular responses during sub-maximal exercise in humans after 14 days of head-down tilt bed rest and inactivity. Eur J Appl Physiol 104:909–918

Cardus D, McTaggart WG, Campbell S (1991) Progress in the development of an artificial gravity simulator (AGS). Physiologist 34(Supplement):S224–S225

Cavagna GA, Thys H, Zamboni A (1976) The sources of external work in level walking and running. J Physiol Lond 262:639–657

Cavagna GA, Willems PA, Heglund NV (1998) Walking on Mars. Nature 393:636

Clément G, Pavy-Le Traon A (2004) Centrifugation as a countermeasure during actual and simulated microgravity: a review. Eur J Appl Physiol 92:235–248

de Boer MD, Seynnes OR, di Prampero PE, Pisot R, Mekjavic IB, Biolo G, Narici MV (2008) Effect of 5 weeks horizontal bed rest on human muscle thickness and architecture of weight bearing and non- weight bearing muscles. Eur J Appl Physiol 104:401–407

De Martino E, Hides J, Elliott JM, Hoggarth M, Zange J, Lindsay K, Debuse D, Winnard A, Beard D, Cook JA, Salomoni SE, Weber T, Scott J, Hodges PW, Caplan N (2021) Lumbar muscle atrophy and increased relative intramuscular lipid concentration are not mitigated by daily artificial gravity after 60-day head-down tilt bed rest. J Appl Physiol 131:356–368

Desplanches D, Hoppeler H, Mayet MH, Denis C, Claassen H, Ferretti G (1998) Effects of bed rest on deltoid muscle morphology and enzymes. Acta Physiol Scand 162:135–140

Di Giovanni C, Chambers RM (1964) Physiologic and psychologic aspects of the gravity spectrum. New Engl J Med 270:88–94

di Prampero PE (2000) Cycling on Earth, in space, on the Moon. Eur J Appl Physiol 82:345–360

di Prampero PE, Antonutto G (1996) Effects of microgravity on muscle power: some possible countermeasures. ESA Symposium Proceedings "Space Station Utilisation", ESOC, Darmstadt, Germany, 30 Sept–2 Oct 1996 (SP-385, December 1996), pp 103-106;

di Prampero PE, Lazzer S, Antonutto G (2009) Human powered centrifuges on the Moon or Mars. Microgravity Sci Technol 21:209–215

di Prampero PE, Narici MV (2003) Muscles in microgravity: from fibres to human motion. J Biomech 36:403–412

Edgell H, Grinberg A, Gagné N, Beavers KR, Hughson RL (2012) Cardiovascular responses to lower body negative pressure before and after 4 h of head-down bed rest and seated control in men and women. J Appl Physiol 113:1604–1612

Ferretti G, Antonutto G, Denis C, Hoppeler H, Minetti AE, Narici MV, Desplanches D (1997) The interplay of central and peripheral factors in limiting maximal O_2 consumption in man after prolonged bed rest. J Physiol Lond 501:677–686

Ferretti G, Berg HE, Minetti AE, Moia C, Rampichini S, Narici MV (2001) Maximal instantaneous power after prolonged bed rest in humans. J Appl Physiol 90:431–435

Ferretti G, Girardis M, Moia C, Antonutto G (1998) The effects of prolonged bed rest on cardiovascular oxygen transport during submaximal exercise in humans. Eur J Appl Physiol 78:398–402

Ferretti G, Iellamo F, Pizzinelli P, Kenfack MA, Lador F, Lucini D, Porta A, Narkiewicz K, Pagani M (2009) Prolonged head down bed rest-induced inactivity impairs tonic autonomic regulation while sparing oscillatory cardiovascular rhythms in healthy humans. J Hypertens 27:551–561

Fischer CL, Johnson PC, Berry CA (1967) Red blood cell mass and plasma volume changes in manned space flight. J Am Med Assoc 200:579–583

Fischer D, Arbeille P, Shoemaker JK, O'Leary DD, Hughson RL (2007) Altered hormonal regulation and blood flow distribution with cardiovascular deconditioning after short-duration head down bed rest. J Appl Physiol 103:2018–2025

Fitts RH, Riley DR, Widrick JJ (2001) Functional and structural adaptations of skeletal muscle to microgravity. J Exp Biol 204:3201–3208

Formenti F, Minetti AE, Borrani F (2015) Pedaling rate is an important determinant of human oxygen uptake during exercise on the cycle ergometer. Physiol Rep 3:1–10

Francescato MP, Girardis M, di Prampero PE (1995) Oxygen cost of internal work during cycling. Eur J Appl Physiol 72:51–57

Gaesser GA, Brooks GA (1975) Muscular efficiency during steady-rate exercise: effects of speed and work rate. J Appl Physiol 38:1132–1139

Gauer OH, Henry JP (1963) Circulatory basis of fluid volume control. Physiol Rev 43:423–481

Girardis M, Linnarsson D, Moia C, Pendergast DR, Ferretti G (1999) Oxygen cost of dynamic leg exercise on a cycle ergometer: effects of gravity acceleration. Acta Physiol Scand 166:239–246

Goodman LS, Banks RD, Grissett JD, Saunders PL (2000) Heart rate and blood pressure responses to +Gz following varied-duration −Gz. Aviat Space Environ Med 71:137–141

Goswami N, Bruner M, Xu D, Bareille MP, Beck A, Hinghofer-Szalkay H, Blaber AP (2015a) Short-arm human centrifugation with 0.4 g at eye and 0.75 g at heart level provides similar cerebrovascular and cardiovascular responses to standing. Eur J Appl Physiol 115:1569–1575

Goswami N, Evans J, Schneider S, von der Wiesche M, Mulder E, Rössler A, Hinghofer-Szalkay H, Blaber AP (2015b) Effects of individualized centrifugation training on orthostatic tolerance in men and women. PLoS One 10:e0125780

Grandl W (2007) Lunar Base 2015 Stage I. Preliminary design study. Acta Astronautica 60:554–560

Greenleaf JE, Gundo DP, Watenpaugh DE, Mulenburg GM, McKenzie MA, Looft-Wilson R, Hargens AR (1997) Cycle-powered short radius (1.9 m) centrifuge: effect of exercise versus passive acceleration on heart rate in humans. NASA Technical Memorandum. 110433

Hargens AR, Bhattacharya R, Schneider SM (2013) Space physiology VI: exercise, artificial gravity, and countermeasure development for prolonged space flight. Eur J Appl Physiol 113: 2183–2192

Hargens AR, Richardson S (2009) Cardiovascular adaptations, fluid shifts, and countermeasures related to space flight. Respir Physiol Neurobiol 169:S30–S33

Hargens AR, Watenpaugh DE (1996) Cardiovascular adaptation to spaceflight. Med Sci Sports Exerc 28:977–982

Hickes F (1902) Lucian's true histories. An English translation. AH Bullen, London

Hughson RL, Helm A, Durante M (2018) Heart in space: effect of the extraterrestrial environment on the cardiovascular system. Nat Rev Cardiol 15:167–180

Johnson WH (1961) Some vestibular problems in space flight. Ann Otol Rhinol Laryngol 70:777–784

Johnson WH, Ireland PE (1966) Preparation of man for travel into space. Proc Roy Soc Med 59: 277–280

Kawakami Y, Akima H, Kubo K, Muraoka Y, Hasegawa H, Kouzaki M, Imai M, Suzuki Y, Gunji A, Kanehisa H, Fukunaga T (2001) Changes in muscle size, architecture, and neural activation after 20 days of bed rest with and without resistance exercise. Eur J Appl Physiol 84: 7–12

Lackner JR, Di Zio P (2000) Artificial gravity as a countermeasure in long-duration space flight. J Neurosci Res 15:169–176

Lackner JR, Graybiel A (1986) The effective intensity of Coriolis, cross-coupling stimulation is gravitoinertial force dependent: implications for space motion sickness. Aviat Space Environ Med 57:229–235

Laing C, Green DA, Mulder E, Hinghofer-Szalkay H, Blaber AP, Rittweger J, Goswami N (2020) Effect of novel short-arm human centrifugation-induced gravitational gradients upon cardiovascular responses, cerebral perfusion and g-tolerance. J Physiol Lond 598:4237–4249

Layne CS, Mulavara AP, McDonald PV, Pruett CJ, Kozlovskaya IB, Bloomberg JJ (2001) Effect of long-duration spaceflight on postural control during self-generated perturbations. J Appl Physiol 90:997–1006

Lazzer S, Plaino L, Antonutto G (2011) The energetics of cycling on Earth, Moon and Mars. Eur J Appl Physiol 111:357–366

Lecheler L, Paulke F, Sonnow L, Limper U, Schwarz D, Jansen S, Klussmann JP, Tank J, Jordan J (2021) Gravity and mastoid effusion. Am J Med 134:e181–e183

Margaria R (1966) La locolozione umana in subgravità. Atti Acc Naz Lincei 78:5–18

Margaria R (1975) Fisiologia muscolare e meccanica del movimento. Mondadori EST, Milano, Italy, pp 138–143

Margaria R, Cavagna GA (1964) Human locomotion in subgravity. Aerospace Med 35:1140–1146

Martin R, Hautier C, Bedu M (2002) Effect of age and pedalling rate on cycling efficiency and internal power in humans. Eur J Appl Physiol 86:245–250

Milesi S, Capelli C, Denoth J, Hutchinson T, Stüssi E (2000) Effects of 17 days bedrest on the maximal voluntary isometric torque and neuromuscular activation of the plantar and dorsal flexors of the ankle. Eur J Appl Physiol 82:197–205

Minetti AE (2011) Bioenergetics and biomechanics of cycling: the role of "internal work.". Eur J Appl Physiol 111:323–329

Moriggi M, Vasso M, Fania C, Capitanio D, Bonifacio G, Salanova M, Blottner D, Rittweger J, Felsenberg D, Cerretelli P, Gelfi C (2010) Long term bed rest with and without vibration exercise countermeasures: effects on human muscle protein dysregulation. Proteomics 10: 3756–3774

Mulder ER, Gerrits KH, Rittweger J, Felsenberg D, Stegeman DF, de Haan A (2008) Characteristics of fast voluntary and electrically evoked isometric knee extensions during 56 days of bed rest with and without exercise countermeasure. Eur J Appl Physiol 103:431–440

Narici MV, Kayser B, Barattini P, Cerretelli P (1997) Changes in electrically evoked skeletal muscle contractions during 17-day spaceflight and bed rest. Int J Sports Med 18(suppl 4):S290–S292

Nicogossian AE, Mohler SR, Gazenko OG, Grigorjev AI (eds) (1996) Space biology and medicine. Joint U.S./Russia publication in five volumes. AIAA, Reston VA, USA/Nauka Press, Moscow, Russia

Noordung H (1929) Das Problem der Befahrung des Weltraums: der Raketen-Motor. Richard Carl Schmidt, Berlin

Norsk P (2014) Blood pressure regulation IV: adaptive responses to weightlessness. Eur J Appl Physiol 114:481–497

Norsk P (2020) Adaptation of the cardiovascular system to weightlessness: surprises, paradoxes and implications for deep space missions. Acta Physiol 228:e13434

Pendergast DR, Olszowka AJ, Rokitka MA, Farhi LE (1987) Gravitational force and cardiovascular system. In: Dejours P (ed) Comparative physiology of environmental adaptations. Karger, Basel, pp 15–26

Porcelli S, Marzorati M, Lanfranconi F, Vago P, Pišot R, Grassi B (2010) Role of skeletal muscles impairment and brain oxygenation in limiting oxidative metabolism during exercise after bed rest. J Appl Physiol 109:101–111

Recktenwald MR, Hodgson JA, Roy RR, Riazanski S, McCall GE, Kozlovskaya I, Washburn DA, Fanton JW, Edgerton VR (1999) Effects of spaceflight on rhesus quadrupedal locomotion after return to 1 G. J Neurophysiol 81:2451–2463

Reeves ND, Maganaris CN, Ferretti G, Narici MV (2005) Influence of 90-day simulated microgravity on human tendon mechanical properties and the effect of resistive countermeasures. J Appl Physiol 98:2278–2286

Rejc E, di Prampero PE, Lazzer S, Grassi B, Simunic B, Pisot R, Antonutto G, Narici M (2015a) Maximal explosive power of the lower limbs before and after 35 days of bed rest under different diet energy intake. Eur J Appl Physiol 115:429–436

Rejc E, di Prampero PE, Lazzer S, Grassi B, Simunic B, Pisot R, Antonutto G, Narici M (2015b) A 35-day bed rest does not alter the bilateral deficit of the lower limbs during explosive efforts. Eur J Appl Physiol 115:1323–1330

Rittweger J, Bareille MP, Clément G, Linnarsson D, Paloski WH, Wuyts F, Zange J, Angerer O (2015). Short-arm centrifugation as a partially effective musculoskeletal countermeasure during 5-day head-down tilt bed rest--results from the BRAG1 study. Eur J Appl Physiol 115: 1233–1244

Saibene F, Minetti AE (2003) Biomechanical and physiological aspects of legged locomotion. Eur J Appl Physiol 88:297–316

Salanova M, Gelfi C, Moriggi M, Vasso M, Viganò A, Minafra L, Bonifacio G, Schiffl G, Gutsmann M, Felsenberg D, Cerretelli P, Blottner D (2014) Disuse deterioration of human skeletal muscle challenged by resistive exercise superimposed with vibration: evidence from structural and proteomic analysis. FASEB J 28:4748–4763

Salvadego D, Keramidas ME, Brocca L, Domenis R, Mavelli I, Rittweger J, Eiken O, Mekjavic IB, Grassi B (2016) Separate and combined effects of a 10-d exposure to hypoxia and inactivity on oxidative function in vivo and mitochondrial respiration ex vivo in humans. J Appl Physiol 121: 154–163

Salvadego D, Keramidas ME, Kölegård R, Brocca L, Lazzer S, Mavelli I, Rittweger J, Eiken O, Mekjavic IB, Grassi B (2018) PlanHab(*): hypoxia does not worsen the impairment of skeletal muscle oxidative function induced by bed rest alone. J Physiol Lond 596:3341–3355

Salvadego D, Lazzer S, Marzorati M, Porcelli S, Rejc E, Šimunič B, Pišot R, di Prampero PE, Grassi B (2011) Functional impairment of skeletal muscle oxidative metabolism during knee-extension exercise after bed rest. J Appl Physiol 111:1719–1726

Strempel S, Schroeder C, Hemmersbach R, Boese A, Tank J, Diedrich A, Heer M, Luft FC, Jordan J (2008) Norepinephrine transporter inhibition alters the hemodynamic response to hypergravitation. J Appl Physiol 104:756–760

Verne J (1865) De la Terre à la Lune, trajet direct en 97 heures 20 min. Hetzel, Paris

Vinetti G, Ferretti G, Hostler D (2022) Effects of water immersion on the internal power of cycling. Med Sci Sports Exerc 54(3):530–535

Webb P (1966) Weight loss in men in space. Science 155:558–560

Widrick JJ, Knuth ST, Norenberg KM, Romatowski JG, Bain JL, Riley DA, Karhanek M, Trappe SW, Trappe TA, Costill DL, Fitts RH (1999) Effect of a 17 day spaceflight on contractile properties of human soleus muscle fibres. J Physiol Lond 516:915–930

Yuan P, Koppelmans V, Reuter-Lorenz P, De Dios Y, Gadd N, Riascos R, Kofman I, Bloomberg J, Mulavara A, Seidler RD (2018) Change of cortical foot activation following 70 days of head-down bed rest. J Neurophysiol 119:2145–2152

Zamparo P, Antonutto G, Capelli C, Girardis M, Sepulcri L, di Prampero PE (1997) Effects of elastic recoil on maximal explosive power of the lower limbs. Eur J Appl Physiol 75:289–297

Zamparo P, Minetti AE, di Prampero PE (2002a) Interplay among the changes of muscle strength, cross-sectional area and maximal explosive power: theory and facts. Eur J Appl Physiol 88:193–202

Zamparo P, Pendergast DR, Termin B, Minetti AE (2002b) How fins affect the economy and efficiency of human swimming. J Exp Biol 205:2665–2676

Zange Z, Müller K, Schuber M, Wackerhage H, Hoffmann U, Günter RW, Adam G, Neuerburg JM, Sinitsyn VE, Bacharev AO, Belichenko OI (1997) Changes in calf muscle performance, energy metabolism and muscle volume caused by long term stay on space station Mir. Int J Sports Med 4:S308–S309

Zhang LF (2001) Vascular adaptation to microgravity: what have we learned? J Appl Physiol 91: 2415–2430

Zhang LF (2013) Region-specific vascular remodeling and its prevention by artificial gravity in weightless environment. Eur J Appl Physiol 113:2873–2895

Chapter 12
A School Goes into Depth

Guido Ferretti

Abstract This chapter is on breath-holding and breath-hold diving. The involvement of the School of Milano in diving is a consequence of the strong connection with Buffalo. Therefore, an analysis of the work carried out at Buffalo is initially presented. This includes the early studies on alveolar gas composition after breath-holding and the projects on the physiology of Korean diving women. The School of Milano entered the game with the Maiorca study, carried out in collaboration with Buffalo. Concepts such as gas exchange and energy expenditure during diving, the cardiovascular responses during diving, including the first demonstration of the occurrence of a diving response in humans, and the respiratory adaptation to breath-hold diving are discussed. An analysis of the limits of deep breath-hold diving is performed, including the effects of lung volumes. Finally, the most recent studies on cardiovascular dynamics during breath-holding and the role of baroreflexes are discussed.

G. Ferretti (✉)
Department of Molecular and Translational Medicine, University of Brescia, Brescia, Italy
e-mail: Guido.Ferretti@unibs.it

G. Ferretti (ed.), *Exercise, Respiratory and Environmental Physiology*, Perspectives in Physiology, https://doi.org/10.1007/978-3-031-19197-8_12

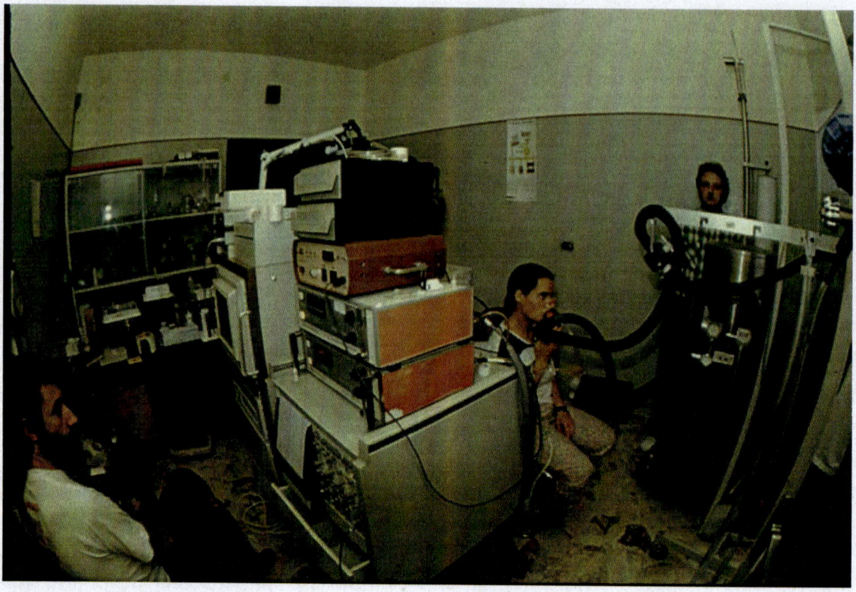

Physiological laboratory in Siracusa, Sicily, June 1986. From left to right, Guido Ferretti, Rossana Maiorca (breathing from a spirometer), and Bruno Grassi. With the kind permission of Mario Costa, Siracusa

12.1 Introduction

The study of diving, whether breath-hold diving or scuba diving, is not part of the ancient tradition of the School of Milano. Neither Margaria nor his predecessors, except Herlitzka, were interested in diving physiology. Only Emilio Agostoni (1929–2021), when he was in Rochester, performed a few experiments on breath-holding. The involvement of the School of Milano in the study of breath-hold diving physiology started in Rochester, and went on in Buffalo.

Notwithstanding a few exceptions, the most remarkable of which was the pioneering study by Gito Teruoka (1889–1966) on the women divers of Japan, (Teruoka 1932), it is a matter of fact that Wallace Fenn (1893–1971), Arthur Otis (1913–2008), and Hermann Rahn (1912–1990) (Fig. 12.1) initiated the systematic study of breath-holding and breath-hold diving in humans. This last in particular looked at breath-holding as a tool for analyzing alveolar gas composition in dynamic conditions by means of the O_2–CO_2 diagram (Rahn and Fenn 1955).

When, in 1956, he moved to Buffalo to become chairman of the Department of Physiology, Rahn hired Leon Farhi (1923–2003) and Edward Lanphier (1922–1999). Lanphier came in 1958 carrying a hyperbaric chamber. The three of

Fig. 12.1 From left to right, Arthur Otis, Hermann Rahn, and Wallace Fenn, portrayed in 1963 at a meeting of the American Physiological Society. With the kind permission of the American Physiological Society

them, with the financial support of the US Navy, put together the plans and applications to establish an environmental physiology laboratory, which eventually became the Center for Research in Special Environments (CRESE). Lanphier was in charge of the hyperbaric chamber. The Department kept expanding in the following years, to cover several fields of physiology useful to medical training. A picture of Buffalo Physiology staff in 1962 is shown in Fig. 12.2.

Lanphier was a very religious man and decided, in 1974, to leave Buffalo and enter a Seminary in Wisconsin. He completed his studies and joined the University of Wisconsin in Madison two years later. He moved the old chamber from Buffalo and established a very productive research program in Madison. He was at the University during the week and managed a rural church that he was the minister for, where he preached on Sundays and Holidays. When Lanphier left Buffalo, Claes Lundgren (1931–2022) came to take over the new chamber and eventually became director of CRESE until his retirement, when Dave Pendergast succeeded to him. A sketch of the large research installations at CRESE in the late 1970s is shown in Chap. 6, Fig. 6.1. The Center, although in a new, smaller form, is still active under the leadership of David Hostler, although in a different Department from the historical Department of Physiology, and is still supported by the US Navy.

Shortly after Lanphier's arrival, Rahn established an exchange agreement with Yonsei University, Seoul, South Korea, which led Suk-Ki Hong (1928–1999) to Buffalo. This international agreement set the conditions for the organization of several renowned scientific expeditions for the study of Korean diving women.

Fig. 12.2 Physiology staff, State University of New York at Buffalo, June 1962. From left to right, back row: BG Covino, CV Paganelli, WK Noell, LE Farhi, JW Boylan; front row: DW Rennie, EH Lanphier, B Bishop, BG Howell, H Rahn. With the kind permission of Dave Pendergast, Buffalo

The first took place in 1961 and the results were collectively presented at the first International Symposium on the Physiology of Breath-Hold Diving, which was organized in Tokyo, Japan, in 1965 (Rahn and Yokoyama 1965).

Claes Lundgren indeed brought the School of Milano into this game. In the mid-1980s, Lundgren entered in touch with Enzo Maiorca (1931–2016), an outstanding competitive breath-hold diver from Siracusa, Sicily, who established several world records in free diving depth and who was the first human to dive to a 100 m under controlled conditions (Fontanebianche, Siracusa, July 30th, 1988). Lundgren wished to organize an expedition to study him and his two daughters, competitive divers as well, during free diving. Paolo Cerretelli accepted to collaborate. Several expeditions to Siracusa took place under his leadership. Guido Ferretti and Bruno Grassi were involved in those studies. More details on this collaboration are given in Sect. 12.5-7. The former took the flag and developed the study of the physiology of breath-holding and diving inside the tradition of the School of Milano.

In this chapter, we aim at following the breath-holding path that, starting from Buffalo, entered the land of the School of Milano and continues nowadays in Brescia. We start from an analysis of the historical works on alveolar gas composition during breath-holding and the construction of the breaking point curve. We then describe the Buffalo work on the Korean diving women, tell the story of the work on the Maiorca family, analyze the physiological limits of breath-hold diving, to the understanding of which the energetic mood of Milano contributed substantially. We finally summarize the most recent studies on the cardiovascular responses to breath-holding within the School of Milano.

12.2 Alveolar Gas Composition During Breath-Holding

Wallace Fenn and Hermann Rahn created the O_2–CO_2 diagram when they were working at Rochester during World War II for the US Army. The history and the meaning of that diagram were told several times (Curran-Everett 2006; Farhi 1990; Ferretti 2018; Subramani et al. 2011; West 2012). Several biographical sketches of the protagonists of that story were published (Fenn 1962; Pappenheimer 1996; Rahn 1979); and the diagram's creators themselves brilliantly summarized the early results in celebrated publications (Fenn et al. 1950; Otis 1964; Rahn and Farhi 1964; Rahn and Fenn 1955). Since that fundamental diagram is not at the core of this chapter, we resignedly address the interested reader to those references. What counts here is that Fenn and Rahn used the O_2–CO_2 diagram as a tool to analyze alveolar gas composition patterns in several special conditions. One of them is breath-holding.

During breath-holding, no exchange of air between the external environment and the alveoli takes place, but pulmonary gas exchange still goes on: thus, carbon dioxide accumulates in the alveoli and oxygen leaves the alveoli through blood. Therefore, during apnea, the partial pressure of oxygen in alveolar air (P_AO_2) decreases, whereas that of carbon dioxide (P_ACO_2) increases. This is an easy prediction: in fact, this concept received the first experimental evidence early in the twentieth century, well before the creation of the O_2–CO_2 diagram (Douglas and Haldane 1909).

However, breath-holding is a dynamic condition in which the equilibria characterizing respiratory steady state are broken. Carbon dioxide is a very soluble gas in tissues and in plasma, whereas oxygen is not. For instance, at sea level at 37 °C, the coefficient of solubility of carbon dioxide (0.515 mL of gas per ml of human plasma per mmHg) is by far higher than that of oxygen (0.0214 mL of gas per ml of human plasma per mmHg) (Christmas and Bassingthwaighte 2017). Hence, the patterns of P_AO_2 and P_ACO_2 on an O_2–CO_2 diagram during breath-holding cannot follow the isopleth for resting gas exchange ratio.

So, the history of the use of the O_2–CO_2 diagram in the analysis of alveolar gas composition during breath-holding started with the wish to identify the P_AO_2 and P_ACO_2 at the breaking point of apneas of maximal duration. The conjecture was that, as we start holding our breath, most of the carbon dioxide produced by metabolism accumulates in blood and in lung parenchyma, so that gas exchange ratio falls to 0.1 or below, describing on an O_2–CO_2 diagram the early part of what Rahn and Fenn (1955) called a hypoventilation loop (see below). This pattern is followed until the stimuli generated by hypoxemia and hypercapnia become so strong that we must interrupt apnea. This defines the volitional breaking point of apnea. Otis et al. (1948) found that, for apneas carried out at rest, P_ACO_2 was only slightly higher, whereas P_AO_2 was by far lower at the volitional breaking point than during quiet breathing at rest.

Otis et al. (1948) modified also the inspired gas mixture. When the subjects breathed oxygen before apnea, no hypoxemia occurred: the attainment of the breaking point was due only to the increase in P_ACO_2, which could reach higher

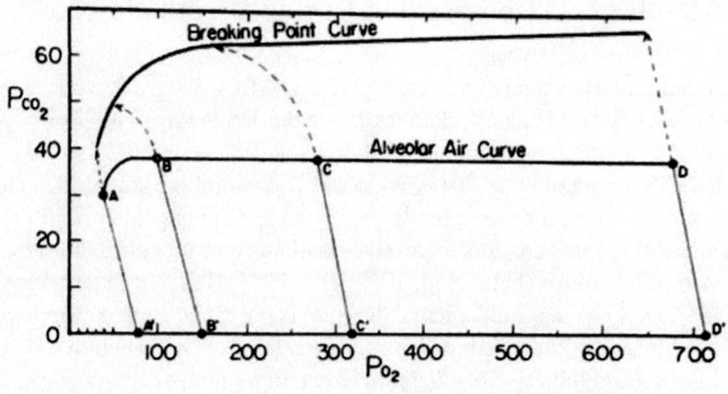

Fig. 12.3 The first representation of an apnea breaking point curve on an O_2–CO_2 diagram. From Rahn and Fenn 1955

values than in resting apneas carried out when subjects were breathing air, because hypoxemia did not interfere with respiratory control. Further manipulation of inspired air composition led to combinations of alveolar gas compositions that, on the O_2–CO_2 diagram, describe a curve similar to that reported in Fig. 12.3, which they called the apnea breaking point curve.

Otis et al. (1946) had already reported that performance in hypoxia is related to alveolar gas composition. On this basis, they defined areas of normal and reduced visual performance on the O_2–CO_2 diagram, taken as an analog of consciousness. They identified zones of severe acapnia, severe hypoxia, anoxic collapse, carbon dioxide narcosis. Alveolar gas compositions lying in those areas are incompatible with consciousness. The breaking point curve of apnea during air breathing falls at the boundaries of the severe hypoxia region. All the alveolar gas compositions below and to the right of the breaking point curve are compatible with wakefulness during breath-holding. Loss of consciousness may occur when, especially during diving, alveolar gas compositions above and to the left of the breaking point curve are attained. It always does occur when the area of anoxic collapse is reached.

However, Otis et al. (1948) did not investigate the activity of respiratory muscles at the volitional breaking point, possibly because of yet unresolved technical difficulties in those days, although the study of their electrical activity could contribute to our understanding of the respiratory drive during breath-holding. Agostoni (1963) was the first to assess the activity of the diaphragm, which he considered the main inspiratory muscle. He also pointed out that the onset of its activity is independent of posture (Massion et al. 1960) and is not masked by that of other respiratory muscles. He observed that the onset of diaphragmatic activity during breath-holding of maximal duration precedes the attainment of the volitional breaking point. He pointed out that, when the respiratory stimuli attain a certain intensity, they overcome the voluntary inhibition of inspiratory muscle activity. From this point on, only the voluntary closure of the airways can prevent breathing, and this, up to the volitional breaking point. Finally, Agostoni (1963) remarked the primary role of

P_ACO_2 in determining the onset of diaphragmatic activity, especially during oxygen breathing, and the subsidiary role of P_AO_2 during hypoxic apneas: when the subject became hypoxic, a lower P_ACO_2 was sufficient to start diaphragmatic contractions. He reported the combination of P_AO_2 and P_ACO_2 at the onset of diaphragmatic activity on the O_2–CO_2 diagram, designing a curve similar in shape to the breaking point curve of Otis et al. (1948), yet displaced downward and rightward with respect to it (Fig. 12.3).

Agostoni's finding generated the current concepts of an easy-going phase and of a struggle phase in maximal breath-holds, the latter being defined as the phase during which the urge to breathe induces progressively stronger involuntary breathing movements (Dejours et al. 1965; Lin 1982; Parkes 2006; Schagatay et al. 2000). More importantly, Agostoni's work set the basis for further experiments, which led Lin et al. (1974) to define another breath-holding curve, which they called the physiological breaking point curve. These authors used the same experimental approach as Otis et al. (1948), yet they interrupted the breath-holds at the onset of the first diaphragmatic contraction. The physiological breaking point curve lies below the breaking point curve of Otis et al. (1948), has the same shape and is similar to that of Agostoni (1963), although more accurately determined.

The two breaking point curves somehow reflect the characteristics of oxygen binding to hemoglobin. The steepest part of the breaking point curves, on the left side of the diagram, corresponds to the steep portion of the oxygen equilibrium curve, where hypoxemia develops. Its slope informs on the relative strength of the hypoxic and the hypercapnic ventilatory drives that lead to end an apnea. As we move rightward along the two curves, these flatten, since we approach the flat part of the oxygen equilibrium curve, hypoxemia is not present anymore, and only the hypercapnic stimulus remains, which determines the apnea breaking point. A selection of data of alveolar gas composition at the end of maximal dry breath-holds is reported in Table 12.1.

The time course of alveolar gas composition from apnea start to the volitional breaking point was also investigated. Whereas P_AO_2 falls rapidly and dramatically, the increase in P_ACO_2 is smooth and modest. Coherently, the volume of alveolar oxygen taken up by blood is much higher than the volume of carbon dioxide added to the alveoli from lung capillary blood. In fact, the latter volume tends to zero during breath-holds, because of carbon dioxide retention, as long as most of the carbon dioxide that is produced by metabolism, is stored in rapidly exchanging tissues and does not stay into the alveoli (Hong et al. 1971; Lanphier and Rahn 1963; Paulev 1969; Tibes and Stegemann 1969). It is noteworthy that the carbon dioxide storage capacity of elite breath-hold divers (Ferretti et al. 1991) resulted twice as high as that of non-divers (Klocke and Rahn 1959; Mithoefer 1959). At least in part, this may be due to larger lung volumes in the former.

When a steady-state condition is broken in the respiratory system, as occurs, e.g., at the beginning of either hyperventilation or hypoventilation, alveolar gas composition, when plotted on an O_2–CO_2 diagram, shifts out of the steady-state isopleth for gas exchange ratio and describes typical patterns that appear in Fig. 12.4 (Rahn and Otis 1949; Rahn and Fenn 1955). As already pointed out, these patterns are called

Table 12.1 Alveolar gas composition at the end of maximal dry breath-holds carried out at sea level on subjects who were breathing ambient air before breath-holding, from different classical studies. In all studies, breath-holds were performed starting from the total lung capacity, except in Otis et al. (1948), in which the initial lung volume was the functional residual capacity. (Modified after Ferretti 2001)

Study	Subjects	Duration	P_AO_2	P_ACO_2
Divers				
Hong et al. 1971	$n = 5$	240	30.6	55.6
Ferretti et al. 1991	A	270	29.3	53.3
	B	300	28.0	46.6
	C	240	40.6	45.6
Non-divers				
Otis et al. 1948	$n = 8$	50	52.1	50.2
Craig Jr 1961	$n = 12$	146	58.0	46.0
Tibes and Stegemann 1969	$n = 6$	165	50.0	45.0
Lin et al. 1974	$n = 5$	160	61.9	53.5
Sterba and Lundgren 1988	$n = 5$	93	72.5	43.3
Ferretti et al. 1991	$n = 9$	150	45.6	50.2
Feiner et al. 1995	$n = 17$	150	51.7	42.9

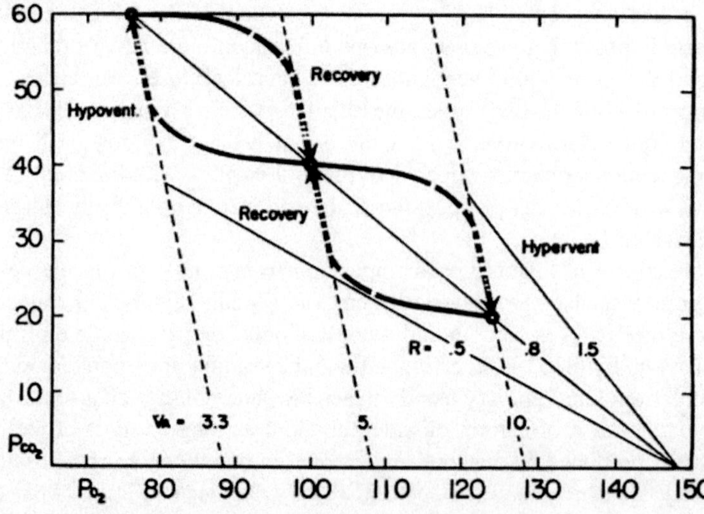

Fig. 12.4 Theoretical alveolar pathway starting at a normal resting alveolar gas composition, when alveolar ventilation is suddenly increased from 5 to 10 L min^{-1}, followed by recovery (lower-right loops), and when alveolar ventilation is suddenly reduced from 5 to 3.3 L min^{-1} (upper-left loops), followed by recovery. (Reproduced from Rahn and Fenn 1955)

hyperventilation or hypoventilation loops. They reflect differences in the kinetics of adjustment of oxygen and carbon dioxide stores. It is not insensate to affirm that the patterns of alveolar gas composition during maximal dry breath-holds follow the most extreme hypoventilation loop, that for an alveolar ventilation of 0. This hypoventilation loop, however, cannot be completed up to the new steady state, but only the first, flatter part of it is covered, until the attainment of the volitional breaking point curve. At that point, the respiratory drive is so strong, that the apnea is necessarily interrupted. Recently, Taboni et al. (2020) analyzed the dynamics of alveolar gas composition during dry breath-holding, treated the corresponding alveolar pathway as a hypoventilation loop, developed an exponential model of it (Farhi and Rahn 1955), and found a close correspondence between the theoretical curve and the experimental data.

12.3 The Korean Diving Women

Humans practiced professional breath-hold diving in many countries for centuries. The aim was military sabotage, marine rescue activities, sea harvesting of sponges, pearl shells, and seafood. This practice has declined in the twentieth century. Scuba divers now carry out rescue and military activities more efficiently than breath-hold divers. Sponge or pearl divers have disappeared from Greece, Australia, and Polynesia. Around 1990, 16,000 diving women (called Hae-Nyo in Korea, Ama in Japan) were reckoned in South Korea, instead of the 30,000 figure reckoned in the mid-1960s (Ferretti and Costa 2003). Most of the Ama ($\sim$70%) live on the Cheju Island, some 60 miles south of the southern tip of Korea. Some 10,000 Ama are still active nowadays. Many of them do not collect seafood anymore, but accompany tourists who wish to dive in their sea: they earn much more money this way than by their traditional activity.

Japanese practitioner Gito Teruoka, who chaired the Japanese Institute for the Science of Labor since the 1920s, pioneered the study of Ama. Teruoka's interests concerned the occupational health, the risk and the exhausting physical activity of those women. He described the diving operations and patterns, their seasonal variations, and the diving equipment of Ama. He was the first to measure their alveolar gas composition at the end of dives. He also tried an estimate of gas exchange during diving, in order to evaluate the hardness of their work, and discussed the possible effects of increased nitrogen partial pressure during the dive (Teruoka 1932). He performed an accurate evaluation of their health condition and collected several demographic data on them. The results of Teruoka remained forgotten for 25 years. Yet Hermann Rahn was aware of them, thanks to Wallace Fenn, who noticed Teruoka's article when they planned their first experiments on breath-holding in the 1940s. It was in 1960 that Kichinosuke Tatai, from the Institute of Public Health, Tokyo, eventually introduced Rahn to the Ama of the Shima Peninsula, where he could watch them diving. His acquaintance to Suk-Ki Hong, who was starting an intensive program on the physiology of the Korean Ama, gave origin to several scientific expeditions to that purpose.

The work of Teruoka had a remarkable impact on Rahn's vision of breath-hold diving, affected his planning of the expedition deeply, and therefore contributed strongly to the rise of interest in the physiology of breath-hold diving in humans in those years. The first International Symposium on the Physiology of Breath-Hold Diving was dedicated to Gito Teruoka. Rahn himself paid a tribute to him at that symposium (Rahn 1965). The article of Teruoka was fully reprinted in the proceedings of that symposium, including several pictures that Teruoka took during his studies.

Hong moved to Buffalo. Numerous studies, mostly on Ama, addressed potential features of physiological adaptations to diving. Diving patterns and techniques, alveolar gas exchange, lung volumes, heart rate (f_H), arterial blood gases, regional blood flows, basal metabolism were studied. Don Rennie (1925–1992) was also involved for the study of temperature regulation under cold stress. In fact, we owe much of our knowledge of the physiology of diving populations to the numerous field studies that Hermann Rahn and Suk-Ki Hong organized since the early 1960s for the investigation of Korean Ama. The results have been summarized in several books or review articles (Hong and Rahn 1967; Hong et al. 1986; Lin and Hong 1996; Rahn and Yokoyama 1965).

Historically, the Ama dived all year round, wearing cotton suits, which got wet during the dives, so that their body was in contact with water. In summer, sea temperature is warm (average surface water temperature around Cheju Island in August is 27 °C), being much colder in winter (average surface water temperature in January is 10 °C). Under these conditions, the thermal stress is very strong. Ama wearing cotton suits were reported to undergo a 2 °C decrease in oral temperature during a working period in August (70 min). In January, although the working period was reduced to 15 min, the decrease in oral temperature attained 4 °C (Kang et al. 1963). Ama are not fat: their mean subcutaneous fat thickness and their percentage of body fat were similar to those of Korean non-diving women of the same age (Rennie et al. 1962).

The minimal water temperature at which a human can stay immersed for three hours without shivering is called "critical water temperature." At the end of three hours at critical water temperature, the thermal insulation of the body attains its maximum, whereas skin blood is reduced to practically zero (Burton and Bazett 1936). The critical water temperature of Korean Ama, with a mean fat thickness of 2.2 mm, was about 27–28 °C, without visible shivering or any increase in oxygen consumption at the end of the third hour of resting immersion (Rennie et al. 1962). Non-diving Korean women of similar mean fat thickness had a critical water temperature of about 32 °C. This reduction in critical water temperature was associated with an increase in thermal insulation of the body, which may result from either stronger peripheral (muscular) vasoconstriction or more effective counter-current heat exchange, in Ama than in non-diving Korean women (Ferretti and Costa 2003). Hormonal and vascular adaptation to heat was also observed. In fact, the Ama showed an increased basal metabolic rate, a stronger thermogenic effect of catecholamines, a reduced heat loss through the limbs at equal limb blood flow, and a greater vasomotor response of the hand to local cooling (Hong et al. 1969a, 1969b; Kang et al. 1963, 1970; Paik et al. 1972).

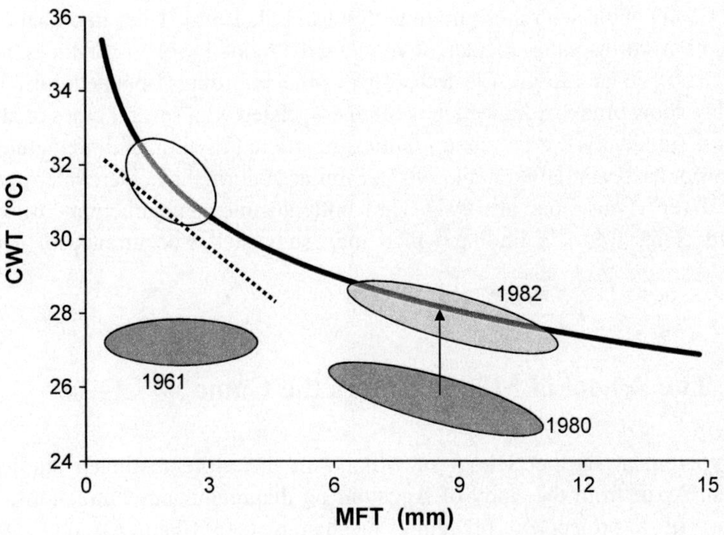

Fig. 12.5 Critical water temperature (CWT) as a function of mean fat thickness (MFT). The black curve describes the relation for Caucasians. The dotted line refers to scuba divers from the Hawaii. The white area identifies Korean non-diving women. The gray areas refer to Ama, investigated, respectively, in 1961 and 1980. These areas lie well below the CWT curve for Caucasians, demonstrating acclimatization to cold. The clear gray area concerns the same Ama studied in 1980, restudied in 1982 after they had been using neoprene wet suit during the dives for two years, during which they clearly lost cold acclimatization. (From Ferretti and Costa 2003)

The thermal adaptation of the Ama to cold could result either from an adaptive physiological phenomenon or from a genetic modification that took place over centuries of diving. That the former was the case became clear with the introduction of neoprene wet suits instead of the traditional cotton suits. Neoprene suits provide an additional thermal insulation on the body surface. When the Ama started using them, they showed a loss of thermal adaptation within a couple of years, as demonstrated by the elevation of critical water temperature (see Fig. 12.5) (Park et al. 1983). Similarly, Greek sponge divers from the Island of Kalymnos, Dodecanese, in the Aegean Sea, started to use neoprene suits in the early 1970s. When their thermal adaptation was studied, some ten years later, their shivering threshold was equal to that of European non-diving controls (Veicsteinas and Rennie 1982). If they were adapted to cold in the past, their adaptation was completely lost.

Although their lung volumes are not as big as a diver could wish, Ama have 15% higher vital capacity than non-diving Korean or Japanese women (Hong et al. 1963; Song et al. 1963). Their ventilatory response to hypoxia is normal and their ventilatory response to hypercapnia is not blunted, despite the long time spent at high carbon dioxide pressure during the dives (Song et al. 1963). The data of alveolar gas composition confirm the finding, by Teruoka (1932), of a low P_AO_2 at the end of a dive (Hong et al. 1963). The data of arterial gas composition, obtained 30 years later, are fully compatible with these findings (Qvist et al. 1993). Their maximal oxygen consumption corresponds to that of moderately trained women of the same age (Hong et al. 1969a, 1969b). Electrocardiographic recordings obtained during dives invariably show bradycardia, which is often associated with several types of arrhythmia, most frequently P wave abnormalities, junctional rhythm, and premature atrial and ventricular beats (Hong et al. 1967). Similar abnormalities were later observed on elite divers (Ferrigno et al. 1991). They undergo splenic contraction during dives (Hurford et al. 1990), a finding which may suggest the occurrence of a diving response.

12.4 The School of Milano Enters the Game

The involvement of the School of Milano in the aforementioned studies was marginal. Apart from the study of Agostoni on diaphragmatic contractions, which was part of a project on breathing mechanics (see Chap. 8), only Arsenio Veicsteinas (1944–2017) had the opportunity to participate with a significant role in a study on breath-hold divers. It was him, in fact, who managed the participation of Don Rennie in the expedition to Greece for the investigation of the thermal responses of Greek sponge divers, mentioned above. In those days, Veicsteinas was at Buffalo as invited assistant professor, working with Don Rennie on temperature regulation. Veicsteinas was in good terms with Piergiorgio Data (1939–2005), who was Professor of Physiology at the Medical School in Chieti, Italy, and who was active in diving physiology and medicine. Data, who was organizing an expedition to Greece for the study of diving, invited Veicsteinas to join him, and this proposed Rennie to come with him to the islands of sponge divers to study their thermal responses. Therefore, it happened that they found themselves on the shore of the Island of Kalymnos, Greece, to study the thermal insulation and the shivering threshold of those divers immersed to the neck in the water of the Aegean Sea (Veicsteinas and Rennie 1982). The temperature regulation project in Buffalo represented also the initiation to physiology of Guido Ferretti. Of the three further ensuing papers (Veicsteinas et al. 1982; Ferretti et al. 1986; Ferretti et al. 1989), one happened to be the MD thesis of Guido Ferretti, who started, quite amazingly, as a Buffalo temperature regulation man. The eclectic tradition is still enduring within the School.

Notwithstanding the study on Kalymnos sponge divers, which resulted from the individual initiative of an enterprising man, the systematic involvement of the School of Milano in diving physiology started a few years later. Massimo Ferrigno, then a young practitioner, fond of diving, breath-hold diver himself, later a successful anesthesiologist, was in Kalymnos with Data's expedition. He was already in touch with Veicsteinas and had already visited Buffalo with him. In Greece, he met Rennie, who managed a salary for him at the hyperbaric chambers at CRESE, with Lundgren. As a breath-hold diver, he knew Enzo Maiorca, who was then the most famous and celebrated competitive breath-hold diver worldwide. He contacted him, on behalf of Lundgren, for a project on extreme breath-hold diving. He invited him to visit Buffalo on the occasion of an international workshop on breath-hold diving, supported by the Undersea and Hyperbaric Medical Society, which took place at CRESE in 1985 (Lundgren and Ferrigno 1987). Eventually, Enzo Maiorca and his two daughters, Patrizia and Rossana (1960–2005), accepted, but asked to be studied in the sea, during actual dives. That would have implied an expedition to Siracusa and Lundgren needed support in Europe for that. Therefore, he contacted Paolo Cerretelli in Geneva, who accepted to collaborate, took over the organization of the expedition in Europe and assigned that task to Guido Ferretti. Maiorca wanted that his trusted doctor, Mario Costa, a reputed anesthesiologist from Siracusa, be involved in the planning of the study. Bruno Grassi participated as well. Three expeditions to Siracusa took place in the second half of the 1980s. Then, a travel to Buffalo for experiments in the hyperbaric chambers followed, in 1990.

That project was the door through which the School of Milano entered the submarine world. Margaria's cultural approach emerged clearly in the study of the energetics of deep breath-hold diving. In those days, Maiorca and his daughters used a free diving technique, wherein a weight, running along a rope that was vertically suspended in the sea from the support boat, pulled the diver down to the planned depth during descent. The weight consisted of a platform on which the diver could either stand or hold on. During ascent, a gas-filled balloon, which an assisting Scuba diver forwarded to her/him at depth, pulled the diver up to the sea surface. The diver released the balloon just before emersion (Majorca 1987). The balloon was replaced by an inflatable suit for the 1988 100-m record dive. Such a technique was expected to reduce the energy cost of diving drastically. Nevertheless, Umberto Pelizzari, an Italian diver who attained the depth of 150 m in 1999, further improved the technique, by using hydrodynamic weights during descent and an improved suit that could be more easily inflated with gas during ascent (Pelizzari 2005). With these improvements, Pelizzari increased the mean diving speed to 1.4 m s^{-1}, which is 40% higher than that of Maiorca during his record dive to 100 m.

For completeness, we remind that Maiorca performed a few hyperventilation maneuvers before diving, consisting of series of deep breaths at about 90% of his vital capacity at a frequency of 4–5 min^{-1} over a period of 8 min. The last breath before submerging was a deep inspiration, so that the dive started at a lung volume close to the total lung capacity (TLC).

12.5 Gas Exchange and Energy Expenditure During Diving

The P_AO_2 and P_ACO_2 values measured before and at the end of the experimental dives performed by the three Maiorcas in the Ionian Sea before Siracusa appear in Fig. 12.6 on an O_2–CO_2 diagram, together with the data obtained during dry breath-holds up to the maximal duration on the same divers (Ferretti et al. 1991). At the end of the dives, P_AO_2 decreased to 30–40 mmHg, while P_ACO_2 attained 47 mmHg, independent of depth and duration. The alveolar volume was remarkably decreased at the end of the dive, because of a net oxygen uptake, with little carbon dioxide

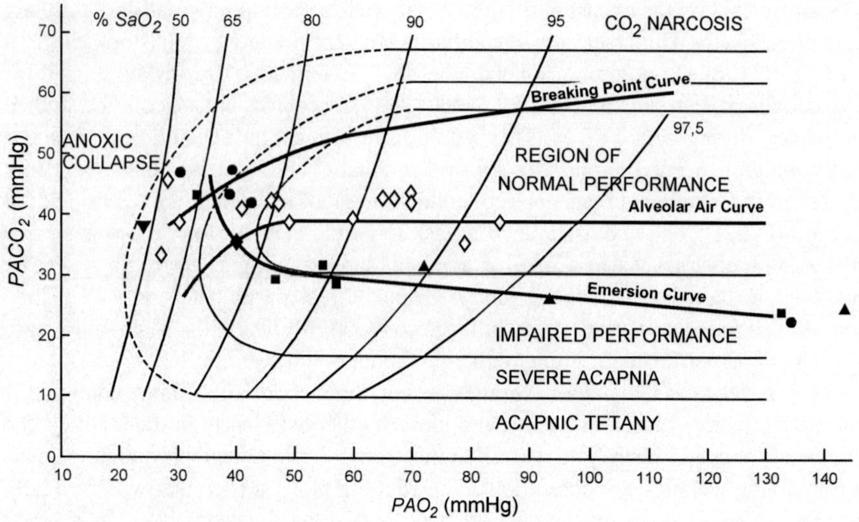

Fig. 12.6 Alveolar air compositions at the end of dry breath-holds of increasing duration up to the maximal are shown on an O_2–CO_2 diagram. P_AO_2 and P_ACO_2, alveolar oxygen and carbon dioxide partial pressures, respectively; S_aO_2, arterial oxygen saturation. The normal alveolar air curve and the apnea breaking point curve are reported. Isopleths for S_aO_2 are also given, as are the lines defining regions on normal or impaired visual performance (Rahn and Fenn 1955). Filled symbols refer to extreme (triangles, Ferretti et al. 1991) or leisure (squares, Hong et al. 1971) divers. Open symbols refer to non-diver controls, from various sources in the literature. The lower-right values below the alveolar air curve reflect pre-dive hyperventilation. Upper-left values refer to maximal breath-holds. (From Ferretti 2001)

addition. Therefore, the gas exchange ratio was close to zero. Peak lactate concentration at the end of the dives ranged between 1.71 and 6.46 mM, partly depending on the depth of the dive.

The alveolar gas compositions observed at the end of the dives lie on a curve, which the authors called "emersion curve." This curve can be approximated to a straight line down to a P_AO_2 of about 45 mmHg. For lower P_AO_2 values, the line bends upward, to intercept the volitional breaking point curve in the same region as for the dry breath-holds.

Ferretti et al. (1991) computed the rate energy expenditure ($\dot{E}$) during the dives. They determined the oxygen uptake from the lungs during the dive, based on the alveolar gas composition and lung volume at the start and the end of each dive. They estimated the changes in tissue oxygen stores, essentially related to oxygen bound to myoglobin, (200 mL, and independent of depth in 20–100 m range, Olszowka and Rahn 1987). To this aim, they assumed equilibrium between tissue and end-capillary oxygen partial pressure (PO_2) in muscle, a myoglobin concentration of 5 mg g^{-1} (Jansson et al. 1982), and an oxygen partial pressure at which myoglobin saturation is 0.5 (P_{50}) of 5 mmHg (Gayeski and Honig 1983). The energy provided by anaerobic glycolysis was calculated from blood lactate values, assuming an energy equivalent of blood lactate accumulation of 3 mL mM^{-1} kg^{-1} (di Prampero 1981). The contribution of anaerobic alactic metabolism is included in the volume of oxygen exchanged above resting metabolic rate in the recovery after the dive. With these premises, Ferretti et al. (1991) obtained values for $\dot{E}$ during the dives ranging between 0.5 and 0.9 L min^{-1}. These values corresponded to some 20–30% of the maximum oxygen uptake of the divers, indicating that $\dot{E}$ during free breath-hold diving is low, and similar to that of walking at the optimal speed or of pedaling on a cycle ergometer at a power of 25–50 W (Chap. 5).

At such metabolic powers, there is no blood lactate accumulation during exercise in humans breathing air at sea level. The blood lactate increase during diving, despite the low metabolic rate, could follow only a marked peripheral vasoconstriction during the dive, most likely in muscles and skin. If this is so, these results could suggest the hypothesis that a diving response may occur in extreme divers during deep breath-hold dives.

12.6 The Concept of Diving Response

In 1940, Per Scholander (1905–1980) demonstrated a peculiar series of events during free dives in diving birds (Scholander 1940). He found extreme bradycardia, associated with a dramatic increase in arterial blood pressure. He called such a response the *Diving Response*. Several authors investigated the diving response afterward. The description of the phenomenon, which was observed in several species, including diving mammals, was subsequently refined and better focused. Nowadays, we can summarize the concept as follows: a strong peripheral vasoconstriction, perhaps of sympathetic origin, occurs, with consequent decrease in cardiac output and increase in arterial blood pressure. Hypertension may be corrected by an extreme bradycardia, perhaps of baroreflex origin, although hypoxic stimulation of peripheral chemoreceptors may also play a role. Cardiac output becomes so low, that it is directed preferentially, if not exclusively, to vital organs, namely heart and brain. This implies thorough redistribution of regional blood flows, so that muscle and skin blood flows fall practically to zero. Therefore, lactate may accumulate in unperfused muscles. Finally, blood oxygen availability is increased, because spleen contraction at the beginning of a dive enhances blood hemoglobin concentration. This phenomenon was discussed in detail in several review articles (Andersen 1966; Blix and Folkow 1983; Butler and Jones 1997; Gooden 1994; Kooyman and Ponganis 1998; Lin 1982; Lindholm and Lundgren 2009; Zapol 1996).

Does a diving response occur also in humans? The most common observation in humans, suggesting such a hypothesis, is the bradycardia reported during face immersion experiments. A decrease in f_H was found in every study on subjects who held their breath at the water surface (Brick 1966; Craig Jr 1963; Harding et al. 1965; Manley 1990), even during exercise (Asmussen and Kristiansson 1968). Later, an increase in arterial blood pressure was also observed, whereas some studies report a fall of cardiac output (Bjertnaes et al. 1984; Lin et al. 1983), while others do

Table 12.2 Summary of the data, obtained on the Maiorcas in the Buffalo hyperbaric chamber, on which the first description of a diving response in humans relied. f_H, heart rate; $\dot{Q}$, cardiac output; Q_{st}, stroke volume of the heart; TPR total peripheral resistance, L_a blood lactate. (Data From Ferrigno et al. 1997)

Subject	Condition	F_H (min^{-1})	Q' (L min^{-1})	Q mL	P_{syst} (mmHg)	P_{diast} (mmHg)	TPR (mmHg min^{-1} L)	L_a (mM)
1	Control	70	5.6	80	145	90	19.3	0.93
1	**Dive**	38	**4.0**	**105**	**290**	**150**	**49.2**	**3.97**
2	Control	78	6.4	82				1.92
2	**Dive**	**28**	**2.8**	**100**				**3.05**
3	Control	86	8.8	102	115	75	10.0	1.29
3	**Dive**	**40**	**4.2**	**105**	**280**	**200**	**54.0**	**5.30**

not (Heistad et al. 1968; Hong et al. 1971). The combination of these results suggests an increase in peripheral vascular resistance, possibly due to peripheral vasoconstriction, with consequent reduction of peripheral blood flows (Bjertnaes et al. 1984; Heistad et al. 1968; Lin et al. 1983). Later, it was shown that these effects were accentuated when the subjects performed their apneas starting from a lower lung volume than their TLC (Andersson and Schagatay 1998), and when they underwent specific breath-hold training protocols (Schagatay et al. 2000).

The isolated results mentioned here above, although suggestive of a possible diving response, did not provide a demonstration of its occurrence in humans. The first comprehensive description of a complete diving response in humans came from the Maiorca study and was obtained in the hyperbaric chambers at Buffalo (Table 12.2) (Ferrigno et al. 1997). An extreme peripheral vasoconstriction was accompanied by a dramatic increase in arterial blood pressure, that in one case reached 280/200 mmHg during descent. The ensuing strong stimulation of arterial baroreceptors caused an extreme drop of f_H, with values lower than 30 min^{-1}. An increase in stroke volume did not compensate for bradycardia, so that cardiac output fell to less than 3 L min^{-1} in correspondence with the minimum f_H. This was enough to ensure perfusion of vital organs, thanks to redistribution of blood flow. In fact, Ferrigno et al. (1997) did not measure peripheral blood flows in their study. However, a very high arterial blood pressure, associated with a very low cardiac output, indicated a dramatic increase in total peripheral resistance during the dive. These findings strongly suggest occurrence of extreme peripheral vasoconstriction. Since lactate accumulated in blood, indicating occurrence of anaerobic metabolism, despite the low metabolic rate, we can speculate that skeletal muscle became unperfused during the dives.

Lactate accumulation was considered as the consequence of an oxygen preserving mechanism (Ferretti 2001). Unperfused muscles do not receive oxygen. Only the oxygen bound to myoglobin can sustain their aerobic metabolism. This represents a small amount, so that muscles have to shift rapidly to anaerobic metabolism. Therefore, the rate at which P_AO_2 and P_aO_2 fall during a dive is reduced. Coherently, during breath-holding with exercise, the stronger is the bradycardia and the hypertension, the slower is the reduction of arterial oxygen saturation (S_aO_2) (Lindholm et al. 1999), as does pulmonary oxygen uptake (Lindholm and Linnarsson 2002).

More recently, an increase in blood hemoglobin concentration, due to spleen contraction in the early part of a dive, has been added to the picture of the human diving response (Bakovic et al. 2003; Espersen et al. 2002; Schagatay et al. 2001), although the phenomenon had already been observed on Ama (Hurford et al. 1990). This mechanism increases blood oxygen stores and thus prolongs breath-hold duration.

12.7 Respiratory Adaptation to Breath-Hold Diving

Figure 12.6 shows also the patterns followed by alveolar gas composition during dry apneas at rest in extreme divers. It appears that some of the values, obtained after maximal apneas, lie beyond (leftward of) the volitional breaking point curve (Ferretti et al. 1991; Lin et al. 1974), at the boundaries of the extreme anoxia region. This was not the case for non-diving controls. The same was observed on the divers studied by Hong et al. (1971). To give an idea of the consequences of this observation, consider that Ferretti (2001) analyzed the alveolar gas composition at the end of maximal breath-holds with respect to the classical ventilatory response curves in hypoxia and hypercapnia (Lloyd et al. 1958; Rebuck et al. 1977). He reported that alveolar gas composition at the volitional breaking point of apnea in elite divers would generate a combined hypoxic and hypercapnia respiratory drive at rest leading to some ten times greater ventilation than during quiet breathing in normoxia, instead of seven times as at the volitional breaking point curve.

The finding of alveolar gas compositions beyond the volitional breaking point curve at the end of maximal breath-holds resumed the hypothesis that breath-hold divers may be characterized by a blunted ventilatory response to hypoxia, hypercapnia, or both. This hypothesis was not new, it was simply forgotten. Indeed, Masuda et al. (1981, 1982) had already reported that Ama have a blunted ventilatory response to both hypoxia and hypercapnia. Similar results were obtained also on synchronized swimmers, underwater hockey players, submarine escape tower trainers, and Royal Navy divers (see Ferretti 2001).

Grassi et al. (1994) determined the ventilatory response to breathing oxygen and carbon dioxide on the Maiorca family. The results of that study appear in Fig. 12.7. In all cases, except when breathing room air, and at all P_ACO_2 levels, steady-state ventilation was lower in the divers than in the controls, indicating a blunted ventilatory response to carbon dioxide in divers. Their ventilatory response to carbon dioxide was even lower than that of individuals, whom Lambertsen (1960) defined as having a low carbon dioxide sensitivity.

In contrast, Grassi et al. (1994), who applied the chemical denervation technique of Dejours et al. (1959), demonstrated no differences in ventilatory responses to hypoxia between divers and controls. These results are not surprising, as long as

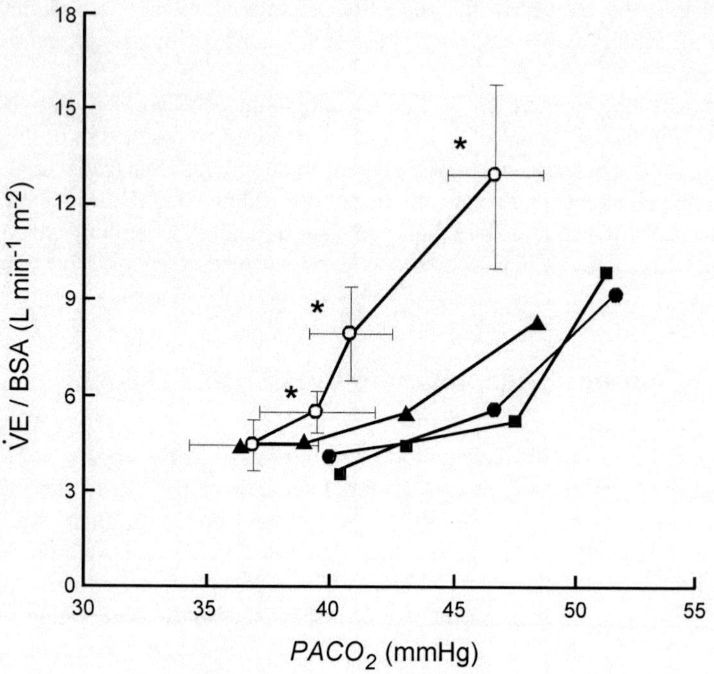

Fig. 12.7 Ventilatory response to carbon dioxide breathing in elite divers (filled symbols) and in non-diver controls (open symbols). On the *y*-axis, expired ventilation ($\dot{V}_E$) is expressed relative to the body surface area (BSA). P_ACO_2 is the alveolar carbon dioxide partial pressure. At any P_ACO_2, $\dot{V}_E$ is lower in divers than in controls. (From Grassi et al. 1994)

during deep dives the brain is not exposed to hypoxia except during the final few meters before emersion, whereas they are exposed to extreme hypercapnia during most of the dive.

The weak point of these studies is that the subjects were all members of the same family, so that we cannot know whether those results unveil an adaptive phenomenon or a complex genetic trait of that family. Nevertheless, the interest for the role of genetics in adaptation to diving has recently gained momentum, as evidence for local genomic selection to improve hypoxia tolerance via spleen size and contraction has been obtained (Ilardo et al. 2018).

12.8 Of the Maximal Diving Depth

The energetic imprinting of the School of Milano drove rapidly the attention toward an analysis of the maximal diving depth and the limits in breath-hold diving. Historically, the first widely accepted hypothesis was an application of gas laws. The assumption was that the TLC could not be compressed below the residual volume (RV). Thus, the ratio of these two volumes would have shown the extent of possible lung volume compression during a dive, and thus the depth limit. Let us take the Ama as an example: Song et al. (1963) obtained a RV of 1.14 L and a TLC of 4.58 L. This indicates that the TLC of Ama can be compressed up to 4 times, and therefore that Ama can tolerate a pressure of 4 ATA. Their maximal diving depth would be 30 m, to be compared with actual diving depths of Ama around 20 m.

Until the early 1960s, diving records were compatible with this assumption. Bob Croft's dive to 73 m in 1968 challenged this hypothesis, but his TLC was so high (9.1 L), that his TLC/RV ratio was above 7 (Schaefer et al. 1968). Croft could exert particularly high expiratory pressures, which might have contributed to reduce his RV during the dives. So, it was concluded that at the 70 m depth his TLC was still larger than his RV, but he got very close to the limits.

Continuous record improvements, however, led to refutation of the TLC/RV ratio hypothesis: lung volumes were not limiting the maximal breath-hold diving depth. The first who formulated the alternative hypothesis of blood shift was Craig Jr (1968): he postulated that the negative pressures attained in the thorax at great depths could recall blood from the extremities to the thorax, thereby increasing the central blood volume. Craig's hypothesis was consolidated by the observation that mere water immersion to the neck may displace up to 700 mL of blood into the thorax (Arborelius Jr et al. 1972). More recently, radiographic images taken at depth using underwater radiological equipment showed smaller lung images, elevation of the diaphragmatic dome, and engorgement of lung blood vessels and increased transverse diameter of the heart (Morelli and Data 1991). Because of blood displacement, the circumference of the chest decreases less than that of the abdomen during a deep dive (Warkander et al. 1996). Whatever one may think of the blood shift hypothesis and of the associated evidence, when the record attained 150 m, it became clear that also the blood shift hypothesis was insufficient to predict diving depth limits.

Let us now perform an axiomatic revolution in our vision of the depth limits of a breath-hold dive, and state that the limit is not set by pressure, as implicitly assumed in the past, but by energy availability. Therefore, let us consider a diving human as a closed system, unable to take energy from the surrounding environment: only the energy that is stored in the body before the dive can be used to sustain resting and exercise metabolic rate, and not even all of it.

The speed that a human can attain, when covering a given distance with a given locomotion mode, is equal to the ratio of the sustaining metabolic power to the energy cost of locomotion (C) (see Chaps. 5 and 6). In diving, however, what matters is to cover the longest possible distance, not a predetermined distance at the fastest possible speed, as in athletics or swimming competitions. This allows getting rid of time, so that we can state that the distance (L) that a diving human can cover, is equal to the ratio of the available metabolic energy (E) to C. Therefore, d is directly proportional to the E stored in the diver's body and inversely proportional to the C of diving (Fig. 12.8). In Fig. 12.8, the slope of the straight lines reported in the top panel is equal to C, whereas in the right panel each hyperbola is an isopleth for E. When moving in water, C is equal to:

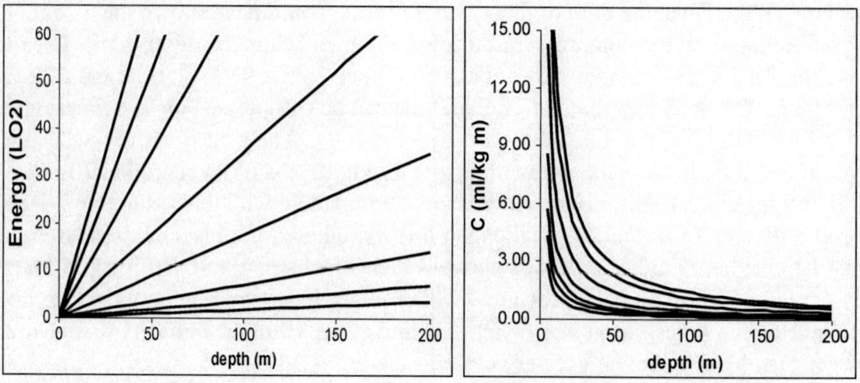

Fig. 12.8 An analytical representation of Eq. 12.2. Reported lines are theoretical. **Left panel**: Energy as a function of depth (note that the total distance covered during the dive is twice the depth). The slope of the straight lines is equal to the energy cost *C*. **Right panel**: *C* as a function of depth. The curvature of the hyperbolas is equal to the overall amount of energy available to the body at the start of the dive

$$C = \frac{D}{\eta} \tag{12.1}$$

where D is the drag of water (see Chap. 6) and η is the mechanical efficiency of diving. If this is so, then:

$$L = \frac{E}{C} = \frac{E\eta}{D} \tag{12.2}$$

Equation 12.2 tells that the maximal diving depth can be improved by: (i) increasing the amount of energy that is stored in the body at the start of the dive; (ii) increasing the mechanical efficiency of diving or of underwater swimming, and (iii) reducing the drag. A reduction of D can be obtained by improving the hydrodynamics of diving, as Pelizzari did for his diving record to 150 m. An increase of η has been achieved by improving the diving technique, taking care of reducing useless movements during the dive to a minimum.

Concerning E, the maximal anaerobic lactic capacity ($E_{La}{}^{max}$) and the maximal anaerobic alactic capacity ($E_{PC}{}^{max}$) dictate the limits of anaerobic metabolisms. Assuming an energy equivalent for blood lactate accumulation of 3 mL of O_2 per kg of body mass per mM of lactate (di Prampero 1981; di Prampero and Ferretti 1999; for a discussion of this issue, see Chap. 3), the former is proportional to the maximal blood lactate concentration, which is essentially invariant. The latter depends on muscle phosphocreatine concentration.

Concerning aerobic metabolism, since there is no oxygen uptake from the external environment, aerobic capacity becomes proportional to the amount of oxygen that is present in the body stores: lungs, blood and tissues. Therefore, we can write that the amount of energy from aerobic sources (Eo_2) is the sum of three terms:

$$Eo_2 = E_L o_2 + E_B o_2 + E_t o_2 \tag{12.3}$$

where suffixes L, B, and t designate the oxygen stores in lungs, blood, and tissues (mostly muscle), respectively. $E_B o_2$ depends on blood hemoglobin concentration, $E_t o_2$ on muscle myoglobin concentration. The latter is practically invariant (Olszowka and Rahn 1987). The former may vary in a dive, if there is spleen contraction (Bakovic et al. 2003; Espersen et al. 2002; Schagatay et al. 2001). Thus, on a first approximation, Eq. 12.3 can be simplified as follows:

$$Eo_2 = E_L o_2 + K \tag{12.4}$$

Equation 12.4 tells that, during a breath-hold dive, there is a linear relationship between Eo_2 and $E_L o_2$ at the start of a dive. Since also E_{La} and E_{PC} are invariant, also the overall amount of energy in the body at the start of the dive varies linearly with Eo_2, so that a diver may increase breath-hold duration and depth by acting on his lung volume at the start of the dive.

Ferretti (2001) simulated alveolar gas patterns in a 150 m breath-hold dive performed at TLC. He assumed that the diver hyperventilated before the dive, as reported by Ferretti et al. (1991), and had a C like that computed for the Maiorca family during free dives (Ferretti et al. 1991). This simulation showed that Enzo Maiorca (TLC = 7.8 L) and Bob Croft (TLC = 9.1 L), at emersion from a 150 m free dive, would have a hypothetical P_AO_2 values of 26.1 and 34.6 mmHg, and a P_ACO_2 of 41.1 and 39.7 mmHg, respectively. These figures are still compatible with consciousness. Higher P_AO_2 values are obtained the larger is the TLC. A diver with a 10 L TLC would emerge from a 150 m dive carried out as described here above with a P_AO_2 of 61 and a P_ACO_2 of 37.5 mmHg, thus in a very safe region of normal visual performance, far enough from the hypoxic collapse boundary.

Concerning anaerobic lactic metabolism, it is noteworthy that the P_AO_2 at emersion is higher, the higher is the rate of blood lactate accumulation. Ferretti (2001) predicted that Enzo Maiorca or Bob Croft could not emerge conscious from a 150 m dive, if their rate of lactate accumulation was 40 instead of 45 $\mu M\ s^{-1}$. The occurrence of a marked diving response, leading to a substantial contribution of anaerobic lactic energy sources to the overall energy expenditure during the dive, is necessary for a record dive. Moreover, Ferretti (2001) predicted that it would be impossible to dive to 150 m and emerge conscious (or even alive), if the diver moves at a speed of 0.9 $m\ s^{-1}$ (speed of Enzo Maiorca during his 100 m world record dive). In fact, the diving speed would be too low, thus prolonging the breath-hold duration to times incompatible with consciousness upon emersion.

12.9 Lung Volumes of Elite Breath-Hold Divers

A corollary of the energetic conjecture on the limits in breath-hold diving depth is the importance of having a high lung volume at the start of the dive. Coherently, the duration of a maximal resting breath-hold is longer, the bigger is the starting lung volume (Andersson and Schagatay 1998; Chapin 1955; Stanley et al. 1975). It is also longer, if the dive is performed with lungs filled with oxygen rather than air (Klocke and Rahn 1959; Otis et al. 1948; Rodbard 1947). Conversely, hypoxia decreases breath-hold duration (Otis et al. 1948; Rahn et al. 1953; Rodbard 1947). These effects follow from Eq. 12.4. Moreover, an increase in metabolic rate above resting accelerates the rate at which oxygen stores are depleted during a breath-hold, and thus reduces the duration of a maximal breath-hold (Åstrand 1960; Craig and Cummings 1958; Rodbard 1947).

Equation 12.4, and the ensemble of this data, allow the prediction that elevated lung volumes represent an advantage for elite competitive breath-hold diving, because they increase $E_{\mathrm{L}}\mathrm{o}_2$, and thus improve the diving depth. As a reference, consider that the TLC of healthy adult European males ranges between 5.4 and 7.9 L, being some 15% lower in females, and depends on body mass (Jouasset 1960). The TLC of Ama appears to be quite low, lower than the European standard (Hong et al. 1963). The 7.8 L TLC of Enzo Maiorca reported above is close to the uppermost limit of healthy Europeans (Ferretti et al. 1991). The first TLC measurement well above such uppermost limit was obtained on Bob Croft, whose TLC, as reported above, was 9.1 L (Schaefer et al. 1968). A TLC above 9 L was observed also on Francisco Ferreira from Cuba, one of the most performing breath-hold divers of the 1990s (Warkander et al. 1996). TLC values above 9 L in elite breath-hold divers have become a common finding in the last 20 years. At least 25 values above 9 L, and 4 above 10 L, can be reckoned in the literature (Ferretti et al. 2012; Lindholm and Nyren 2005; Loring et al. 2007; Patrician et al. 2021; Walterspacher et al. 2011). That top-level breath-hold divers have a TLC 10–25% higher than the upper limit of the normal range has become a widespread notion.

In addition, to increase their lung volume above the TLC, and thus their $E_{\mathrm{L}}\mathrm{o}_2$, several divers practice oropharyngeal breathing (lung packing) prior to diving. By doing so, they can increase their TLC by further 1.5–2.0 L (Lemaître et al. 2010; Lindholm and Nyren 2005). A woman diver was even able to introduce 2.6 L of air in her lung above her TLC by lung packing (Schipke et al. 2015). Lung packing may increase airway pressure up to 90 cmH_2O (Loring et al. 2007), limits venous return (Novalija et al. 2007) and thus reduces cardiac output by up to 40% (Eichinger et al. 2010). Therefore, the divers become hypotensive and tachycardic. Whether such elevated lung volumes are the results of genetic selection or of long-term training is yet to be determined.

12.10 Dynamics of Cardiovascular Responses to Dry Breath-Holding

The possibility of investigating the dynamics of cardiovascular responses to apnea has been restricted for a long time, for lack of available technology. Indeed, historically, only the dynamics of f_H during breath-holding was studied, whether in dry conditions or during immersion or diving. Bradycardia is an almost unanimous and universal observation during breath-holding studies and is considered a key component of the diving response. But, when does bradycardia occur during breath-holding? Is it a continuous phenomenon? Or does it characterize specific periods during maximal breath-holds? A beat-by-beat analysis of several cardiovascular variables is needed to better understand the characteristics, the origin and the meaning, not only of bradycardia, but of the overall cardiovascular responses to breath-holding.

In the Maiorca study (Ferrigno et al. 1997), invasive beat-by-beat recordings of arterial blood pressure were obtained, while cardiac output was continuously measured by impedance cardiography, during simulated free dives in a hyperbaric chamber. A comparison of the time course of the recorded variables showed that hypertension preceded bradycardia, so that the latter appeared as a baroreflex response attempting to correct the extreme elevation in arterial blood pressure. Eventually, the dynamics of the f_H fall is not a continuous phenomenon, but follows patterns that are more complex.

Dynamic studies of the cardiovascular system cannot be carried out during dives. Recently, the construction of a 40-m deep swimming pool in Padova, Italy, has improved the logistic context for such studies, but the technological means are still insufficient. Marabotti et al. (2008, 2009) could obtain Doppler echocardiography data at a depth of 5 and 10 m below surface, but those were not dynamic beat-by-beat measures. They nevertheless documented a slightly lower cardiac output during apnea at depth with respect to the water surface, due to lower stroke volume. They suggested a possible role of chest squeeze in the cardiovascular response at depth.

Perini et al. (2008) had been the first to describe the kinetics of the cardiovascular responses to dry apnea at rest. They identified three phases, two of which, the first (φ_1), lasting in all cases about 30 s, and the third (φ_3), are dynamic phases and are characterized by opposite changes in mean arterial pressure and f_H, suggesting possible baroreflex modulation of arterial pressure, with a second phase (φ_2) between them, characterized by cardiovascular steady state. The observed patterns were the same when apneas were carried out with the body immersed in cool water (Perini et al. 2010). Therefore, the φ_1 dynamics is not a consequence of the diving reflex (Gooden 1994; Manley 1990). Other authors then analyzed cardiovascular responses at various times during dry breath-holding (Costalat et al. 2013, 2015; Tocco et al. 2012, 2013).

Perini et al. (2008, 2010) hypothesized that the transition between φ_2 and φ_3 corresponds to the attainment of the physiological breaking point of apnea and the onset of diaphragmatic contractions (Agostoni 1963; Cross et al. 2013; Lin et al.

1974). If this is the case, the duration of φ_2 should be directly proportional to the size of and inversely proportional to the rate of body oxygen consumption.

This statement has several corollaries. In fact, φ_2 duration ought to be shorter at exercise than at rest, because of higher metabolic rate. It ought to vary with Eo_2, being shorter in hypoxia than in normoxia, and longer in hyperoxia than in normoxia. It should also be shorter when apnea starts from the functional residual capacity (FRC) rather than from TLC. The first corollary was shown to be the case indeed, and so much that even at 50 W exercise, a light power indeed, a steady φ_2 was not visible anymore, perhaps being covered by φ_1 (Sivieri et al. 2015), as suggested by its re-appearance in oxygen apneas carried out at exercise (Taboni et al. 2018). The second corollary was supported by the data obtained during oxygen breathing by Fagoni et al. (2015). However, in this latter study, φ_2 duration during oxygen apneas was only 30% higher than during apneas in air, although Eo_2 was five times larger in the former than in the latter case: this means that φ_2 duration is not directly proportional to Eo_2 at any given metabolic rate. If this is so, a linear relationship between φ_2 duration and Eo_2 would not cross the origin of the axis and the hyperbola describing the inverse relationship between φ_2 duration and metabolic rate would be translated.

Since Otis et al. (1948) analyzed the alveolar gas composition at the volitional breaking point of apnea, peripheral and central chemoreflexes have been under scrutiny as possible determinants of apnea breaking points (Fitz-Clarke 2018; Foster and Sheel 2005; Lin 1982). In this context, the breaking point of apnea and the onset of diaphragmatic contractions would result from the combined effects of hypoxia and hypercapnia. During oxygen apneas, hemoglobin remains fully saturated until volitional exhaustion and no hypoxia occurs. If this is so, the attainment of the physiological breaking point in oxygen apneas would depend on carbon dioxide accumulation and subsequent hydrogen ion stimulation of central chemoreceptors only. The sensitivity of the ventilatory response to carbon dioxide is much higher than that of the ventilatory responses to hypoxia (Nielsen and Smith 1952). Modest increases in P_ACO_2 may suffice to elicit diaphragmatic contractions.

The model of the cardiovascular responses to apnea proposed by Costalat et al. (2015) suggests that the temporal evolution of f_H in the final part of a maximal apnea is characterized by a linear decrease, corresponding to the time course of f_H in φ_3. Costalat et al. (2017) did not observe this fall of f_H in apneas at exercise, a condition in which baroreflex sensitivity around the baroreflex operating point is greatly decreased (Bringard et al. 2017; Kardos et al. 2001). This simple fact might well suffice to explain the flatness of the time course of f_H in φ_3 at exercise, if the f_H response is a baroreflex attempt at correcting blood pressure changes.

At the beginning of φ_1, when apnea is performed from elevated lung volume, there is a sudden fall of stroke volume and of arterial blood pressure (Andersson and Schagatay 1998; Palada et al. 2007; Perini et al. 2008). This may be caused by an immediate increase in central venous pressure due to the high lung volumes at which apneas are performed (Andersson and Schagatay 1998; Fagoni et al. 2017). The pressure fall is opposed by a concomitant increase in f_H, which we attribute to baroreflex activation (Fagoni et al. 2015; Taboni et al. 2021). If this interpretation is correct, φ_1 should disappear if apneas are performed from FRC rather than from TLC. The duration of φ_1 is invariant and independent of the metabolic rate and of Eo_2 (Fagoni et al. 2015; Sivieri et al. 2015). The physiological mechanisms behind φ_1 are unrelated to those behind φ_2 are φ_3.

Taboni et al. (2021) have analyzed the dynamics of arterial baroreflexes during φ_1. They showed that, during the first seconds of apnea, the R–R interval (reciprocal of f_H) decreases following a baroreflex curve, possibly mediated by withdrawal of vagal tone, as occurs during exercise (Bringard et al. 2017). Then, after the attainment of a minimum in mean blood pressure, baroreflex resets in both resting and exercise apneas, so that the operating point is displaced toward higher blood pressure levels in φ_2. The pattern of baroreflex resetting suggests that it is not due to the action of a central command mechanism. Taboni et al. (2021) speculated on the possible role of enhanced sympathetic activation in baroreflex resetting.

12.11 Conclusions: How Buffalo Influenced Milano

The discussion of the dynamics of the cardiovascular responses to breath-holding that we have represented in the previous paragraph demonstrates the deep influence that Buffalo and Hermann Rahn heritage had on the School of Milano in the field of breath-holding and breath-hold diving physiology. Milano had a similar influence on Buffalo in the field of swimming physiology, as shown in Chap. 6.

The influence was however much deeper than one could say from what we discussed so far. It attained the deepest level of physiological thinking, wherein Buffalo pervaded the spirit of the School of Milano and vice versa (the relationship was definitely biunivocal), to reshape a common approach to the study of physiology and a common way of thinking about science and physiology. The result was more than a mere sum of two different yet complementary histories. A clear example of what we mean is provided by a recent study (Taboni et al. 2020), on the patterns followed by the alveolar gas composition during dry breath-holding in air and in oxygen on the O_2–CO_2 diagram, which were treated as corresponding to the hypoventilation loop for an alveolar ventilation of 0.

Concerning the paths followed by the development of physiological thinking, they are shaped by a mixture of history, cultural *milieu* in which individuals grow, cultural exchange between Schools, psychological mood pervading a scientific environment, ability of absorbing and integrating in the vision of a School the results of others. Buffalo and Milano had different histories, but shared a vision of research and of science and were characterized by a similar, though obviously not identical, cultural *milieu*: this greatly facilitated the scientific exchanges between the two Schools. The Buffalo–Milano connection was one of the most fruitful and fertile relationships between Schools in physiology, and this, not much in terms of number of common publications, yet elevated, but, mostly and most importantly, for the common and integrated approach to science that shaped the mind of people from both Schools. This is the best heritage that Rodolfo Margaria and Hermann Rahn have left to us and to the future generations.

References

Agostoni E (1963) Diaphragm activity during breath holding: factors related to its onset. J Appl Physiol 18:30–36

Andersen HT (1966) Physiological adaptations in diving vertebrates. Physiol Rev 46:212–243

Andersson J, Schagatay E (1998) Effects of lung volume and involuntary breathing movements on the human diving response. Eur J Appl Physiol 77:19–24

Arborelius M Jr, Balldin UI, Lilja B, CEG L (1972) Hemodynamic changes in man during immersion with the head above water. Aerospace Med 43:592–598

Asmussen E, Kristiansson NG (1968) The "diving bradycardia" in exercising man. Acta Physiol Scand 73:527–535

Åstrand PO (1960) Breath holding during and after muscular exercise. J Appl Physiol 15:220–224

Bakovic D, Valic Z, Eterovic D, Vukovic I, Obad A, Marinovic-Terzic I, Dujic Z (2003) Spleen volume and blood flow response to repeated breath-hold apneas. J Appl Physiol 95:1460–1466

Bjertnaes L, Hange A, Kjekshus J, Soyland E (1984) Cardiovascular responses to face immersion and apnea during steady-state muscle exercise. Acta Physiol Scand 120:605–612

Blix AS, Folkow B (1983) Cardiovascular adjustments to diving in mammals and birds. In: Shephard JT, Abboud FM (eds) Handbook of Physiology. The cardiovascular system III, vol 3. American Physiological Society, sect. 2, Bethesda, MD, pp 917–945

Brick I (1966) Circulatory responses to immersing the face in water. J Appl Physiol 21:33–36

Bringard A, Adami A, Fagoni N, Fontolliet T, Lador F, Moia C, Tam E, Ferretti G (2017) Dynamics of the RR-interval versus blood pressure relationship at exercise onset in humans. Eur J Appl Physiol 117:619–630

Burton AC, Bazett HC (1936) A study of the average temperature of the tissues, and of the exchanges of heat and vasomotor responses in man by means of a bath calorimeter. Am J Phys 117:36–54

Butler PJ, Jones DR (1997) Physiology of diving of birds and mammals. Physiol Rev 77:837–899

Chapin JL (1955) Relationship between lung volume and breath-holding breaking point. J Appl Physiol 8:88–90

Christmas KM, Bassingthwaighte JB (2017) Equations for O_2 and CO_2 solubilities in saline and plasma: combining temperature and density dependences. J Appl Physiol 122:1313–1320

Costalat G, Coquart J, Castres I, Joulia F, Sirost O, Clua E, Lemaître F (2017) The oxygen-conserving potential of the diving response: a kinetic-based analysis. J Sports Sci 35:678–687

Costalat G, Coquart J, Castres I, Tourny C, Lemaître F (2013) Hemodynamic adjustments during breath-holding in trained divers. Eur J Appl Physiol 113:2523–2529

Costalat G, Pichon A, Joulia F, Lemaître F (2015) Modeling the diving bradycardia: toward an oxygen-conserving breaking point? Eur J Appl Physiol 115:1475–1484

Craig AB Jr (1961) Causes of loss of consciousness during underwater swimming. J Appl Physiol 16:583–586

Craig AB Jr (1963) Heart rate responses to apneic underwater diving and to breath holding in man. J Appl Physiol 18:854–862

Craig AB Jr (1968) Depth limits in breath-hold diving (an example of Fennology). Respir Physiol 5: 14–22

Craig FN, Cummings EG (1958) Breath holding during exercise. J Appl Physiol 13:30–34

Cross TJ, Breskovic T, Sabapathy S, Zubin Maslov P, Johnson BD, Dujic Z (2013) Respiratory muscle pressure development during breath-holding in apnea divers. Med Sci Sports Exerc 45: 93–101

Curran-Everett D (2006) A classic learning opportunity from Fenn, Rahn, and Otis (1946): the alveolar gas equation. Adv Physiol Educ 30:58–62

Dejours P, Girard F, Labrousse Y, Teillac A (1959) Etude de la régulation de la ventilation de repos chez l'homme en haute altitude. Rev Franc Etud Clin Biol 4:115–127

Dejours P, Puccinelli R, Armand J, Dicharry M (1965) Concept and measurement of ventilatory sensitivity to carbon dioxide. J Appl Physiol 20:890–897

di Prampero PE (1981) Energetics of muscular exercise. Rev Physiol Biochem Pharmaco 89:143–222

di Prampero PE, Ferretti G (1999) The energetics of anaerobic muscle metabolism: a reappraisal of older and recent concepts. Respir Physiol 118:103–115

Douglas CG, Haldane JS (1909) The causes of periodic or Cheyne-stokes breathing. J Physiol Lond 38:401–419

Eichinger M, Walterspacher S, Scholz T, Tetzlaff R, Puderbach M, Tetzlaff K, Kopp-Schneider A, Ley S, Choe K, Kauczor HU, Sorichter S (2010) Glossopharyngeal insufflation and pulmonary hemodynamics in elite breath hold divers. Med Sci Sports Exerc 42:1688–1695

Espersen K, Frandsen H, Lorentzen T, Kanstrup IL, Christensen NJ (2002) The human spleen as an erythrocyte reservoir in diving-related interventions. J Appl Physiol 92:2071–2079

Fagoni N, Sivieri A, Antonutto G, Moia C, Taboni A, Bringard A, Ferretti G (2015) Cardiovascular responses to dry resting apneas in elite divers while breathing pure oxygen. Respir Physiol Neurobiol 219:1–8

Fagoni N, Taboni A, Vinetti G, Bottarelli S, Moia C, Bringard A, Ferretti G (2017) Alveolar gas composition during maximal and interrupted apneas in ambient air and pure oxygen. Respir Physiol Neurobiol 235:45–51

Farhi LE (1990) World War II and respiratory physiology: the view from Rochester, New York. J Appl Physiol 69:1565–1570

Farhi LE, Rahn H (1955) Gas stores of the body and the unsteady state. J Appl Physiol 7:472–484

Feiner JR, Bickler PE, Severinghaus JW (1995) Hypoxic ventilatory response predicts the extent of maximal breath-holds in man. Respir Physiol 100:213–222

Fenn WO (1962) Born fifty years too soon. Annu Rev Physiol 24:1–10

Fenn WO, Rahn H, Otis AB (1950) Respiratory system. Annu Rev Physiol 12:179–204

Ferretti G (2001) Extreme human breath-hold diving. Eur J Appl Physiol 84:254–271

Ferretti G (2018) The O_2-CO_2 diagram. A revolution in respiration physiology. pH 2:36–57

Ferretti G, Costa M (2003) Diversity in and adaptation to breath-hold diving in humans. Comp Biochem Physiol 136:205–213

Ferretti G, Costa M, Ferrigno M, Grassi B, Marconi C, Lundgren CEG, Cerretelli P (1991) Alveolar gas composition and exchange during deep breath-hold diving and dry breath holds in elite divers. J Appl Physiol 70:794–802

Ferretti G, Costa M, Moroni R, Ranieri P, Butti F, Sponsiello N (2012) Lung volumes of extreme breath-hold divers. Sport Sci Health 7:55–59

Ferretti G, Veicsteinas A, Rennie DW (1986) Regional heat flows of resting and exercising men immersed in cool water. J Appl Physiol 64:1239–1248

Ferretti G, Veicsteinas A, Rennie DW (1989) Conductive and convective heat flows of exercising humans in cold water. J Appl Physiol 67:2473–2480

Ferrigno M, Ferretti G, Ellis A, Warkander D, Costa M, Cerretelli P, Lundgren CEG (1997) Cardiovascular changes during deep breath-hold dives in a pressure chamber. J Appl Physiol 83:1282–1290

Ferrigno M, Grassi B, Ferretti G, Costa M, Marconi C, Cerretelli P, Lundgren CEG (1991) Electrocardiogram during deep breath-hold dives by elite divers. Undersea Biomed Res 18: 81–91

Fitz-Clarke JR (2018) Breath-hold diving. Compr Physiol 8:585–630

Foster GE, Sheel AW (2005) The human diving response, its function, and its control. Scand J Med Sci Sports 15:3–12

Gayeski TEJ, Honig CR (1983) Direct measurement of intracellular O_2 gradients: role of convection and myoglobin. Adv Exp Med Biol 159:613–621

Gooden BA (1994) Mechanism of the human diving response. Integr Physiol Behav Sci 29:6–16

Grassi B, Ferretti G, Costa M, Ferrigno M, Panzacchi A, Lundgren CEG, Marconi C, Cerretelli P (1994) Ventilatory responses to hypercapnia and hypoxia in elite breath-hold divers. Respir Physiol 97:323–332

Harashima S (1965) Gito Teruoka: his life and work. In: Rahn H, Yokoyama T (eds) Physiology of breath-hold diving and the ama of Japan. DC, NAS-NRC Publ, Washington, p 1341

Harding PE, Roman D, Whelan RF (1965) Diving bradycardia in man. J Physiol 181:401–409

Heistad DD, Aboud FM, Eckstein JW (1968) Vasoconstrictor response to simulated diving in men. J Appl Physiol 25:542–549

Hong SK, Kim PK, Pak HK, Kim JK, Yoo MJ, Rennie DW (1969a) Maximal aerobic power of Korean women divers. Fed Proc 28:1284–1288

Hong SK, Lee CK, Kim JK, Song SH, Rennie DW (1969b) Peripheral blood flow and heat flux of Korean women divers. Fed Proc 28:1143–1148

Hong SK, Lin YC, Lally DA, Yim BJB, Kominami N, Hong PW, Moore TO (1971) Alveolar gas exchange and cardiovascular function during breath holding with air. J Appl Physiol 30:540–547

Hong SK, Rahn H (1967) The diving women of Korea and Japan. Sci Am 216:34–43

Hong SK, Rahn H, Kang DH, Song SH, Kang BS (1963) Diving pattern, lung volumes and alveolar gas of the Korean diving women (ama). J Appl Physiol 18:457–465

Hong SK, Rennie DW, Park YS (1986) Cold acclimatization and deacclimatization in Korean women divers. Exercise Sports Sci Rev 14:231–268

Hong SK, Song SH, Kim PK, Suh CS (1967) Seasonal observations on the cardiac rhythm during diving in the Korean ama. J Appl Physiol 23:18–22

Hurford WE, Hong SK, Park YS, Ahn DW, Shiraki K, Mohri M, Zapol WM (1990) Splenic contraction during breath-hold diving in the Korean ama. J Appl Physiol 69:932–936

Ilardo MA, Moltke I, Korneliussen TS, Cheng J, Stern AJ, Racimo F, de Barros DP, Sikora M, Seguin-Orlando A, Rasmussen S et al (2018) Physiological and genetic adaptations to diving in sea nomads. Cell 173:569–580

Jansson E, Sylvén C, Nordevang E (1982) Myoglobin in the quadriceps femoris muscle of competitive cyclists and untrained men. Acta Physiol Scand 114:627–629

Jouasset D (1960) Normalisation des épreuves fonctionnelles respiratoires dans les pays de la Communauté Européenne du Charbon et de l'Acier. Poumon Coeur 10:1145–1159

Kang BS, Han DS, Paik KS, Park YS, Kim JK, Kim CS, Rennie DW, Hong SK (1970) Calorigenic action of norepinephrine in the Korean women divers. J Appl Physiol 29(6–9):1970

Kang BS, Song SH, Suh CS, Hong SK (1963) Changes in body temperature and basal metabolic rate of the ama. J Appl Physiol 18:483–488

Kardos A, Watterich G, de Menezes R, Csanady M, Casadei B, Rudas L (2001) Determinants of spontaneous baroreflex sensitivity in a healthy working population. Hypertension 37:911–916

Klocke FJ, Rahn H (1959) Breath holding after breathing oxygen. J Appl Physiol 14:689–693

Kooyman GL, Ponganis PJ (1998) The physiological basis of diving to depth: birds and mammals. Annu Rev Physiol 60:19–32

Lambertsen CJ (1960) Carbon dioxide and respiration in acid base homeostasis. Anesthesiology 21: 642–651

Lanphier EH, Rahn H (1963) Alveolar gas exchange during breath-holding with air. J Appl Physiol 18:478–482

Lemaître F, Clua E, Andréani B, Castres I, Chollet D (2010) Ventilatory function in breath-hold divers: effect of glossopharyngeal insufflation. Eur J Appl Physiol 108:741–747

Lin YC (1982) Breath hold diving in terrestrial mammals. Exerc Sports Sci Rev 10:270–307

Lin YC, Hong SK (1996) Hyperbaria: breath-hold diving. In: Fregly MJ, Blatteis CM (eds) Handbook of physiology. Environmental physiology. Oxford University Press, New York, pp 979–995

Lin YC, Lally DA, Moore TO, Hong SK (1974) Physiological and conventional breath-hold breaking points. J Appl Physiol 37:291–296

Lin YC, Shida KK, Hong SK (1983) Effects of hypercapnia, hypoxia and rebreathing on circulatory response to apnea. J Appl Physiol 54:172–177

Lindholm P, Linnarsson D (2002) Pulmonary gas exchange during apnea in exercising men. Eur J Appl Physiol 86:487–491

Lindholm P, Lundgren CEG (2009) The physiology and pathophysiology of human breath-hold diving. J Appl Physiol 106:284–292

Lindholm P, Nyren S (2005) Studies on inspiratory and expiratory glossopharyngeal breathing in breath-hold divers employing magnetic resonance imaging and spirometry. Eur J Appl Physiol 94:646–651

Lindholm P, Sundblad P, Linnarsson D (1999) Oxygen-conserving effects of apnea in exercising men. J Appl Physiol 87:2122–2127

Lloyd BB, Jukes MGM, Cunningham DJC (1958) The relation between the alveolar oxygen pressure and the respiratory response to carbon dioxide in man. Q J Exp Physiol 43:214–227

Loring SH, O'Donnell CR, Butler JP, Lindholm P, Jacobson F, Ferrigno M (2007) Transpulmonary pressures and lung mechanics with glossopharyngeal insufflation and exsufflation beyond normal lung volumes in competitive breath-hold divers. J Appl Physiol 102:841–846

Lundgren CGE, Ferrigno M (1987) The physiology of breath-hold diving. Undersea and Hyperbaric Medical Society, Bethesda MD, USA

Majorca E (1987) Depth records: practical considerations. In: Lundgren CEG, Ferrigno M (eds) The physiology of breath—hold diving. Undersea and Hyperbaric Medicine Society, Bethesda, MD, pp 291–294

Manley L (1990) Apnoeic heart rate responses in humans. A review Sports Med 9:286–310

Marabotti C, Belardinelli A, L'Abbate A, Scalzini A, Chiesa F, Cialoni D, Passera M, Bedini R (2008) Cardiac function during breath-hold diving in humans: an echocardiographic study. Undersea Hyperb Med 35:83–90

Marabotti C, Scalzini A, Cialoni D, Passera M, Ripoli A, L'Abbate A, Bedini R (2009) Effect of depth and chest volume on cardiac output during breath-hold diving. Eur J Appl Physiol 106: 683–689

Massion J, Meulders M, Colle J (1960) Fonction posturale des muscles respiratoires. Arch Int Physiol Biochim 68:314–326

Masuda Y, Yoshida A, Hayashi F, Sasaki K, Honda Y (1981) The ventilatory responses to hypoxia and hypercapnia in the Ama. Jpn J Physiol 31:187–197

Masuda Y, Yoshida A, Hayashi F, Sasaki K, Honda Y (1982) Attenuated ventilatory responses to hypercapnia and hypoxia in assisted breath-hold divers (funado). Jpn J Physiol 32:327–336

Mithoefer JC (1959) Lung volume restriction as a ventilatory stimulus during breath-holding. J Appl Physiol 14:701–705

Morelli L, Data PG (1991) Thoracic radiological changes during deep breath-hold diving (abstract). In: Data PG, Lahiri S (eds) Proceedings of the workshop on the physiology of deep breath-hold diving. Chieti, Italy, pp 28–30

Nielsen M, Smith H (1952) Studies on the regulation of respiration in acute hypoxia; with an appendix on respiratory control during prolonged hypoxia. Acta Physiol Scand 24:293–313

Novalija J, Lindholm P, Loring SH, Diaz E, Fox JA, Ferrigno M (2007) Cardiovascular aspects of glossopharyngeal insufflation and exsufflation. Undersea Hyperb Med 34:415–423

Olszowka AJ, Rahn H (1987) Breath-hold diving. In: Sutton JR, Houston CS, Coates G (eds) Hypoxia and cold. New York, Praeger, pp 417–428

Otis AB (1964) Quantitative relationships in steady-state gas exchange. In: Fenn WO, Rahn H (eds) Handbook of physiology. Respiration. American Physiological Society, Washington, DC, pp 681–698

Otis AB, Rahn H, Epstein MA, Fenn WO (1946) Performance as related to alveolar air. Am J Phys 146:207–221

Otis AB, Rahn H, Fenn WO (1948) Alveolar gas changes during breath holding. Am J Phys 152: 674–686

Paik KS, Kang BS, Han DS, Rennie DW, Hong SK (1972) Vascular responses of Korean Ama to hand immersion in cold water. J Appl Physiol 32:446–450

Palada I, Eterovic D, Obad A, Bakovic D, Valic Z, Ivancev V, Lojour M, Shoemaker JK, Dujic Z (2007) Spleen and cardiovascular function during short apneas in divers. J Appl Physiol 103: 1958–1963

Pappenheimer J (1996) Hermann Rahn: July 5, 1912-June 23, 1990. Biogr Mem Natl Acad Sci 69: 242–267

Park YS, Rennie DW, Lee SI, Park YD, Paik KS, Kang DH, Suh DJ, Lee SH, Hong SY, Hong SK (1983) Time course of deacclimatization to cold water immersion in Korean women divers. J Appl Physiol 54:1708–1716

Parkes MJ (2006) Breath-holding and its break point. Exp Physiol 91:1–15

Patrician A, Gasho C, Spajic B, Caldwell HG, Bakovic-Kramaric D, Barak O, Drviš I, Dujic Z, Ainslie PN (2021) Case studies in physiology: breath-hold diving beyond 100 meters – cardiopulmonary responses in world-champion divers. J Appl Physiol 130:1345–1350

Paulev PE (1969) Respiratory and cardiovascular effects of breath-holding. Acta Physiol Scand Suppl 324:1–116

Pelizzari U (2005) Profondamente, 2nd edn. Mondadori, Milano

Perini R, Gheza A, Moia C, Sponsiello N, Ferretti G (2010) Cardiovascular time course during prolonged immersed static apnea. Eur J Appl Physiol 110:277–283

Perini R, Tironi A, Gheza A, Butti F, Moia C, Ferretti G (2008) Heart rate and blood pressure time courses during prolonged dry apnea in breath-hold divers. Eur J Appl Physiol 104:1–7

Qvist J, Hurford WE, Park YS, Radermacher P, Falke KJ, Ahn DW, Guyton GP, Stanek KS, Hong SK, Weber RE, Zapol WM (1993) Arterial blood gas tensions during breath-hold diving in the Korean ama. J Appl Physiol 75:285–293

Rahn H (1965) Gito Teruoka: his contribution to the physiology of the Ama. In: Physiology of breath-hold diving and the Ama of Japan, vol 1341. NAS-NRC Publ, Washington, DC, pp 9–10

Rahn H (1979) Wallace Osgood Fenn: August 27,1893-September 20,1971. Biogr Mem Natl Acad Sci 50:140–173

Rahn H, Bahnson HT, Muxworthy JF, Hagen JM (1953) Adaptation to high altitude: changes in breath-holding time. J Appl Physiol 6:154–157

Rahn H, Farhi LE (1964) Ventilation, perfusion and gas exchange – the $\dot{V}_A/\dot{Q}$ concept. In: Fenn WO, Rahn H (eds) Handbook of Physiology. Respiration. American Physiological Society, Washington, DC, pp 735–766

Rahn H, Fenn WO (1955) A graphical analysis of the respiratory gas exchange, The O_2-CO_2 diagram. American Physiological Society, Washington, DC

Rahn H, Otis AB (1949) Continuous analysis of alveolar gas composition during work, hyperpnea, hypercapnia and anoxia. J Appl Physiol 1:717–724

Rahn H, Yokoyama T (1965) Physiology of breath-hold diving and the Ama of Japan. Washington, DC, NAS-NRC Publ 1341

Rebuck AS, Slutsky AS, Mahutte CK (1977) A mathematical expression to describe the ventilatory response to hypoxia and hypercapnia. Respir Physiol 31:107–116

Rennie DW, Covino BG, Howell BJ, Song SH, Kang BS, Hong SK (1962) Physical insulation of Korean diving women. J Appl Physiol 17:961–966

Rodbard S (1947) The effect of oxygen, altitude and exercise on breath-holding time. Am J Phys 150:142–148

Schaefer KE, Allison RD, Dougherty JH Jr, Carey CR, Walker R, Yost F, Parker D (1968) Pulmonary and circulatory adjustments determining the limits of depth in breath-hold diving. Science 162:1020–1023

Schagatay E, Andersson JP, Hallen M, Palsson B (2001) Selected contribution: role of spleen emptying in prolonging apneas in humans. J Appl Physiol 90:1623–1629

Schagatay E, van Kampen M, Emanuelsson S, Holm B (2000) Effects of physical and apnea training on apneic time and the diving response in humans. Eur J Appl Physiol 82:161–169

Schipke JD, Kelm M, Siegmund K, Muth T, Sievers B, Steinerd S (2015) Lung packing in breath hold-diving: an impressive case of pulmo–cardiac interaction. Respir Med Case Rep 16:120–121

Scholander PF (1940) Experimental investigations on the respiratory function in diving mammals and birds, vol 22. Norske Videnskaps-Akademie Oslo, Hvalradets Skrifter, pp 1–131

Sivieri A, Fagoni N, Bringard A, Capogrosso M, Perini R, Ferretti G (2015) A beat-by-beat analysis of cardiovascular responses to dry resting and exercise apneas in elite divers. Eur J Appl Physiol 115:119–128

Song SH, Kang DH, Kang BS, Hong SK (1963) Lung volumes and ventilatory responses to high CO_2 and low O_2 in the Ama. J Appl Physiol 18:466–470

Stanley NN, Altose MD, Kelsen SG, Ward CF, Cherniack NS (1975) Changing effect of lung volume on respiratory drive in man. J Appl Physiol 38:768–773

Sterba JA, Lundgren CE (1988) Breath-hold duration in man and the diving response induced by face immersion. Undersea Biomed Res 15:361–375

Subramani S, Kanthakumar P, Maneksh D, Sidharthan A, Rao SV, Parasuraman V, Tharion E (2011) O_2-CO_2 diagram as a tool for comprehension of blood gas abnormalities. Adv Physiol Educ 35:314–320

Taboni A, Fagoni N, Fontolliet T, Grasso G, Moia C, Vinetti A, Ferretti G (2020) Breath holding as an example of extreme hypoventilation: experimental testing of a new model describing alveolar gas pathways. Exp Physiol 105:2216–2225

Taboni A, Vinetti G, Bruseghini P, Camelio S, D'Elia M, Moia C, Ferretti G, Fagoni N (2018) Cardiovascular responses to dry apneas at exercise in air and in pure oxygen. Respir Physiol Neurobiol 255:17–21

Taboni A, Vinetti G, Fontolliet T, Grasso G, Tam E, Moia C, Ferretti G, Fagoni N (2021) Baroreflex responses during dry resting and exercise apneas in air and pure oxygen. Eur J Appl Physiol 121:539–547

Teruoka G (1932) Die Ama und ihre Arbeit. Arbeitsphysiologie 5:239–251

Tibes U, Stegemann J (1969) Behavior of the end tidal respiratory gas pressure, O_2 uptake and CO_2 output following simple apnea in water, on land and apneic diving. Pflugers Arch 311:300–311

Tocco F, Crisafulli A, Melis F, Porru C, Pittau G, Milia R, Concu A (2012) Cardiovascular adjustments in breath-hold diving: comparison between divers and non-divers in simulated dynamic apnea. Eur J Appl Physiol 112:543–554

Tocco F, Marongiu F, Pinna M, Roberto S, Pusceddu M, Angius L, Migliaccio GM, Milia R, Concu A, Crisafulli A (2013) Assessment of circulatory adjustments during underwater apnea in elite divers by means of a portable device. Acta Physiol 207:290–298

Veicsteinas A, Ferretti G, Rennie DW (1982) Superficial shell insulation in resting and exercising men in cold water. J Appl Physiol 52:1557–1564

Veicsteinas A, Rennie DW (1982) Thermal insulation and shivering threshold in Greek sponge divers. J Appl Physiol 52:845–850

Walterspacher S, Scholz T, Tetzlaff K, Sorichter S (2011) Breath-hold diving: respiratory function on the longer term. Med Sci Sports Exerc 43:1214–1219

Warkander DE, Ferrigno M, Lundgren CEG, McCoy K (1996) Some physiological parameters in a breathhold dive to 107 msw (351 fsw) in the ocean. Undersea Hyperb Med 23:575

West JB (2012) The physiological legacy of the Fenn, Rahn and Otis school. Am J Phys 303: L1845–L1851

Zapol WM (1996) Diving physiology of the Weddel seal. In: Fregly MJ, Blatteis CM (eds) Handbook of physiology. Environmental physiology. Oxford University Press, New York, pp 1049–1056